Lecture Notes in Computer Science 1379
Edited by G. Goos, J. Hartmanis and J. van Leeuwen

Springer
Berlin
Heidelberg
New York
Barcelona
Budapest
Hong Kong
London
Milan
Paris
Santa Clara
Singapore
Tokyo

Tobias Nipkow (Ed.)

Rewriting Techniques and Applications

9th International Conference, RTA-98
Tsukuba, Japan, March 30 – April 1, 1998
Proceedings

Springer

Series Editors

Gerhard Goos, Karlsruhe University, Germany
Juris Hartmanis, Cornell University, NY, USA
Jan van Leeuwen, Utrecht University, The Netherlands

Volume Editor

Tobias Nipkow
Institut für Informatik, Technische Universität München
D-80290 München, Germany
E-mail: nipkow@informatik.tu-muenchen.de

Cataloging-in-Publication data applied for

Die Deutsche Bibliothek - CIP-Einheitsaufnahme

Rewriting techniques and applications : 9th international conference ; proceedings / RTA-98, Tsukuba, Japan, March 28 - April 1, 1998. Tobias Nipkow (ed.). - Berlin ; Heidelberg ; New York ; Barcelona ; Budapest ; Hong Kong ; London ; Milan ; Paris ; Santa Clara ; Singapore ; Tokyo : Springer, 1998
(Lecture notes in computer science ; Vol. 1379)
ISBN 3-540-64301-X

CR Subject Classification (1991): F.4, D.3, F.3.2, I.1, I.2.2-3

ISSN 0302-9743
ISBN 3-540-64301-X Springer-Verlag Berlin Heidelberg New York

Typesetting: Camera-ready by author
SPIN 10631976 06/3142 – 5 4 3 2 1 0 Printed on acid-free paper

Preface

This volume contains the papers presented at the *9th International Conference on Rewriting Techniques and Applications* (RTA-98) held in Tsukuba, Japan, March 30 – April 1, 1998, at the University of Tsukuba. RTA is the major forum for the presentation of new research in all aspects of rewriting, including equational logic, λ-calculus, theorem proving, and functional programming. This is the first time that RTA was held outside Europe or North America. It is only fitting that RTA-98 took place in Japan, as there is a large Japanese research community working on rewriting, which has made many important contributions.

There were 61 submissions from Belgium (1), Denmark (0.5), France (12.17), Germany (17.33), Israel (0.5), Italy (1.5), Japan (10), Mexico (0.67), The Netherlands (0.5), Portugal (1.33), Spain (2), Sweden (1.1), UK (4.5), and USA (7.9), of which the program committee selected 22 for presentation. In addition, there were invited talks by Hubert Comon, Jan Willem Klop, and Gordon Plotkin.

This year, for the first time, the program committee awarded a *Best Student Paper* prize. It went to Johannes Waldmann of the University of Jena, Germany, for his paper *Normalization of S-Terms is Decidable.* In this paper Johannes Waldmann shows that in Combinatory Logic it is decidable if a ground term built from the combinator S alone has a normal form, thus settling a problem that had been open for a long time.

In addition to the usual presentation of research papers, RTA-98, also for the first time, featured two workshops.

Many people helped to make RTA-98 a success. It is a particular pleasure to thank Aart Middeldorp and Tetsuo Ida for hosting RTA-98 in Tsukuba, and especially for the support they provided (financial and otherwise) right from the start. The program committee and 113 additional referees worked hard to guarantee a high-quality program. Paliath Narendran helped with the electronic advertising. Isabelle Gnaedig handled the poster. Werner Buchert and Gabriele Turner took care of a lot of text processing. RTA-98 was organized in cooperation with the Japan Society for Software Science and Technology. The following sponsors provided financial support: Chubu Electric Company, Electrotechnical Laboratories, International Information Science Foundation, Japan Society for the Promotion of Science, Kayamori Foundation of Information Science Advancement, Japanese Ministry of Education, Science, Sports and Culture (Grant-in-Aid for Scientific Research on Priority Areas "Research on the Principles for Constructing Software with Evolutionary Mechanisms"), Okawa Foundation for Information and Telecommunications, Technical University of Munich, University of Tsukuba (Center for Tsukuba Advanced Research Alliance and Institute of Information Sciences and Electronics). I am very grateful to all of them.

München, February 1998

Tobias Nipkow
RTA-98 Program Chair

Conference Organization

Program Chair

Tobias Nipkow (München)

Local Arrangements Chairs

Aart Middeldorp and Tetsuo Ida (Tsukuba)

Program Committee

Jürgen Avenhaus (Kaiserslautern)
Roberto Di Cosmo (Paris)
Harald Ganzinger (Saarbrücken)
Deepak Kapur (Albany)
Tobias Nipkow (München)
Michio Oyamaguchi (Tsu)
Franz Baader (Aachen)
Maribel Fernández (Paris)
Michael Hanus (Aachen)
Aart Middeldorp (Tsukuba)
Vincent van Oostrom (Amsterdam)
Ralf Treinen (Paris)

Organizing Committee

Hubert Comon (Cachan)
Harald Ganzinger (Saarbrücken)
Jieh Hsiang (Taipei)
Klaus Madlener (Kaiserslautern)
Paliath Narendran (Albany)
Yoshihito Toyama (Tatsunokuchi)

External Referees

M. Alpuente
I. Alouini
J.-M. Andreoli
T. Aoto
F. Barbanera
C. Beierle
I. Bethke
R. Bloo
A. Bockmayr
M. Chakravarty
P. Ciancarini
H. Comon
E. Contejean
A. Corradini
P.-L. Curien
J. Denzinger
N. Dershowitz
G. Dowek
I. Durand
R. Echahed
C. Ferdinand
M. Ferreira
I. Gent
A. Geser
N. Ghani
J. Giesl
A. Goerdt
J. Goubault-Larrecq
B. Gramlich
S. Guerrini
M. Harris
N. Heintze
R. Hennicker
A. Herzig
S. Hirokawa
D. Hofbauer
S. Hölldober
D. Holt
A.-A. Holzbacher
U. Hustadt
K. Indermark
F. Jacquemard
J.-P. Jouannaud
S. Kahrs
Y. Kaji
G. Kemper
D. Kesner
J.W. Klop
M. Kurihara
J. Levy
B. Löchner
S. Lucas
D. Lugiez
I. Mackie
K. Madlener
C. Marché
D. Marchignoli
A. Masini
R. Matthes
D. McAllester
R. McNaughton
P.-A. Melliés
C. Meyer
M. Müller
C. Munoz
P. Narendran
J. Niehren
R. Nieuwenhuis
M. Ogawa
E. Ohlebusch
Y. Ohta
S. Okui
F. Otto
L. Pacholski
V. Padovani
A. Piperno
A. Podelski
C. Prehofer
L. Puel
F. van Raamsdonk
C. Ringeissen
K. Rose
A. Rubio
M. Rusinowitch
M. Sakai
M. Schmidt-Schauß
K. Schulz
A. Schürr
H. Seidl
D. Seipel
G. Sénizergues
G. Sivakumar
J. Steinbach
M. Stone
R. Strandh
V. Strehl
J. Stuber
P. Thiemann
R. Thomas
S. Tison
A. Togashi
M. Tommasi
Y. Toyama
M. Veanes
S. Vorobyov
U. Waldmann
M. Walicki
C. Weidenbach
A. Weiermann
R. Wilhelm
C.-P. Wirth
M. Wolf
T. Yamada

Table of Contents

Origin Tracking in Term Rewriting

J.W. Klop

CWI and Vrije Universiteit, Amsterdam

Abstract. The notion of descendants or residuals with its inverse notion of ancestors is classical in the theory of rewriting, both in first order term rewriting and in higher-order rewriting, such as lambda calculus. Recently this classical notion has been given much attention. On the one hand, the notion has been studied in an abstract, axiomatic way, in order to isolate the essential properties of the descendant concept. On the other hand descendants were studied in a very concrete way, inspired by practical considerations such as error recovery in program executions and program slicing.

In the latter endeavour the emphasis is on tracing back the symbols constituting an expression to their 'causes' in an earlier expression in the rewrite sequence. This method is also known as origin tracking. The corresponding descendant notion is a refinement of the classical one.

In our talk we present some basic properties of origin tracking. It will be apparent that various labeled versions of the rewrite systems concerned are important to establish such properties.

We treat three (theoretical) applications of the origin tracking technique: first, a simplified proof of the classical theorem of Huet and Le'vy about needed reduction; second, another proof of the Genericity Lemma in pure lambda calculus; third, a simple proof of Berry's sequentiality theorem for lambda calculus.

The third application actually takes place in the setting of infinitary lambda calculus, where Boehm trees are normal forms; and Berry's theorem will easily generalize to lazy lambda trees (or Levy-Longo trees), and to Berarducci trees.

Simultaneous Critical Pairs and Church-Rosser Property

Satoshi Okui *

Department of Information Engineering
Faculty of Engineering
Mie University
Tsu 514, Japan

Abstract. We introduce *simultaneous critical pairs*, which account for *simultaneous* overlapping of several rewrite rules. Based on this, we introduce a new CR-criterion widely applicable to arbitrary left-linear term rewriting systems. Our result extends the well-known criterion given by Huet (1980), Toyama (1988), and Oostrom (1997) and incomparable with other well-known criteria for left-linear systems.

1 Introduction

Church-Rosser (CR for short) property is one of the most important properties of rewrite systems. For (finite) strongly normalising systems, this property is decidable by testing joinability of *critical pairs* whereas for arbitrary systems, undecidable even if they are *weak* CR. For this reason, many researchers have been interested in various (decidable) sufficient conditions ensuring this property. We refer to these conditions as *CR-criteria*.

To obtain CR-criteria, the critical-pair-based approach is still available as far as *left-linear* systems are concerned (Note that non-left-linear systems may not be CR even in the absence of critical pairs). The general joinability requirement is too general to ensure CR property so that the point of research in this direction is to discover suitable joinability requirements that are as general as possible. Rosen observed in [18] that left-linear systems that have only trivial critical pairs are CR. Huet gave in his seminal paper on CR property the first non-trivial condition admitting *parallel-closed* critical pairs [5]. After that, Toyama relaxed this criterion, showing that *overlay* critical pairs allow a much general joinability requiment [20].

Recently, two important contributions were added. One is a study of Oostrom. He elucidated relevance of *developments* in study of CR-criteria, showing that in the Huet-Toyama criterion a parallel rewrite step can be replaced with a more general complete development step [16] (See Fig. 1 (a) where $\multimap$ stands for a complete development step). The other is a study of Gramlich. He considered a generalization of critical pairs, demonstrating a new widely applicable CR-criterion based on *parallel* critical pairs [4] (See Fig. 1 (b) where $\nparallel\!\!\to$ stands for a parallel rewrite step).

* `mailto:okui@cs.info.mie-u.ac.jp`, `http://www.cs.info.mie-u.ac.jp/~okui/`

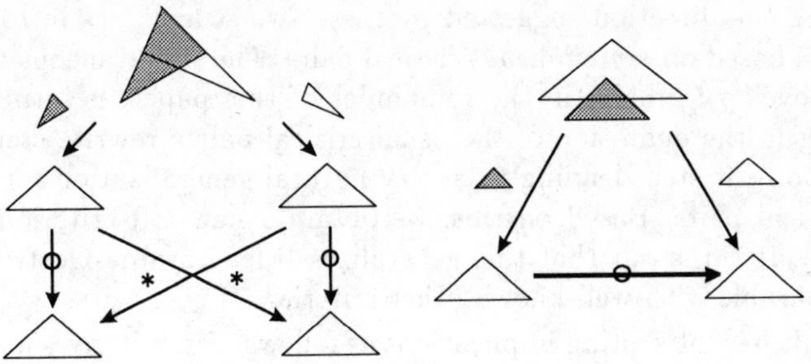

(a) Huet-Toyama-Oostrom criterion

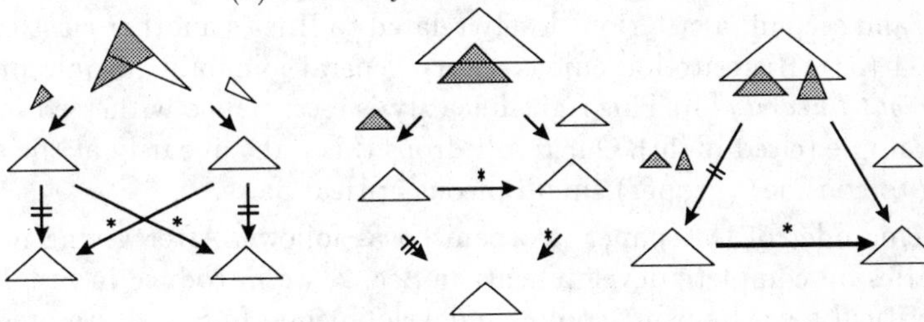

(b) Gramlich's criterion

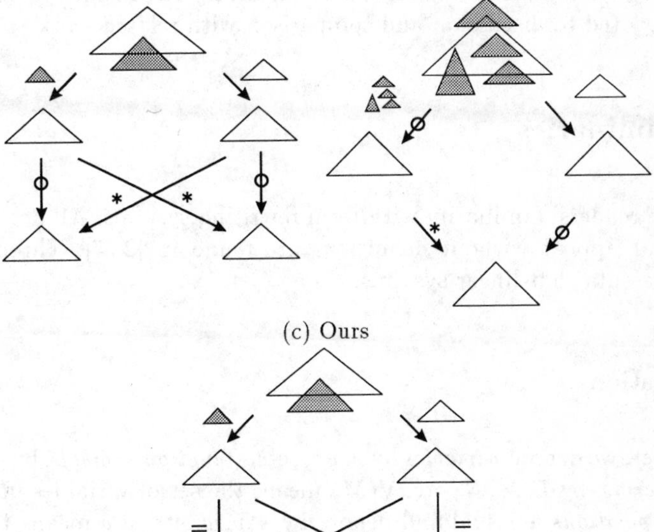

(c) Ours

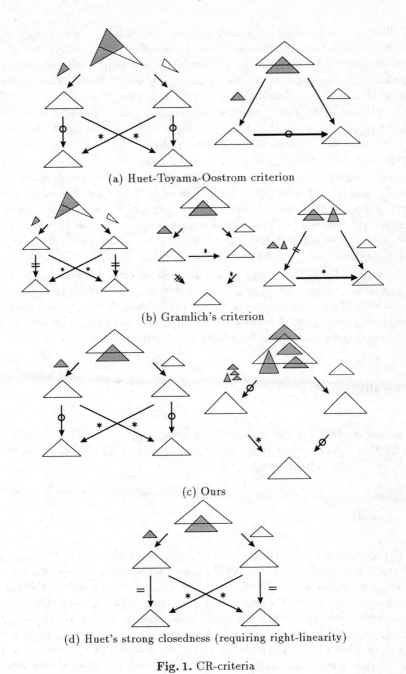

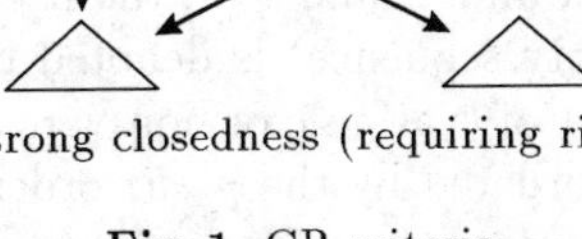

(d) Huet's strong closedness (requiring right-linearity)

Fig. 1. CR-criteria

Following the direction suggested by these two studies, we introduce a new CR-criterion based on *simultaneous* critical pairs. The simultaneous critical pair, which we develop formally in the remainder of this paper, is intuitively given by replacing in the definition of the usual critical pair a rewrite step to a complete development step, leading to a very natural generalization for Gramlich's parallel critical pairs. Based on this, we obtain a new CR-criterion illustrated in Fig. 1 (c). It turns out that this generalizes Huet-Toyama-Oostrom criterion and incomparable with well-known other criteria.

Our result has interesting implications as follows. First, it gives a partial solution to a well-known open problem related to Huet's parallel-closedness (Problem 13 in [3]), and second, a criterion closely related to Huet's another classical criterion (Fig. 1 (d)). His criterion enjoys a very general joinability requirement but requires *right-linearity* [5]. This right-linearity is essential as witnessed in Lévy's counterexample (cited in [5]). Our result drops this right-linearity at the expense of having to consider (proper) simultaneous critical pairs.

The remainder of this paper is organized as follows. After giving necessary preliminaries on complete developments in Sec. 2, we introduce in Sec. 3 simultaneous critical pairs by using complete developments. In Sec. 4, we present our criterion and show that it implies CR for arbitrary left-linear systems. Finally, Sec. 5 is devoted to discussion and comparison with related work.

2 Preliminaries

We assume readers' familiarity with term rewriting systems. All necessary basic notions that appear without definitions are found in [2, 7]. Throughout this paper, we assume left-linear systems.

2.1 Notation

In this paper, we denote *variables* by x, y, z, etc., *function symbols*, by f, g, h, a, b, c, etc., and *terms*, by L, M, N, etc. $\mathrm{V}(M)$ means the set of variables in a term M. We define *positions* as usual and denote by u, v, w, etc. $u.v$ means the position given by concatenation of u and v, and $v \backslash u$, the position w such that $v.w = u$. The root position (the empty sequence) is denoted by ε. $\mathrm{Pos}(M)$ ($\mathrm{PosV}(M)$ or $\mathrm{PosF}(M)$) stands for the set of (variable or non-variable) positions in M respectively. These are partially ordered by the *prefix* ordering ($u \leq v \;\Leftrightarrow u.w = v$ for some w). We denote *substitutions* by σ, θ, ρ, etc., and the *domain* of a substitution σ, by $\mathrm{D}(\sigma)$. $M\sigma$ stands for the *application* of a substitution σ to a term M. We give *composition* of substitutions simply by juxtaposition; $M(\sigma\rho) = (M\sigma)\rho$. The set of most general unifiers of terms M and N is denoted by $\mathrm{mgu}(M, N)$. The length of a rewrite sequence A, the height of a term M (as a tree), and the number of elements in a finite set S are denoted by $|A|$, $|M|$, and $|S|$ respectively.

2.2 Finite Development Theorem

In this paper, *developments* play a central role. Roughly speaking, a development of a set U of redexes of a term M is a rewrite sequence issued from M such that in each step one of the redexes that "descends" from U is contracted. For *orthogonal* systems or *λ-calculus* this propagation of redexes can be formalized by the *residual*, a well-established notion to keep track of propagation of redex positions. However, this is not enough for overlapping systems because several rewrite rules are applicable to a redex position in general. To formalize developments precisely so as to make "finite development theorem" hold, we also keep track of propagation of rewrite rules. We follow the approach of Boudol [1].

Definition 1. A *redex pattern* is a pair $\langle u, \alpha \to \beta \rangle$ of some position u and some rewrite rule $\alpha \to \beta$. For any term M, the set of redex patterns *of* M is defined as $\mathrm{Red}(M) = \{\langle u, \alpha \to \beta \rangle \mid \exists\, \sigma\ \alpha\sigma = M|_u\ \}$.

We denote redex patterns as $\bar{u}, \bar{v}$, etc. The position u and/or the right hand side β of rewrite rule may be omitted when irrelevant or clear from context. Clearly, we have $\mathrm{Red}(M) \subseteq \mathrm{Red}(M\sigma)$ for any substitution σ. For $U \subseteq \mathrm{Red}(M)$ and some position $v \in \mathrm{PosF}(M)$, we define $U|_v = \{\langle v \backslash u, \alpha \to \beta \rangle \mid \langle u, \alpha \to \beta \rangle \in U\ \&\ u \geq v\}$.

We say two redex patterns $\langle u_1, \alpha_1 \to \beta_1 \rangle$ and $\langle u_2, \alpha_2 \to \beta_2 \rangle$ are *connected* if either $u_1 \geq u_2\ \&\ u_1 \backslash u_2 \in \mathrm{PosF}(\alpha_2)$ or $u_2 \geq u_1\ \&\ u_2 \backslash u_1 \in \mathrm{PosF}(\alpha_1)$.

Definition 2. A set of redex patterns is *simultaneous* (or *consistent*) if it consists of pairwise unconnected elements.

Clearly, $U|_v$ is simultaneous if U is. Two sets U and V of redex patterns are said to be *compatible* (written as $U \uparrow V$) if any $\bar{u} \in U$ and $\bar{v} \in V$ are unconnected. The following definition is similar to the usual definition of residuals [6] except for specifying employed rewrite rules explicitly.

Definition 3. Let $\bar{u} = \langle u, \alpha_1 \to \beta_1 \rangle$ be a redex pattern of M and $A\colon M \overset{\langle v,\, \alpha_2 \to \beta_2 \rangle}{\to} N$ a rewrite step. The set $\bar{u} \backslash A$ of *residuals* of $\bar{u}$ by A is defined as follows.

$$\bar{u} \backslash A = \begin{cases} \{\bar{u}\} \text{ if } u \not\geq v \\ \emptyset \quad \text{if } u \geq v \text{ and } v \backslash u \in \mathrm{PosF}(\alpha_2) \\ \{\langle v.w'.((v.w) \backslash u), \alpha_1 \to \beta_1 \rangle \mid w' \text{ is a position of the variable } x \text{ in } \beta_2\} \\ \quad \text{if } u \geq v.w \text{ for the position } w \text{ of some variable, say } x, \text{ of } \alpha_2. \end{cases}$$

The above definition is extended to $U \backslash A$ for a set U of redex patterns of M by $U \backslash A = \bigcup_{\bar{u} \in U} \bar{u} \backslash A$. For extension to arbitrary rewrite sequences, we need some care; we do not have $U \backslash M \to N \subseteq \mathrm{Red}(N)$ in general because our present setting allows overlapping systems. For a rewrite sequence $A (= A_1; \ldots; A_n)$, we define $U \backslash A = U \backslash A_1 \backslash \ldots \backslash A_n$ only when $U \backslash A_1 \backslash \ldots \backslash A_i \uparrow \bar{u}_{i+1}$ for $i = 1, \ldots, n-1$ (where $\bar{u}_i$ is the redex pattern contracted in the step A_i). This condition is satisfied when A is a development introduced below.

Definition 4. Let $U(\subseteq \mathrm{Red}(M))$ be a simultaneous set of redex patterns. A rewrite sequence $A: M \overset{*}{\rightarrow} N$ is called a *development* of U on M if

1. $|A| = 0$, otherwise,
2. $A = A_1; A_2(|A_1| = 1)$, an element, say $\bar{u}$, of U is contracted in A_1, and A_2 is a development of $U \backslash A_1$.

Moreover, an infinite rewrite sequence is called development of U on M if every finite prefix is a development of U on M.

A finite development $A: M \overset{*}{\rightarrow} N$ of U on M is called a *complete* development of U (written as $M \overset{U}{\twoheadrightarrow} N$) if $U \backslash A = \emptyset$. Note that a complete development of U on M is also a development of V on M for any simultaneous V such that $U \subseteq V \subseteq \mathrm{Red}(M)$ and we have $V \backslash A = (V - U) \backslash A$. The most important fact on developments is the following *finite development theorem*. We refer to a result of Boudol [1] without a proof:

Theorem 5. *For any term M and any simultaneous set $U \subseteq \mathrm{Red}(M)$,*

1. *every development of U on M is finite, and can be extended to a complete development, and*
2. *all complete developments of U on M end with the same term.*

This theorem allows us to write $U \backslash M \overset{V}{\twoheadrightarrow} N$ simply as $U \backslash V$. Complete development sequences are defined just as in the case of rewrite sequences: we denote $M_0 \overset{U_0}{\twoheadrightarrow} M_1 \overset{U_1}{\twoheadrightarrow} \ldots \overset{U_{n-1}}{\twoheadrightarrow} M_n$ as $M_0 \overset{*}{\twoheadrightarrow} M_n$. This induces a reflexive transitive relation. Note that $\overset{*}{\twoheadrightarrow} = \overset{*}{\rightarrow}$.

To end this chapter, we list some facts concerning residuals and complete developments that are used in the rest of this paper. The first one is known as the *permutation lemma*.

Lemma 6 [1]. *For any simultaneous sets $U, V \subseteq \mathrm{Red}(M)$ such that $U \uparrow V$ we have*

$$N_1 \overset{U}{\twoheadleftarrow} M \overset{V}{\twoheadrightarrow} N_2 \;\Rightarrow\; N_1 \overset{V \backslash U}{\twoheadrightarrow} M' \overset{U \backslash V}{\twoheadleftarrow} N_2.$$

The next one, which is equivalent to the above permutation lemma, is frequently used later.

Lemma 7. *For any simultaneous sets $U, V \subseteq \mathrm{Red}(M)$ such that $U \uparrow V$ we have*

$$M \overset{U \cup V}{\twoheadrightarrow} N \quad \textit{iff} \quad M \overset{U}{\twoheadrightarrow} L \overset{V \backslash U}{\twoheadrightarrow} N.$$

The last one is most involved. It states in essence that in some cases $\bar{u} \backslash A$ is determined only by the initial and the final terms of A.

Lemma 8. [2] *Let A and B be rewrite sequences from a linear term M to a term M', σ a substitution, N a term, and $v \in \mathrm{Pos}(N)$. We can regard A and B as rewrite sequences from $N[M\sigma]_v$ to $N[M'\sigma]_v$. We then have $\bar{u} \backslash A = \bar{u} \backslash B (\subseteq \mathrm{Red}(N[M'\sigma]_v))$ for any $\bar{u} \in \mathrm{Red}(N[M\sigma]_v) - \{\langle v.v', r\rangle \mid \langle v', r\rangle \in \mathrm{Red}(M)\}$.*

[2] This fact is found in [16] (without a proof).

Proof. Instead of a tedious formal proof, we present here only an intuitive proof sketch. It follows from $\bar{u} \notin \{\langle v.v', r\rangle \mid \langle v', r\rangle \in \text{Red}(M)\}$ that $\bar{u}\backslash A, \bar{u}\backslash B (\subseteq \text{Red}(N[M'\sigma]_v))$ is well defined. Let $V_1 = \{\langle v', r\rangle \in \text{Red}(N) \mid v' \not\geq v\}$ and $V_2 = \{\langle v.v', r\rangle \mid \langle v', r\rangle \in \text{Red}(M\sigma) - \text{Red}(M)\}$ temporary. If $V_1 = V_2 = \emptyset$ the result is trivial. Otherwise, $\bar{u}$ is in V_1 or in V_2. In the former case, we obtain $\bar{u}\backslash A = \{\bar{u}\} = \bar{u}\backslash B$. Consider the later case. If $\text{V}(M')$ is empty the result is trivial. Otherwise, we can find a variable $x \in \text{V}(M')$. Since rewrite sequences do not introduce any new variable, x must appear in M. Furthermore, x appears in M only once. Hence, from the definition of residuals we know that redex patterns in $M'\sigma$ below the position of x in M' are residuals of redex patterns in $M\sigma$ below the unique position of x in M, from which we are able to infer the conclusion. □

3 Simultaneous Critical Pairs

In this section, we formally introduce our *simultaneous* critical pairs. For this, we need to formulate "simultaneous" overlapping. This is not difficult because we assume left-linearity. First, we state two auxiliary facts.

Definition 9. Let M, N be terms and u a position. We say M overlaps with N *at* u if $u \in \text{PosF}(N)$ and $M\sigma = N|_u\sigma$ for some substitution σ. Let M_1, M_2 be terms and u_1, u_2 positions. We say $\langle u_1, M_1\rangle$ and $\langle u_2, M_2\rangle$ *overlap* if either M_1 overlaps with M_2 at $u_2\backslash u_1$ or M_2 overlaps with M_1 at $u_1\backslash u_2$.

Lemma 10. *Suppose $\langle u_1, M_1\rangle$ and $\langle u_2, M_2\rangle$ are unconnected, and $\langle u, M\rangle$ and $\langle u_i, M_i\rangle$ overlap for $i = 1, 2$. If $u_1 \leq u_2$ there exists some variable position v of M_1 such that $u_1.v \leq u_2$.*

Proof. The proof is by induction on $|M_1|$. First, consider the case $|M_1| = 1$. We have $u_1 \geq u$. M_1 is a constant or a variable. Suppose it is a constant. Since $\langle u_1, M_1\rangle$ and $\langle u, M\rangle$ overlap, $u\backslash u_1$ is a position of a constant or a variable in M. From this and $u_2 > u_1$ we have $u_2 > u$ and $u\backslash u_2 \notin \text{Pos}(M)$. This contradicts the assumption of $\langle u_2, M_2\rangle$ and $\langle u, M\rangle$ overlapping so that we take ε as v.

In case $|M_1| > 1$, we distinguish two cases according to $u_1 \geq u$ or $u_1 < u$. Suppose $u_1 \geq u$. Since $u_1 < u_2$, there exists a natural number i such that $u_1.i \leq u_2$. Since $\text{PosF}(M)$ is lower closed, $u\backslash u_2 \in \text{PosF}(M)$ implies $u\backslash(u_1.i) \in \text{PosF}(M)$. Namely, $u\backslash u_1$ is a position of a function symbol with arity more than i in M, thus so is ε in M_1. Hence, we have $i \in \text{Pos}(M_1)$. We see $\langle u_1.i, M_1|_i\rangle$ and $\langle u, M\rangle$ overlap and $\langle u_1.i, M_1|_i\rangle$ and $\langle u_2, M_2\rangle$ are unconnected. This allows as to apply induction hypothesis, so it follows from $u_1.i \leq u_2$ that there exists some $v' \in \text{PosV}(M_1|_i)$ such that $u_1.i.v' \leq u_2$. Let $v = i.v'$. Since $v \in \text{PosV}(M_1)$ we obtain the desired result. Finally, we treat the case $u_1 < u$. We see $\langle u, M_1|_{u_1\backslash u}\rangle$, and $\langle u, M\rangle$ overlap and $\langle u, M_1|_{u_1\backslash u}\rangle$ and $\langle u_2, M_2\rangle$ are unconnected. Therefore, we can apply induction hypothesis; from $u \leq u_2$ we obtain $v' \in \text{PosV}(M_1|_{u_1\backslash u})$ such that $u.v' \leq u_2$. Let $v = (u_1\backslash u).v'$. Since $u.v' = u_1.v$ and $v \in \text{PosV}(M_1)$, we are done. □

In the above lemma, the assumption that $\langle u, M\rangle$ and $\langle u_i, M_i\rangle$ overlap for $i = 1, 2$ is necessary. For example, consider $\langle u_1, M_1\rangle = \langle \varepsilon, f(a)\rangle$ and $\langle u_2, M_2\rangle = \langle 2, b\rangle$. Although they are unconnected and $u_1 \le u_2$, we can not find a variable position $v \in \mathrm{PosV}(M_1)$ such that $u_1.v \le u_2$.

Lemma 11. *Let M and N are linear terms such that* $\mathrm{V}(M) \cap \mathrm{V}(N) = \emptyset$. *If M and N are unifiable there exists some $\sigma \in \mathrm{mgu}(M, N)$ with the following properties:*

1. *$D(\sigma) \subseteq \mathrm{V}(M) \cup \mathrm{V}(N)$.*
2. *$M\sigma$ and $N\sigma$ are linear.*
3. *$M|_u\sigma = N|_u$ for any $u \in \mathrm{PosV}(M) \cap \mathrm{Pos}(N)$.*

The boring and rather straightforward proof of this lemma is omitted. We only note here that most general unifiers given by the well-known Martelli-Montanari style unification algorithm [10] fulfill the above requirements, so it is quite easy to find such most general unifiers.

The above lemmata serve for reducing a simultaneous overlap to repetition of usual overlaps. We give the following inductive definition:

Definition 12. Consider a left-linear system. Let M be a linear term, v a position, and U a (finite) simultaneous set of redex patterns such that $\langle v, M\rangle (= \bar{v})$ and $\langle u, \alpha\rangle$ overlap for any $\langle u, \alpha\rangle$ in U. Suppose, without loss of generality, no variable appears in $U \cup \{\bar{v}\}$ twice. We define a *simultaneous overlap* of U and $\bar{v}$ (notation $\langle U; \bar{v}\rangle$) by induction on $|U|$. In case $U = \emptyset$ we simply define $\langle U; \bar{v}\rangle$ as $\bar{v}$. Otherwise, let $\langle u, \alpha\rangle$ be a minimal[3] element of U, $U' = U - \{\langle u, \alpha\rangle\}$, and according to $v \le u$ or $u < v$ we define $\langle v', M'\rangle$ as $\langle v, M\sigma\rangle$ or $\langle u, \alpha\sigma\rangle$ respectively, where σ is a substitution in $\mathrm{mgu}(M|_{v\backslash u}, \alpha)$ or in $\mathrm{mgu}(\alpha|_{u\backslash v}, M)$ given by Lemma 11. We learn from Lemma 10 and Lemma 11 that M' is linear, $\langle v', M'\rangle$ and $\langle u', \alpha'\rangle$ overlap for any $\langle u', \alpha'\rangle$ in U' [4], and no variable appears twice. Hence, we define $\langle U; \langle v, M\rangle\rangle$ as $\langle U'; \langle v', M'\rangle\rangle$.

Note that the definition of $\langle U; \bar{v}\rangle$ is somewhat ambiguous as for several possibilities of $\sigma_1, \ldots, \sigma_n$ (and of minimal elements in U). This, however, does not give rise to any problem because they are variants of each other.

Hereafter, we use $SP(M)$ to refer the set $\{\langle u, N\rangle \mid \exists \sigma N\sigma = M|_u\}$. We now introduce the notion of simultaneous critical pairs as follows.

Definition 13. Consider a left-linear system $\mathcal{R}$. Let $\bar{v} = \langle v, \alpha \to \beta\rangle$ be a redex pattern and $U = \{\langle u_1, \alpha_1 \to \beta_1\rangle, \ldots, \langle u_n, \alpha_n \to \beta_n\rangle\} (n \ge 1)$ a simultaneous set of redex patterns such that $\langle v, \alpha \to \beta\rangle$ and $\langle u_i, \alpha_i \to \beta_i\rangle$ overlap for all $1 \le i \le n$ and that at least one of $u_1, \ldots, u_n, v$ is ε. Suppose $\mathrm{V}(\alpha), \mathrm{V}(\alpha_1), \ldots, \mathrm{V}(\alpha_n)$ are pairwise disjoint and $\bar{v} \notin U$. We can consider a simultaneous overlap $\langle U; \bar{v}\rangle$. Let us denote this as $\langle u, M\rangle$. We then obtain a complete development $M \twoheadrightarrow P$ of $U|_u$

[3] We define $\langle u_1, M_1\rangle \le \langle u_2, M_2\rangle$ as $u_1 \le u_2$.

[4] Since $\langle u, \alpha\rangle$ is minimal, we have $u \perp u'$ or $u.v \le u'$ for some $v \in \mathrm{PosV}(\alpha)$ (Lemma 10). In both cases, Lemma 11 garantees that $\langle v', M'\rangle$ and $\langle u', \alpha'\rangle$ overlap.

and a rewrite step $M \to Q$ using $\{\bar{v}\}|_u$. We call the pair $\langle P, Q\rangle$ of P and Q a *simultaneous critical pair* of $\mathcal{R}$.

A typical simultaneous critical pair is the one we have already seen in Fig. 1 (b) where the elements of U are represented pictorially by the gray triangles and $\bar{v}$ the white triangle. A concrete example is given below:

Example 1. The following left-linear system

$$\mathcal{R}_{AC} = \begin{cases} x + (y + z) \to (x + y) + z & (a) \\ x + y \to y + x & (c) \end{cases}$$

possesses the following simultaneous critical pairs:

$$\begin{array}{lll}
(1)\ \langle (y+z)+x,\ (x+y)+z\rangle & \text{from } \langle \varepsilon, (c)\rangle & \text{and } \langle \varepsilon, (a)\rangle \\
(2)\ \langle x+(z+y),\ (x+y)+z\rangle & \text{from } \langle 2, (c)\rangle & \text{and } \langle \varepsilon, (a)\rangle \\
(3)\ \langle x+((x'+y')+z'),\ (x+x')+(y'+z')\rangle & \text{from } \langle 2, (a)\rangle & \text{and } \langle \varepsilon, (a)\rangle \\
(4)\ \langle (z+y)+x,\ (x+y)+z\rangle & \text{from } \{\langle \varepsilon, (c)\rangle, \langle 2, (c)\rangle\} & \text{and } \langle \varepsilon, (a)\rangle \\
(5)\ \langle ((x'+y')+z')+x,\ (x+x')+(y'+z')\rangle & \text{from } \{\langle \varepsilon, (c)\rangle, \langle 2, (a)\rangle\} & \text{and } \langle \varepsilon, (a)\rangle
\end{array}$$

Among them, (1)-(3) are also regarded as usual critical pairs and (4)-(5) are proper simultaneous critical pairs. Note that for (1)-(3) we have to consider the reverse ones so that this system totally has eight simultaneous critical pairs.

The above system has only finitely many simultaneous critical pairs. This is generally true for arbitrary finite left-linear systems. To see this, we recall the finite development theorem for left-linear systems (Theorem 5). This theorem determines a unique (modulo renaming) P (and Q) for a pair $\langle U, \bar{v}\rangle$ required in Definition 13. Hence, what we have to know is that the number of such pairs $\langle U, \bar{v}\rangle$ is finite. This is true due to the requirement of $U \cup \{\bar{v}\}$ containing at least one ε.

To end this section, we present two more lemmata used in the next section.

Lemma 14. *Let M_1, M_2 be linear terms and u_1, u_2 positions such that $u_1 \backslash u_2 \in \mathrm{PosF}(M_1)$. Suppose $\mathrm{V}(M_1) \cap \mathrm{V}(M_2) = \emptyset$ and $\langle u_1, M_1\rangle, \langle u_2, M_2\rangle \subseteq \mathrm{SP}(M)$ for some M. We see that M_2 overlaps with M_1 at $u_1 \backslash u_2$ and $\langle u_1, M_1\sigma\rangle \in \mathrm{SP}(M)$ for any σ in $\mathrm{mgu}(M_1|_{u_1\backslash u_2}, M_2)$.*

The proof is easy and omitted. Using this, we obtain the following:

Lemma 15. *Let M be a linear term, v a position, and U a simultaneous set of redex patterns such that $\langle v, M\rangle (= \bar{v})$ and $\langle u, \alpha\rangle$ overlap for any $\langle u, \alpha\rangle$ in U. Suppose no variable appears in $U \cup \{\bar{v}\}$ twice. We then have*

$$U \subseteq \mathrm{SP}(N)\ \&\ \bar{v} \in \mathrm{SP}(N) \quad \Rightarrow \quad \langle U; \bar{v}\rangle \in \mathrm{SP}(N)$$

for any term N.

Proof. The proof is by induction on $|U|$. The case $|U| = 0$ is trivial. Suppose $|U| > 0$. Let $\langle u, \alpha \rangle$ be a minimal element of U and $U' = U - \{\langle u, \alpha \rangle\}$. Lemma 14 ensures that either α overlaps with M or M overlaps with α. According to $v \leq u$ or $u < v$ we define $\langle v', M' \rangle$ as $\langle v, M\sigma \rangle$ or $\langle u, \alpha\sigma \rangle$ respectively, where σ is in $\mathrm{mgu}(M|_{v\backslash u}, \alpha)$ or in $\mathrm{mgu}(\alpha|_{u\backslash v}, M)$ given by Lemma 11. We have $\langle v', M' \rangle \in \mathcal{M}$ by Lemma 14. We learn from Lemma 10 and Lemma 11 that M' is linear, $\langle v', M' \rangle$ and $\langle u', \alpha' \rangle$ overlap for any $\langle u', \alpha' \rangle$ in U', and no variable appears twice. Therefore, the result follows by induction hypothesis. □

4 Strong Closedness

In this section, we introduce based on the notion of simultaneous critical pairs our CR-criteria for arbitrary left-linear systems.

Definition 16. A simultaneous critical pair $\langle P, Q \rangle$ is *strongly closed* [5] *(with respect to complete development* $\multimap$*)* if

$$P \xrightarrow{*} M \multimapinv Q$$

for some term M.

It is powerful to allow arbitrary rewrite sequences. We list below some examples.

Example 2. Recall Example 1. It is not difficult to see $P \xrightarrow{*} Q$ for all the eight simultaneous critical pairs $\langle P, Q \rangle$ of $\mathcal{R}_{AC}$. Hence $\mathcal{R}_{AC}$ is strongly closed wrt $\multimap$.

Example 3 [4]. Consider the left-linear (but neither terminating nor right-linear) system

$$\mathcal{R}_G = \begin{cases} f(a, a) \to b & (a) \\ a \to a' & (b) \\ f(a', x) \to f(x, x) & (c) \\ f(x, a') \to f(x, x) & (d) \\ f(a', a') \to b & (e) \\ b \to f(a', a') & (f) \end{cases}$$

This system was given in [4] as an example such that its CR property can not be ensured by any existing criteria (given before [4]). $\mathcal{R}_G$ has the following (at most) eleven simultaneous critical pairs (we also count the reverse ones for (1)-(5)):

(1) $\langle f(a', a'),\ f(a', a') \rangle$	from $\langle \varepsilon, (c) \rangle$	and $\langle \varepsilon, (d) \rangle$
(2) $\langle f(a', a),\ b \rangle$	from $\langle 1, (b) \rangle$	and $\langle \varepsilon, (a) \rangle$
(3) $\langle f(a, a'),\ b \rangle$	from $\langle 2, (b) \rangle$	and $\langle \varepsilon, (a) \rangle$
(4) $\langle f(a', a'),\ b \rangle$	from $\langle \varepsilon, (c) \rangle$	and $\langle \varepsilon, (e) \rangle$
(5) $\langle f(a', a'),\ b \rangle$	from $\langle \varepsilon, (d) \rangle$	and $\langle \varepsilon, (e) \rangle$
(6) $\langle f(a', a'),\ b \rangle$	from $\{\langle 1, (b) \rangle,\ \langle 2, (b) \rangle\}$	and $\langle \varepsilon, (a) \rangle$

[5] This name is coined by Huet [5]. For left-linear systems without proper simultaneous critical pairs, his definition and ours coincide except for the difference of $\to$ and $\multimap$.

(1) is trivial. For (2) and (3) we have $P \rightarrow \cdot \leftarrow Q$. For (4)-(6) we have $P \rightarrow Q$. Hence, $\mathcal{R}_G$ is strongly closed (wrt $\mathrel{\circ\!\!\!\rightarrow}$).

We will see more examples in Sec.5.

What we would like to state is that the strong closedness implies CR property. For this, it suffices to show the lemma presented below. The key idea of proving this is to develop *all* redex patterns that make overlaps at first. The rest part then follows by the permutation lemma.

Lemma 17. *Consider a left-linear strongly closed (wrt* $\mathrel{\circ\!\!\!\rightarrow}$*) system. For any term* M*,* N_1*, and* N_2 *there exists some term* N *such that* $N_1 \mathrel{\leftarrow\!\!\!\circ} M \rightarrow N_2$ *implies* $N_1 \overset{*}{\rightarrow} N \mathrel{\leftarrow\!\!\!\circ} N_2$*.*

Proof. Suppose $N_1 \overset{U}{\mathrel{\leftarrow\!\!\!\circ}} M \overset{v}{\rightarrow} N_2$. We will show that $N_1 \overset{*}{\mathrel{\circ\!\!\!\rightarrow}} N \mathrel{\leftarrow\!\!\!\circ} N_2$ for some N (This outline is shown in Fig. 2). In case $U \uparrow \bar{v}$, we immediately obtain N such that $N_1 \mathrel{\circ\!\!\!\rightarrow} N \mathrel{\leftarrow\!\!\!\circ} N_2$ by appeal to Lemma 6. Hereafter, we assume $U \not\uparrow \bar{v}$. Let $U' (\subseteq U)$ be the set of all redex patterns that make overlaps with $\bar{v}$, i.e. $U' = \{\langle u, \alpha' \rightarrow \beta' \rangle \mid \exists \sigma \; \alpha'\sigma = \alpha|_{v \backslash u}\sigma \text{ or } \alpha\sigma = \alpha'|_{u \backslash v}\sigma\}$. We can consider a simultaneous overlap $\langle U'; \bar{v} \rangle$. We denote this as $\langle u, L \rangle$. Suppose $\bar{v} \notin U'$. Since U' is non-empty, we can consider a simultaneous critical pair $\langle P, Q \rangle$ such that

$$P \overset{U'|_u}{\mathrel{\leftarrow\!\!\!\circ}} L \overset{\{v\}|_u}{\rightarrow} Q.$$

Since $U' \subseteq \mathrm{SP}(M)$ and $\bar{v} \in \mathrm{SP}(M)$, we have $\langle U'; \bar{v} \rangle \in \mathrm{SP}(M)$ by Lemma 15. Namely, there exists some substitution ρ such that $L\rho = M|_u$. By Lemma 7 we have

$$M \overset{U'}{\mathrel{\circ\!\!\!\rightarrow}} M_1 \overset{U \backslash U'}{\mathrel{\circ\!\!\!\rightarrow}} N_1$$

for some M_1. Since $M_1 = M[P\rho]_u$ and $N_2 = M[Q\rho]_u$, it follows from the strong closedness that

$$M_1 \overset{U_1}{\mathrel{\circ\!\!\!\rightarrow}} \ldots \overset{U_n}{\mathrel{\circ\!\!\!\rightarrow}} M_{n+1} \overset{U_0}{\mathrel{\leftarrow\!\!\!\circ}} N_2 \qquad (1)$$

for some terms $M_2, \ldots M_{n+1}$ and simultaneous sets $U_0, \ldots, U_n$. On the other hand, in case $\bar{v} \in U'$ we immediately obtain $M_1 \overset{U \backslash v}{\mathrel{\leftarrow\!\!\!\circ}} N_2$ by Lemma 7. So, in any case we have (1).

Clearly, $\langle u, L \rangle$ and each element of $U - U'$ are unconnected, so that $\langle u, P \rangle$ and each element of $(U - U') \backslash U'$ are also unconnected. On the other hand, the elements of U_1 appear only in the place of P. From this, we can reason that $U \backslash U' (= (U - U') \backslash U') \uparrow U_1$. (This situation is shown in Fig. 3 where the black triangles represent the elements of U or $(U - U') \backslash U'$ and the gray triangles the elements of U_1.) Therefore, we can apply Lemma 6 to obtain

$$M_2 \overset{(U \backslash U') \backslash U_1}{\mathrel{\circ\!\!\!\rightarrow}} N_2' \overset{U_1 \backslash (U \backslash U')}{\mathrel{\leftarrow\!\!\!\circ}} N_1$$

for some term N_2'. We have $(U \backslash U') \backslash U_1 \uparrow U_2$ by the same reasoning. Therefore, repeating this process n times yields a complete development $M_{n+1} \mathrel{\circ\!\!\!\rightarrow} N_{n+1}'$ of

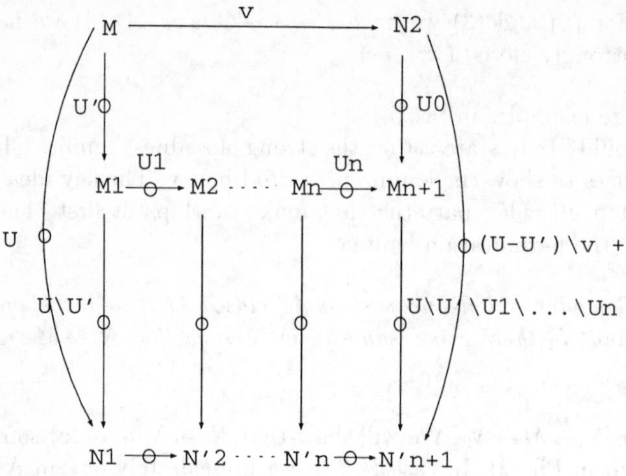

Fig. 2. Proof outline of Lemma 17

$U\backslash U'\backslash U_1\backslash\ldots\backslash U_n$ and a complete development sequence $N_1 \twoheadrightarrow N'_2 \ldots \twoheadrightarrow N'_{n+1}$, hence a rewrite sequence $N_1 \xrightarrow{*} N'_{n+1}$. By Lemma 8 we have

$$U\backslash U'\backslash U_1\backslash\ldots\backslash U_n = (U - U')\backslash U'\backslash U_1\backslash\ldots\backslash U_n = (U - U')\backslash\bar{v}\backslash U_0.$$

So, by Lemma 7 we obtain a complete development step

$$N_2 \overset{(U-U')\backslash v \,\cup\, U_0}{\twoheadrightarrow} N'_{n+1}.$$

Taking N'_{n+1} as N, we are done. □

Our proof heavily relies on the notion of the complete development, which simplifies our proof. The idea of using complete developments in study of CR-criteria is not new and goes back to Oostrom [16]. Actually, our proof of Lemma 17 is inspired from his proof.

Now, our main result is obtained from the above lemma by easy diagram chasing:

Theorem 18. *Left-linear systems are CR if every simultaneous critical pair is strongly closed (wrt* $\twoheadrightarrow$*).*

As special cases, we obtain the following decidable sufficient conditions for CR:

Corollary 19. *Left-linear systems are CR if for any simultaneous critical pair* $\langle P, Q\rangle$ *we have* $P \twoheadrightarrow \cdot \twoheadleftarrow Q$.

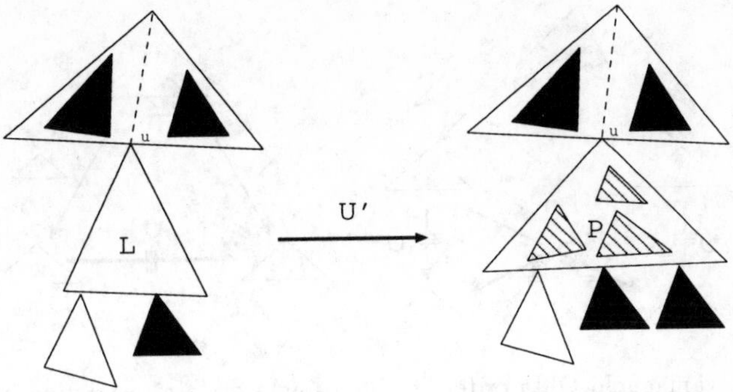

Fig. 3. Picture of $U \backslash U' \uparrow U_1$

Corollary 20. *Left-linear systems are CR if for any simultaneous critical pair* $\langle P, Q \rangle$ *we have* $P \mathrel{-\!\!\circ\!\!\!\rightarrow} Q$ *or* $P \mathrel{\leftarrow\!\!\!\circ\!-} Q$.

It should be noted that our result partially answers an open problem related to Huet's parallel-closedness [5], which enquires whether it is possible to reverse the direction of the parallel step in Huet's parallel-closedness [3]. This is known to be a hard problem. To the best of the author's knowledge, only a partial solution is given by Oyamaguchi and Ohta, who have shown that a left-linear system is CR if its critical pairs are closed by *upside parallel* rewrite step in the reverse direction [17] (see Fig. 4. [6]). A significance of our result is giving another partial solution for this problem, where "partial" means that we have to consider not only usual critical pairs but also proper simultaneous ones. Apart from this, our result is more general than [17].

Our result is also compared with Huet's another classical CR-criterion called *strong closedness*. As mentioned in Sec.1, it always requires *right-linearity* [5]. Another significance of our result is dropping this right-linearity at the cost of considering proper simultaneous critical pairs.

5 Related Work

Besides the studies of Oostrom and Glamlich that we have already mentioned, closely related to ours is an unpublished work of Toyama [19]. His criterion is given by replacing in our criterion complete development steps with parallel rewrite steps (see Fig. 5) but additionally requires $\mathrm{V}_V(N) \subseteq \mathrm{V}_U(M)$ where $\mathrm{V}_W(L)$ means the set of variable positions in L below some $w \in W$. This subsumes the result of [4], which was done independently of [19]. Interestingly, our result drops this additional condition, which is essential in his criterion, at the expense of considering nested positions.

[6] In Fig. 4 $\overset{U}{\mathrel{+\!\!\!+\!\!\!+\!\!\!\rightarrow}}$ stands for the union of the parallel rewrite relation $\overset{U}{\mathrel{+\!\!\!+\!\!\!\rightarrow}}$ and $\overset{\varepsilon}{\leftarrow}$

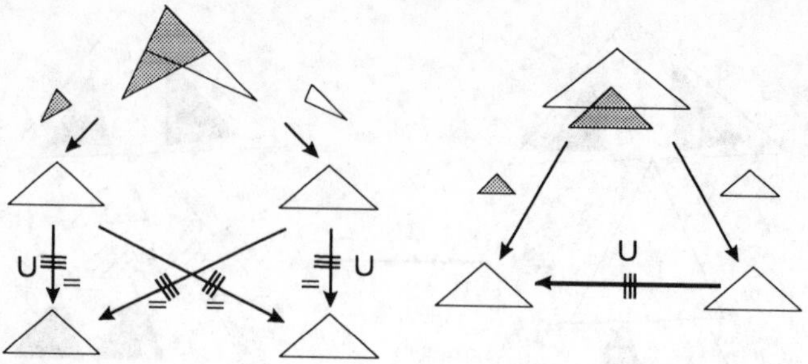

Fig. 4. Oyamaguchi-Ohta criterion (imposing a certain restriction on U [17])

In spite of the difference of parallel and complete development steps, [4], [19], and ours share the same split [21]. One of the most important differences is that we have more freedom for manipulating rewrite steps (by using Lemma 7) since complete development steps are defined for possibly nested positions, while parallel rewrite steps, only for disjoint positions. This is why the additional condition [19] becomes unnecessary in our criterion.

Gramlich has compared in [4] Huet-Toyama's parallel-closedness [5, 20], Huet's strong closedness [5], and his own parallel critical pair criterion, showing that they are incomparable one another. We extend this comparison by incorporating criteria given by Oostrom, Oyamaguchi and Ohta, and Toyama (We omit Huet's strong closedness as it always assumes right-linearily). In Table 1, $\mathcal{R}_1$ [12], $\mathcal{R}_2$ and $\mathcal{R}_3$ refer to the following left-linear CR systems:

$$\mathcal{R}_1 = \begin{cases} f(g(g(x)))\to a \\ f(g(h(x)))\to b \\ f(h(g(x)))\to b \\ f(h(h(x)))\to c \\ g(x)\to h(x) \\ a\to b \\ b\to c \end{cases} \mathcal{R}_2 = \begin{cases} a\to c \\ b\to c \\ f(a,b)\to d \\ f(x,c)\to f(c,c) \\ f(c,x)\to f(c,c) \\ d\to f(a,c) \\ d\to f(c,b) \end{cases} \mathcal{R}_3 = \begin{cases} f(x)\to x \\ f(x)\to f(f(x)) \end{cases}$$

From this table, we can observe that these three criteria are incomparable one another.

On the other hand, our result extends the Huet-Toyama-Oostrom criterion. That is because any left-linear system satisfying the criterion has diamond property wrt $\circ\!\!\!\rightarrow$, thereby, strongly closed wrt $\circ\!\!\!\rightarrow$[7].

In this paper, we have restricted ourselves to first-order rewrite systems although *higher-order* and *combinatory* rewrite systems are important current research topics [8, 13, 9]. Mayr and Nipkow described in [11] that some well-known results in first-order systems, including CR-property of orthogonal systems, are

[7] This is pointed out for the author by an anonymous referee.

Table 1. Incomparable criteria

	$\mathcal{R}_1$	$\mathcal{R}_2$	$\mathcal{R}_3$
(i) Toyama-Gramlich [19, 4]	○	×	×
(ii) Oyamaguchi-Ohta [17]	×	○	×
(iii) ours	×	×	○

○ ... satisfied × ... not satisfied

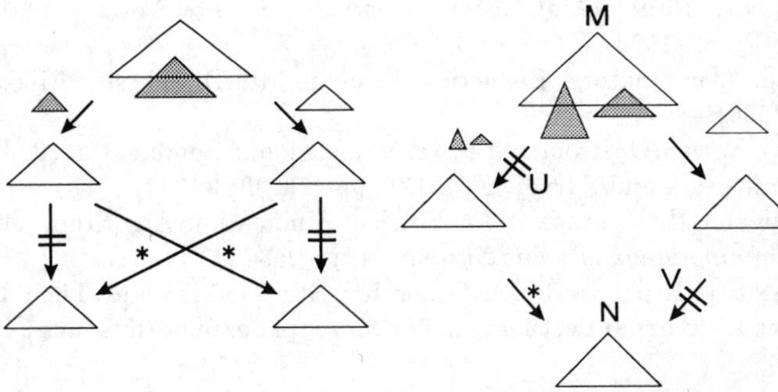

Fig. 5. Toyama's criterion (requiring a certain restriction)

lifted to (orthogonal) *pattern rewriting systems* (PRS's). (See also [14].) In [16], Oostrom also has presented his result for PRS's. Therefore, it is natural to ask whether our result also holds for PRS's. This paper lacks consideration on this. The author expects this is true as our proof essentially relies only on the permutation lemma and the finite development theorem, which hold for left-linear PRS's as in [15].

Acknowledgments. The author is grateful to Michio Oyamaguchi, Aart Middeldorp, Bernhard Gramlich, and anonymous referees for their valuable comments and advice. The author is also grateful to the participants in the TRS meeting held in Tsukuba for their discussion.

References

1. G. Boudol: Computational Semantics of Term Rewriting Systems, In M. Nivat and J.C. Reynolds, eds., *Algebraic Method in Semantics*, pp.169–236, Cambridge University Press (1985).
2. N. Dershowitz and J.-P. Jouannaud: Rewrite Systems, In J. van Leeuwen, ed., *Handbook of Theoretical Computer Science*, Vol.B, pp.243–320, The MIT Press (1990).

3. N. Dershowitz, J.-P. Jouannaud, and J. W. Klop: Open Problems in Rewriting, In *Proceedings of RTA-91* (LNCS 488), pp.445–456, Springer (1991).
4. B. Gramlich: Confluence without termination via parallel critical pairs, In *Proceedings of CAAP-96* (LNCS 1059), pp.211–225, Springer (1996).
5. G. Huet: Confluent Reductions: Abstract Properties and Applications to Term Rewriting Systems, *JACM*, 27, pp.797–821,1980.
6. G. Huet and J.J. Lévy: Computations in Orthogonal Rewrite Systems, I and II, In J.-L. Lassez and G. Plotkin, eds., *Computational Logic, Essays in Honor of Alan Robinson*, pp.396–443, MIT Press, 1991.
7. J. W. Klop: Term Rewriting Systems, In S. Abramsky, D.M. Gabbay, and T.S.E. Maibaum, eds., *Handbook of Logic in Computer Science*, Vol.2, pp.2–116, Oxford University Press (1992).
8. J. W. Klop: Combinatory Reduction Systems, Ph.D. Thesis, Rijksuniversiteit Utrecht (1980).
9. J. W. Klop, V. van Oostrom and F. van Raamsdonk: Combinatory Reduction Systems, Introduction and Survey, *TCS* 121, pp.279–308 (1993)
10. A. Martelli and U. Montanari: An Efficient Unification Algorithm, *ACM Trans. on Programming languages and Systems*, 4, pp.258–282 (1982).
11. R. Mayr and T. Nipkow: Higher-Order Rewrite Systems and Their Confluence, `http://www4.informatik.tu-muenchen.de/nipkow/pubs/hrs.html`, To appear in *TCS*.
12. A. Middeldorp: Personal Communication (1997).
13. T. Nipkow: Higher-Order Critical Pairs, In Proceedings of *LICS-91*, pp.342–349 (1991).
14. T. Nipkow: Orthogonal Higher-Order Rewrite Systems are Confluent, In Proceedings of *TLCA-93* (LNCS 664), pp.306–317 (1993)
15. V. van Oostrom: Development Closed Critical Pairs, In HOA-95 Selected Papers (LNCS 1074), pp. 185–200 (1995).
16. V. van Oostrom: Developing Developments, *TCS*, 175.1, pp.159–181 (1997).
17. M. Oyamaguchi and Y. Ohta: A New Parallel Closed Condition for Church-Rosser of Left-Linear Term Rewriting Systems, In *Proceedings of RTA-97* (LNCS 1232), pp.187–201 (1997).
18. B. K. Rosen: Tree Manipulating Systems and Church-Rosser Theorems, *JACM*, 20, pp.160–187 (1973).
19. Y. Toyama: On the Church-Rosser Property of Term Rewriting Systems (Japanese), *Technical Report*, 17672, NTT ECL (1981).
20. Y. Toyama: Commutativity of Term Rewriting Systems, In Fuchi and L. Kott eds., *Programming of Future Generation Computer*, Vol.II, pp.393–407, North-Holland (1988).
21. Y. Toyama: Personal Communication (1997).

Church-Rosser Theorems for Abstract Reduction Modulo an Equivalence Relation

Enno Ohlebusch

University of Bielefeld, Technische Fakultät
P.O. Box 100131, 33501 Bielefeld, Germany
enno@TechFak.Uni-Bielefeld.DE

Abstract. A very powerful method for proving the Church-Rosser property for abstract rewriting systems has been developed by van Oostrom. In this paper, his technique is extended in two ways to abstract rewriting modulo an equivalence relation. It is shown that known Church-Rosser theorems can be viewed as special cases of the new criteria. Moreover, applications of the new criteria yield several new results.

1 Introduction

An abstract reduction[1] system (ARS) is just a set of objects and a sequence of binary relations on it. ARSs are called general replacement systems in [Ros73,Sta75]. Abstract rewriting comprises several kinds of rewriting like term-, string-, graph-, and conditional rewriting. Thus a repetition of similar definitions and concepts is avoided by stating them once and for all on an abstract level (for ARSs, that is). In essence, a rewriting relation models non-deterministic computations. A key property in the theory of rewriting is the Church-Rosser property which guarantees that the results of such computations are unique. Church-Rosser theorems for abstract rewriting were given for instance by Hindley [Hin64], Rosen [Ros73], Staples [Sta75], Huet [Hue80], Geser [Ges90], and van Oostrom [Oos94a]. The reader is referred to the survey of Klop [Klo92] for details. In this paper, we study ARSs which are additionally equipped with an equivalence relation $\sim$ on the set of objects. Such an ARS is called Church-Rosser if every two objects which are convertible have reducts that are equivalent. It is well-known that for abstract reduction modulo $\sim$ the Church-Rosser property does not coincide with confluence. This is in sharp contrast to pure abstract rewriting. Sethi's [Set74] Church-Rosser theorem for bounded ARSs has subsequently been improved by Huet [Hue80]. Further Church-Rosser theorems for abstract rewriting modulo $\sim$ were obtained in the context of completion modulo equations; see e.g. [JM84,JK86]. In this paper, we will give several new sufficient conditions for the Church-Rosser property for abstract rewriting modulo $\sim$ and review known methods. This is essentially done by generalizing the powerful method devised by van Oostrom [Oos94a].

[1] The terms reduction and rewriting are used synonymously throughout the paper.

2 Preliminaries

Definition 1. An *abstract reduction system* $\mathcal{A} = (A, \langle \to_\alpha \rangle_{\alpha \in I}, \sim)$ is a structure consisting of a set of objects A, a sequence of relations $\to_\alpha$ on A, and an equivalence relation $\sim$ on A. A relation $\to_\alpha$ is said to be a *reduction* relation *labeled* by α. The reduction relation of $\mathcal{A}$ is the union of its constituent reduction relations: $\to_{\mathcal{A}} = \bigcup_{\alpha \in I} \to_\alpha$. When the ARS is clear from the context, it will be suppressed. Two ARSs $\mathcal{A} = (A, \langle \to_\alpha \rangle_{\alpha \in I}, \sim)$ and $\mathcal{B} = (A, \langle \to_\beta \rangle_{\beta \in J}, \sim)$ are *reduction equivalent* if $\to_{\mathcal{A}} = \to_{\mathcal{B}}$.

The symbols $\leftarrow$, $\to^=$, $\to^+$, and $\to^*$ denote the inverse, the reflexive closure, the transitive closure, and the reflexive transitive closure of $\to$, respectively. We use $\to_\alpha \cdot \to_\beta$ to denote the composition of $\to_\alpha$ and $\to_\beta$. The relation $\to^* \cdot {}^*\!\leftarrow$ ($\to^* \cdot \sim \cdot {}^*\!\leftarrow$) is called *joinability* (*modulo* $\sim$) and denoted by $\downarrow$ ($\downarrow_\sim$). We further define $\mapsto\!\!\!\!\!\leftarrow\ = \to \cup \leftarrow \cup \sim$, $\approx\ = (\mapsto\!\!\!\!\!\leftarrow)^*$, and $\to_\sim\ = \sim \cdot \to \cdot \sim$. The relation $\approx$ is called *conversion.*

If $a \to b$, then we speak of a *reduction step* from a to b. An element $a \in A$ is in *normal form* if there is no element $b \in A$ with $a \to b$; it *has a normal form* if $a \to^* b$ for some normal form b. The ARS $\mathcal{A}$ is called *normalizing* (weakly normalizing–WN) if every term has a normal form. $\mathcal{A}$ is *terminating* (strongly normalizing–SN) if there is no infinite reduction sequence w.r.t. $\to$. $\mathcal{A}$ is *terminating modulo* $\sim$ (SN$\sim$) if there is no infinite reduction sequence w.r.t. $\to_\sim$.

The label of a finite reduction sequence is the string of the labels of its constituent reduction steps. The Greek letters σ, τ, μ, ν etc. will be used to denote strings. The concatenation of two strings σ and τ is denoted by $\sigma\tau$.

The rest of this section is copied from [Oos94a]. The reader should consult [Oos94a] or [Oos94b] for more details and intuitive explanations.

Definition 2. A (*general*) *multiset* is a collection in which elements are allowed to occur more than once or even infinitely often. A *finite* multiset has finitely many different elements which occur finitely often. A *set* is a multiset in which elements occur either not at all or infinitely often.

To distinguish between set comprehension and finite multiset comprehension, braces will be used to denote the former and square brackets to denote the latter. For example $[\alpha, \beta]$ denotes the finite multiset with exactly one occurrence of both α and β, whereas $\{\alpha\}$ denotes the set multiset with infinitely many occurrences of α.

The following (in)equalities illustrate the differences between finite and set multisets, as well as sum and union: $[\alpha] \uplus [\alpha] = [\alpha, \alpha] \neq [\alpha] = [\alpha] \cup [\alpha]$, $\{\alpha\} \uplus \{\alpha\} = \{\alpha, \alpha\} = \{\alpha\} = \{\alpha\} \cup \{\alpha\}$, $[\alpha, \alpha] - [\alpha] = [\alpha]$, $\{\alpha\} - [\alpha, \alpha] = \{\alpha\}$, and $[\alpha, \alpha] - \{\alpha\} = \emptyset$.

The multiset $[\sigma]$ of labels of a string σ is the sum of all label occurrences in it, so in particular we have $[\sigma\tau] = [\sigma] \uplus [\tau]$. For example, if we have digits as labels, $[132343] = [1, 3, 2, 3, 4, 3]$.

In the sequel, $\prec$ denotes a strict partial order on the set of labels I.

Definition 3. (1) The *down-set* $\curlyvee\alpha$ of α is defined by $\curlyvee\alpha = \{\beta \mid \beta \prec \alpha\}$. This is extended to multisets and strings by defining $\curlyvee M = \bigcup_{\alpha \in M} \curlyvee\alpha$ and $\curlyvee\sigma = \curlyvee[\sigma]$. For example, $\curlyvee 2 = \curlyvee[0,2] = \curlyvee 212 = \{0,1\}$.

(2) The *(standard) multiset extension* (denoted by $\prec_{mul}$) of the partial order $\prec$ is defined by

$$M \prec_{mul} N \quad \text{if} \ \exists\, X, Y, Z : M = Z \uplus X, N = Z \uplus Y, X \subseteq \curlyvee Y \text{ and } Y \neq \emptyset.$$

Furthermore, $\preceq_{mul}$ will be used to denote the reflexive closure of $\prec_{mul}$.

Definition 4. *The (lexicographic maximum) measure* grades strings by finite multisets and is denoted by $|\cdot|$. It is defined inductively by $|\varepsilon| = \emptyset$, where ε denotes the empty string, and $|\alpha\sigma| = [\alpha] \uplus (|\sigma| - \curlyvee\alpha)$.

The measure of a rewrite sequence $a_0 \to_{\alpha_0} a_1 \to_{\alpha_1} \dots a_m \to_{\alpha_m} a_{m+1}$ is the measure of its labels, i.e., $|\alpha_0\alpha_1 \dots \alpha_m|$.

Intuitively, the lexicographic maximum measure takes the multiset of elements which are maximal (in the $\succ$ ordering) with respect to the elements to their left in the string. Operationally, one can think of filtering out the noise before proceeding to the right. For instance, taking the measures of the strings of digits 132343 and 211 yields $|132343| = [1,3,3,4]$ and $|211| = [2]$. Taking the measure of the rewrite sequence $a \to_2 b \to_1 c$ yields $|a \to_2 b \to_1 c| = |21| = [2]$.

3 Properties of ARSs

The following definitions describe basic properties of ARSs. Suppose that $\vdash\!\dashv$ is a symmetric relation on A with $\vdash\!\dashv^* = \sim$. Note that $\sim^* = \sim$. Throughout this section, $\vdash\!\dashv$ indeed coincides with $\sim$. A (partial) analysis of the relationships between the last six properties can be found in the next section.

Definition 5. Let $\mathcal{A} = (A, \langle \to_\alpha \rangle_{\alpha \in I}, \sim)$ be an ARS.

1. We write $\Diamond(\to^* \cdot \sim)$ and say that the *diamond property* holds for $\to^* \cdot \sim$ if $\sim \cdot {}^*\!\leftarrow \cdot \to^* \cdot \sim \ \subseteq \ \downarrow_\sim$.
2. $\mathcal{A}$ is *Church-Rosser modulo* $\sim$ (CR$\sim$) if $\approx \ \subseteq \ \downarrow_\sim$.
3. $\mathcal{A}$ is *almost Church-Rosser modulo* $\sim$ (ACR$\sim$) if ${}^*\!\leftarrow \cdot \sim \cdot \to^* \ \subseteq \ \downarrow_\sim$.
4. $\mathcal{A}$ is *confluent modulo* $\sim$ (CON$\sim$) if ${}^*\!\leftarrow \cdot \to^* \ \subseteq \ \downarrow_\sim$.
5. $\mathcal{A}$ is *locally confluent modulo* $\sim$ (LCON$\sim$) if $\leftarrow \cdot \to \ \subseteq \ \downarrow_\sim$.
6. $\mathcal{A}$ is *strongly* LCON$\sim$ (SLCON$\sim$) if $\leftarrow \cdot \to \ \subseteq \to^= \cdot \sim \cdot {}^*\!\leftarrow$.
7. $\mathcal{A}$ is *coherent with* $\vdash\!\dashv$ (COH$\vdash\!\dashv$) if $\vdash\!\dashv \cdot \to^* \ \subseteq \ \downarrow_\sim$.
8. $\mathcal{A}$ is *locally coherent with* $\vdash\!\dashv$ (LCOH$\vdash\!\dashv$) if $\vdash\!\dashv \cdot \to \ \subseteq \ \downarrow_\sim$.
9. $\mathcal{A}$ is *strongly coherent with* $\vdash\!\dashv$ (SCOH$\vdash\!\dashv$) if $\vdash\!\dashv \cdot \to^* \cdot \sim \ \subseteq \ \downarrow_\sim$.
10. $\mathcal{A}$ is *compatible with* $\vdash\!\dashv$ (COM$\vdash\!\dashv$) if $\vdash\!\dashv \cdot \to^* \ \subseteq \ \to^* \cdot \sim$.
11. $\mathcal{A}$ is *strongly compatible with* $\vdash\!\dashv$ (SCOM$\vdash\!\dashv$) if $\vdash\!\dashv \cdot \to \ \subseteq \ \to^= \cdot \sim$.
12. $\mathcal{A}$ is *locally commuting with* $\vdash\!\dashv$ (LCMU$\vdash\!\dashv$) if $\vdash\!\dashv \cdot \to \ \subseteq \ \to^+ \cdot \sim$.

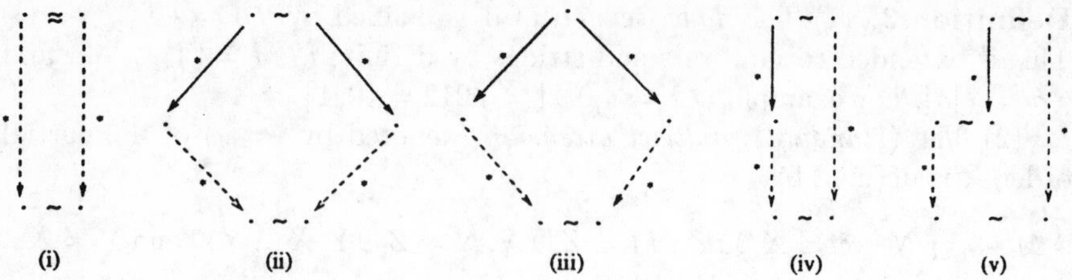

Fig. 1. Properties modulo $\sim$

Figure 1 illustrates the properties (i) CR$\sim$, (ii) ACR$\sim$, (iii) CON$\sim$, (iv) COH$\sim$, and (v) SCOH$\sim$. For instance, $\mathcal{A}$ is CON$\sim$ if $\forall a, b, c \in A$ with $a \;{}^*\!\!\leftarrow b \rightarrow^* c$, there are $d, e \in A$ such that $a \rightarrow^* d \sim e \;{}^*\!\!\leftarrow c$. The term "confluent modulo $\sim$" for property CON$\sim$ is taken from [Ave95]. Note that property ACR$\sim$ is also called "confluence modulo $\sim$" in [Hue80]. In order to avoid the same name for two different properties, we use the notion "almost Church-Rosser modulo $\sim$" for the latter. Note that ACR$\sim$ implies confluence of $\rightarrow/\sim$ in $A/\sim$ but *not* vice versa; cf. [PPE94]. The term "locally commuting with $\vdash\!\dashv$" stems from [JM84]. Figure 2 depicts the relationships between the properties defined above. Most of the implications are trivial. The non-trivial implications are proven in the remainder of this section.

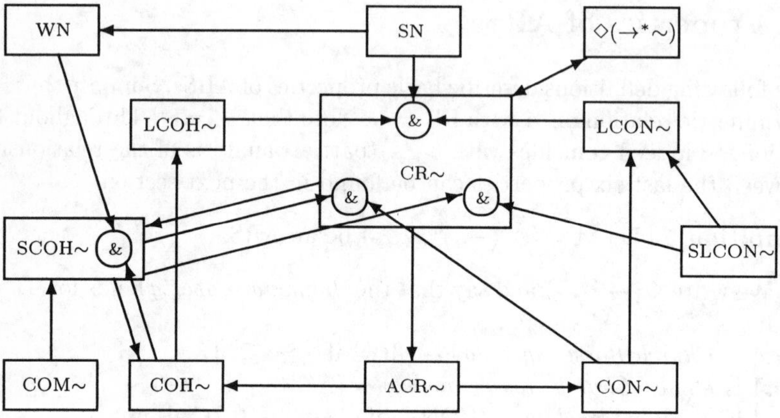

Fig. 2. Connections between the properties.

Proposition 6. *Let $\mathcal{A}$ be an ARS. The following statements are equivalent.*

1. *$\mathcal{A}$ is CR$\sim$.*
2. *$\diamond(\rightarrow^* \cdot \sim)$ holds.*
3. *$\mathcal{A}$ is CON$\sim$ and SCOH$\sim$.*

Proof. Apparently, (1) implies (2) and (2) implies (3). We prove that (3) implies (1). Let $a \overset{k}{\models\!\dashv} b$ be given, i.e., k is the number of $\models\!\dashv$-steps in the conversion $a \approx b$. We show by induction on k that there are $c, d \in A$ such that $a \to^* c \sim d \,{}^*\!\!\leftarrow b$. If $k = 0$, then $a = b$ and the claim holds. Consider $a \overset{k}{\models\!\dashv} b' \models\!\dashv b$. According to the inductive hypothesis, there are $c, d \in A$ such that $a \to^* c \sim d \,{}^*\!\!\leftarrow b'$. We further proceed by case analysis.
(i) $b' \sim b$: Since $c \sim d \leftarrow b' \sim b$ and $\mathcal{A}$ is SCOH$\sim$, there exist $e, f \in A$ such that $c \to^* e \sim f \,{}^*\!\!\leftarrow b$. That is, $a \to^* e \sim f \,{}^*\!\!\leftarrow b$.
(ii) $b' \to b$: For $d \,{}^*\!\!\leftarrow b' \to b$ there are $e, f \in A$ such that $d \to^* e \sim f \,{}^*\!\!\leftarrow b$ because $\mathcal{A}$ is CON$\sim$. Since $\mathcal{A}$ is SCOH$\sim$, it follows from $c \sim d \to^* e \sim f$ that there are $g, h \in A$ with $a \to^* c \to^* g \sim h \,{}^*\!\!\leftarrow f \,{}^*\!\!\leftarrow b$.
(iii) $b' \leftarrow b$: Trivial. □

Lemma 7. *For every ARS $\mathcal{A}$ the following statements hold.*

1. *If $\mathcal{A}$ is WN and COH$\sim$, then it is SCOH$\sim$.*
2. *If $\mathcal{A}$ is WN, CON$\sim$, and COH$\sim$, then it is CR$\sim$.*
3. *If $\mathcal{A}$ is WN and ACR$\sim$, then it is CR$\sim$.*

Proof. (1) Consider $a \sim b \to^* c \sim d$. Let c' be a normal form of c. It follows from COH$\sim$ in combination with the fact that c' is irreducible that there is an $e \in A$ such that $a \to^* e \sim c'$. Analogously, there is an $f \in A$ such that $d \to^* f \sim c'$. All in all, $a \to^* e \sim f \,{}^*\!\!\leftarrow d$.
(2) By (1), $\mathcal{A}$ is SCOH$\sim$. Thus, it is CR$\sim$ by Proposition 6.
(3) Direct consequence of (2) because ACR$\sim$ implies CON$\sim$ and COH$\sim$. □

Statements (2) and (3) of the preceding lemma are well-known. (2) can be found in [Ave95], Satz 4.1.6 and (3) in [Hue80], Lemma 2.6. The fact that ACR$\sim$ does not imply CR$\sim$ is also well-known; see [Hue80]. The counterexample in Fig. 3(i) was given by Aart Middeldorp; note that the ARS lacks WN and SCOH$\sim$. A counterexample showing that the combination of CON$\sim$ and COH$\sim$ does not imply ACR$\sim$ is depicted in Fig. 3(ii).

We next come to the first extension of van Oostrom's result.

Definition 8. Let $\to_\alpha$ and $\to_\beta$ be two relations on A. We say that $\to_\alpha$ subcommutes with $\to_\beta$ modulo $\sim$ if ${}_\alpha\!\!\leftarrow \cdot \to_\beta \subseteq \to_\beta^= \cdot \sim \cdot {}_\alpha^=\!\!\leftarrow$. Moreover, $\to_\alpha$ commutes with $\to_\beta$ modulo $\sim$ if $\to_\alpha^*$ subcommutes with $\to_\beta^*$ modulo $\sim$.

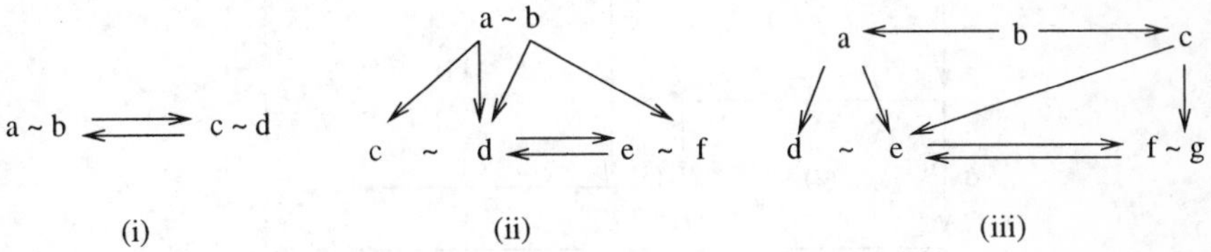

Fig. 3. Counterexamples

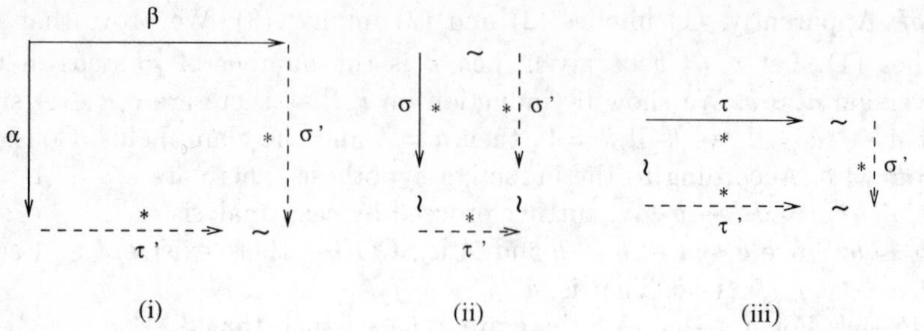

Fig. 4. Assumptions of Theorem 9.

Theorem 9. *Let $\mathcal{A} = (A, \langle \to_\alpha \rangle_{\alpha \in I}, \sim)$ be an ARS and let $\succ$ be a well-founded partial order on I. Let $I_v, I_h \subseteq I$, $\to_v = \bigcup_{\alpha \in I_v} \to_\alpha$, and $\to_h = \bigcup_{\beta \in I_h} \to_\beta$.*

1. *If $\forall \alpha \in I_v$, $\beta \in I_h$, $\sigma \in I_v^*$, $\tau \in I_h^*$ the diagrams in Fig. 4 hold so that $|\beta| \succeq_{mul} |\tau'|$, $\sigma' \in I_v^*$, $\tau' \in I_h^*$, then $\to_v$ commutes with $\to_h$ modulo $\sim$.*
2. *If $\to_\mathcal{A} = \to_v = \to_h$, then $\mathcal{A}$ is CR$\sim$.*

Proof. (1) Let $||\sigma||$ denote the length of σ, i.e., the number of labels in the string σ and let $>$ denote the natural order on $\mathbb{N}$. Let $>_{lex}$ be defined by $(|\tau|, ||\sigma||) >_{lex} (|\tau'|, ||\sigma'||)$ if $|\tau| \succ_{mul} |\tau'|$ or $|\tau| = |\tau'|$ and $||\sigma|| > ||\sigma'||$, where $\tau, \tau' \in I_h^*$ and $\sigma, \sigma' \in I_v^*$. So first the horizontal labels are compared and then the length of the vertical ones. Since $\succ$ is well-founded and $>_{lex}$ is the lexicographic product of $\succ_{mul}$ and $>$, $>_{lex}$ is well-founded too. Now we proceed by induction on $>_{lex}$. The proof is illustrated in Fig. 5. The rectangle (1) exists by assumption; cf. Fig. 4(i). Since $|\beta| \succeq_{mul} |\nu|$ and $||\alpha\sigma|| > ||\sigma||$, the inductive hypothesis is applicable in (2). (3) is true because of Fig. 4(ii). Then one can apply the inductive hypothesis again in (4) because $|\beta\tau| \succ_{mul} |\tau|$. The proof of (1) is concluded by applying Fig. 4(iii) in (5).
(2) Evidently, $\mathcal{A}$ is SCOH$\sim$. By (1), $\to_\mathcal{A}$ is CON$\sim$ too. Hence $\mathcal{A}$ is CR$\sim$ by Proposition 6. □

Note that in the above theorem, the order $\succ_{mul}$ can be replaced by any well-founded order $\gg$ on strings of labels satisfying the property $\beta\tau \gg \tau$.

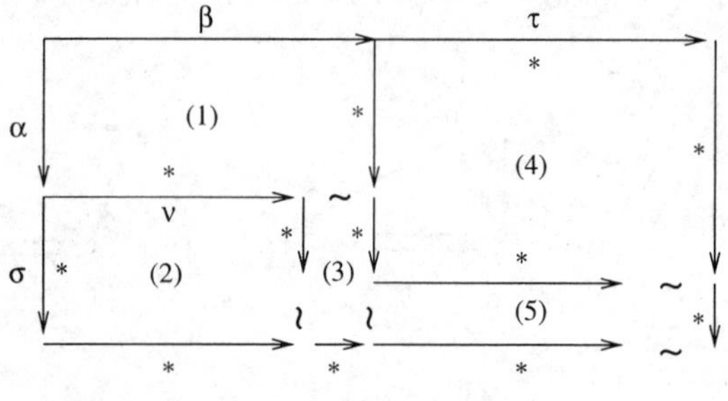

Fig. 5. Proof of Theorem 9.

A well-known result for pure abstract rewriting states that strong confluence ($\leftarrow \cdot \rightarrow \subseteq \rightarrow^{=} \cdot {}^{*}\!\leftarrow$) implies confluence. The modulo analogue to strong confluence is the property $\leftarrow \cdot \rightarrow \subseteq \rightarrow^{=} \cdot \sim \cdot {}^{*}\!\leftarrow$. This property, however, does not imply CON∼ as witnessed by the example in Fig. 3(iii); note that the system lacks SCOH∼. That's why we call the above property strong local confluence modulo ∼ instead of strong confluence modulo ∼. Since CON∼ is not a consequence of SLCON∼, the following result does not follow directly from Proposition 6.

Corollary 10. *If the ARS $\mathcal{A}$ is SLCON∼ and SCOH∼, then it is CR∼.*

Proof. Using Theorem 9, CON∼ can be shown as in the proof of [Oos94a], Corollary 4.6. Then the claim follows from Proposition 6. □

The following proposition owing to Huet [Hue80] is the modulo analogue to Newman's Lemma (a simpler proof than the original one given in [Hue80] can be found in [Oos94b]). It shows that in the presence of termination local versions of CON∼ and COH∼ are sufficient to infer CR∼.

Proposition 11. *If $\mathcal{A}$ is SN, LCON∼, and LCOH∼, then it is CR∼.*

Proof. We show that $\mathcal{A}$ is CON∼ and SCOH∼. In order to show that if (i) $a \; {}^{*}\!\leftarrow b \rightarrow^{*} d$ or (ii) $a \sim b \rightarrow^{*} c \sim d$, then $a \downarrow_{\sim} d$, we use induction on the well-founded order $\rightarrow^{+}$, i.e., we assume that (i) and (ii) hold for all $e \in A$ with $b \rightarrow^{+} e$. It is not difficult to verify (i) along the lines of the proof of Theorem 9. We prove (ii). If $b = c$, then the claim holds vacuously. So suppose $b \rightarrow c' \rightarrow^{*} c$. Since $\mathcal{A}$ is LCOH∼, there are $e, f \in A$ such that $a \rightarrow^{*} f \sim e \; {}^{*}\!\leftarrow c'$; see Fig. 6(1). Then the proposition follows by applying the inductive hypothesis thrice; see Fig. 6(2)-(4). □

4 Local Decreasingness

In this section, we consider ⊣⊢ instead of ∼. So in the sequel an ARS is a structure $\mathcal{A} = (A, \langle \rightarrow_{\alpha} \rangle_{\alpha \in I}, \vdash\!\dashv)$, where ⊣⊢ is a symmetric relation on A with

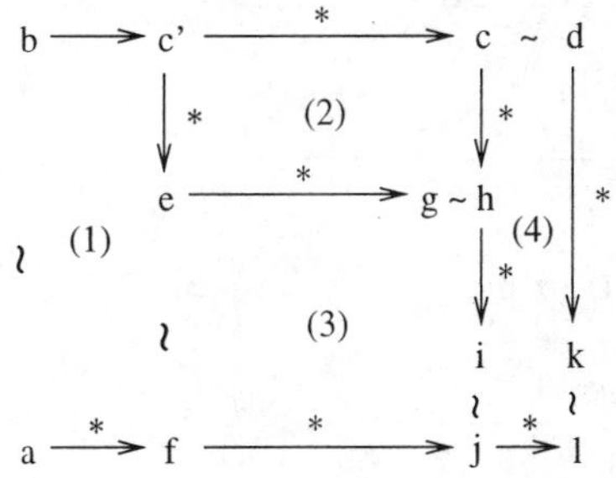

Fig. 6. Proof of Proposition 11.

$\vdash\!\dashv^* = \sim$. The next lemma shows that compatibility is a stronger property than coherence (for space reasons we refrain from giving a more detailed analysis of the relationships between the last six properties of Definition 5).

Lemma 12. *For every ARS $\mathcal{A}$ the following holds.*

1. COM$\vdash\!\dashv$ *is equivalent to* COM$\sim$.
2. SCOM$\vdash\!\dashv$ $\Rightarrow$ COM$\vdash\!\dashv$ $\Rightarrow$ SCOH$\vdash\!\dashv$ $\Rightarrow$ COH$\vdash\!\dashv$ $\Rightarrow$ LCOH$\vdash\!\dashv$,

Proof. (1) By induction on k one shows that $\vdash\!\dashv \cdot \rightarrow^* \subseteq \rightarrow^* \cdot \sim$ implies $\overset{k}{\vdash\!\dashv} \cdot \rightarrow^* \subseteq \rightarrow^* \cdot \sim$.
(2) We only prove SCOM$\vdash\!\dashv$ $\Rightarrow$ COM$\vdash\!\dashv$ because the remaining implications obviously hold. In fact, we show SCOM$\vdash\!\dashv$ $\Rightarrow$ COM$\sim$. To this end, we first show $a \sim b \rightarrow c \Rightarrow a \rightarrow^= d \sim c$ by induction on k in $a \overset{k}{\vdash\!\dashv} b$. The base case $k = 0$ clearly holds. Suppose $a \vdash\!\dashv b_1 \overset{k}{\vdash\!\dashv} b \rightarrow c$. According to the inductive hypothesis, we have $a \vdash\!\dashv b_1 \rightarrow^= b_2 \sim c$. Since $\mathcal{A}$ is SCOM$\vdash\!\dashv$ it follows $a \rightarrow^= b_3 \sim b_2 \sim c$. Hence $a \rightarrow^= b_3 \sim c$. We next show $a \sim b \rightarrow^* c \Rightarrow a \rightarrow^* d \sim c$ by induction on k in $b \rightarrow^k c$. Again, the base case $k = 0$ is true. So suppose $a \sim b \rightarrow b_1 \rightarrow^k c$. By the above result, it follows $a \rightarrow^= b_2 \sim b_1 \rightarrow^k c$. Now the inductive hypothesis implies $a \rightarrow^= b_2 \rightarrow^* b_3 \sim c$. Thus, $a \rightarrow^* b_3 \sim c$. □

We next present the main theorem of this paper. It constitutes the aforementioned second (and proper) generalization of van Oostrom's result to rewriting modulo $\sim$. As in Theorem 9, it is required that the constituent relations of the ARS under consideration are ordered by a well-founded order. The proof of this theorem is postponed to Sect. 5. Its consequences are treated first.

Definition 13. Let $\mathcal{A} = (A, \langle \rightarrow_\alpha \rangle_{\alpha \in I}, \vdash\!\dashv)$ be an ARS and $\succ$ a well-founded order on I. For $\alpha, \beta \in I$, we write $LD(\alpha, \beta)$ and $LD(\alpha)$ if the respective diagram in Fig. 7 holds. Here γ stands for $\curlyvee\alpha \cup \curlyvee\beta$. Such diagrams are called *locally decreasing*.

Main Theorem 14. *Let $\mathcal{A} = (A, \langle \rightarrow_\alpha \rangle_{\alpha \in I}, \vdash\!\dashv)$ be an ARS. If there is a well-founded partial order $\succ$ on I such that for all $\alpha, \beta \in I$ properties $LD(\alpha, \beta)$ and $LD(\alpha)$ hold, then $\mathcal{A}$ is CR$\sim$.*

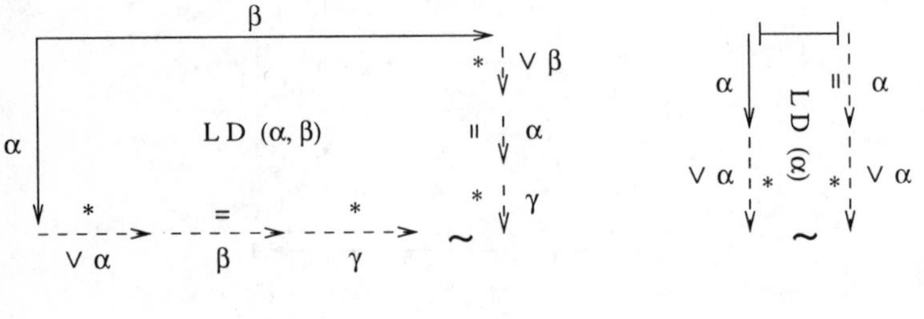

Fig. 7. Locally decreasing diagrams.

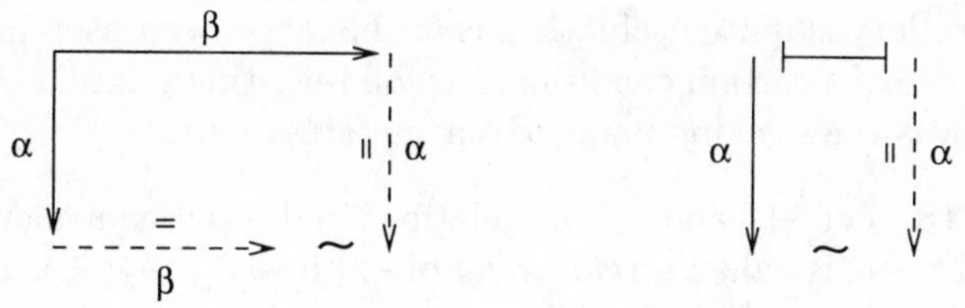

Fig. 8. Subcommutation modulo $\sim$ and SCOM$\vdash\!\dashv$.

The main theorem has many consequences. Among them are Corollaries 15, 17, and 19, which are extensions to rewriting modulo $\sim$ of results owing to Hindley-Rosen (see [Ros73], Theorem 3.5), Rosen (see [Ros73], Theorem 3.8), and Staples (see [Sta75], Sect. 3), respectively.

Corollary 15. *Let $\mathcal{A} = (A, \langle \to_\alpha \rangle_{\alpha \in I}, \vdash\!\dashv)$ be an ARS. If, for all $\alpha, \beta \in I$, $\to_\alpha$ subcommutes with $\to_\beta$ modulo $\sim$ and $\to_\alpha$ is* SCOM$\vdash\!\dashv$, *then $\mathcal{A}$ is* CR$\sim$.

Proof. Let $\prec$ be the empty order on I. Then we have $LD(\alpha, \beta)$ and $LD(\alpha)$ for all $\alpha, \beta \in I$; see F ig. 8. Thus, the corollary follows from Theorem 14. □

If the constituent relations are transitive and reflexive, then subcommutation modulo $\sim$ coincides with commutation modulo $\sim$ and SCOM$\vdash\!\dashv$ coincides with COM$\vdash\!\dashv$ (the lemma of Hindley-Rosen is formulated in terms of commutation).

Definition 16. Let $\mathcal{A} = (A, \langle \to_1, \to_2 \rangle, \vdash\!\dashv)$ be an ARS. We say that $\to_2$ requests $\to_1$ modulo $\sim$ if ${}_1^*\!\leftarrow \cdot \to_2^* \subseteq \to_2^* \cdot \to_1^* \cdot \sim \cdot {}_1^*\!\leftarrow$ and $\vdash\!\dashv \cdot \to_2^* \subseteq \to_2^* \cdot \to_1^* \cdot \sim \cdot {}_1^*\!\leftarrow$; see Fig. 9 and cf. [Ros73], Definition 3.7.

Corollary 17. *Let $\mathcal{A} = (A, \langle \to_1, \to_2 \rangle, \vdash\!\dashv)$. If $\to_1$ and $\to_2$ are* CON$\sim$, *$\to_2$ requests $\to_1$ modulo $\sim$, and $\to_1$ is* COM$\vdash\!\dashv$, *then $\mathcal{A}$ is* CR$\sim$.

Proof. Define $\to_\alpha = \to_1^*$ and $\to_\beta = \to_2^*$. So $\to_\alpha$ and $\to_\beta$ are transitive and reflexive relations. Take as order $\alpha \prec \beta$. $LD(\alpha, \alpha)$ and $LD(\beta, \beta)$ hold because $\to_1$ and $\to_2$ are CON$\sim$. $LD(\alpha, \beta) = LD(\beta, \alpha)$ and $LD(\beta)$ hold since $\to_2$ requests $\to_1$ modulo $\sim$. Finally, $LD(\alpha)$ is true because $\to_1$ is COM$\vdash\!\dashv$. So the claim is also a corollary to Theorem 14. □

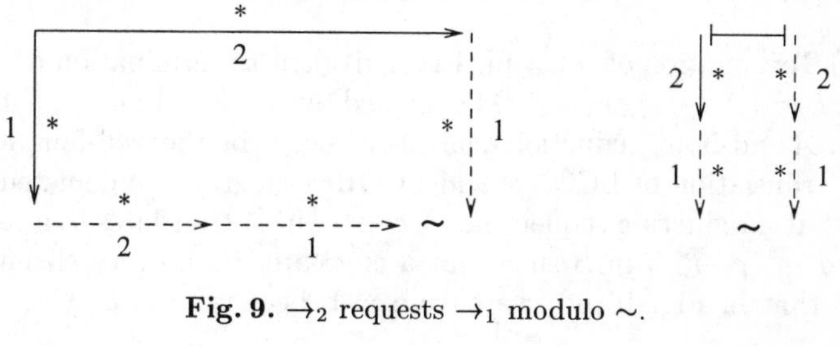

Fig. 9. $\to_2$ requests $\to_1$ modulo $\sim$.

The next corollary is quite useful. It has for instance been used in [Ohl97] to show confluence of a certain conditional graph rewriting relation. As a matter of fact, this was the starting point of our investigations.

Definition 18. Let $\rightarrow_1$ and $\rightarrow_2$ be relations and $\sim$ an equivalence relation on A. Relation $\rightarrow_1$ is called a *refinement* of $\rightarrow_2$ if $\rightarrow_2 \subseteq \rightarrow_1^*$; it is said to be a *compatible refinement* of $\rightarrow_2$ *modulo* $\sim$ if it is a refinement of $\rightarrow_2$ and for all $a \rightarrow_1^* b$ there are $c, d \in A$ such that $a \rightarrow_2^* c \sim d \; {}_2^*\!\leftarrow b$; cf. [Sta75], Sect. 3.

Corollary 19. *Let* $\mathcal{A}_1 = (A, \rightarrow_1, \vdash\dashv)$ *be a compatible refinement of* $\mathcal{A}_2 = (A, \rightarrow_2, \vdash\dashv)$ *modulo* $\sim$*. If* $\rightarrow_2$ *is* CON$\sim$ *and* $\rightarrow_1$ *is* COM$\vdash\dashv$*, then* $\rightarrow_1$ *is* CR$\sim$*.*

Proof. There are at least four different proofs of this corollary. The first one proceeds by showing that the corollary is also a consequence of Theorem 14. The second one is by extending Staples' original proof ([Sta75], Sect. 3). The third proof uses Staples' original result ([Sta75], Sect. 3) and was suggested by Aart Middeldorp. Finally, in the fourth proof one shows that the diamond property holds for the relation $\rightarrow_1^* \cdot \sim$ (cf. Proposition 6). This proof was suggested to the author by Vincent van Oostrom. Since it is the shortest one, it is stated here. So consider a divergence $a \sim \; {}_1^*\!\leftarrow \; \rightarrow_1^* \; \sim b$. It follows

$$\begin{array}{lll}
a \sim \; \rightarrow_2^* \; \sim \; {}_2^*\!\leftarrow \qquad \rightarrow_2^* \; \sim \; {}_2^*\!\leftarrow \; \sim b & & \text{(compatibility)} \\
a \sim \; \rightarrow_2^* \; \sim \; \rightarrow_2^* \; \sim \; {}_2^*\!\leftarrow \; \sim \; {}_2^*\!\leftarrow \; \sim b & & (\rightarrow_2 \text{ is CON}\sim) \\
a \sim \; \rightarrow_1^* \; \sim \; \rightarrow_1^* \; \sim \; {}_1^*\!\leftarrow \; \sim \; {}_1^*\!\leftarrow \; \sim b & & \text{(refinement)} \\
\qquad a \rightarrow_1^* \; \sim \; {}_1^*\!\leftarrow b & & (\rightarrow_1 \text{ is COM}\sim) \; \square
\end{array}$$

It is not difficult to prove the following converse of the preceding corollary. Let $\mathcal{A}_1 = (A, \rightarrow_1, \vdash\dashv)$ be a compatible refinement of $\mathcal{A}_2 = (A, \rightarrow_2, \vdash\dashv)$ modulo $\sim$. If $\rightarrow_1$ is CON$\sim$ and $\rightarrow_2$ is COM$\vdash\dashv$, then $\rightarrow_2$ is CR$\sim$.

Corollary 20 is closely related to Proposition 11. In its first statement (owing to Huet [Hue80]), SN$\sim$ cannot be weakened to SN as Example 21 (see [Ave95], Beispiel 4.1.8) shows. However, if we replace LCOH$\vdash\dashv$ with the stronger property LCMU$\vdash\dashv$, then SN is sufficient. This fact is made precise in the second statement which is due to Jouannaud and Munoz [JM84].

Corollary 20. *Let* $\mathcal{A}$ *be an ARS.*

1. *If* $\mathcal{A}$ *is* SN$\sim$, LCON$\sim$, *and* LCOH$\vdash\dashv$, *then it is* CR$\sim$.
2. *If* $\mathcal{A}$ *is* SN, LCON$\sim$, *and* LCMU$\vdash\dashv$, *then it is* CR$\sim$.

Proof. (1) Termination of $\rightarrow_\sim$ in A is equivalent to termination of $\rightarrow / \sim$ in $A/\sim$. Let $\mathcal{B} = (A, \langle \rightarrow_{\mathbf{a}} \rangle_{\mathbf{a} \in A/\sim}, \vdash\dashv)$ be defined by: $\forall a, b \in A : a \rightarrow_{\mathbf{a}} b$ iff $a \rightarrow b$. Evidently, $\mathcal{A}$ and $\mathcal{B}$ are reduction equivalent. Let $\succ$ be the well-founded order $\rightarrow_\sim^+$. The translation of LCON$\sim$ and LCOH$\vdash\dashv$ diagrams is depicted in Fig. 10. Note that a reduction sequence $a \rightarrow c \rightarrow^* d$ in $\mathcal{A}$ translates to a reduction sequence $a \rightarrow_{\mathbf{a}} c \rightarrow_\sigma^* d$ in $\mathcal{B}$ such that $\mathbf{a}$ is greater than every element in σ. Also note that in Fig. 10(iv), we have $\mathbf{a} = \mathbf{b}$ because $a \sim b$. Thus (ii) and

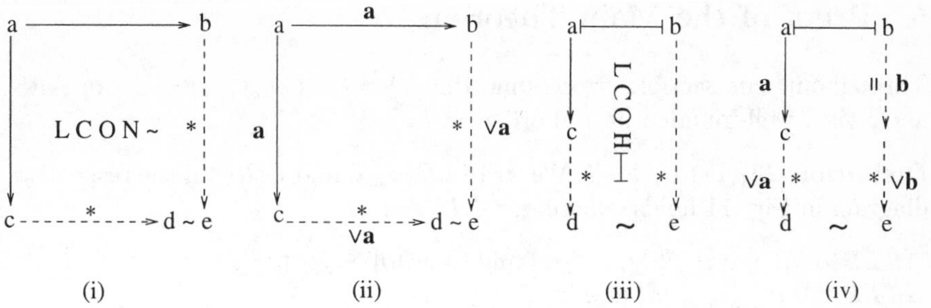

Fig. 10. Translation of LCON$\sim$ and LCOH$\vdash\!\dashv$ diagrams.

(iv) are locally decreasing diagrams and $\mathcal{A}$ is CR$\sim$ by Theorem 14.
(2) The combination of SN and LCMU$\vdash\!\dashv$ implies SN$\sim$; see [JM84], Theorem 14. Now the claim follows from (1) because LCMU$\vdash\!\dashv$ implies LCOH$\vdash\!\dashv$. □

Example 21. Let the ARS $\mathcal{A}$ be defined by $a \leftarrow b \overset{\vdash\!\dashv}{\leftarrow} c \overset{\vdash\!\dashv}{\rightarrow} d \rightarrow e$. $\mathcal{A}$ is SN, LCON$\sim$, and LCOH$\vdash\!\dashv$ but not CR$\sim$ (in fact, it lacks LCOH$\sim$).

The following result of Jouannaud and Kirchner [JK86] is the basis of Knuth-Bendix completion modulo equations.

Proposition 22. *Let* $\mathcal{A} = (A, \langle \rightarrow, \rightarrow_i \rangle, \vdash\!\dashv)$ *be an ARS, where* $\rightarrow \subseteq \rightarrow_i \subseteq \rightarrow_\sim$ *(*$\rightarrow_i$ *is an intermediate relation, that is). Note that* $\rightarrow_{\mathcal{A}} = \rightarrow \cup \rightarrow_i = \rightarrow_i$*. If* $\mathcal{A}$ *is* SN$\sim$, LCOH$\vdash\!\dashv$, *and* $\forall a, b, c \in A$ *with* $a \leftarrow b \rightarrow_i c$ *there are* $d, e \in A$ *such that* $a \rightarrow_i^* d \sim e \; {}_i^*\!\!\leftarrow c$*, then* $\mathcal{A}$ *is* CR$\sim$.

Proof. In order to use Corollary 20, one has to show LCON$\sim$. This indeed follows from the premises; the proof is the same as the direct proof of CR$\sim$ in [JK86], Theorem 5. □

It is relatively simple to show that Sethi's Theorem 2.3 [Set74] is also a direct consequence of Theorem 14. We don't give the details here because the mentioned theorem is already subsumed by Corollary 20; cf. [Hue80]. Sethi also provided an application of his theorem to covering matrices. In contrast to the aforementioned, Corollary 10 and Proposition 11 aren't consequences of Theorem 14. This is because $LD(\alpha)$ cannot be ensured. Yet another Church-Rosser theorem is worth mentioning here, namely Theorem 1 of [PPE94] which is useful in studying the lazy partial λ-calculus. In that theorem, the symmetric relation $\vdash\!\dashv$ is the union of some relation $\rightsquigarrow$ and its inverse. Thus it is also not covered by Theorem 14.

We finally remark that in Theorem 14 it is possible to distinguish between vertical and horizontal reductions (as in Theorem 9). We refrained from doing so because all the corollaries can be proven without this distinction. Note, however, that with this distinction Theorem 14 is a proper generalization of van Oostrom's main result; see [Oos94a], Main Theorem 3.7.

5 Proof of the Main Theorem

Throughout this section, we assume that $\mathcal{A} = (A, \langle \rightarrow_\alpha \rangle_{\alpha \in I}, \vdash\!\dashv)$ is an ARS and $\succ$ is a well-founded partial order on I.

Definition 23. Let $\alpha, \beta \in I$. We write $LD(\alpha,\beta)$ and $LD(\alpha)$ if the respective diagram in Fig. 11 holds, where $\sigma, \tau \in I^*$ and

1. $LD(\alpha,\beta)$: $|\alpha| \uplus |\beta| \succeq_{mul} |\alpha\tau|$ and $|\alpha| \uplus |\beta| \succeq_{mul} |\beta\sigma|$,
2. $LD(\alpha)$: $|\alpha| \succeq_{mul} |\sigma|$ and $|\tau| \subseteq \curlyvee\alpha$.

It is fairly simple to show that the preceding definition is equivalent to Definition 13. That's why the above diagrams are called locally decreasing.

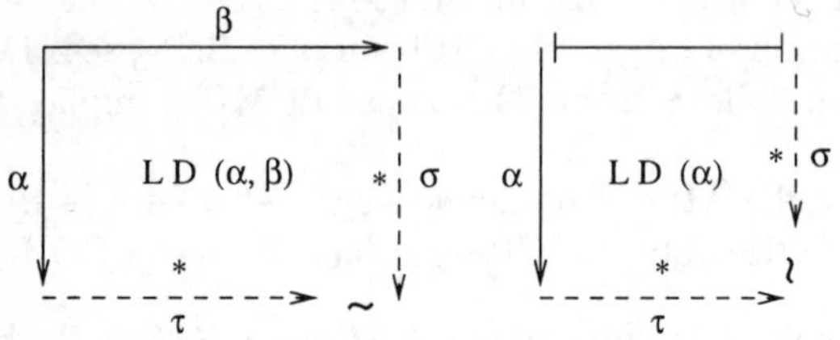

Fig. 11. Local decreasingness.

Definition 24. The diagram in Fig. 12(i) is *decreasing modulo* (DM) if

$$|\sigma| \uplus |\tau| \succeq_{mul} |\sigma\tau'| \text{ and } |\sigma| \uplus |\tau| \succeq_{mul} |\tau\sigma'| \,.$$

Figure 12(ii) and (iii) shows the cases in which $\tau = \varepsilon$ and $\sigma = \varepsilon$, respectively. Note that if there are no $\sim$-steps at all in the diagram, then we have a decreasing diagram as defined in [Oos94a,Oos94b]. By this observation, the following two lemmata do not need an extra proof because the proofs of the corresponding lemmata for decreasing diagrams carry over immediately.

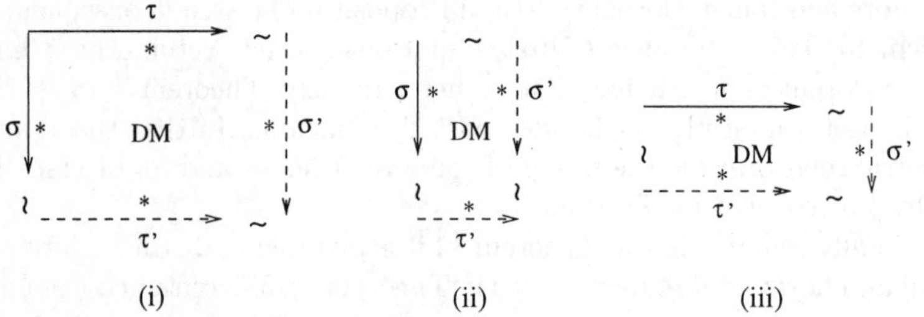

Fig. 12. Decreasing diagram modulo (DDM).

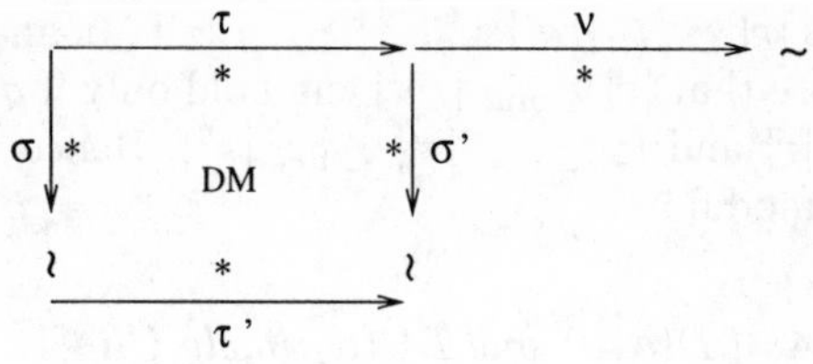

Fig. 13. Pasting is hypothesis decreasing.

Lemma 25. *([Oos94a], Lemma 3.6) In the situation of Fig. 13, if $\tau \neq \varepsilon$, then the inequality $|\sigma| \uplus |\tau\nu| \succ_{mul} |\sigma'| \uplus |\nu|$ holds.* □

Lemma 26. *([Oos94a], Lemma 3.5) If we paste together two decreasing diagrams modulo DM_1 and DM_2 as depicted in Fig. 14(i), where $\tau \neq \varepsilon$, then the resulting diagram in Fig. 14(ii) is also a decreasing diagram modulo.* □

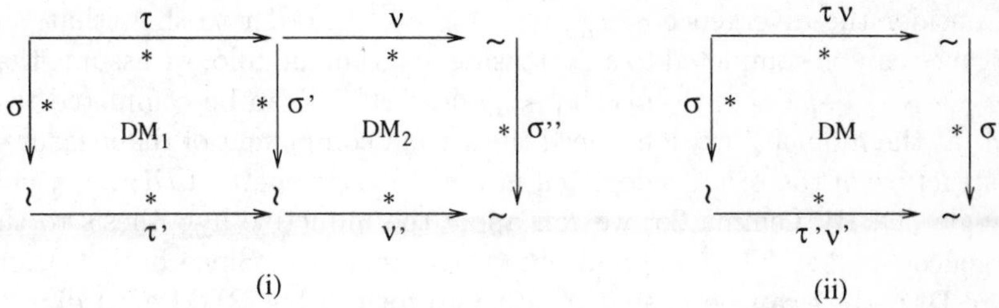

Fig. 14. Pasting preserves decreasingness modulo.

Lemma 27. *If we paste together two decreasing diagrams modulo DM_1 and DM_2 as depicted in Fig. 15(i), then the resulting diagram in Fig. 15(ii) is also a decreasing diagram modulo.*

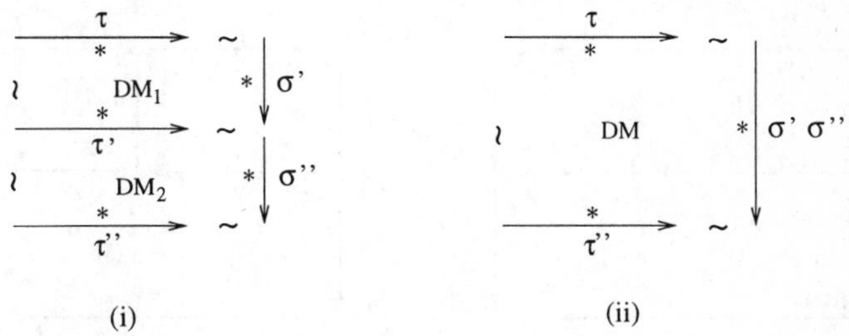

Fig. 15. Pasting preserves decreasingness modulo, part 2.

Proof. DM_1 implies $|\tau| \succeq_{mul} |\tau\sigma'|$ and $|\tau| \succeq_{mul} |\tau'|$ (an analogous statement holds for DM_2). Note that $|\tau| \succeq_{mul} |\tau\sigma'|$ can hold only if $\sigma' \subseteq \curlyvee\tau$. It follows $|\tau\sigma'\sigma''| = |\tau\sigma''| = |\tau|$ and $|\tau| \succeq_{mul} |\tau'| \succeq_{mul} |\tau''|$. Hence the resulting diagram is decreasing modulo. □

Main Theorem 14. *$LD(\alpha, \beta)$ and $LD(\alpha)$ imply CR$\sim$.*

Proof. Let $>_{lex} \subseteq (I^* \times \mathbb{N}) \times (I^* \times \mathbb{N})$ be the lexicographic product of $\succ_{mul}$ and the natural order on $\mathbb{N}$. Note that $>_{lex}$ is well-founded. The measure of a divergence $b \sim \cdot \, {}_{\sigma}^{*}\!\!\leftarrow a \rightarrow_{\tau}^{*} \cdot \sim c$ is defined to be the pair $(|\sigma| \uplus |\tau|, k)$, where $k = 0$ if either $b \sim a \sim c$ or $b \sim \cdot \, {}_{\sigma}^{+}\!\!\leftarrow a \rightarrow_{\tau}^{+} \cdot \sim c$; otherwise we can either write $b \sim \cdot \, {}_{\sigma}^{+}\!\!\leftarrow a \overset{k}{\vdash\!\dashv} c$ or $b \overset{k}{\vdash\!\dashv} a \rightarrow_{\tau}^{+} \cdot \sim c$ which defines k.

By Proposition 6, it is sufficient to prove $\Diamond(\rightarrow^* \cdot \sim)$. In order to show this, we proceed by induction on the well-founded order $>_{lex}$ and distinguish between the following cases: (i) $\sigma = \varepsilon = \tau$, (ii) $\sigma \neq \varepsilon \neq \tau$, and (iii) $\sigma = \varepsilon \neq \tau$ (the case $\sigma \neq \varepsilon = \tau$ is symmetric).
(i) This case holds trivially.
(ii) Consider the divergence $\sim \cdot \, {}_{\alpha\sigma}^{+}\!\!\leftarrow \cdot \rightarrow_{\beta\tau}^{+} \cdot \sim$. In order to show that this divergence can be completed to a decreasing diagram modulo, we assume that every $\sim \cdot \, {}_{\mu}^{*}\!\!\leftarrow \cdot \rightarrow_{\nu}^{*} \cdot \sim$ with $|\mu| \uplus |\nu| \prec_{mul} |\alpha\sigma| \uplus |\beta\tau|$ can be completed to a DDM (at the moment, we don't need the second component of the measure). The structure of the proof is depicted in Fig. 16(i). Property $LD(\alpha, \beta)$ yields rectangle (1). By Lemma 25, we can apply the inductive hypothesis to the divergence $\sim \cdot \, {}_{\mu}^{*}\!\!\leftarrow \cdot \rightarrow_{\tau}^{*} \cdot \sim$ and obtain rectangle (2). Since both (1) and (2) are DM, they can be pasted together to form a big DDM according to Lemma 26. A renewed application of (the mirrored version of) Lemma 25 to the divergence $\sim \cdot \, {}_{\sigma}^{*}\!\!\leftarrow \cdot \rightarrow_{\nu\tau'}^{*} \cdot \sim$ yields rectangle (3). With the aid of Lemma 26, we paste the big DDM and diagram (3) together, so that the whole diagram is decreasing modulo.
(iii) Consider $\overset{k}{\vdash\!\dashv} \cdot \vdash\!\dashv \cdot \rightarrow_{\beta\tau}^{+} \cdot \sim$, where $k \in \mathbb{N}$. The structure of this proof can be found in Fig. 16(ii). Property $LD(\beta)$ yields rectangle (1). By Lemma 25, we can apply the inductive hypothesis to the divergence $\sim \cdot \, {}_{\mu}^{*}\!\!\leftarrow \cdot \rightarrow_{\tau}^{*} \cdot \sim$

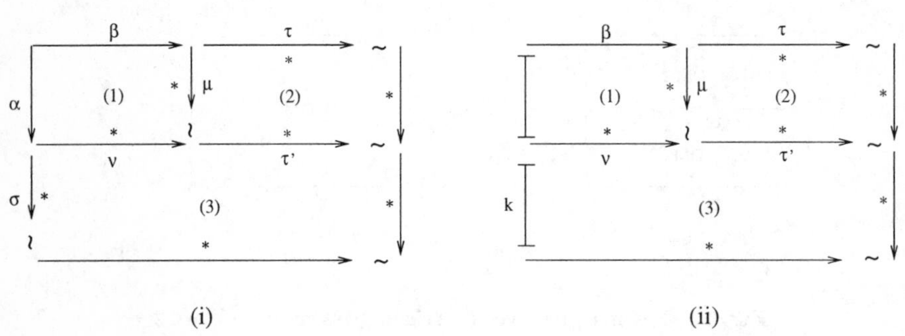

Fig. 16. Proof of Theorem 14

and obtain rectangle (2). Since both (1) and (2) are DM, they can be pasted together to form a big DDM according to Lemma 26. In particular, this implies $|\beta\tau| \succeq_{mul} |\nu\tau'|$. Thus, $(|\beta\tau|, k+1) >_{lex} (|\nu\tau'|, k)$ and the inductive hypothesis is applicable to $\overset{k}{\vdash\!\dashv} \cdot \rightarrow^*_{\nu\tau'} \cdot \sim$. Therefore, rectangle (3) is a DDM. Now Lemma 27 shows that the whole diagram is decreasing modulo. □

Acknowledgements: In the submitted version of this paper, the proof of the main theorem consisted of showing CON$\sim$ and SCOH$\sim$ which is much more complicated than showing the diamond property for $\rightarrow^* \cdot \sim$. It was Vincent van Oostrom who observed that $\diamond(\rightarrow^* \cdot \sim)$ is equivalent to CR$\sim$ (see Proposition 6) and that the proof of Theorem 14 can considerably be simplified by this observation. I am thus most grateful to him. Furthermore, I would like to thank Aart Middeldorp and the anonymous referees for valuable comments on the previous version of the paper.

References

[Ave95] J. Avenhaus. *Reduktionssysteme.* Springer Verlag, 1995 (in German).

[Ges90] A. Geser. *Relative Termination.* PhD thesis, Universität Passau, 1990.

[Hin64] R. Hindley. *The Church-Rosser Property and a Result in Combinatory Logic.* PhD thesis, University of Newcastle-upon-Tyne, 1964.

[Hue80] G. Huet. Confluent Reductions: Abstract Properties and Applications to Term Rewriting Systems. *J. of the ACM* **27***(4)*, pages 797–821, 1980.

[JK86] J.-P. Jouannaud and H. Kirchner. Completion of a Set of Rules Modulo a Set of Equations. *SIAM J. on Computing* **15***(4)*, pages 1155–1194, 1986.

[JM84] J.-P. Jouannaud and M. Munoz. Termination of a Set of Rules Modulo a Set of Equations. In *Proceedings of the 7th International Conference on Automated Deduction*, pages 175–193. LNCS **170**, 1984.

[Klo92] J.W. Klop. Term Rewriting Systems. In S. Abramsky, D. Gabbay, and T. Maibaum, editors, *Handbook of Logic in Computer Science*, volume 2, pages 1–116. Oxford University Press, 1992.

[Ohl97] E. Ohlebusch. Conditional Graph Rewriting. In *Proceedings of the 6th International Conference on Algebraic and Logic Programming*, pages 144–158. LNCS **1298**, 1997.

[Oos94a] V. van Oostrom. Confluence by Decreasing Diagrams. *Theoretical Computer Science* **126***(2)*, pages 259–280, 1994.

[Oos94b] V. van Oostrom. *Confluence for Abstract and Higher-Order Rewriting.* PhD thesis, Vrije Universiteit te Amsterdam, 1994.

[PPE94] R. Pino Pérez and Ch. Even. An Abstract Property of Confluence Applied to the Study of the Lazy Partial Lambda Calculus. In *Proceedings of the 3rd International Symposium on Logical Foundations of Computer Science*, pages 278–290. LNCS **813**, 1994.

[Ros73] B.K. Rosen. Tree-Manipulating Systems and Church-Rosser Theorems. *J. of the ACM* **20***(1)*, pages 160–187, 1973.

[Set74] R. Sethi. Testing for the Church-Rosser Property. *J. of the ACM* **21**, pages 671–679, 1974.

[Sta75] J. Staples. Church-Rosser Theorems for Replacement Systems. In J. Crosley, editor, *Algebra and Logic*, pages 291–307. Lecture Notes in Mathematics **450**, 1975.

Automatic Monoids Versus Monoids with Finite Convergent Presentations*

Friedrich Otto[1], **Andrea Sattler-Klein**[2], **Klaus Madlener**[2]

[1] Fachbereich Mathematik/Informatik, Universität Kassel, D-34109 Kassel
otto@theory.informatik.uni-kassel.de

[2] Fachbereich Informatik, Universität Kaiserslautern, D-67653 Kaiserslautern
{sattler,madlener}@informatik.uni-kl.de

Abstract. Due to their many nice properties *groups with automatic structure* (*automatic groups*) have received a lot of attention in the literature. The multiplication of an automatic group can be realized through finite automata based on a regular set of (not necessarily unique) representatives for the group, and hence, each automatic group has a tractable word problem and low derivational complexity. Consequently it has been asked whether corresponding results also hold for *monoids with automatic structure*. Here we show that there exist finitely presented monoids with automatic structure that cannot be presented through finite and convergent string-rewriting systems, thus answering a question in the negative that is still open for the class of automatic groups. Secondly, we present an automatic monoid that has an exponential derivational complexity, which establishes another difference to the class of automatic groups. In fact, both our example monoids are bi-automatic. In addition, it follows from the first of our examples that a monoid which is given through a finite, noetherian, and weakly confluent string-rewriting system need not have *finite derivation type*.

1 Introduction

In recent years the computational aspect has become more and more prominent in combinatorial group and semigroup theory. Although in general not much information on the algebraic structure presented can be extracted from a finite presentation, various methods have been used successfully in certain instances.

One approach is based on the notion of a *convergent presentation*. A monoid-presentation $(\Sigma; R)$ is called *convergent* if the string-rewriting system R is convergent, that is, the reduction relation induced by R is both noetherian and confluent. The presentations of this form are of particular interest, because the reduction relation induced by a convergent string-rewriting system R yields a unique irreducible string for each congruence class of the Thue congruence generated by R. Hence, if a monoid M has a finite presentation of this form, then

* This work was supported by the Deutsche Forschungsgemeinschaft (Projekte Ma 1208/5-1 und Ot 79/4-1).

it has a decidable *word problem*. In fact, for the class of finite convergent presentations, the *uniform word problem* is decidable, which implies that for this class various decision problems can be solved that are undecidable in general. For example, the problem of deciding whether the monoid presented is actually a group is decidable for this class [Ott86, BoOt93].

Gilman [Gil84] considers groups that are given through finite, monadic, and convergent string-rewriting systems. His approach is based on performing computations with regular sets. Corresponding considerations had already led Book to the result that for monoid-presentations involving finite, monadic, and confluent string-rewriting systems, all those properties of Thue congruences are decidable that can be expressed through *linear sentences* [Boo82b, Boo83].

As an alternative to the rewrite approach Epstein et al developed the notion of *groups with automatic structure* during the 1980's [Eps92]. A finitely generated group G has an *automatic structure* if it has a finite set of generators Σ and a regular set $S \subseteq \Sigma^*$ of representatives, which, however, need not be unique, such that the following tasks can be performed by finite state acceptors:

(1.) Given two elements $u, v \in S$, decide whether or not they both represent the same element of the group G.
(2.) For $a \in \Sigma$, if two elements $u, v \in S$ are given, decide whether or not ua and v represent the same element of the group G.

Thus, an automatic structure provides a regular set of representatives for the group considered such that the group operation can be realized on these representatives through the finite automata of (2.), which accordingly are called the *multiplier automata*.

Although the definition mentions explicitly a set of generators Σ for the group considered, it turns out that the existence of an automatic structure is independent of the finite set of generators chosen. Further, the word problem for each automatic group can be solved in quadratic time, and each automatic group has quadratic *derivational complexity*, that is, its *Dehn function* has a quadratic upper bound. Finally, each automatic group is finitely presented and it satisfies the homological finiteness condition FP_∞ [Alo92], and hence, each automatic group has *finite derivation type* [CrOt96].

If the group G has a finite convergent presentation $(\Sigma; R)$, then the set of irreducible strings IRR(R) is a regular set of unique representatives for the group G, and so also condition (1.) above is satisfied. But in general there need not exist multiplier automata in this situation. And indeed, there exist finitely presented groups that do admit finite convergent presentations, but that do not have an automatic structure [Ger92]. However, it is still not known whether each automatic group admits a finite convergent presentation.

Recently, the notion of automatic structure has been generalized to monoids [Hud96]. *Automatic monoids* do also have word problems that are decidable in quadratic time, but it turns out that for monoids the existence of an automatic structure does indeed depend on the actually chosen set of generators. Also an automatic monoid need not be finitely presented [CRRT97].

Here we also consider the class of automatic monoids. We present some examples showing that further properties do not carry over from automatic groups to automatic monoids. First we present a monoid N that has a finite, noetherian, and weakly confluent presentation as well as an infinite left-regular, weight-reducing, and confluent presentation. This monoid has an automatic structure, but it does not have finite derivation type. Hence, it does not admit any finite convergent presentation. This shows that the class of monoids that have a finite convergent presentation and the class of automatic monoids are incomparable under inclusion. For the class of monoids this provides a negative anwer to the open problem mentioned above. Further, it proves that the property of having a finite, noetherian, and weakly confluent presentation is not sufficient to imply that the monoid considered has finite derivation type.

Our second example consists of a finitely presented monoid that is automatic, but that has an exponential Dehn function. In fact, both our example monoids are *bi-automatic*.

This paper is structured as follows. In the next section we restate some definitions and notation regarding monoid-presentations and string-rewriting systems, and we present the definition of an automatic structure for a finitely generated monoid in full detail. In Sections 3 and 4 we present the two example monoids mentioned above. The paper closes with a short discussion of the consequences of our results and some open problems.

2 Automatic monoids

First we restate some definitions regarding string-rewriting systems and monoid-presentations in order to establish notation. For additional information the reader is asked to consult the literature, where the monograph by Book and Otto [BoOt93] is our main reference on string-rewriting systems, and the monograph by Hopcroft and Ullmann [HoUl79] is our main reference on automata theory. Then we give the definition of an automatic structure for a monoid, and we restate some recent results on automatic monoids.

Let Σ be a finite alphabet. Then Σ^+ denotes the set of all non-empty strings over Σ, and $\Sigma^* := \Sigma^+ \cup \{\lambda\}$ denotes the set of all strings over Σ including the empty string λ. For $u, v \in \Sigma^*$, the concatenation of u and v is simply written as uv, and exponents are used to abbreviate strings, that is, $u^0 := \lambda$, $u^1 := u$, and $u^{n+1} := u^n u$ for all $u \in \Sigma^*$ and integers $n \geq 1$. For $u \in \Sigma^*$, the *length* of u is denoted by $|u|$.

For $\Gamma \subset \Sigma$, $\pi_\Gamma : \Sigma^* \to \Gamma^*$ denotes the projection, that is, π_Γ is the morphism that is induced by the mapping $a \mapsto a$ $(a \in \Gamma)$ and $b \to \lambda$ $(b \in \Sigma \smallsetminus \Gamma)$. Further, the *$\Gamma$-length* of a string $u \in \Sigma^*$ is defined as $|u|_\Gamma := |\pi_\Gamma(u)|$.

A *string-rewriting system* R on Σ is a subset of $\Sigma^* \times \Sigma^*$, the elements of which are called (*rewrite*) *rules*. Usually these rules will be written in the form $\ell \to r$ to improve readability. By dom(R) we denote the *domain* of R, which is the set dom$(R) := \{\ell \in \Sigma^* \mid \exists r \in \Sigma^* : (\ell \to r) \in R\}$. The system R is called *length-reducing* if $|\ell| > |r|$ holds for each rule $(\ell \to r)$ of R.

A string-rewriting system R on Σ induces several binary relations on Σ^*, the simplest of which is the *single-step reduction relation* $\rightarrow_R$:

$$u \rightarrow_R v \text{ iff } \exists x, y \in \Sigma^* \exists (\ell \rightarrow r) \in R : u = x\ell y \text{ and } v = xry.$$

Its reflexive and transitive closure $\rightarrow_R^*$ is the *reduction relation* induced by R. If $u \rightarrow_R^* v$, then u is an *ancestor* of v and v is a *descendant* of u (mod R). A string u is called *reducible* (mod R), if there exists a string v such that $u \rightarrow_R v$ holds; otherwise, u is *irreducible*. By IRR(R) we denote the set of all irreducible strings, and by RED(R) we denote the set of all reducible strings (mod R). Obviously, $\text{RED}(R) = \Sigma^* \cdot \text{dom}(R) \cdot \Sigma^*$ and $\text{IRR}(R) = \Sigma^* \smallsetminus \text{RED}(R)$. Thus, if R is finite or left-regular, then RED(R) and IRR(R) are both regular sets. Here a system R is called *left-regular* if dom(R) is a regular set.

The reflexive, symmetric, and transitive closure $\leftrightarrow_R^*$ of $\rightarrow_R$ is a congruence on Σ^*, the *Thue congruence* generated by R. For $u \in \Sigma^*$, $[u]_R := \{v \in \Sigma^* \mid u \leftrightarrow_R^* v\}$ is the *congruence class* of u (mod R). The set $M_R := \{[u]_R \mid u \in \Sigma^*\}$ of congruence classes is a monoid under the operation $[u]_R \circ [v]_R = [uv]_R$ with identity $[\lambda]_R$. It is the factor monoid $\Sigma^* / \leftrightarrow_R^*$, and it is uniquely determined by Σ and R. Accordingly, the ordered pair $(\Sigma; R)$ is called a *(monoid-) presentation* of M_R with *generators* Σ and *defining relations* R. A monoid M is called *finitely generated* if it has a presentation $(\Sigma; R)$ with a finite set of generators Σ, and it is called *finitely presented* if it has a finite presentation.

For $u, v \in \Sigma^*$, if $u \leftrightarrow_R^* v$, then there exists a *derivation*

$$u = w_0 \leftrightarrow_R w_1 \leftrightarrow_R \ldots \leftrightarrow_R w_m = v,$$

where $\leftrightarrow_R$ denotes the symmetric closure of the single-step reduction relation $\rightarrow_R$. The derivation above is said to be of *length* m. Let $d_R : \Sigma^* \times \Sigma^* \rightarrow \mathbb{N}$ be the function defined by

$$d_R(u,v) := \begin{cases} \min\{m \mid \exists w_0, w_1, \ldots, w_m \in \Sigma^* : \\ \qquad u = w_0 \leftrightarrow_R w_1 \leftrightarrow_R \ldots \leftrightarrow_R w_m = v\}, & \text{if } u \leftrightarrow_R^* v, \\ 0, & \text{otherwise,} \end{cases}$$

that is, $d_R(u,v)$ is the length of the shortest derivation from u to v, if $u \leftrightarrow_R^* v$ holds. This function is called the *derivational complexity* of the string-rewriting system R.

Further, for $n \in \mathbb{N}$, we define

$$\delta_R(n) := \max\{d_R(u,v) \mid u, v \in \Sigma^*, |u| + |v| \leq n\}.$$

Then $\delta_R : \mathbb{N} \rightarrow \mathbb{N}$ is the *isoperimetric function* (or *Dehn function*) of R.

Although the Dehn function δ_R is defined explicitly for the string-rewriting system R (or the presentation $(\Sigma; R)$), it is essentially an invariant of the monoid M_R. For two functions $f, g : \mathbb{N} \rightarrow \mathbb{N}$, we write $f \preceq g$ if there are positive integers α, β, and γ such that $f(n) \leq \alpha \cdot g(\beta \cdot n) + \gamma \cdot n$ holds for all $n \in \mathbb{N}$, and we call two functions $f, g : \mathbb{N} \rightarrow \mathbb{N}$ *equivalent* ($f \sim g$), if $f \preceq g$ and $g \preceq f$ both hold.

Proposition 1. [MaOt85, Pri95] *Let* $(\Sigma_1; R_1)$ *and* $(\Sigma_2; R_2)$ *be two finite presentations of the same monoid, that is,* $M_{R_1} \cong M_{R_2}$. *Then the Dehn functions* δ_{R_1} *and* δ_{R_2} *are equivalent.*

We are in particular interested in presentations involving string-rewriting systems of certain restricted forms. A string-rewriting system R on Σ is called

- *noetherian* if there is no infinite sequence of reduction steps of the form $u_0 \to_R u_1 \to_R u_2 \to_R \cdots \to_R u_i \to_R u_{i+1} \to_R \cdots$;
- *weight-reducing* if there exists a *weight-function* $\varphi : \Sigma \to \mathbb{N}_+$ such that $\varphi(\ell) > \varphi(r)$ holds for each rule $(\ell \to r)$ of R, where φ is extended to a morphism from Σ^* to $\mathbb{N}$;
- *confluent*, if, for all $u, v, w \in \Sigma^*$, $u \to_R^* v$ and $u \to_R^* w$ imply that v and w have a common descendant (mod R);
- *convergent*, if it is noetherian and confluent.

If R is convergent, then each congruence class $[u]_R$ contains a unique irreducible string u_0, which thus can serve as the normal form for each string in $[u]_R$. Since u_0 can be obtained from u by a finite sequence of reduction steps, the *word problem* is decidable for each finite convergent string-rewriting system.

A presentation $(\Sigma; R)$ involving a convergent string-rewriting system R is called a *convergent presentation.* Since the decidability of the word problem is an invariant of finitely generated monoids, we have the following result.

Proposition 2. *If a monoid has a finite convergent presentation, then it has a decidable word problem.*

It is easily seen that a weight-reducing string-rewriting system is noetherian. In fact, if R is a finite, weight-reducing, and confluent string-rewriting system on Σ, then the word problem for R can be solved in linear time [BoOt93], and $\delta_R(n) \sim n$, that is, the Dehn function of R is linearly bounded.

Finally, we come to the main definition of this section, that is, the definition of an automatic structure for a monoid-presentation.

Let Σ be a finite alphabet. We will be interested in certain subsets of $\Sigma^* \times \Sigma^*$ that are accepted by *finite state acceptors* (fsas). To make this precise we proceed as follows.

Let $\#$ be an additional symbol. We define a finite alphabet $\Sigma_\#$ as

$$\Sigma_\# := ((\Sigma \cup \{\#\}) \times (\Sigma \cup \{\#\})) \smallsetminus \{(\#, \#)\}.$$

This alphabet is called the *padded extension* of Σ. A mapping $\nu : \Sigma^* \times \Sigma^* \to \Sigma_\#^*$ is then defined as follows:

if $u := a_1 a_2 \cdots a_n$ and $v := b_1 b_2 \cdots b_m$, where $a_1, \ldots, a_n, b_1, \ldots, b_m \in \Sigma$, then

$$\nu(u, v) := \begin{cases} (a_1, b_1)(a_2, b_2) \cdots (a_m, b_m)(a_{m+1}, \#) \cdots (a_n, \#), & \text{if } m < n, \\ (a_1, b_1)(a_2, b_2) \cdots (a_m, b_m), & \text{if } m = n, \\ (a_1, b_1)(a_2, b_2) \cdots (a_n, b_n)(\#, b_{n+1}) \cdots (\#, b_m), & \text{if } m > n. \end{cases}$$

A subset $L \subseteq \Sigma^* \times \Sigma^*$ is called *synchronously regular*, if $\nu(L) \subseteq \Sigma_{\#}^*$ is accepted by some finite state acceptor.

An *automatic structure* for a finitely generated monoid-presentation $(\Sigma; R)$ consists of a fsa W over Σ, a fsa $M_=$ over $\Sigma_{\#}$, and fsas M_a $(a \in \Sigma)$ over $\Sigma_{\#}$ satisfying the following conditions:

(0.) $L(W) \subseteq \Sigma^*$ is a complete set of (not necessarily unique) representatives for the monoid M_R, that is, $L(W) \cap [u]_R \neq \emptyset$ holds for each $u \in \Sigma^*$,
(1.) $L(M_=) = \{\nu(u,v) \mid u, v \in L(W) \text{ and } u \leftrightarrow_R^* v\}$, and
(2.) for all $a \in \Sigma$, $L(M_a) = \{\nu(u,v) \mid u, v \in L(W) \text{ and } ua \leftrightarrow_R^* v\}$.

The fsa W is called the *word acceptor*, $M_=$ is the *equality recognizer*, and the M_a $(a \in \Sigma)$ are the *multiplier automata* for the automatic structure. The language $C := L(W)$ is a set of representatives for the monoid M_R, the language $L_= := L(M_=)$ is the *equality language*, and $L_a := L(M_a)$ is the *multiplication language* for a. A monoid-presentation is called *automatic* if it admits an automatic structure, and a monoid is called *automatic* if it has an automatic presentation.

Groups with automatic structure have been investigated thoroughly. It is known that the word problem of an automatic group can be solved in quadratic time, that its Dehn function is bounded from above by a polynomial of degree 2, that the existence of an automatic structure is independent of the chosen set of generators, and that each automatic group is finitely presented. Large classes of groups have been identified that are automatic, and various closure properties have been established for the class of automatic groups. For a detailed treatment of this class of groups see the monograph by Epstein [Eps92].

Until recently the automatic monoids did not receive much attention. This may partly be due to the fact that they do enjoy only a few of the nice properties that the automatic groups have.

Proposition 3. [CRRT97]

(a) The word problem for an automatic monoid can be solved in quadratic time.
(b) For monoids the existence of an automatic structure depends on the chosen set of generators.
(c) An automatic monoid is not necessarily finitely presented.

Finally, an automatic structure for a monoid-presentation $(\Sigma; R)$ is a *bi-automatic structure* if there exist fsas $\overline{M}_a$ $(a \in \Sigma)$ over $\Sigma_{\#}$ such that,

(3.) for all $a \in \Sigma$, $L(\overline{M}_a) = \{\nu(u,v) \mid u, v \in L(W) \text{ and } au \leftrightarrow_R^* v\}$.

For $a \in \Sigma$, ${}_aL := L(\overline{M}_a)$ is the *left-multiplication language* for a.

For bi-automatic groups the conjugacy problem is decidable [Eps92], but it actually is an open problem whether or not each automatic group is in fact bi-automatic. Bi-automatic groups have been considered in detail by Gersten and Short [GeSh91].

3 Automatic monoids and convergent presentations

There exist finitely presented monoids, in fact groups, that admit finite convergent presentations, but that do not have an automatic structure [Ger92]. On the other hand, it is currently not known whether each automatic group admits a finite convergent presentation. Here we answer this question in the negative for the more general class of automatic monoids.

Let $\Sigma := \{a, b, c, d, A, W\}$, and let R be the following finite string-rewriting system on Σ:

$$R := \{abbc \rightarrow abc, da \rightarrow Aabb, WA \rightarrow \lambda\}.$$

By N we denote the monoid that is given through the presentation $(\Sigma; R)$.

In this section we will establish various properties of the monoid N. We begin with some simple observations.

Lemma 4.

(a) The string-rewriting system R is weight-reducing.
(b) The string-rewriting system R is not confluent.

Let R^∞ denote the string-rewriting system

$$R^\infty := R \cup \{ab^{2n}c \rightarrow ab^{2n-1}c \mid n \geq 1\}.$$

Lemma 5. *The string-rewriting system R^∞ is weight-reducing, confluent, left-regular, and equivalent to R.*

Using the left-regular convergent string-rewriting system R^∞ each string can be reduced to its irreducible descendant. Hence, the word problem is decidable for the monoid N. Observe, however, that the results of Ó'Dúnlaing [Ó'Dú83] do not apply to the system R^∞, since the left-regular systems considered by Ó'Dúnlaing are of a more restricted form than R^∞. Nevertheless, the standard method of reducing a string to its irreducible descendant by applying only leftmost reduction steps, which Book originally developed for finite, length-reducing, and confluent string-rewriting systems [Boo82a], can be adjusted to apply to R^∞. Here it is important to notice that a string $uab^{2n-1}c$ is irreducible mod R^∞, whenever the string uab^{2n} is. Hence, after reducing the prefix $uab^{2n}c$ of a string w to $uab^{2n-1}c$ the reduction process can proceed without scanning a suffix of $uab^{2n-1}c$. This yields the following result.

Corollary 6. *There is a linear-time algorithm that, given a string $w \in \Sigma^*$ as input, determines the irreducible descendant w_0 of w mod R^∞.*

In particular, this gives the following.

Corollary 7. *The word problem for the monoid N can be solved in linear time.*

The subsystem $R_0 := \{da \rightarrow Aabb, WA \rightarrow \lambda\}$ of R is convergent. Further, for each $w_1 \in \mathrm{IRR}(R_0)$, if $w_1 \rightarrow^*_{R^\infty \setminus R_0} w_2$, then $w_2 \in \mathrm{IRR}(R_0)$. Hence, for each string $w \in \Sigma^*$, there is a reduction sequence of the form

$$w \rightarrow^*_{R_0} w_1 \rightarrow^*_{R^\infty \setminus R_0} w_0 \in \mathrm{IRR}(R^\infty).$$

Thus, w_1 is of the form $w_1 = u_0 ab^{2n_1} cu_1 \cdots u_{m-1} ab^{2n_m} cu_m$, where $u_0, u_1, \ldots, u_m \notin \Sigma^* \cdot \{ab^{2n}c \mid n \geq 1\} \cdot \Sigma^*$, and $w_0 = u_0 ab^{2n_1-1} cu_1 \cdots u_{m-1} ab^{2n_m-1} cu_m$.

Since R^∞ is weight-reducing, the reduction $w \rightarrow^*_{R_0} w_1$ is of length at most $c \cdot |w|$ for some constant $c \geq 1$, and $|w_1| \leq |w| + 2 \cdot |w|_d \leq 3 \cdot |w|$. Further, we have the following technical result, which is easily verified by induction. Here, for $m \in \mathbb{N}$, $\leftrightarrow^m_R$ denotes the m-fold iteration of the relation $\leftrightarrow_R$.

Lemma 8. *For all* $n \geq 1$, $ab^{2n}c \leftrightarrow^{4n-3}_R ab^{2n-1}c$.

Thus, in R the above reduction sequence from w to w_0 can be simulated as follows:

$$\begin{aligned} w \rightarrow^*_{R_0} \; & w_1 = u_0 ab^{2n_1} cu_1 \cdots u_{m-1} ab^{2n_m} cu_m \\ \leftrightarrow^{4n_1-3}_R \; & u_0 ab^{2n_1-1} cu_1 \cdots u_{m-1} ab^{2n_m} cu_m \\ \cdots \; & \\ \leftrightarrow^{4n_m-3}_R \; & u_0 ab^{2n_1-1} cu_1 \cdots u_{m-1} ab^{2n_m-1} cu_m = w_0, \end{aligned}$$

that is, this derivation has length at most

$$c \cdot |w| + \sum_{i=1}^{m} (4n_i - 3) \leq c \cdot |w| + 4 \cdot |w_1| \leq (c + 12) \cdot |w|.$$

Since $w_0 \in \mathrm{IRR}(R^\infty)$ is the unique representative of $[w_0]_R$, this gives the following result on the Dehn function δ_R for N.

Corollary 9. *δ_R is linearly bounded, that is, $\delta_R(n) \leq c_0 \cdot n$ for some integer constant $c_0 \geq 1$.*

A string-rewriting system S is called *confluent on* $[w]_S$ for some string w if, for all $x, y, z \in [w]_S$ satisfying $x \rightarrow^*_S y$ and $x \rightarrow^*_S z$, y and z have a common descendant (mod S). The system S is called *weakly confluent* if it is confluent on $[r]_S$ for each rule $(\ell \rightarrow r) \in S$ [MNOZ93].

Although the system R is not confluent, it has at least the following weaker property.

Lemma 10. *The string-rewriting system R is weakly confluent.*

Based on the infinite convergent string-rewriting system R^∞ we now establish the existence of an automatic structure for the presentation $(\Sigma; R)$.

Theorem 11. *The monoid N is automatic.*

Proof. We consider the finite presentation $(\Sigma; R)$. Since R^∞ is left-regular, convergent, and equivalent to R, the set $C := \mathrm{IRR}(R^\infty)$ is a complete set of unique representatives for the monoid N, and in addition C is a regular set. Hence, there exist a finite state acceptor W over Σ and a finite state acceptor $M_=$ over $\Sigma_\#$ such that $L(W) = C$ and $L(M_=) = \{\nu(u,v) \mid u,v \in C$ and $u \leftrightarrow_R^* v\} = \{\nu(u,u) \mid u \in C\}$. It remains to verify the existence of the multiplier automata M_s for $s \in \Sigma$.

(1.) $\boldsymbol{s := a}$: Let $u \in C$. Then u can uniquely be written as $u = u_1(Wd)^m$ for some $u_1 \notin \Sigma^* \cdot \{Wd\}$ and some $m \geq 0$. Hence, $ua = u_1(Wd)^m a \to_{R^\infty}^* u_1ab^{2m}$. If u_1 does not end in d, then $u_1ab^{2m} \in C$; otherwise, $u_1 = u_2d$ for some $u_2 \notin \Sigma^* \cdot \{W\}$, and $u_1ab^{2m} = u_2dab^{2m} \to_{R^\infty} u_2Aab^{2m+2} \in C$. Hence, the multiplier automaton M_a is to accept the language L_a, where

$$\begin{aligned} L_a := &\{\nu(u,ua) \mid ua \in C\} \cup \\ &\{\nu(u_1(Wd)^m, u_1ab^{2m}) \mid u_1 \in C, u_1 \notin \Sigma^* \cdot \{d\}, m \geq 1\} \cup \\ &\{\nu(u_2d(Wd)^m, u_2Aab^{2m+2}) \mid u_2 \in C, u_2 \notin \Sigma^* \cdot \{W\}, m \geq 0\}, \end{aligned}$$

which is easily seen to be regular.

(2.) $\boldsymbol{s \in \{b,d,W\}}$: For each $u \in C$, also $ub, ud, uW \in C$. Hence, L_b, L_d, and L_W are easily seen to be regular.

(3.) $\boldsymbol{s := c}$: Let $u \in C$. If $u = u_1ab^{2n}$ for some $n \geq 1$, then $uc = u_1ab^{2n}c \to_{R^\infty} u_1ab^{2n-1}c \in C$. Thus,

$$\begin{aligned} L_c = &\{\nu(u_1ab^{2n}, u_1ab^{2n-1}c) \mid u_1 \in C, u_1 \notin \Sigma^* \cdot \{d\}, n \geq 1\} \cup \\ &\{\nu(u,uc) \mid uc \in C\}, \end{aligned}$$

which is regular.

(4.) $\boldsymbol{s := A}$: Let $u \in C$. If $u = u_1W$, then $uA = u_1WA \to_{R^\infty} u_1 \in C$; otherwise, $uA \in C$. Thus

$$L_A = \{\nu(u_1W, u_1) \mid u_1W \in C\} \cup \{\nu(u,uA) \mid uA \in C\},$$

which is certainly regular.

Thus, C and the automata M_s $(s \in \Sigma)$ for the languages L_s constitute an automatic structure for the presentation $(\Sigma; R)$ of N. □

Actually, using the same case analysis the following stronger result can be established.

Corollary 12. *The monoid N is bi-automatic.*

Despite all the nice properties that the monoid N has, we will show in the following that it does not admit a finite convergent presentation, since it does not satisfy one of the conditions that are necessary for admitting a finite convergent presentation. See the recent paper by Kobayashi and Otto [OtKo97] for an overview on these conditions. One of these conditions is the property of having finite derivation type, which was introduced by Squier in [SOK94]. It can be shown that the monoid N does not have finite derivation type.

This property is defined for a given monoid-presentation $(\Sigma; S)$ based on an infinite graph $\Gamma(\Sigma; S) := (V, E, \sigma, \tau, ^{-1})$ that is associated with this presentation (for details see e.g. [SOK94, CrOt94, OtKo97]). This graph represents the single-step reduction relation $\rightarrow_S$ induced by the string-rewriting system S.

On the set $P(\Gamma(\Sigma; S))$ of all paths in $\Gamma(\Sigma; S)$ certain equivalence relations are considered that are called *homotopy relations.* The collection of all homotopy relations on $P(\Gamma(\Sigma; S))$ is closed under arbitrary intersection. Since the set $P^{(2)}(\Gamma(\Sigma; S))$ of all pairs of parallel paths is itself a homotopy relation, each subset $B \subseteq P^{(2)}(\Gamma(\Sigma; S))$ is contained in a smallest homotopy relation $\simeq_B$ on $P(\Gamma(\Sigma; S))$, which is accordingly called the homotopy relation *generated* by B. Here two paths p and q are called *parallel*, if they have a common initial vertex and a common terminal vertex.

The monoid-presentation $(\Sigma; S)$ is said to have *finite derivation type*, FDT for short, if there exists a finite subset $B \subseteq P^{(2)}(\Gamma(\Sigma; S))$ which generates $P^{(2)}(\Gamma(\Sigma; S))$ as a homotopy relation.

Squier has established the following results on the property FDT.

Proposition 13. [SOK94]

(a) If $(\Sigma_1; S_1)$ and $(\Sigma_2; S_2)$ are two finite presentations of the same monoid, then $(\Sigma_1; S_1)$ has FDT *if and only if $(\Sigma_2; S_2)$ has* FDT.
(b) If a monoid M has a finite convergent presentation, then M has FDT.

Because of part (a) of this proposition the property FDT is actually a property of finitely presented monoids.

We claim that the monoid N does not have FDT. In [SOK94] Squier presents a finitely presented monoid with a decidable word problem that does not have FDT. Following the proof given by Squier the following result can be established. The proof, which is technically involved, can be found in a technical report by two of the authors [OtSa97].

Theorem 14. *The monoid N does not have* FDT.

Because of Proposition 13(b) this result has the following consequence.

Corollary 15. *The monoid N does not have a finite convergent presentation.*

Hence, we see that the class of monoids that have a finite convergent presentation and the class of finitely presented monoids that are (bi)-automatic are incomparable under inclusion. Further, this result shows that a monoid with an easily decidable word problem need not have finite derivation type even if it has a finite, noetherian, and weakly confluent presentation. Hence, Proposition 13(b) cannot be strengthened to this class of presentations.

4 Automatic monoids and Dehn functions

As mentioned before the Dehn function of an automatic group is bounded by a quadratic function. The Dehn function of the automatic monoid N considered in the previous section is even bounded by a linear function. Here, however, we will show that Dehn functions of automatic monoids are in general not bounded by polynomial functions at all. For this we present a finitely presented (bi-)automatic monoid the Dehn function of which grows exponentially.

Let $\Sigma = \{0, 1, ¢, \$\}$, and let R consist of the following 4 rules:

$$0\$ \to 1\$,\ 1\$ \to ¢\$,\ 0¢ \to 10,\ 1¢ \to ¢0.$$

Lemma 16. *The system R is convergent.*

Proof. If we order the alphabet Σ by taking $\$ > 0 > 1 > ¢$, then we see that with respect to the length-lexicographical ordering $>_{\text{lex}}$ induced by $>$, $\ell >_{\text{lex}} r$ holds for each rule $(\ell \to r) \in R$. Hence, R is noetherian.

The left-hand sides of the rules of R do not overlap. Hence, R does not admit any critical pairs, that is, R is also confluent. □

Thus, the set $\text{IRR}(R)$ of irreducible strings is a regular set of unique representatives for the monoid M presented by $(\Sigma; R)$. Obviously, $\text{IRR}(R) = \{\$, ¢\}^* \cdot \{0, 1\}^*$. Thus, for all $n \geq 1$, $¢^n\$$ is the only irreducible element in the set $\{0, 1, ¢\}^n \cdot \{\$\}$. Since R is convergent and length-preserving, this immediately yields the following.

Lemma 17. *For all $u \in \{0, 1, ¢\}^*$, we have $u\$ \to_R^* ¢^{|u|}\$$.*

Based on this observation we now establish the following result.

Lemma 18. *The presentation $(\Sigma; R)$ is bi-automatic.*

Proof. We take C to be the set $\text{IRR}(R)$. Then

$$\begin{aligned} L_= &= \{\nu(w_1, w_2) \mid w_1, w_2 \in IRR(R),\ w_1 \leftrightarrow_R^* w_2\} \\ &= \{\nu(w, w) \mid w \in \text{IRR}(R)\}, \text{ which is obviously regular.} \end{aligned}$$

For $u \in \text{IRR}(R)$, we have $u0, u1 \in \text{IRR}(R)$. Hence,

$$\begin{aligned} L_0 &= \{\nu(w_1, w_2) \mid w_1, w_2 \in \text{IRR}(R),\ w_1 0 \leftrightarrow_R^* w_2\} \\ &= \{\nu(u, u0) \mid u \in \text{IRR}(R)\}, \text{ and} \end{aligned}$$

$$L_1 = \{\nu(u, u1) \mid u \in \text{IRR}(R)\}, \text{ which are regular, too.}$$

Now let $u = u_1 u_2$, where $u_1 \in \{\$, ¢\}^*$ and $u_2 \in \{0, 1\}^*$. Then

$$u¢ \to_R^* \begin{cases} u_1¢, & \text{if } u_2 = \lambda, \\ u_1¢0^i, & \text{if } u_2 = 1^i \quad (i \geq 1), \\ u_1u_3 10^{i+1}, & \text{if } u_2 = u_3 01^i\ (i \geq 0), \end{cases}$$

and

$$u\$ \rightarrow_R^* \begin{cases} u_1\$, & \text{if } u_2 = \lambda, \\ u_1 ¢^{|u_2|}\$, & \text{if } u_2 \neq \lambda \text{ (cf. Lemma 17).} \end{cases}$$

It is easily seen that the sets $L_¢$ and $L_\$$ are regular.

Further, $¢u, \$u \in \mathrm{IRR}(R)$,

$$0u \rightarrow_R^* \begin{cases} 0u_2, & \text{if } u_1 = \lambda, \\ 1^i 0 u_2, & \text{if } u_1 = ¢^i \quad (i \geq 1), \\ ¢ u_1 u_2, & \text{if } |u_1|_\$ > 0 \text{ (cf. Lemma 17),} \end{cases}$$

and

$$1u \rightarrow_R^* \begin{cases} 1u_2, & \text{if } u_1 = \lambda, \\ ¢ 1^{i-1} 0 u_2, & \text{if } u_1 = ¢^i \quad (i \geq 1), \\ ¢ u_1 u_2, & \text{if } |u_1|_\$ > 0 \text{ (cf. Lemma 17).} \end{cases}$$

Hence, also the sets ${}_¢L$, ${}_\$L$, ${}_0L$, and ${}_1L$ are regular. Thus, $(\Sigma; R)$ is indeed bi-automatic. □

There are no overlaps between the left-hand sides of the rules of R. Thus, if $u = u_1 \ell u_2$ for some $u_1, u_2 \in \Sigma^*$ and some rule $(\ell \rightarrow r) \in R$, and if $u = w_0 \rightarrow_R w_1 \rightarrow_R \ldots \rightarrow_R w_m \in \mathrm{IRR}(R)$ is a reduction sequence, then there exists an index $k \geq 0$ such that the following conditions are satisfied:

$$w_k = v_1 \ell v_2,\ w_{k+1} = v_1 r v_2, \text{ and } u_i \rightarrow_R^* v_i,\ i = 1, 2.$$

Based on this observation the following result can be derived.

Lemma 19. *Let $u \in \Sigma^*$ and $u_0 \in \mathrm{IRR}(R)$ such that $u \leftrightarrow_R^* u_0$. Then there exists an integer $n_u \in \mathbb{N}$ such that each reduction sequence $u = w_0 \rightarrow_R w_1 \rightarrow_R \cdots \rightarrow_R u_0$ has length n_u.*

For $n \in \mathbb{N}$, let $f(n)$ denote the length of the reduction sequences that reduce the string $0^n\$$ to its irreducible representative $¢^n\$$.

Lemma 20. *$f(n) = 2^{n+2} - 2n - 4$ for all $n \geq 1$.*

Proof. First we describe reduction sequences from $0^n\$$ to $¢^n\$$ inductively:

$n = 1:$ $\quad 0\$ \rightarrow_R 1\$ \rightarrow_R ¢\$$, that is, $f(1) = 2$.

$$n \rightarrow n+1: 0^{n+1}\$ = 00^n\$ \rightarrow_R^{f(n)} 0¢^n\$ \rightarrow_R^n 1^n 0\$ \rightarrow_R^2 1^n ¢\$ \rightarrow_R^n ¢0^n\$$$
$$\rightarrow_R^{f(n)} ¢¢^n\$ = ¢^{n+1}\$, \text{ that is, } f(n+1) = 2 \cdot f(n) + 2n + 2.$$

Based on this recursion we obtain the statement about $f(n)$ by induction:

$$n = 1: \quad f(1) = 2 = 2^3 - 2 - 4.$$

$$\begin{aligned} n \rightarrow n+1: f(n+1) &= 2 \cdot f(n) + 2n + 2 \\ &= 2 \cdot (2^{n+2} - 2n - 4) + 2n + 2 \\ &= 2^{n+3} - 2(n+1) - 4. \end{aligned}$$ □

Finally, we complete our investigation of the presentation $(\Sigma; R)$ by proving the following result on the distance function d_R.

Lemma 21. *$d_R(u\$, ¢^n\$) = n_{u\$}$ for all $u \in \{0,1,¢\}^n$, $n \geq 0$.*

Proof. By Lemma 17 $u\$ \leftrightarrow_R^* ¢^n\$$ for all $u \in \{0,1,¢\}^n$, $n \geq 0$, and by Lemma 19, $u\$ = v_0 \rightarrow_R \ldots \rightarrow_R v_m = ¢^n\$$ implies that $m = n_{u\$}$.

Now let $u\$ = w_0 \leftrightarrow_R \ldots \leftrightarrow_R w_\ell = ¢^n\$$ be a derivation of minimal length from $u\$$ to $¢^n\$$. We claim that this derivation is in fact a reduction sequence, which then yields $\ell = d_R(u\$, ¢^n\$) = n_{u\$}$.

Assume to the contrary that, for some index j, we have $w_j \leftarrow_R w_{j+1}$, and let this index j be chosen maximal. Then $u\$ = w_0 \leftrightarrow_R^j w_j \leftarrow_R w_{j+1} \rightarrow_R w_{j+2} \rightarrow_R \ldots \rightarrow_R w_\ell = ¢^n\$$. Observe that $j+1 < \ell$, since $w_\ell = ¢^n\$$ is irreducible. Since R is convergent, there is a reduction sequence

$$w_j = g_0 \rightarrow_R g_1 \rightarrow_R \ldots \rightarrow_R g_k = ¢^n\$.$$

Hence,

$$w_{j+1} \rightarrow_R w_{j+2} \rightarrow_R \ldots \rightarrow_R w_\ell = ¢^n\$$$

and

$$w_{j+1} \rightarrow_R w_j \rightarrow_R g_1 \rightarrow_R \ldots \rightarrow_R g_k = ¢^n\$$$

are two reduction sequences from w_{j+1} to $¢^n\$$. By Lemma 19 they both have the same length $\ell - j - 1$, which implies that $k = \ell - j - 2$. Hence, $u\$ = w_0 \leftrightarrow_R^j w_j \rightarrow_R g_1 \rightarrow_R \ldots \rightarrow_R g_k = ¢^n\$$ is a derivation from $u\$$ to $¢^n\$$ that has length $j + k = \ell - 2$. This obviously contradicts the choice of the original derivation. Thus, the derivations of minimal length from $u\$$ to $¢^n\$$ are in fact reduction sequences. □

Combining Lemma 20 and Lemma 21 we obtain the following result.

Corollary 22.

a) $d_R(0^n\$, ¢^n\$) = 2^{n+2} - 2n - 4$ for all $n \geq 1$.

b) $\delta_R(2n+2) \geq 2^{n+2} - 2n - 4$ for all $n \geq 1$.

Thus, we can summarize our results as follows.

Theorem 23. *There exists a finite convergent monoid-presentation that is bi-automatic such that the corresponding Dehn function grows exponentially.*

5 Concluding remarks

While the monoids with automatic structure share some of the nice properties of the groups with automatic structure [CRRT97], we have seen that this is not true for the rate of growth of their Dehn functions. While the Dehn function of an automatic group only grows quadratically, that of an automatic monoid may grow exponentially. This is true even if the finitely presented monoid considered is bi-automatic, and if it is given through a finite convergent presentation.

On the other hand, we have seen that a bi-automatic monoid may not have a finite convergent presentation even if it is finitely presented, has an easily decidable word problem, and has a linearly bounded Dehn function. However, the question remains open whether each automatic group has a finite convergent presentation.

The monoid N considered in Section 3 is given through a finite, weight-reducing, and weakly confluent string-rewriting system, from which we obtained an infinite, weight-reducing, and confluent string-rewriting system that is left-regular. Since $abb \leftrightarrow_R^* Wda$ holds, but abb is not congruent to any string of shorter length, we see that no length-reducing and confluent string-rewriting system is equivalent to R. If, however, we introduce three additional letters f, g, and h as short forms for the strings bc, da, and dW, respectively, then we obtain an infinite left-regular, length-reducing, and confluent string-rewriting system T^∞ on $\Gamma := \Sigma \cup \{f, g, h\}$ such that $(\Gamma; T^\infty)$ is another presentation of the monoid N.

As pointed out by an anonymous referee the notion of automatic structure may be generalized to the setting of term-rewriting systems, using (bottom-up) tree automata. In this setting the equivalence problem for ground terms should be easily decidable for automatic structures, but at this point it is not clear what other properties the ground term algebras with automatic structure may have.

References

[Alo92] J.M. Alonso. Combings of groups. In G. Baumslag and C.F. Miller III, editors, *Algorithms and Classification in Combinatorial Group Theory*, Math. Sciences Research Institute Publ. 23, pages 165–178. Springer-Verlag, New York, 1992.

[Boo82a] R.V. Book. Confluent and other types of Thue systems. *J. Association Computing Machinery*, 29:171–182, 1982.

[Boo82b] R.V. Book. The power of the Church-Rosser property in string-rewriting systems. In D.W. Loveland, editor, *6th Conference on Automated Deduction*, Lecture Notes in Computer Science 138, pages 360–368. Springer-Verlag, Berlin, 1982.

[Boo83] R.V. Book. Decidable sentences of Church-Rosser congruences. *Theoretical Computer Science*, 24:301–312, 1983.

[BoOt93] R.V. Book and F. Otto. *String-Rewriting Systems*. Springer-Verlag, New York, 1993.

[CRRT97] C.M. Campbell, E.F. Robertson, N. Ruškuc, and R.M. Thomas. *Automatic Semigroups*. Technical Report No. 1997/29, Department of Mathematics and Computer Science, University of Leicester, 1997.

[CrOt94] R. Cremanns and F. Otto. Finite derivation type implies the homological finiteness condition FP_3. *Journal of Symbolic Computation*, 18:91–112, 1994.

[CrOt96] R. Cremanns and F. Otto. For groups the property of having finite derivation type is equivalent to the homological finiteness condition FP_3. *Journal of Symbolic Computation*, 22:155–177, 1996.

[Eps92] D.B.A. Epstein. *Word Processing In Groups.* Jones and Bartlett Publishers, 1992.

[Ger92] S.M. Gersten. Dehn functions and l1-norms of finite presentations. In G. Baumslag and C.F. Miller III, editors, *Algorithms and Classification in Combinatorial Group Theory,* Math. Sciences Research Institute Publ. 23, pages 195–224. Springer-Verlag, New York, 1992.

[Gil84] R.H. Gilman. Computations with rational subsets of confluent groups. In J. Fitch, editor, *Proc. EUROSAM 84,* Lecture Notes in Computer Science 174, pages 207–212. Springer-Verlag, Berlin, 1984.

[GeSh91] S.M. Gersten and H.B. Short. Rational subgroups of biautomatic groups. *Annals of Mathematics,* 134:125–158, 1991.

[HoUl79] J.E. Hopcroft and J.D. Ullman. *Introduction to Automata Theory, Languages, and Computation.* Addison-Wesley, Reading, M.A., 1979.

[Hud96] J.F.P. Hudson. Regular rewrite systems and automatic structures. In J. Almeida, G.M.S. Gomes, and P.V. Silva, editors, *Semigroups, Automata and Languages,* pages 145–152. World Scientific, Singapure, 1996.

[MaOt85] K. Madlener and F. Otto. Pseudo-natural algorithms for the word problem for finitely presented monoids and groups. *Journal of Symbolic Computation,* 1:383–418, 1985.

[MNOZ93] K. Madlener, P. Narendran, F. Otto, and L. Zhang. On weakly confluent monadic string-rewriting systems. *Theoretical Computer Science,* 113:119–165, 1993.

[Ó'Dú83] C. Ó'Dúnlaing. Infinite regular Thue systems. *Theoretical Computer Science,* 25:171–192, 1983.

[Ott86] F. Otto. On deciding whether a monoid is a free monoid or is a group. *Acta Informatica,* 23:99–110, 1986.

[OtKo97] F. Otto and Y. Kobayashi. Properties of monoids that are presented by finite convergent string-rewriting systems - a survey. In D.Z. Du and K. Ko, editors, *Advances in Algorithms, Languages and Complexity,* pages 226–266. Kluwer Academic Publ., Dordrecht, 1997.

[OtSa97] F. Otto and A. Sattler-Klein. *Some remarks on finitely presented monoids with automatic structure.* Mathematische Schriften Kassel No. 9/97, Fachbereich Mathematik/Informatik, Universität Kassel, 1997.

[Pri95] S.J. Pride. Geometric methods in combinatorial semigroup theory. In J. Fountain, editor, *Proc. of Int. Conf. on Semigroups, Formal Languages, and Groups,* pages 215–232. Kluwer Academic Publ., Dordrecht, 1995.

[SOK94] C.C. Squier, F. Otto, and Y. Kobayashi. A finiteness condition for rewriting systems. *Theoretical Computer Science,* 131:271–294, 1994.

Decidable and Undecidable Second-Order Unification Problems*

Jordi Levy

Institut d'Investigació en Intel·ligència Artificial
Consejo Superior de Investigaciones Científicas
http://www.iiia.csic.es/~levy

Abstract. There is a close relationship between word unification and second-order unification. This similarity has been exploited for instance for proving decidability of monadic second-order unification. Word unification can be easily decided by transformation rules (similar to the ones applied in higher-order unification procedures) when variables are restricted to occur at most twice. Hence a well-known open question was the decidability of second-order unification under this same restriction. Here we answer this question negatively by reducing simultaneous rigid E-unification to second-order unification. This reduction, together with an inverse reduction found by Degtyarev and Voronkov, states an equivalence relationship between both unification problems.
Our reduction is in some sense reversible, providing decidability results for cases when simultaneous rigid E-unification is decidable. This happens, for example, for one-variable problems where the variable occurs at most twice (because rigid E-unification is decidable for just one equation). We also prove decidability when no variable occurs more than once, hence significantly narrowing the gap between decidable and undecidable second-order unification problems with variable occurrence restrictions.

1 Introduction

Word unification [Mak77,Sch91], linear second-order unification [Lev96], context unification [Com93,SS95] and second-order unification [Pie73] are closely related problems. The relationship between word unification and linear second-order unification becomes clear when we codify a word unification problem, like $F \cdot a \cdot G \stackrel{?}{=} G \cdot a \cdot F$, as a linear second-order unification problem $\lambda x.F(a(G(x))) \stackrel{?}{=} \lambda x.G(a(F(x)))$. The relationship between word unification and second-order unification is not so clear, but was used, for instance, to prove decidability of monadic second-order unification [Far88]. Despite their similarities, word unification is decidable [Mak77], second-order unification is undecidable [Gol81], and the question is open for linear second-order unification and context unification (although it is conjectured to be decidable).

* This work was partially supported by the project MODEL (TIC97-0579-C02-01) funded by the CICYT, and the ESPRIT Basic Research Actions CCL and CONSOLE

Decidability of word unification was an open question for a long time and its proof [Mak77] involves a lot of technicalities. However, it is very easy to prove that it is decidable when no variable occurs more than twice in a problem. The same main ideas were used to prove that linear second-order unification and context unification are decidable when no variable occurs more than twice [Lev96]. Thus, the arising question is, are these ideas applicable to second-order unification? The answer is no. We prove this undecidability result by reduction of another undecidable unification problem: simultaneous rigid E-unification [GRS87,DV96]. This reduction, together with a inverse reduction found by Degtyarev and Voronkov [DV95], states a close relationship between both unification problems. More precisely, proves that both problems are polynomial-time equivalent. Based on the preliminary version of this paper, Veanes [Vea98] also proves some of the results we prove. Other related results and ideas appeared in [Sch97].

Our reduction is in some sense reversible, providing decidability results for cases when simultaneous rigid E-unification is decidable. This happens, for example, for one (second-order) variable problems where the variable occurs at most twice, since (non-simultaneous) rigid E-unification is decidable [GNPS88].

This paper proceeds as follows. In section 2 we introduce all the unification problems we will deal with and some preliminary definitions and notation. In section 3 we prove undecidability of second-order unification, when variables are restricted to occur at most twice, by reduction from simultaneous rigid E-unification. We started our research trying to prove the decidability of the problem. The difficulties we found to achieve this purpose suggested us how we could, in fact, prove undecidability, and which undecidable problem we had to chose. However, since simultaneous rigid E-unification is decidable for one equation, we prove in section 4 decidability for one second-order variable problems. Additionally, in section 4, we also prove decidability for problems where variables occur at most once. This closes the gap between decidable and undecidable second-order unification problems w.r.t. variable occurrence restrictions.

2 Preliminary Definitions

We assume that the reader is familiar with unification problems, second-order typed λ-calculus and related topics. Variables are denoted by capital letters ($X, Y, Z \ldots$ when they are first-order, and $F, G \ldots$ when they are second-order variables), constants are denoted by lower case letters ($a, b \ldots$ when they are 0-ary constants, and $f, g \ldots$ for functions), terms by $t, u, v, w \ldots$ and substitutions by Greek letters $\sigma, \rho, \theta \ldots$. Substitutions are represented by finite sets of variable-term pairs, like $\sigma = [X_1 \mapsto t_1] \cdots [X_n \mapsto t_n]$. The application of a substitution σ to a term t is represented by $\sigma(t)$. Notation $t|_p$ represents subterm at position p of t, and $t[u]_p$ represents term t where subterm at position p has been replaced by u. We assume that any term is second-order typed and is written in $\beta\eta$-long normal form, i.e. any term has the form $\lambda \overline{x} \,.\, a(t_1, \ldots, t_n)$ where $\overline{x}$ is a (possibly empty) list of first-order bound variables, a may be a (at most) third-order constant, a second-order free variable or a first-order bound variable (in this later case $n = 0$), and t_i are also second-order terms in normal form.

2.1 Word Unification

It is easy to describe a complete[1] (non-terminating) procedure for word unification in terms of *transformation rules* [GS90]. Any state of the process is represented by a pair $\langle S, \sigma \rangle$, where S is the problem and σ the substitution computed until that moment. We proceed by applying a substitution ρ, that transforms the pair into a new one $\langle \rho(S), \rho \circ \sigma \rangle$ where $\rho(S)$ can be later simplified. At some point, more than a rule can be applicable, thus the procedure is not deterministic. We distinguish two kinds of words, *rigid* words (when they start by a constant, like $a \cdot w$) and *flexible* words (when they start by a variable, like $X \cdot w$). Therefore, we have three kinds of equations. For each kind of equation, the set of applicable transformations are as follows:

Rigid-rigid equations, like $a \cdot w_1 \stackrel{?}{=} a \cdot w_2$. Only **simplification rule** is applicable

$$\langle \{a \cdot w_1 \stackrel{?}{=} a \cdot w_2\} \cup S, \sigma \rangle \Rightarrow \langle \{w_1 \stackrel{?}{=} w_2\} \cup S, \sigma \rangle$$

Rigid-flexible equations, like $a \cdot w_1 \stackrel{?}{=} X \cdot w_2$. We can apply two different rules:

Projection rule to instantiate the variable on the head by the empty word $\rho = [X \mapsto \epsilon]$.

Imitation rule to instantiate $\rho = [X \mapsto a \cdot X']$. A fresh variable X' is introduced, and the equation is transformed into a rigid-rigid equation $a \cdot \rho(w_1) \stackrel{?}{=} a \cdot X' \cdot \rho(w_2)$ that is later simplified into $\rho(w_1) \stackrel{?}{=} X' \cdot \rho(w_2)$

Flexible-flexible equations, like $X \cdot w_1 \stackrel{?}{=} Y \cdot w_2$.

If $X = Y$, we can simplify the equation by removing both occurrences of the variable to get $w_1 \stackrel{?}{=} w_2$.

Otherwise we can instantiate one of the variables $\rho = [X \mapsto Y \cdot X']$, introducing a new fresh variable X'. The equation is transformed into $Y \cdot X' \cdot \rho(w_1) \stackrel{?}{=} Y \cdot \rho(w_2)$ and simplified into $X' \cdot \rho(w_1) \stackrel{?}{=} \rho(w_2)$.

If no variable occurs more than twice, after instantiating and simplifying equations, no transformation rule increases the size of the problem (in term of number of symbols). Since, there are finitely many unification problems of a given size (up to variable renaming), we can easily prove decidability of the problem [SS95] under this two-occurrences restriction. In fact, although word unification is infinitary[2], we can prove that, under this restriction, there exists a *finite* representation of the (maybe infinite) set of unifiers. For instance, the problem $X \cdot a \stackrel{?}{=} a \cdot X$ has infinitely many most general unifiers $[X \mapsto a \cdot \dots \cdot a]$, but we can represent all them by a regular expression $[X \mapsto \epsilon] \circ [X \mapsto a \cdot X]^*$.

[1] Notice that this procedure computes all most general unifiers, but not only most general unifiers.

[2] We can have infinitely many most general unifiers for a given unification problem.

2.2 Second-order Unification

Pietrzykowski [Pie73] was the first to describe a complete second-order unification procedure. The rules that this procedure uses are quite similar to the rules we have described for word unification. We also distinguish between rigid and flexible second-order normal terms. Given a term in normal form $\lambda\overline{x}\,.\,a(t_1,\ldots,t_n)$, if a is a constant or a bound variable, the term is said to be rigid, and flexible if a is a free variable.

A second-order unification problem is a finite set $\{t_1 \stackrel{?}{=} u_1,\ldots,t_n \stackrel{?}{=} u_n\}$ of pairs of second-order terms. For all our purposes, we can assume that our unification problems do not contain third-order constants or λ-bindings, i.e. we can assume that any term is first-order typed and has the form $a(t_1,\ldots,t_n)$, where a is a second-order free variable or second-order constant, and t_i are also first-order terms. Goldfarb [Gol81] proved that second-order unification, even under this restriction, is undecidable.

If we are only interested in deciding if a problem has a solution or not, and not in finding *all* its most general unifiers, we can simplify Pietrzykowski's procedure (notice that all flexible-flexible equations are solvable). These decision procedures for unification problems are called *pre-unification* procedures. Huet [Hue75] was the first to describe a pre-unification procedure for typed λ-calculus. The set of transformation rules for second-order pre-unification can be easily derived from either Huet's higher-order pre-unification procedure or from Pietrzykowski's second-order unification procedure. This set is as follows:

Simplification rule.

$$\langle\{a(t_1,\ldots,t_n) \stackrel{?}{=} a(u_1,\ldots,u_n)\}\cup S,\sigma\rangle \;\Rightarrow\; \langle \bigcup_{i\in[1..n]} \{t_i \stackrel{?}{=} u_i\}\cup S,\sigma\rangle$$

Projection rule. If we have a rigid-flexible equation like $\lambda\overline{x}\,.\,F(t_1,\ldots,t_n) \stackrel{?}{=} \lambda\overline{x}\,.\,g(u_1,\ldots,u_m)$ we can instantiate

$$[F\mapsto \lambda x_1\ldots\lambda x_n\,.\,x_i] \qquad \text{for some } i\in[1..n]$$

Imitation rule. Or, we can instantiate

$$[F\mapsto \lambda x_1\ldots\lambda x_n\,.\,g(F_1'(x_1,\ldots,x_n),\ldots,F_m'(x_1,\ldots,x_n))]$$

Proposition 1. *The procedure based on the previous transformation rules is a sound and complete pre-unification procedure for second-order unification.*

The previous proposition ensures semi-decidability of second-order unification, so when we say "undecidable" or "not decidable" we always mean semi-decidable.

2.3 Simultaneous Rigid E-Unification

Simultaneous rigid E-unification was introduced in [GRS87] in order to extend the tableau method, the method of matings and other proof methods to first-order logic with equality. After some faulty proofs of its decidability, it was proved to be undecidable in [DV96]. The (non-simultaneous) *rigid E-unification problem* can be formulated as follows. Given a finite set of first-order equations $\{t_i \cong u_i \mid i \in [1..n]\}$ and an equation $v \cong w$, decide if there exists a ground[3] substitution θ such that the formula

$$\theta(t_1) = \theta(u_1) \wedge \cdots \wedge \theta(t_n) = \theta(u_n) \Rightarrow \theta(v) = \theta(w)$$

is provable in first-order logic with equality. An instance of the problem is denoted by $t_1 \cong u_1 \wedge \cdots \wedge t_n \cong u_n \vdash_\forall v \cong w$, and is called a *rigid equation. Simultaneous rigid E-unification* is formalised as the problem of finding a simultaneous solution for a finite set of rigid equations.

For simplicity, we will also introduce a different notion of rigid unification called **rigid O-unification**. Given a rigid O-unification problem $\bigwedge_{i\in[1..n]} t_i \subseteq u_i \vdash_\forall v \subseteq w$, we say it is solvable if there exists a ground substitution θ such that the formula $\left(\bigwedge_{i\in[1..n]} \theta(t_i) \subseteq \theta(u_i)\right) \Rightarrow \theta(v) \subseteq \theta(w)$ is provable in first-order logic with a monotonic pre-order relation, i.e. without considering the symmetry rule.

Since a rewriting system defines a monotonic pre-order relation, we can reformulate rigid O-unification as follows. Given a term rewriting system $\{t_i \to u_i \mid i \in [1..n]\}$, and a pair of terms v and w, decide if there exists a substitution θ such that $\theta(v) \to^* \theta(w)$ using the ground rewriting system $\{\theta(t_i) \to \theta(u_i) \mid i \in [1..n]\}$.

It is easy to prove decidability of rigid O-unification from decidability of rigid E-unification, and to prove undecidability of *simultaneous* rigid O-unification from undecidability of *simultaneous* rigid E-unification.

Proposition 2. *Rigid E-unification is reducible to rigid O-unification. Simultaneous rigid E-unification is reducible to simultaneous rigid E-unification.*

Proof: Replace every rigid equation $\bigwedge_{i\in[1..n]} t_i \cong u_i \vdash_\forall v \cong w$ by the rigid inclusion $\bigwedge_{i\in[1..n]}(t_i \subseteq u_i \wedge u_i \subseteq t_i) \vdash_\forall v \subseteq w$ ■

3 Reducing Simultaneous Rigid E-Unification to Second-Order Unification

In this section we reduce simultaneous rigid O-unification to second-order unification where second-order typed variables are restricted to occur at most twice in the unification problem. This reduction is based in the following main lemma.

[3] Requiring θ to be ground is not relevant since, if there exist a non-ground solution, then there exists also a ground solution.

Lemma 1 (Main Lemma). *The rigid equation over the signature* $\langle \Sigma, \mathcal{X} \rangle$

$$t_1 \subseteq u_1 \wedge \cdots \wedge t_m \subseteq u_m \vdash_\forall v \subseteq w \tag{1}$$

has a solution if, and only if, the following second-order equation

$$F(a(b,v), u_1, \ldots, u_m) \stackrel{?}{=} a(F(b, t_1, \ldots, t_m), w) \tag{2}$$

together with the following set of equations

$$\left.\begin{array}{l} X \stackrel{?}{=} G_x(f_1(\overrightarrow{Y_x}), \ldots, f_N(\overrightarrow{Y_x})) \\ b \stackrel{?}{=} G_x(b, \ldots, b) \end{array}\right\} \quad \forall X \in \mathcal{X} \,.\, \exists i \in [1..m] \,.\, u_i = X \tag{3}$$

have a solution.
Where we have assumed $\overrightarrow{Y_x} \notin \mathcal{X}$ *are lists of first-order variables of the appropriate length,* $a, b \notin \Sigma$ *and* $\Sigma = \{f_1, \ldots, f_N\}$ *is a finite signature.*[4]

Example 1. The following rigid inclusion $c \subseteq X \vdash_\forall f(c,c) \subseteq f(d,e)$, where $\Sigma = \{c, d, e, f\}$ and $\mathcal{X} = \{X\}$, is solvable if, and only if, the following set of second-order equations is solvable.

$$\begin{array}{l} F(a(b, f(c,c)), X) \stackrel{?}{=} a(F(b,c), f(d,e)) \\ X \stackrel{?}{=} G_x(c, d, e, f(Y_x^1, Y_x^2)) \\ b \stackrel{?}{=} G_x(b, b, b, b) \end{array}$$

In this case both systems are unsolvable. However, notice that the first equation $F(a(b, f(c,c)), X) \stackrel{?}{=} a(F(b,c), f(d,e))$ alone, has a solution

$$[F \mapsto \lambda x, y \,.\, y][X \mapsto a(c, f(d,e))]$$

To avoid this problem with variables occurring as the right-hand side of a premise, like in this case X, we require equations (3).

Proof of Main Lemma:
Implication $\Rightarrow$
Let θ be a solution of the rigid equation (1). Without lose of generality, we can assume that θ is ground and the signature Σ only contains constant symbols from equation (1). We can derive $\theta(v) \subseteq \theta(w)$ from $\bigwedge_{i \in [1..m]} \theta(t_i) \subseteq \theta(u_i)$ using only reflexivity, transitivity and monotonicity inference rules for the $\subseteq$ binary relation.

Using Birkhoff's theorem, we can prove that there exist a sequence of terms $s_1, \ldots, s_{k+1}$ such that $\theta(v) = s_1$ and $\theta(w) = s_{k+1}$ and for any $j \in [1..k]$ there exist an $i_j \in [1..m]$ and a position p_j of s_j such that

$$\begin{array}{l} s_{j+1} = s_j[\theta(u_{i_j})]_{p_j} \\ s_j|_{p_j} = \theta(t_{i_j}) \end{array}$$

[4] We can consider Σ as the set of constants $f_1, \ldots, f_N$ occurring in the original rigid E-unification problem.

i.e. we can rewrite s_j into s_{j+1} using $t_{i_j} \to u_{i_j}$ as a rewriting rule at position p_j.

Define the second-order substitution σ as follows:

$$\begin{aligned}\sigma(X) &= \theta(X) \quad \text{for } X \in \mathcal{X} \\ \sigma(F) &= \lambda x_0, x_1, \ldots, x_m \,.\, a(\ldots a(a(x_0, s_1[x_{i_1}]_{p_1}), s_2[x_{i_2}]_{p_2}) \ldots, s_k[x_{i_k}]_{p_k})\end{aligned}$$

It is a straightforward exercise to prove that this substitution σ is a solution of the second-order equation (2).

For any variable X satisfying $\exists j \in [1..m] \,.\, u_i = X$ define σ_x as follows. Let $f_k \in \Sigma$ be the constant such that $\theta(X) = f_k(s_1, \ldots, s_p)$, for some terms $s_1, \ldots, s_p$, where $p = \text{arity}(f_k)$, (regard that θ is ground and the signature Σ only contains constant symbols from equation (1)), then

$$\begin{aligned}\sigma_x(\overrightarrow{Y_x}) &= \overrightarrow{s} \\ \sigma_x(G_x) &= \lambda x_0, x_1, \ldots, x_N \,.\, x_k\end{aligned}$$

It is also straightforward to prove that $\sigma_x \circ \sigma$ satisfies equation (3) for X.

As far as these substitutions σ and σ_x, for any X, have disjoint domains, the composition of all them $\sigma \circ \circ_{x \in \mathcal{X}} \sigma_x$ solves all the second-order equations (2) and (3).

Implication $\Leftarrow$

Suppose that equation (2) is solvable. We can only apply the imitation or projection rules, and by completeness of the pre-unification procedure, one of the two problems that we obtain has to be solvable.

If we apply the projection rule, it has to be necessarily $[F \mapsto \lambda x_0, \ldots, x_m \,.\, x_0]$. Any other projection function would lead to $u_i \stackrel{?}{=} a(t_i, w)$, and this equation has no solution unless u_i contains a (which is not the case because $a \notin \Sigma$) or u_i is a variable. If $u_i = X$, we have equations (3) written for X. The only possible solutions for the second one of these equations $b \stackrel{?}{=} G_x(b, \ldots, b)$ are $[G_x \mapsto \lambda x_1, \ldots, x_n \,.\, x_k]$, for some $k \in [1..N]$, and $[G_x \mapsto \lambda x_1, \ldots, x_n \,.\, b]$. Applying any of this substitutions to the rest of equations $X \stackrel{?}{=} a(t_i, w)$ and $X \stackrel{?}{=} G_x(f_1(\overrightarrow{Y_x}), \ldots, f_N(\overrightarrow{Y_x}))$ results in the following unsolvable problems $\{X \stackrel{?}{=} a(t_i, w), X \stackrel{?}{=} f_k(\overrightarrow{Y_x})\}$, for some $k \in [1..N]$, or $\{X \stackrel{?}{=} a(t_i, w), X \stackrel{?}{=} b\}$. Therefore, we can conclude that, after instantiating F by the only possible projection function, we obtain $a(b, v) \stackrel{?}{=} a(b, w)$. Now, by simplification we obtain $v \stackrel{?}{=} w$.

If we apply the imitation rule, we obtain:

$$\begin{aligned}F_1(a(b, v), u_1, \ldots, u_m) &\stackrel{?}{=} a(F_1(b, t_1, \ldots, t_m), F_2(b, t_1, \ldots, t_m)) \\ F_2(a(b, v), u_1, \ldots, u_m) &\stackrel{?}{=} w\end{aligned}$$

where F_1 and F_2 are both fresh variables.

If this system is solvable, then the first equation –which is quite similar to the original one– has to be also solvable. We can repeat the same argument for this equation. Iterating this argument, we can conclude that: there exist a $k \geq 0$ such that, after applying k times the imitation rule to the first equation, and later the projection and the simplification rules, the system we get is solvable. The system will be:

$$
\begin{aligned}
& v \stackrel{?}{=} F_1(b, t_1, \dots, t_m) \\
& F_1(a(b,v), u_1, \dots, u_m) \stackrel{?}{=} F_2(b, t_1, \dots, t_m) \\
& F_2(a(b,v), u_1, \dots, u_m) \stackrel{?}{=} F_3(b, t_1, \dots, t_m) \\
& \dots \\
& F_k(a(b,v), u_1, \dots, u_m) \stackrel{?}{=} w
\end{aligned}
$$

for $k > 0$, and $u \stackrel{?}{=} w$ for $k = 0$.

Solvability of one of these system ensures that there exist a ground substitution σ such that $\sigma(v)$ can be rewritten into $\sigma(F_1(a(b,v), u_1, \dots, u_m))$ in one parallel rewriting step, and $\sigma(F_1(a(b,v), u_1, \dots, u_m))$ can be rewritten into $\sigma(F_2(a(b,v), u_1, \dots, u_m))$, etc. using ground rewriting rules: $\sigma(b) \to \sigma(a(b,v))$, $\sigma(t_1) \to \sigma(u_1)$, ... , $\sigma(t_m) \to \sigma(u_m)$. Again, Birkhoff's theorem proves that we can deduce:

$$\sigma(b) \subseteq \sigma(a(b,v)) \wedge \sigma(t_1) \subseteq \sigma(u_1) \wedge \cdots \wedge \sigma(t_m) \subseteq \sigma(u_m) \ \vdash \ \sigma(v) \subseteq \sigma(w)$$

in first-order logic with a monotonic pre-order relation $\subseteq$. Therefore, the rigid equation:

$$b \subseteq a(b,v) \wedge t_1 \subseteq u_1 \wedge \cdots \wedge t_m \subseteq u_m \ \vdash_\forall \ v \subseteq w$$

has a solution. As far as a and b do not occur in t_i, u_i, v and w, we do not need to assume $b \subseteq a(b,v)$ in that derivation. Therefore, we can also prove that the rigid equation (1) has a solution. ■

Theorem 1. *There is an effective method that reduces simultaneous rigid E-unification to second-order unification, where second-order variables are restricted to occur at most twice in a unification problem and no equation contains more than one second-order variable.*

Proof: We have already seen that simultaneous rigid E-unification is reducible to simultaneous rigid O-unification. Now, lemma 1 can be easily extended to reduce simultaneous rigid O-unification to second-order unification.

Suppose we have a system of n rigid equations, over a first-order signature $\langle \Sigma, \mathcal{X} \rangle$,

$$t_1^i \subseteq u_1^i \wedge \cdots \wedge t_{m_i}^i \subseteq u_{m_i}^i \vdash_\forall v^i \subseteq w^i$$

for $i \in [1..n]$.

We define a new second-order signature

$$\langle \Sigma \cup \{a, b\}, \mathcal{X} \cup \bigcup_{\substack{x \in \mathcal{X} \\ j \in [1..N]}} \{Y_x^j\} \cup \bigcup_{i \in [1..n]} \{F^i\} \cup \bigcup_{x \in \mathcal{X}} \{G_x\} \rangle$$

where, apart from the constants and variables we already had, we have introduced a new constant symbol b, a new binary function symbol a, N first-order variables for each variable $X \in \mathcal{X}$, where $N = \max\{arity(f) \mid f \in \Sigma\}$, a second-order variable F^i, with arity $m_i + 1$, for each rigid equation of the system, and a second-order variable G_x with arity cardinality of Σ, for each variable $X \in \mathcal{X}$.

We can effectively construct a second-order unification problem, containing the following second-order equation for each rigid equation:

$$F^i(a(b, v^i), u_1^i, \ldots, u_{m_i}^i) \stackrel{?}{=} a(F^i(b, t_1^i, \ldots, t_{m_i}^i), w^i)$$

and the following equations

$$\begin{aligned} X &\stackrel{?}{=} G_x(f_1(\overrightarrow{Y_x}), \ldots, f_N(\overrightarrow{Y_x})) \\ b &\stackrel{?}{=} G_x(b, \ldots, b) \end{aligned}$$

for any variable $X \in \mathcal{X}$ satisfying $\exists i \in [1..n] . \exists j \in [1..m_i] . u_j^i = X$.

An extension of main lemma can be used to prove the equivalence between this system and the original simultaneous rigid equations. Notice that any second-order variable F^i or G_x in the equations occurs only twice and there are not equations containing more than one second-variable. ■

Since there exist a reduction [DV95] from second-order unification to simultaneous rigid E-unification we have the following corollary.

Corollary 1. *Second-order unification, simultaneous rigid E-unification, and second-order unification where second-order variables are restricted to occur at most twice and equations to do not contain more than one second-order variable, are all three equivalent.*
The second-order unification problem is undecidable, even if we restrict second-order variables to occur at most twice, and equations to do not contain more than one second-order variable.

4 Decidability Results

In this section we prove decidability of second-order unification problems only containing one second-order variable, which occurs at most twice. The impossibility to prove this result for more than one variable suggested us how to prove undecidability (in the previous section). It also helped us establish a relationship between second-order unification and simultaneous rigid E-unification. Additionally, we also prove decidability for problems not containing any repeated variable.

If we compare second-order unification rules with word unification rules, in section 2, at first sight it seems that the two-occurrences restriction is going to carry over, like in the word unification case. The simplification and projection rules always decrease the size of a problem. However, in this case, application of imitation rule can increase the size of a problem, even if we restrict variables to occur at most twice.

For instance, if we apply imitation rule to

$$F(t_1, t_2) \stackrel{?}{=} g(F(u_1, u_2), v)$$

we obtain a bigger problem (in term of number of symbols)

$$\begin{array}{l} F_1'(t_1, t_2) \stackrel{?}{=} g(F_1'(u_1, u_2), F_2'(u_1, u_2)) \\ F_2'(t_1, t_2) \stackrel{?}{=} v \end{array}$$

Moreover, since some terms are duplicated (like t_1, t_2, u_1, u_2), second-order variables of these terms may occur now more than twice!

We can overcome the second problem by assigning a directed acyclic graph (DAG) to each problem to avoid duplication of terms. In our example we would have:

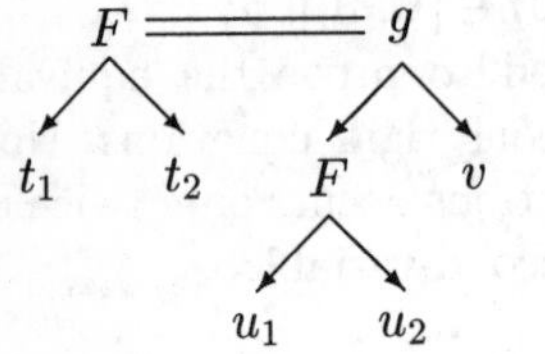

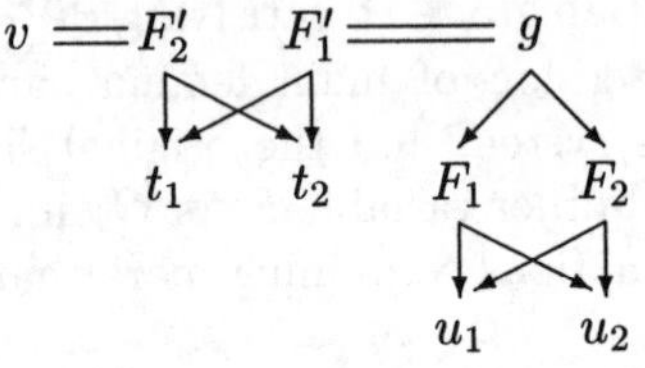

$$F(t_1, t_2) \stackrel{?}{=} g(F(u_1, u_2), v) \qquad \begin{array}{l} F_1'(t_1, t_2) \stackrel{?}{=} g(F_1'(u_1, u_2), F_2'(u_1, u_2)) \\ F_2'(t_1, t_2) \stackrel{?}{=} v \end{array}$$

Then, we can define the *size of a problem* as a pair *(number of constant occurrences, number of variable occurrences)* of its assigned DAG. We compare these pairs using a lexicographic order. Since the simplification rule always removes two constant occurrences, it always decreases the size of the problem. The projection rule removes a variable occurrence and does not increase the number of constant occurrences. However, the imitation rule may increase the number of variable occurrences (although, if no variable occurs more than twice, it never increases the number of constant occurrences). This proves the following lemma.

Lemma 2. *Any infinite transformation sequence contains infinitely many imitation steps.*

To characterise non-terminating transformation sequences we have to study the imitation rule in detail. When we apply the imitation rule to a rigid-flexible equation $F(t_1, \ldots, t_n) \stackrel{?}{=} g(u_1, \ldots, u_m)$, the occurrence of g on the right hand side of the equation is removed, i.e. the equation is replaced by new equations $F_i'(t_1, \ldots, t_n) \stackrel{?}{=} u_i$ for $i \in [1..m]$. And, if there exists another occurrence of F in another term, an occurrence of g is added to this term. Thus, we can see the imitation rule as *moving* constant occurrences from one place to another. This image can help to characterise non-terminating sequences as follows.

Definition 1. *We say that a unification problem is in* **normal form** *if it does not contain any rigid-rigid equation. Notice that, given a solvable unification problem, we can always find an equivalent problem in normal form by repeatedly applying simplification rule.*

We say that two variables F and G are **equivalent** *in an unification problem S, noted $F \cong G$, if for some substitution θ, the normal form of $\theta(S)$ contains a flexible-flexible equation of the form $F(t_1, \ldots, t_n) \stackrel{?}{=} G(u_1, \ldots, u_m)$.*
We say that a variable F is **connected with** *another variable G in a unification problem S, noted $F \leadsto G$, if for some substitution θ, the normal form of $\theta(S)$ contains an equation of the form $F(t_1, \ldots, t_n) \stackrel{?}{=} v$, where v contains the variable G, and it is not in the head of v.*
Let $\stackrel{\cong}{\leadsto}$ denote the relation $\cong^* \circ \leadsto$. We say that a unification problem contains a **variable cycle** *if there is a non-empty sequence of variables such that $F_1 \stackrel{\cong}{\leadsto} F_2 \stackrel{\cong}{\leadsto} \cdots \stackrel{\cong}{\leadsto} F_n$.*

In our example $F(t_1, t_2) \stackrel{?}{=} g(F(u_1, u_2), v)$, we have $F \leadsto F$, therefore, it contains a variable cycle.

Theorem 2. *Any infinite transformation sequence is generated by a problem containing a variable cycle.*
Therefore, it is decidable whether a second-order unification problem not containing variable cycles has a unifier.

Proof: By lemma 2 we know that any infinite sequence contains infinitely many imitation steps. Since initially there are finitely many variables, and when we instantiate one variable we only introduce finitely many new fresh variables, we can conclude that, some variable F of the original problem is involved in an *infinite sequence of chained imitation steps*:

$$\cdots [F \mapsto \lambda \overline{x} \,.\, g_1(\ldots, F_1(\overline{x}), \ldots)] \cdots [F_1 \mapsto \lambda \overline{x} \,.\, g_2(\ldots, F_2(\overline{x}), \ldots)] \cdots$$

Assume F is one of the maximal (w.r.t. the relation $\stackrel{\cong}{\leadsto}$) variables involved in one of such chained sequences. This is always possible unless the relation is cycling.

Firstly, we will prove that some variable G satisfying $F \stackrel{\cong}{\leadsto} G$ is also involved in one of such infinite sequences of chained imitation steps.

At some point of the transformation sequence, the problem contains, at least, one rigid-flexible pair $F(\ldots) \stackrel{?}{=} g_1(t_1, \ldots, t_m)$. Otherwise, the imitation step $[F \mapsto \lambda \overline{x} \,.\, g_1(\ldots F_1(\overline{x}) \ldots)]$ would never been applied. By applying this imitation step to this problem, we replace this equation by a finite set of equations containing $F_1(\ldots) \stackrel{?}{=} \rho(t_i)$ for some $i \in [1..m]$. Notice that an occurrence of the function symbol g_1 is removed when we replace $g_1(t_1, \ldots, t_m)$ by $\rho(t_i)$ on the right-hand side of equations. Therefore, since there are finitely many occurrences of function symbols in $g_1(t_1, \ldots, t_m)$, we can not repeat this process infinitely many times, unless some variable G occurring in t_i is also involved in an infinite sequence of chained imitation steps. If G is a variable of the original problem, we have $F \leadsto G$, and the work is done. Otherwise, let G' be the variable of the original problem that originates the chained sequence where G is involved. We can prove that initially there is an equation containing both F and G', and either $F \leadsto G'$, $F \cong G'$ or $G' \leadsto F$. In the first case the work is done. The last case is not possible because we have assumed that F is maximal. In the second case we can repeat the same reasoning for G'. At some point we have to find a variable G'' such that $F \cong G' \overset{\cong}{\rightsquigarrow} G''$. Otherwise it is not possible to have $F \rightsquigarrow G$ at some point of the transformation sequence.

Now we can repeat the same argument for G, or G' or G''. Since originally there are finitely many variables, this process allows us to construct a cycle $F \overset{\cong}{\rightsquigarrow}^* H \overset{\cong}{\rightsquigarrow}^* \circ \overset{\cong}{\rightsquigarrow} H$ for some variable H of the original problem. Moreover H is involved in an infinite sequence of chained imitation steps. ■

A direct consequence of this theorem is the following decision result.

Corollary 2. *It is decidable whether a second-order unification problem, where no second-order variable occurs more than once, has a unifier.*

Proof: Ii is not difficult to prove that, if no variable is repeated, we can not have any cycle.

We can also prove this result directly. If no variable is repeated, no transformation rule can increase the size of the problem. However, we still have to prove that no transformation rule can duplicate a variable occurrence. This is true if we represent the unification problem as a DAG. ■

If we have multiple occurrences of a variable, we have to deal with infinite transformation sequences. This does not seems easy. We only have been able to do that when the problem only contains a second-order variable and this variable only occurs twice.

Theorem 3. *It is decidable whether a second-order unification problem, containing a single second-order variable and where this variable only occurs twice, has a unifier.*

Proof: The only possible cycle $F \rightsquigarrow F$ is generated if we have an equation $\lambda\overline{x}\,.\,F(t_1,\ldots,t_n) \overset{?}{=} v$ where F occurs in v.

For simplicity, we will assume that only one function symbol g occurs between the root of v and F. Without loss of generality, we will assume that F occurs in the first argument of v. We can have other more complex situations, but they can also be proved to be decidable using the same main ideas. Under these simplifications, we only need to consider the following equation.

$$F(t_1,\ldots,t_n) \overset{?}{=} g(F(u_1,\ldots,u_n),v_1,\ldots,v_m)$$

We can repeat the same argument as in the proof of lemma 1. If this equation has a solution, since it is a rigid-flexible equation, we can obtain another solvable system after applying the imitation rule k many times to the first equation and later the projection rule, for some $k \geq 0$. If we only apply imitation rule, we can generate an infinite transformation sequence. However, if the problem has a solution, there is a *finite* sequence of transformations leading to a set with only flexible-flexible equations. Therefore, at some point we have to apply projection rule to the first equation. The problem is that we can not conjecture when!

After these k many imitation steps, and a projection step, we get a system which is equivalent to the following one.

$$
\begin{aligned}
& t_i \stackrel{?}{=} g(u_i, X_1, \ldots, X_m) \\
& X_1 \stackrel{?}{=} F_0^{(p)}(u_1, \ldots, u_n) \\
& \cdots \\
& X_m \stackrel{?}{=} F_m^{(p)}(u_1, \ldots, u_n) \\
& F_1^{(p)}(t_1, \ldots, t_n) \stackrel{?}{=} F_1^{(p-1)}(u_1, \ldots, u_n) \\
& \cdots \\
& F_m^{(p)}(t_1, \ldots, t_n) \stackrel{?}{=} F_m^{(p-1)}(u_1, \ldots, u_n) \\
& F_1^{(p-1)}(t_1, \ldots, t_n) \stackrel{?}{=} F_1^{(p-2)}(u_1, \ldots, u_n) \\
& \cdots \\
& F_m^{(p-1)}(t_1, \ldots, t_n) \stackrel{?}{=} F_m^{(p-2)}(u_1, \ldots, u_n) \\
& \cdots \\
& F_1^{(1)}(t_1, \ldots, t_n) \stackrel{?}{=} v_1 \\
& \cdots \\
& F_m^{(1)}(t_1, \ldots, t_n) \stackrel{?}{=} v_m
\end{aligned}
$$

for some $i \in [1..n]$.

Applying the same ideas as in section 3, we can prove that solvability of this system is equivalent to solvability of the following rigid equation:

$$\sigma(t_1) \subseteq \sigma(u_1) \wedge \cdots \wedge \sigma(t_n) \subseteq \sigma(u_n) \ \vdash_\forall \ \sigma(v_1) \subseteq \sigma(X_1) \wedge \cdots \wedge \sigma(v_m) \subseteq \sigma(X_m)$$

for some σ being unifier of $t_i \stackrel{?}{=} g(u_i, X_1, \ldots, X_m)$ for some $i \in [1..n]$. There are finitely many of such unifiers. The problem has been reduced to solvability of finitely many instances of a rigid equation, which is decidable.

For more complex cycle situations (not considered in this proof) we get a similar rigid equation. ■

5 Conclusions and Further Work

Since Goldfarb proved the undecidability of the second-order unification problem [Gol81], very few decidable and undecidable subclasses of second-order unification problems have been found. Here we have characterised decidability for classes defined in terms of number of occurrences per variable.

Moreover, we have stated a very close relationship between the simultaneous rigid E-unification and the second-order unification problems. This relationship allows us to translate some decidability/undecidability results from one class of problems to the other.

In [Lev96] we proved that *linear* second-order unification is decidable when no second-order variable occurs more than twice. Here, we have proved that, under this same restriction, second-order unification is undecidable. This establishes a clear difference between this two apparently similar problems. Notice that second-order unification is undecidable, whereas linear second-order unification has been conjectured to be decidable.

Acknowledgements

I would like to acknowledge M. Bonet, A. Rubio, M. Villaret and all the anonymous referees of this paper for their comments and support, and specially to R. Nieuwenhuis for suggesting me the possible relationship between the problem I *tried* to prove decidable and the simultaneous rigid E-unification problem.

References

[Com93] H. Comon. Completion of rewrite systems with membership constraints. Technical report, CNRS and LRI, Université de Paris Sud, 1993.

[DV95] A. Degtyarev and A. Voronkov. Reduction of second-order unification to simultaneous rigid E-unification. Technical Report 109, Computer Science Department, Uppsala University, 1995.

[DV96] A. Degtyarev and A. Voronkov. The undecidability of simultaneous rigid E-unification. *Theoretical Computer Science*, 166(1-2):291–300, 1996.

[Far88] W. M. Farmer. A unification algorithm for second-order monadic terms. *Annals of Pure and Applied Logic*, 39:131–174, 1988.

[GNPS88] J. H. Gallier, P. Narendran, D. Plaisted, and W. Snyder. Rigid E-unification is NP-complete. In *Proc. IEEE Conf. on Logic in Computer Science, LICS'88*, pages 338–346, 1988.

[Gol81] W. D. Goldfarb. The undecidability of the second-order unification problem. *Theoretical Computer Science*, 13:225–230, 1981.

[GRS87] J. H. Gallier, S. Raatz, and W. Snyder. Theorem proving using rigid E-unification: Equational matings. In *Proc. IEEE Conf. on Logic in Computer Science, LICS'87*, pages 338–346, 1987.

[GS90] J. H. Gallier and W. Snyder. Designing unification procedures using transformations: A survey. *Bulletin of the EATCS*, 40:273–326, 1990.

[Hue75] G. Huet. A unification algorithm for typed λ-calculus. *Theoretical Computer Science*, 1:27–57, 1975.

[Lev96] J. Levy. Linear second-order unification. In *7th Int. Conf. on Rewriting Techniques and Applications, RTA'96*, volume 1103 of *LNCS*, pages 332–346, New Jersey, USA, 1996.

[Mak77] G. S. Makanin. The problem of solvability of equations in a free semigroup. *Math. USSR Sbornik*, 32(2):129–198, 1977.

[Pie73] T. Pietrzykowski. A complete mechanization of second-order logic. *J. of the ACM*, 20(2):333–364, 1973.

[Sch91] K. U. Schulz. Makanin's algorithm, two improvements and a generalization. Technical Report CIS-Bericht-91-39, Centrum für Informations und Sprachverarbeitung, Universität München, 1991.

[Sch97] A. Schubert. Second-order unification and type inference for church-style polymorphism. Technical Report TR 97-02(239), Institute of Informatics, Warsaw University, 1997.

[SS95] M. Schmidt-Schauß. Unification of stratified second-order terms. Technical Report 12/94, Johan Wolfgang-Goethe-Universität, Frankfurt, Germany, 1995.

[Vea98] M. Veanes. The relation between second-order unification and simultaneous rigid E-unification. Technical Report MPI-I-98-2-005, Max-Planck Institut für Informatik, 1998.

On the Exponent of Periodicity of Minimal Solutions of Context Equations

Manfred Schmidt-Schauß[1] & Klaus U. Schulz[2]

[1] Fachbereich Informatik, Johann Wolfgang Goethe-Universität, Postfach 11 19 32, D-60054 Frankfurt,Germany
Tel: (+49)69-798-28597, Fax: (+49)69-798-28919,
E-mail: schauss@ki.informatik.uni-frankfurt.de

[2] Center for Information and Language Processing (CIS), University of Munich, Oettingenstr 67, D-80538 München, Germany
Tel: (+49)89-2178-2700, Fax: (+49)89-2178-2701,
E-mail: schulz@cis.uni-muenchen.de

Abstract. Context unification is a generalisation of string unification where words are generalized to terms with one hole. Though decidability of string unification was proved by Makanin, the decidability of context unification is currently an open question. This paper provides a step in understanding the complexity of context unification and the structure of unifiers. It is shown, that if a context unification problem of size d is unifiable, then there is also a unifier with an exponent of periodicity smaller than $\mathcal{O}(2^{1.07d})$. We also prove $\mathcal{NP}$-hardness for restricted cases of the context unification problem and compare the complexity of general context unification with that of general string unification.

1 Introduction

Context unification is a generalisation of string unification. The latter is known to be decidable [11], while the decidability of the first one is currently open. Context unification also generalizes first order unification by a special kind of second order variables (the context variables), which can be instantiated only by (ground) terms with a single distinguished hole. It is well-known that permitting an arbitrary number of (equally named) holes results in an undecidable problem, since this is exactly second order unification [7, 5]. Thus it can be seen as a specialization of higher-order unification(see e.g. [13, 8, 16]), though the methods are different.

Contexts are used in various areas of computer science. Usually, they appear as notation $C[s]$, where $C[.]$ means a context, and s means a term. $C[s]$ is the resulting term after plugging the term s into the hole of $C[.]$. They appear also in the more general form of n-ary contexts like $C[.,\ldots,.]$. Contexts are used for notational convenience in automated deduction, term rewriting and programming languages, in particular program fragments are often expressed as contexts.

Context unification appears as a subproblem in constraint solving with membership constraints [3]. It also appears as a subproblem in some equational unification algorithms [15]. An interesting application is in the field of computer linguistics, where contexts are used to express the semantics of utterances in natural languages [12].

There are some known decidability results on restricted context unification problems. If the signature is restricted to symbols of arity 1 or 0, then second-order unification is decidable [4], which holds also for context unification. If the number of occurrences of every context variable and every variable is restricted to be at most two, then context unification is decidable [10]. The decidability of stratified context unification problems was announced in [14, 10]).

The main contribution of this paper is a generalisation of a result of Kościelski and Pacholski [9], which by itself improves similar results by Makanin and Bulitko [11, 2]. The result says, given a word equation E, there exists a recursive function f in the length d of E such that the maximal number n of periodic repetitions of some non-empty word in a component of a minimal solution of E does not exceed $f(d)$. The number n is called the exponent of periodicity of the solution, hence f gives an upper bound on the exponent of periodicity. This result was the key lemma which was used by Makanin to show decidability of string unification [11]. The bound given in [9] is the smallest known bound. We generalize this result to context unification following the structure of the proof in [9] and obtain the same upper bound.

This results may be useful for future research on decidability of context unification. A further application may be to show an improved upper bound on the complexity of stratified context unification.

2 Terms and Contexts

Let Σ denote a signature, i.e, a set of functions symbols of fixed arity. We shall assume that Σ contains at least one constant. Let Ω (the "hole") denote a new constant not in Σ. We shall also use an infinite set $\mathcal{X}$ of *context variables*, denoted $X, Y, \ldots$.

The set of (ground first order) *terms* built over Σ is defined as usual. A *ground context* is a term built over the extended signature $\Sigma \cup \{\Omega\}$ that has exactly one occurrence of Ω. Given a ground context A and a ground term/context B, we may apply A to B, which means to replace the occurrence of Ω in A by B, obtaining $A[B/\Omega]$. Obviously, the operation of applying ground contexts to ground contexts is associative, with identity Ω. The application of A to B is denoted in the form $A \cdot B$, or AB, for simplicity. Furthermore we define $A^0 := \Omega$ and $A^n := AA^{n-1}$ for $n > 0$. Letters $A, B, \ldots$ will denote ground terms/contexts.

The set of *positions* of a ground term/context is defined as usual. For a ground context A, the position τ of Ω in A is also called the path to Ω. The length of τ is the *depth* of A, denoted $|A|$. A ground context A is called *simple*, iff there is no other ground context B, such that $A = B^n$ for some $n > 1$. A ground context B is called a (proper) *prefix* of a ground context A if there exists a (non-empty) ground context C such that $A = BC$. In this situation, C is called a (proper) *suffix* of A (if B is non-empty).

The set of *context terms* is defined as follows. If $f \in \Sigma$ has arity $n \geq 0$, and if $W_1, \ldots, W_n$ are context terms, then $f(W_1, \ldots, W_n)$ is a context term. If V is a context term, and if X is a context variable, then XV is a context term. *Contexts* are defined like context terms, but have exactly one occurrence of the special constant Ω. As for ground contexts, we may also apply a context U to a context/context term V, obtaining a context (context term) UV. Again, this operation is associative, with identity Ω. The context U *occurs in* the context/context term V if V can be represented in the form $V = W_1 U W_2$, for contexts/context terms W_1, W_2. If U and V are ground, then the position of the root of the marked occurrence of W_2 in $W_1 U W_2$ is called the *position of the hole* of the marked occurrence of U in V. Letters $U, V, W, \ldots$ denote contexts/context terms.

A *context equation* is an equation $W_1 = W_2$ between context terms. The *size* of $W_1 = W_2$ is the total number of occurrences of symbols from $\Sigma \cup \mathcal{X}$ in the equation. An assignment σ of ground contexts to context variables *solves* the context equation $W_1 = W_2$ if $\sigma(W_1) = \sigma(W_2)$ when we apply all ground contexts $\sigma(X)$ to their arguments.

Given a solution σ of the context equation $W_1 = W_2$, the maximal number p such that there exists a variable X in $W_1 = W_2$ such that $\sigma(X)$ can be represented in the form BA^pC for a non-empty context A is called the *exponent of periodicity* of the solution σ. A solution σ of $W_1 = W_2$ is called *minimal* if there is no other solution σ' of $W_1 = W_2$ with $\Sigma_i(size(\sigma'(X_i))) < \Sigma_i(size(\sigma(X_i)))$, where X_i are the context variables occurring in W_1, W_2.

In this paper we want to give an upper bound on the exponent of periodicity of a minimal solution of a context equation.

3 Presentations of ground contexts and terms

In this section we describe a representation of ground contexts/terms where iterated occurrences of the same simple word are factored out. We shall often consider occurrences of the same ground context A in a given term/context. An expression $A^{(i)}$ will always refer to a fixed occurrence of A.

Definition 1. Two occurrences $A^{(1)}$ and $B^{(2)}$ of ground contexts A and B in a ground term/context C are called *aligned* if the holes of $A^{(1)}$ and $B^{(2)}$ are on the same path of C. $A^{(1)}$ and $B^{(2)}$ *overlap* in C if there exists a position in C that belongs to $A^{(1)}$ and $B^{(2)}$, not marking the hole of either occurrence.

Lemma 2. *Let A and B be ground contexts. If, for some ground terms/contexts C and D we have $A^{(1)}D = B^{(1)}A^{(2)}C$, then $A^{(1)}$ and $B^{(1)}$ are aligned in AD. Either A is a prefix of B or vice versa.*

Proof. Assume that $A^{(1)}$ and $B^{(1)}$ were not aligned. Then $A^{(2)}$ would be an occurrence of A properly inside $A^{(1)}$, a contradiction. The rest is obvious. □

Lemma 3. *Let A, B be ground contexts. If $AB = BA$, then there exists a ground context C such that $A = C^m$ and $B = C^n$ for suitable naturals m, n.*

Proof. This can be proved as for words using induction on the depth and using Lemma 2. □

Lemma 4. *Let A, B be ground contexts. Then $ABC = BAD$ for ground terms/contexts C, D implies that $AB = BA$.*

Proof. By induction on the size of AB. If A or B is trivial (i.e., $= \Omega$), or if $A = B$, then the lemma holds. In the other case, by Lemma 2 we may assume without loss of generality that B is a prefix of A, and we can write $A = BA'$. The remaining equation is then $BA'BC = BBA'D$. By cancellation we get the smaller equation $A'BC = BA'D$, which implies $A'B = BA'$ by induction, and hence $AB = BA'B = BBA' = BA$. □

Lemma 5. *Let A be a simple context. Then $AAC = BAD$ for a ground context B such that $|B| \leq |A|$ and ground terms/contexts C, D implies that either $B = A$ or $B = \Omega$.*

Proof. From Lemma 2 we know that B is a prefix of A under the given assumptions. Let $A = BA_1$. Hence we have $BA_1BA_1C = BBA_1D$ and $A_1BA_1C = BA_1D$. By Lemma 4 we get $A_1B = BA_1$. Since A is simple, the result follows now from Lemma 3. □

Definition 6. Let $A^{(1)}$ and $A^{(2)}$ be two aligned occurrences of the ground context A in the ground term/context B. The *distance* of $A^{(1)}$ and $A^{(2)}$ is the length of the path between the roots of both occurrences.

Lemma 7. *Let A be a simple context. Assume that the ground term/context C contains two occurrences of the context AA. If the two occurrences overlap, then they are aligned. Moreover, if both occurrences are different, their distance is at least $|A|$.*

Proof. We may assume that C is a ground term. Assume that the two occurrences $A^{(1)}A^{(2)}$ and $A^{(3)}A^{(4)}$ of AA have root position τ and τ' respectively. Since $A^{(1)}A^{(2)}$ and $A^{(3)}A^{(4)}$ overlap we may assume without loss of generality that τ is a prefix of τ'. This means that there exists a ground context B at position τ and terms D, E such that $BA^{(1)}A^{(2)}D = A^{(3)}A^{(4)}E$. By Lemma 2, either B is a prefix of A or vice versa. If B is a prefix of A, then by Lemma 5 either $B = \Omega$ or $B = A$, and in both cases it follows that $A^{(1)}A^{(2)}$ and $A^{(3)}A^{(4)}$ are aligned. If A is a prefix of $B = AB'$, then $AB'A^{(1)}A^{(2)}D = A^{(3)}A^{(4)}E$. Cancellation yields $B'A^{(1)}A^{(2)}D = A^{(4)}E$. The occurrences $A^{(1)}$ and $A^{(4)}$ are aligned since otherwise $A^{(2)}$ would be properly contained in $A^{(4)}$. It follows that $A^{(1)}A^{(2)}$ and $A^{(3)}A^{(4)}$ are aligned.

Assume that $A^{(1)}A^{(2)}$ and $A^{(3)}A^{(4)}$ are different occurrences. Let B be a ground context such that $BA^{(1)}A^{(2)}D = A^{(3)}A^{(4)}E$. The depth $|B|$ is the distance between $A^{(1)}A^{(2)}$ and $A^{(3)}A^{(4)}$. If $|B| < |A|$, then Lemma 5 yields $B = \Omega$, which is a contradiction. □

Definition 8. A context term/context S is called a *skeleton* if each context variable occurs at most once in S. The *reduct* of S is the ground term/context that is obtained when we replace each occurrence of a context variable in S by Ω. The *order* of a skeleton S is the number of context variables occurring in S. Two skeletons S and S' are *equivalent* if S' can be obtained from S by a bijective renaming of variables. S is *finer* than S' if we obtain S' from S—modulo a bijective renaming of variables—by assigning Ω to some of the variables of S.

In general we shall assume that the variables of a skeleton S are given in a fixed order, and we often write $S(X_1, \ldots, X_u)$ to make variables and order explicit.

Definition 9. If a skeleton S has an occurrence of the context XBX', where B is a ground context, then $|B|$ is called the *distance* between X and X' in S.

Definition 10. Let A be a simple context. A skeleton $S(X_1, \ldots, X_u)$ is called an *A-skeleton* if S has an occurrence of the context AX_iA, for each $1 \leq i \leq u$.

Lemma 11. *Let XCX' be a context occurring in the A-skeleton S, where C is a ground context. Then the distance of X and X' in S is at least $|A|$.*

Definition 12. Let A be a simple context. An A-skeleton S is called an *A-skeleton* for the ground term/context B if B is the reduct of S. An A-skeleton S of B is *maximal* if there is no A-skeleton of B that is finer than S.

Lemma 13. *Let A be a simple context.*

1. *Each ground term/context B has a maximal A-skeleton.*
2. *Two maximal A-skeletons for the same ground term/context B are always equivalent.*

Proof. By Lemma 7, two occurrences of AA in B either do not overlap, or they are aligned and their distance is at least $|A|$. Hence we may independently replace each subcontext AA of B by an expression AXA, using a new variable for each new context. We receive a maximal A-skeleton S for B. If also S' is a maximal A-skeleton for B, each subexpression $AX'A$ of S' reduces to an occurrence of AA in B. Hence S contains a corresponding subexpression AXA. Since S' cannot be finer than S, S and S' are equivalent. □

In view of the previous lemma we shall henceforth talk about *the greatest* A-skeleton of a ground term/context. In the greatest A-skeleton, variables are just used as "named holes". We shall now move to another representation where variables stand for maximal sequences of stable occurrences of A of length $k \geq 0$. An occurrence $A^{(1)}$ of A in the ground context/term C is *stable* if C has a subcontext $A^{(0)}A^{(1)}A^{(2)}$.

Definition 14. Let A be a simple context. An A-skeleton $S(X_1, \ldots, X_u)$ is called *A-stable* if the following conditions are satisfied

1. S does not have any occurrence of the context AA,

2. S does not have any occurrence of a context of the form $X_i A X_j$.

Clearly, if S is A-stable, then every A-subskeleton of S is A-stable.

Definition 15. Let A be a simple context, and let $S(X_1, \ldots, X_u)$ be an A-stable skeleton ($u \geq 0$). Then $plug_{S \leftarrow A}$ is the function that assigns to every u-tuple $(k_1, \ldots, k_u)$ of natural numbers the ground term/context that is obtained from S by replacing each variable X_i by A^{k_i}, for $i = 1, \ldots, u$.

Definition 16. Let A be a simple context. An *A-presentation* of the ground term/context B is a pair $(T(\vec{X}), \vec{k})$ where $T(\vec{X}) = T(X_1, \ldots, X_u)$ is an A-stable A-skeleton of order $u \geq 0$ and $\vec{k} = (k_1, \ldots, k_u)$ is a sequence of natural numbers such that $B = plug_{T \leftarrow A}(k_1, \ldots, k_u)$. B is of *A-order* u if it has an A-presentation of order u.

Let A be a simple context. Let B be a ground term/context and let S be the greatest A-skeleton of B. When we replace each maximal sequence $X_{i_1} A X_{i_2} \ldots A X_{i_{k_i}}$ (where $k_i \geq 0$) occurring in S by a variable X_i, using distinct variables for distinct sequences, then we obviously obtain an A-presentation $(T(\vec{X}), (\vec{k}))$ of B. Conversely, if $(T(\vec{X}), (\vec{k}))$ is an A-presentation of B, and if we replace each variable X_i by an expression $X_{i_1} A X_{i_2} \ldots A X_{i_{k_i}}$, then we obtain the greatest A-skeleton of B. Hence, from Lemma 13 we obtain

Lemma 17. *Let A be a simple context. Then each ground term/context B has an A-presentation. Moreover, two A-presentations for the same ground term/context B are always equivalent.*

In the sequel, we shall talk about *the* A-presentation of a given ground term/context, and we shall often write $T(\vec{X})$ (or T) if the sequence $\vec{k}$ (resp. $\vec{X}$) is not relevant in the given context.

4 Composition of Presentations

In this section we discuss how A-presentations of ground contexts/terms can be formed on the basis of given A-presentations of their subcontexts/terms. From now on, we fix a simple context A. We write $S = gulp(B)$ if the A-skeleton S is the A-presentation of B and $plug_S(.)$ instead of $plug_{S \leftarrow A}(.)$.

Definition 18. An occurrence $A^{(1)}A^{(2)}$ of the context AA in the ground term/context B is called *adaptive* if B has the form $B = A_1 A^{(1)} A^{(2)} B_1$ where the context A_1 is a proper, possibly empty suffix of A. The occurrence $A^{(1)}A^{(2)}$ of AA in the ground context B is called *final* if B has the form $B_1 A^{(1)} A^{(2)} A_2$ where A_2 is a proper, possibly empty prefix of A.

Lemma 19. *Let A be a simple context. For every ground term/context B there is at most one adaptive and at most one final occurrence of AA in B.*

Proof. Assume that $B = A_1^{(1)}A^{(2)}A^{(3)}B_1$ and $B = A_2^{(4)}A^{(5)}A^{(6)}B_2$ mark two adaptive occurrences $A^{(2)}A^{(3)}$ and $A^{(5)}A^{(6)}$ of A in B. Then $A_1^{(1)}$ and $A_2^{(4)}$ have to be aligned in B since otherwise the suffix $A_2^{(4)}$ of A would contain $A^{(2)}$. It follows easily that the two marked occurrences of AA overlap. By Lemma 7, the two marked occurrences of AA are aligned. Since both A_1 and A_2 are proper prefixes of A, the distance of the two occurrences of AA is smaller then $|A|$. Hence the occurrences coincide, by Lemma 7. Uniqueness of final occurrences can be proved in the same way. □

Let T be an A-skeleton with set of variables $\{X_1, \ldots, X_u\}$. The variable X of T is an *adaptive* (resp. *final*) variable of T if T has a subcontext $A^{(1)}XA^{(2)}$ that represents an adaptive (final) occurrence of AA in the reduct of T. By Lemma 19, T has at most one adaptive (final) variable.

Lemma 20. *Let C be a ground context, and let T be the A-presentation of C. If T has an adaptive variable X_0 and a final variable X_u, and if T has at least two variables, then X_0 and X_u are different.*

Proof. Assume that $X_0 = X_u$. Then C can be represented in the form $C = A_1A^{k+2}A_2$ where A_1 is a proper suffix of A and A_2 is a proper prefix of A. It follows easily that all occurrences of the context AA in C belong to the marked sequence and the A-presentation of C has only one variable, contradicting our assumption. □

The previous lemma enables us to use the following *notational convention*: if the skeleton $S(X_1, \ldots, X_u)$ has an adaptive (final) variable, then this is X_1 (resp. X_u). In the sequel, we consider compositions of A-presentations, and the convention will become relevant. We first show how maximal A-skeletons are composed.

Lemma 21. *Let $f \in \Sigma$ be an n-ary function symbol, let $B_1, \ldots, B_n$ be ground terms of orders $v_1, \ldots, v_n$ respectively, let $(g_{1,1}, \ldots, g_{1,v_1}), \ldots, (g_{n,1}, \ldots, g_{n,v_n})$ denote sequences of natural numbers. Let $C_i = \mathrm{plug}_{\mathrm{gulp}(B_i)}(g_{i,1}, \ldots, g_{i,v_i})$, and let S_i denote the greatest A-skeleton of C_i ($i = 1, \ldots, n$). We assume that distinct skeletons do not share variables. Then the greatest A-skeleton S of $f(C_1, \ldots, C_n)$ has either the form $f(S_1, \ldots, S_n)$ or it may be obtained from $f(S_1, \ldots, S_n)$ introducing one additional variable X. In the latter case, S has the form $AXAS'$, for a suitable skeleton S', and at most one of the terms C_i has an occurrence of A.*

Proof. Obviously S may have the form $f(S_1, \ldots, S_n)$. In the other case $f(C_1, \ldots, C_n)$ has an occurrence $A^{(1)}A^{(2)}$ of the context AA that does not exist in any subterm C_i. Hence $A^{(1)}$ must mark the root of $f(C_1, \ldots, C_n)$ which implies that S has the form $AXAS'$. In this situation, exactly one of the subterms C_i is aligned with $A^{(1)}$, and the other subterms cannot have any occurrence of A. □

Lemma 22. *Let $f \in \Sigma$ be an n-ary function symbol, let $B_1, \ldots, B_n$ be ground terms of orders $v_1, \ldots, v_n$ respectively, let $(g_{1,1}, \ldots, g_{1,v_1}), \ldots, (g_{n,1}, \ldots, g_{n,v_n})$ denote sequences of natural numbers, let $\vec{g}_i := (g_{i,1}, \ldots, g_{i,v_i})$. Then either*

1. $f(plug_{gulp(B_1)}(\vec{g}_1), \ldots, plug_{gulp(B_n)}(\vec{g}_n))$
   $= plug_{gulp(f(B_1,\ldots,B_n))}(\vec{g}_1, \ldots, \vec{g}_n)$, *or otherwise*
2. *with one possible exception, all terms* B_i *have order* 0 *(hence* $\vec{g}_i$ *is empty), and* $f(plug_{gulp(B_1)}(\vec{g}_1), \ldots, plug_{gulp(B_n)}(\vec{g}_n))$ *has one of the following forms*
   (a) $plug_{gulp(f(B_1,\ldots,B_n))}(g_{i,1}+1, \ldots, g_{i,v_i})$.
   (b) $plug_{gulp(f(B_1,\ldots,B_n))}(0, g_{i,1}, \ldots, g_{i,v_i})$.

Proof. Let $C_i = plug_{gulp(B_i)}(\vec{g}_i)$, and let S_i denote the greatest A-skeleton of C_i $(i = 1, \ldots, n)$. Note that $f(plug_{gulp(B_1)}(\vec{g}_1), \ldots, plug_{gulp(B_n)}(\vec{g}_n))$ is the reduct of $f(S_1, \ldots, S_n)$. Let us consider the two cases mentioned in Lemma 21. If the greatest A-skeleton S of $f(C_1, \ldots, C_n)$ has the form $f(S_1, \ldots, S_n)$, then its reduct yields $plug_{gulp(f(B_1,\ldots,B_n))}(\vec{g}_1, \ldots, \vec{g}_n)$ and we have Case 1. In the second case mentioned above it follows that at most one of the terms B_i has order > 0. If S has the form $AXAX'S'$, then we are in Case 2 (a). In the other case, we obtain situation (b). □

Consider two ground terms/contexts B and C. If S is the greatest A-skeleton of BC, then S can be described in the form S_1S_2 where S_1 (S_2) describes the subskeleton of S that reduces to B (resp. C). In order to have a unique splitting point we shall use the following convention: if S contains a variable X that marks the splitting point between B and C, then the variable is attached to the lower part, which means that S_2 has the form XAS_2'. With this convention, the decomposition is unique.

Lemma 23. *Let* B_1 *be a ground context and* C_1 *a ground term, of orders* v *and* w *respectively, let* $(g_1, \ldots, g_v)$ *and* $(h_1, \ldots, h_w)$ *denote sequences of natural numbers. Let* $B = plug_{gulp(B_1)}(g_1, \ldots, g_v)$*, let* $C = plug_{gulp(C_1)}(h_1, \ldots, h_w)$*. Let* S (S_B, S_C) *denote the greatest* A*-skeleton of* BC *(resp.* B, C*), let* $S = S_1S_2$ *be the partition of* S *into the subskeletons that reduce to* B *and* C *respectively.*

1. S_1 (S_2) *is finer than, or equivalent to* S_B *(resp.* S_C*).*
2. *If* S_1 *has a variable* X *not occurring in* S_B*, then* S_1 *can be represented in the form* $S_1'XA'$ *where* A' *is a proper and non-empty prefix of* $A = A'A''$ *and* A'' *is a prefix of* C*.* S_1 *can have at most one variable not occurring in* S_B*.*
3. *If* S_2 *has a variable* X *not occurring in* S_C*, then* S_2 *can be represented in the form* $A''XS_3$ *where* A'' *is a proper, possibly empty, suffix of* $A = A'A''$ *and* A' *is a suffix of* B*.* S_2 *can have at most one variable not occurring in* S_C*.*

Proof. Part 1 is obvious since each occurrence of AA in B or C respectively defines a parallel occurrence of AA in BC. The first statement of Part 2 is obvious, the second follows easily from Lemma 11. The first statement of Part 3 is obvious. Assume that S_2 has variables X_1, X_2 not occurring in S_C. Let $A_1X_1S_3$ and $A_2X_2S_4$ denote two representations of S_2 of the form described in Part 3. Then S_2 has representations $A_1X_1A^{(1)}S_5$ and $A_2X_2A^{(2)}S_6$. Hence A_1

and A_2 have to be aligned since otherwise $A^{(2)}$ would be properly contained in A_1. By Lemma 11, $X_1 = X_2$. □

In the sequel, two variables X, X' of a skeleton S are said to have *contact* in S if S contains a subcontext XAX' or $X'AX$.

Lemma 24. *Let B_1 be a ground context and C_1 a ground term, with orders $v \geq 0$ and $w \geq 0$ respectively. Let $(g_1, \ldots, g_v)$ and $(h_1, \ldots, h_w)$ denote sequences of natural numbers. Then one of the following cases holds*[3]*:*

1. $plug_{gulp(B_1)}(g_1, \ldots, g_v) plug_{gulp(C_1)}(h_1, \ldots, h_w) =$ $plug_{gulp(B_1 C_1)}(g_1, \ldots, g_v, h_1, \ldots, h_w)$,
2. $plug_{gulp(B_1)}(g_1, \ldots, g_v) plug_{gulp(C_1)}(h_1, \ldots, h_w) =$ $plug_{gulp(B_1 C_1)}(g_1, \ldots, g_v + c + 2 + h_1, \ldots, h_w)$,
3. $plug_{gulp(B_1)}(g_1, \ldots, g_v) plug_{gulp(C_1)}(h_1, \ldots, h_w) =$ $plug_{gulp(B_1 C_1)}(g_1, \ldots, g_{v-1}, g_v + c + 1, h_1, \ldots, h_w)$,
4. $plug_{gulp(B_1)}(g_1, \ldots, g_v) plug_{gulp(C_1)}(h_1, \ldots, h_w) =$ $plug_{gulp(B_1 C_1)}(g_1, \ldots, g_v, h_1 + c + 1, \ldots, h_w)$,
5. $plug_{gulp(B_1)}(g_1, \ldots, g_v) plug_{gulp(C_1)}(h_1, \ldots, h_w) =$ $plug_{gulp(B_1 C_1)}(g_1, \ldots, g_v, c, h_1, \ldots, h_w)$,
6. $plug_{gulp(B_1)}(g_1, \ldots, g_v) plug_{gulp(C_1)}(h_1, \ldots, h_w) =$ $plug_{gulp(B_1 C_1)}(g_1, \ldots, g_{v-1}, g_v + 1, h_1 + 1, h_2 \ldots, h_w)$,
7. $plug_{gulp(B_1)}(g_1, \ldots, g_v) plug_{gulp(C_1)}(h_1, \ldots, h_w) =$ $plug_{gulp(B_1 C_1)}(g_1, \ldots, g_{v-1}, g_v + 1, 0, h_1, \ldots, h_w)$,
8. $plug_{gulp(B_1)}(g_1, \ldots, g_v) plug_{gulp(C_1)}(h_1, \ldots, h_w) =$ $plug_{gulp(B_1 C_1)}(g_1, \ldots, g_v, 0, h_1 + 1, \ldots, h_w)$,
9. $plug_{gulp(B_1)}(g_1, \ldots, g_v) plug_{gulp(C_1)}(h_1, \ldots, h_w) =$ $plug_{gulp(B_1 C_1)}(g_1, \ldots, g_v, 0, 0, h_1, \ldots, h_w)$,

where $c \in \{0, 1\}$.

Proof. Let $B = plug_{gulp(B_1)}(g_1, \ldots, g_v)$, let $C = plug_{gulp(C_1)}(h_1, \ldots, h_w)$. Let S (S_B, S_C) denote the greatest A-skeleton of BC (resp. B, C), let $S = S_1 S_2$ be the partition of S into the subparts that reduce to B and C respectively. The previous lemma shows that S_1 (resp. S_2) is obtained from S_B (resp. S_C) by introducing at most one new variable.

1. Case 0: $S_1 = S_B$ and $S_2 = S_C$. In this case, the first equation holds.
2. Case 1: S_1 has one additional variable X and $S_2 = S_C$, or vice versa. If X has contact in S to variables X_1 and X_2 of S_1 and S_2 respectively, then X_1 is the final variable of S_B and X_2 is the adaptive variable of S_C. We obtain equation 2 for $c = 0$. In the other case, if X has contact in S to a variable X_1 of S_1 we receive equation 3 with $c = 0$. Otherwise, if X has contact to

[3] In the following expressions, letters g_v (resp. h_1) in sums on the right-hand sides have to be replaced by 0 if $v = 0$ (resp. $w = 0$).

a variable X_2 of S_2 we receive equation 4 with $c = 0$. Otherwise, if X does not have contact to another variable we receive equation 5 with $c = 0$.

3. Case 2: S_1 has one additional variable X_1 and S_2 has one additional variable X_2. We distinguish two subcases.
   (a) X_1 and X_2 have contact in S. We distinguish the four cases, as in Case 1. It is easily seen that we obtain the same equations, with $c = 1$ in each case.
   (b) X_1 and X_2 do not have contact[4]. Again, we distinguish the four cases. For two contacts we obtain equation 6, if only X_1 has contact we obtain equation 7, if only X_2 has contact we obtain equation 8, otherwise we receive equation 9. □

5 Linear Diophantine Systems and Solutions of Context Equations

In this section we reduce the problem of finding an upper bound for the exponent of periodicity of a minimal solution of a context equation to the problem of finding an upper bound of a minimal positive integer solution of a system of linear diophantine equations. The system of linear diophantine equations that will be obtained from a given solution of a context equation of size d has exactly the same form as in the case of word equations of size d. Consequently we will be able to use the results of [9] on systems of linear diophantine equations and we receive the same bound on the exponent of periodicity as in the case of word equations. The results of this section are obtained by a simple adaptation of the results given in [9] for the case of word equations.

In the sequel, $W = W'$ denotes a fixed context equation, and $\mathcal{Y}$ denotes the set of context variables occurring in the equation. We assume that a solution σ of $W = W'$ is given. As before, A denotes a fixed simple context.

Definition 25. If $X \in \mathcal{Y}$ and $\sigma(X)$ has order u, then X is called a variable of order u ($u \geq 0$).

Let $Y \in \mathcal{Y}$ be a variable of order $u \geq 2$, let $T_{\sigma(Y)}(X_1, \ldots, X_u)$ denote the A-presentation of $\sigma(Y)$. On the basis of the notational convention introduced in the previous section, X_1 (resp. X_u) is called a *variable in boundary position*, and the variables $X_2, \ldots, X_{u-1}$ are called *variables in isolated position*. In the sequel we assume that for distinct variables $Y, Y' \in \mathcal{Y}$ the A-presentations of $\sigma(Y)$ and $\sigma(Y')$ respectively always use disjoint sets of variables. The set of all variables occurring in an A-presentation $T_{\sigma(Y)}$ for some $Y \in \mathcal{Y}$ is called the set of *presentation variables* and denoted $\mathcal{Z}$. We choose an ordering $(Z_1, \ldots, Z_t)$ of $\mathcal{Z}$ such that all variables that belong to an A-presentation $T_{\sigma(Y)}$ where Y has order 1 come before all variables in boundary position which in turn come before all variables in isolated position.

---

[4] An example that shows that this case can occur can easily be translated from Example 2.11 on page 675 in [9] for words.

Definition 26. For every context term V with variables in $\mathcal{Y}$ and for every tuple $(h_1, \ldots, h_t)$ of natural numbers we define a term $\langle V\rangle(h_1, \ldots, h_t)$ as follows:

1. if $f \in \Sigma$ is n-ary, and if $V_1, \ldots, V_n$ are context terms, then $\langle f(V_1, \ldots, V_n)\rangle(h_1, \ldots, h_t) := f(\langle V_1\rangle(h_1, \ldots, h_t), \ldots, \langle V_n\rangle(h_1, \ldots, h_t))$,
2. if $Y \in \mathcal{Y}$ and V is a context term, then $\langle YV\rangle(h_1, \ldots, h_t) := T_{\sigma(Y)}(h_{i_1}, \ldots, h_{i_u})\langle V\rangle(h_1, \ldots, h_t)$ where $T_{\sigma(Y)}(Z_{i_1}, \ldots, Z_{i_u})$ is the A-presentation of $\sigma(Y)$.

Note that, by the second clause, h_i gives the number of stable occurrences of A that are plugged in for the presentation variable Z_i, for $i = 1, \ldots, t$. To make this more transparent we may extend the definition and define $\langle Y\rangle(h_1, \ldots, h_t) := T_{\sigma(Y)}(h_{i_1}, \ldots, h_{i_u})$ for $Y \in \mathcal{Y}$.

Definition 27. Given the context equation $W = W'$, we use the following notation:

1. if $i = 0, 1$, then d_i (d'_i) denotes the number of occurrences of variables of order i in W (W'),
2. d_2 (d'_2) denotes the number of occurrences of variables of order ≥ 2 in W (W'), and
3. d_Σ (d'_Σ) denotes the number of occurrences of function symbols in W (W').
4. if $i \in \{0, 1, 2\}$, then $d_i^+ := d_i + d'_i$, and $d_\Sigma^+ := d_\Sigma + d'_\Sigma$.

In the sequel, we describe the sequence $(Z_1, \ldots, Z_t)$ in the form $(\vec{Z}_1, \vec{Z}_2, \vec{Z}_3)$ where $\vec{Z}_1$ denotes the sequence of all presentation variables of order 1, $\vec{Z}_2$ denotes the sequence of all presentation variables in boundary position, and $\vec{Z}_3$ denotes the sequence of all presentation variables in isolated position. Accordingly, sequences of t natural numbers will be denoted in the form $(\vec{h}_1, \vec{h}_2, \vec{h}_3)$.

Lemma 28. *There exists a sequence $(L_1, \ldots, L_l)$ of linear functions*

$$L_i(\vec{h}_1, \vec{h}_2, \vec{h}_3) = \sum_{h_j \in \vec{h}_1} c_{i,j}h_j + \sum_{h_j \in \vec{h}_2} c'_{i,j}h_j + \sum_{h_j \in \vec{h}_3} c''_{i,j}h_j + c_i$$

with non-negative integer coefficients such that

1. $\langle W\rangle(\vec{h}_1, \vec{h}_2, \vec{h}_3) = plug_{gulp(\sigma(W))}(L_1(\vec{h}_1, \vec{h}_2, \vec{h}_3), \ldots, L_l(\vec{h}_1, \vec{h}_2, \vec{h}_3))$,
2. $\sum_{i,j} c_{i,j} = d_1$, $\sum_{i,j} c'_{i,j} = 2d_2$, $c'_{i,j} \leq 1$, *and if $Z_j \in \vec{Z}_2$ is a final variable, then* $c'_{1,j} = 0$,
3. *for all $i = 1, \ldots, l$, the cardinality of $\{j \mid c'_{i,j} > 0\}$ does not exceed 2, and for $i = 1$ it does not exceed 1,*
4. *if $c''_{i,j} > 0$ for some i, j, then $L_i = h_j$ and $i \neq 1$,*
5. $\sum_i c_i \leq 2d_0 + 3d_1 + 3d_2 + d_\Sigma$.

Proof. We use induction on the structure of W.
Case 1: $W = f(V_1, \ldots, V_n)$, where f has arity $n \geq 0$. If f has arity 0, then $\langle W\rangle = f$, and we may choose the empty sequence of linear

functions. Otherwise, by induction hypothesis, for $i = 1, \ldots, n$ there exist sequences of linear functions $(L_{i,1}, \ldots, L_{i,l_i})$ such that $\langle V_i\rangle(\vec{h}_1, \vec{h}_2, \vec{h}_3) = plug_{gulp(\sigma(V_i))}(L_{i,1}(\vec{h}_1, \vec{h}_2, \vec{h}_3), \ldots, L_{i,l_i}(\vec{h}_1, \vec{h}_2, \vec{h}_3))$ and Properties 2-5 hold (interpreting here the numbers d_j and d_Σ relative to the context term V_i). By Definition 26 we have (writing $(\vec{h})$ for $(\vec{h}_1, \vec{h}_2, \vec{h}_3)$)

$$\begin{aligned}\langle W\rangle(\vec{h}) &= f(\ldots, \langle V_i\rangle(\vec{h}), \ldots)\\ &= f(\ldots, plug_{gulp(\sigma(V_i))}(L_{i,1}(\vec{h}), \ldots, L_{i,l_i}(\vec{h})), \ldots).\end{aligned}$$

We distinguish the three cases mentioned in Lemma 22, where $B_i := \sigma(V_i)$. In the first case, $\langle W\rangle(\vec{h})$ takes the form $plug_{gulp(\sigma(W))}(L_{1,1}(\vec{h}), \ldots, L_{n,l_n}(\vec{h}))$. Hence Property 1 holds. It is trivial to verify that also Properties 2-5 hold by induction hypothesis.

In case 2 (a) of Lemma 22, $\langle W\rangle(\vec{h})$ takes the form $plug_{gulp(\sigma(W))}(L_{i,1}(\vec{h}) + 1, \ldots, L_{i,l_i}(\vec{h}))$ and we have Property 1. Properties 2, 3 hold trivially by induction hypothesis. Property 5 holds since the new word W has one additional function symbol. To verify Property 4 note that, by Lemma 21 $L_{i,1}$ gives the position of an adaptive variable. Hence $L_{i,1}$ cannot have the form $c''_{i,j}$. In case 2 (b) of Lemma 22, $\langle W\rangle(\vec{h})$ takes the form $plug_{gulp(\sigma(W))}(0, L_{i,1}(\vec{h}), \ldots, L_{i,l_i}(\vec{h}))$ and we have Property 1. Properties 2-4 hold by induction hypothesis, and (5) since we have a new function symbol in W.

Case 2: $W = YV$. By induction hypothesis there exists a sequence of linear functions $(L_1, \ldots, L_l)$ such that

$$\langle V\rangle(\vec{h}_1, \vec{h}_2, \vec{h}_3) = plug_{gulp(\sigma(V))}(L_1(\vec{h}_1, \vec{h}_2, \vec{h}_3), \ldots, L_l(\vec{h}_1, \vec{h}_2, \vec{h}_3))$$

and Properties 2-5 hold. By Definition 26 we have

$$\langle W\rangle(\vec{h}) = T_{\sigma(Y)}(h_{i_1}, \ldots, h_{i_u}) plug_{gulp(\sigma(V))}(L_1(\vec{h}), \ldots, L_l(\vec{h}))$$

where $T_{\sigma(Y)}(Z_{i_1}, \ldots, Z_{i_u})$ is the A-presentation of $\sigma(Y)$. In other words, $\langle W\rangle(\vec{h})$ has the form $plug_{gulp(\sigma(Y))}(h_{i_1}, \ldots, h_{i_u})\ plug_{gulp(\sigma(V))}(L_1(\vec{h}), \ldots, L_l(\vec{h}))$. We may apply Lemma 24 for $B_1 := \sigma(Y)$ and $C_1 := \sigma(V)$. Since $\sigma(Y)\sigma(V) = \sigma(W)$, Property 1 holds in all cases. Property 2 holds by construction; to see that $c'_{i,j} \leq 1$ note that if we are in Case 2 of Lemma 24, and if $u \geq 2$, then Z_{i_u} is a final variable. To see Property 3, note that if we obtain a new component $h_{i_u} + c + 2 + L_1(\vec{h})$ (Case 2 of Lemma 24), and if $h_{i_u} \in \vec{h}_2$, then $u \geq 2$. Property 4 can be easily verified. To prove that Property 5 holds as well, note that adding a variable of order ≥ 1 will contribute at most 3 to $\sum_i c_i$ (Case 2 of Lemma 24 for $c = 1$), a variable of order 0 can contribute at most 2 (Case 3 of Lemma 24 for $c = 1$). □

Theorem 29. *It $W = W'$ is a context equation of size d, and if p is the exponent of periodicity of a minimal solution of the equation, then $p - 2$ is a coordinate of a minimal solution of a system $\mathcal{M} = \{M_1(\vec{u}, \vec{w}) = v'_1, \ldots, M_m(\vec{u}, \vec{w}) = v'_m)\}$*

of linear diophantine equations with non-negative coefficients $m_{i,j}, m'_{i,k}, m_i$ *such that* $M_i = \sum_{u_j \in \vec{u}} m_{i,j} u_j + \sum_{w_k \in \vec{w}} m'_{i,k} w_k + m_i$ *and moreover*

1. $2\sum_{i,j} m_{i,j} + \sum_{i,k} m'_{i,k} \leq 2d - 4$,
2. $m'_{i,k} \leq 1$,
3. $\sum_i m_i \leq 3d - 5$,
4. $m \leq 2d - 2$.

Given Lemma 28 as prerequisite, for proving Theorem 29 the proof of the corresponding Theorem in [9] can be used without any modifications. For the convenience of the reader, a sketch is given. Assume that σ is a minimal solution of the context equation $W = W'$ with exponent of periodicity p. Let $(\vec{u}, \vec{w}, \vec{v})$ be a sequence of natural numbers such that $\sigma(Y) = \langle Y \rangle(\vec{u}, \vec{w}, \vec{v})$ for all $Y \in \mathcal{Y}$. It follows easily from Definition 26 that $\sigma(W) = \langle W \rangle(\vec{u}, \vec{w}, \vec{v})$ and $\sigma(W') = \langle W' \rangle(\vec{u}, \vec{w}, \vec{v})$. Hence, by Lemma 28 there exist sequences $(L_1, \ldots, L_l)$ and $(L'_1, \ldots, L'_{l'})$ such that

$$\langle W \rangle(\vec{u}, \vec{w}, \vec{v}) = plug_{gulp(\sigma(W))}(L_1(\vec{u}, \vec{w}, \vec{v}), \ldots, L_l(\vec{u}, \vec{w}, \vec{v})),$$
$$\langle W' \rangle(\vec{u}, \vec{w}, \vec{v}) = plug_{gulp(\sigma(W'))}(L'_1(\vec{u}, \vec{w}, \vec{v}), \ldots, L'_{l'}(\vec{u}, \vec{w}, \vec{v})).$$

Since $\sigma(W) = \sigma(W')$, Part 2 of Lemma 17 shows that $l = l'$ and $L_i(\vec{u}, \vec{w}, \vec{v}) = L'_i(\vec{u}, \vec{w}, \vec{v})$ for $i = 1, \ldots l$. By assumption, $p - 2$ is a coordinate of a minimal solution of the system of linear diophantine equations $\mathcal{L} = \{L_1 = L'_1, \ldots, L_l = L'_{l'}\}$. When we identify some variables in $\vec{v}$ that necessarily receive the same value under any solution of $\mathcal{L}$, and when we erase trivial equations afterwards, still each solution of the simplified system will define a solution of $W_1 = W_2$, and still $p - 2$ is a coordinate of a minimal solution of the resulting system. The system $\mathcal{M}$ is obtained in this way, using some trivial further modifications.

Theorem 30. *Let* $W_1 = W_2$ *be a context equation of size* d. *If* p *is the exponent of periodicity of a minimal solution of* $W_1 = W_2$, *then* $p \leq (5d-6)(e^{1/e})^{2d-3}+2 = O(2^{1.07d})$.

Proof. In [9] it is shown that for a minimal solution of the system $\mathcal{M}$ described in Theorem 29 the number $(5d - 6)(e^{1/e})^{2d-3}$ gives an upper bound on the size of each coordinate. □

When translating the bounds given in Lemma 28 into the bounds of Theorems 29, 30 it becomes relevant that every context equation contains at least two constants. In other formulations of context equations, individual variables (denoting first order terms) may be used for building context terms. Clearly individual variables x may always be replaced by a context term of the form Xa, where $a \in \Sigma$ is a constant occurring in the equation that is chosen non-deterministically. Hence we receive a bound of $O(e^{2.14d})$ in this more general situation. Additional arguments would show that the sharper bound $O(e^{1.07d})$ is valid also in this case.

6 Some lower complexity bounds

Proposition 31. *If individual variables may be used in context terms, then solvability of context matching problems where each context variable occurs at most once is $\mathcal{NP}$-hard.*

Proof. A *1-IN-3-SAT* problem is given by a set of clauses $\{C_1, \ldots, C_n\}$ where each clause contains exactly three positive literals. A truth value assignment α solves $\{C_1, \ldots, C_n\}$ if exactly one literal of each clause is mapped to 1 under α. It is well-known that solvability of 1-IN-3-SAT problems is $\mathcal{NP}$-complete [6]. We may encode clause C_i, say, with literals p_i, q_i, r_i, by the context matching equation $\mathcal{E}_i$ of the form $X_i(f(p_i, q_i, r_i)) = g(f(1,0,0), f(0,1,0), f(0,0,1))$. Obviously, the context matching problem $\{\mathcal{E}_1, \ldots, \mathcal{E}_n\}$ has a solution iff $\{C_1, \ldots, C_n\}$ has a solution. □

Proposition 32. *Stratified context unification is $\mathcal{NP}$-hard*

Proof. We use the $\mathcal{NP}$-hardness of the 1-IN-3-SAT problem. Let $q_i, i = 1, \ldots, n$ be the propositional variables, and let the clause set be $\{p_{i,1} \vee p_{i,2} \vee p_{i,3} \mid i = 1, \ldots, m\}$, where $p_{i,j}$ is the propositional variable $q_{\varphi(i,j)}$. For every propositional variable q_i, let $X_i, i = 1, \ldots, n$ be a context variable. For every clause, there is a new context variable $Y_i, i = 1, \ldots, m$. The equations are: $X_i(a) = f(a,a)$ for $i = 1, \ldots, n$, $Y_i((f(a,b)) = g(f(a,b), f(a,b), f(a,b))$ and $Y_i(f(b,a)) = g(X_{\varphi(i,1)}(b), X_{\varphi(i,2)}(b), X_{\varphi(i,3)}(b))$ for $i = 1, \ldots, m$. It is obvious that this context unification problem is stratified [14, 10], since every occurrence of a context variable is either at top level, or it is of the form $g(\ldots, X(.), \ldots)$, and there are no first order variables.

The possible instantiations of $X_i(b)$ are exactly $f(a,b)$ or $f(b,a)$, where we interpret the first as false and the second as true. The possible instantiations for $Y_i(f(b,a))$ are the three terms $\{g(f(b,a), f(a,b), f(a,b)), g(f(a,b), f(b,a), f(a,b)), g(f(a,b), f(a,b), f(b,a))\}$, which corresponds to the three possibilities of yes/no-assignments to propositional variables in the clause with index i. Thus an instance of the 1-IN-3-SAT problem given above is solvable exactly iff the constructed stratified context unification problem is unifiable. □

Proposition 33. *Context unification is at least as hard as string unification with linear constant restrictions.*

Proof. (sketch) Let $s = t, C$ be a unification problem with the linear constant restrictions C [1]. Let d be a constant. Every variable x_i is translated into a context variable X_i, every constant a in the strings s, t will be translated to the context $f_a(\Omega, X_1(d), \ldots, X_k(d))$, where $\{x_1, \ldots, x_k\}$ is the the set of variables, such that a should not occur in the instantiation of x_i. The string equation is translated into the context unification problem $\Gamma := \{Sd = Td, STd = TSd\}$, where S, T are the translations of the terms s, t, respectively.

We have to show that unifiability of $s = t$ is equivalent to the unifiability of Γ. Given a solution of $s = t, C$, a unifier of Γ can be constructed, since the linear constant restrictions are satisfied. For the other direction, let σ be a unifier of Γ. Then Lemma 2 implies that w.l.o.g. $\sigma(S)$ is a prefix of $\sigma(T)$. Since the depths are equal, $\sigma(S) = \sigma(T)$. The rest is obvious. □

Unifiability of string equations with free function symbols can be decided using a non-deterministic polynomial algorithm with an oracle that decides unifiability of string equations together with linear constant restrictions [1]. Thus context unification is at least as hard as general string unification, provided the latter has at least exponential time complexity.

References

1. F. Baader and K. U. Schulz. Unification in the union of disjoint equational theories: Combining decision procedures. *J. Symbolic Computation*, 21:211–243, 1996.
2. V. Bulitko. Equations and inequalities in a free group and free semigroup. *Tul. Gos. Ped. Inst. Učen. Zap. Mat. Kafedr. Vyp 2 Geometr. i Algebra*, pages 242–252, 1970. in Russian.
3. H. Comon. Completion of rewrite systems with membership constraints, part I: Deduction rules and part II: Constraint solving. Technical report, CNRS and LRI, Université de Paris Sud, 1993. to appear in JSC.
4. W. Farmer. A unification algorithm for second order monadic terms. *Annals of Pure and Applied Logic*, 39:131–174, 1988.
5. W. Farmer. Simple second-order languages for which unification is undecidable. *Theoretical Computer Science*, 87:173–214, 1991.
6. M. Garey and D. Johnson. *"Computers and Intractability": A guide to the theory of NP-completeness.* W.H. Freeman and Co., San Francisco, 1979.
7. W. Goldfarb. The undecidability of the second-order unification problem. *Theoretical Computer Science*, 13:225–230, 1981.
8. G. Huet. A unification algorithm for typed λ-calculus. *Theoretical Computer Science*, 1:27–57, 1975.
9. A. Kościelski and L. Pacholski. Complexity of Makanin's algorithms. *Journal of the Association for Computing Machinery*, 43:670–684, 1996.
10. J. Levy. Linear second order unification. In *Proc. of the 7th Int. Conf. on Rewriting Techniques and Applications*, volume 1103 of *LNCS*, pages 332–346, 1996.
11. G. Makanin. The problem of solvability of equations in a free semigroup. *Math. USSR Sbornik*, 32(2):129–198, 1977.
12. J. Niehren, M. Pinkal, and P. Ruhrberg. On equality up-to constraints over finite trees, context unification, and one-step rewriting. In *Proc. of the Int. Conf. on Automated Deduction*, volume 1249 of *LNCS*, pages 34–48, 1997.
13. T. Pietrzykowski. A complete mechanization of second-order type theory. *J. ACM*, 20:333–364, 1973.
14. M. Schmidt-Schauß. Unification of stratified second-order terms. Internal Report 12/94, Fachbereich Informatik, Universität Frankfurt, Germany, 1994.
15. M. Schmidt-Schauß. An algorithm for distributive unification. In *Proc. of the 7th Int. Conf. on Rewriting Techniques and Applications*, volume 1103 of *LNCS*, pages 287–301, 1996.
16. W. Snyder and J. Gallier. Higher-order unification revisited: Complete sets of transformations. *J. Symbolic Computation*, 8:101–140, 1989.

Unification in Extensions of Shallow Equational Theories

Florent Jacquemard*, Christoph Meyer, Christoph Weidenbach**

Max-Planck-Institut für Informatik
Im Stadtwald
66123 Saarbrücken, Germany
email: {florent,meyer,weidenb}@mpi-sb.mpg.de

Abstract. We show that unification in certain extensions of shallow equational theories is decidable. Our extensions generalize the known classes of shallow or standard equational theories. In order to prove decidability of unification in the extensions, a class of Horn clause sets called sorted shallow equational theories is introduced. This class is a natural extension of tree automata with equality constraints between brother subterms as well as shallow sort theories. We show that saturation under sorted superposition is effective on sorted shallow equational theories. So called semi-linear equational theories can be effectively transformed into equivalent sorted shallow equational theories and generalize the classes of shallow and standard equational theories.

1 Introduction

Algorithms to solve unification and word problems in an equational theory play a crucial role in many areas of computer science like automated deduction, logic and functional programming, and symbolic constraint solving. Many algorithms are dedicated to particular theories and often semantic conditions are assumed. In addition, a lot of progress has been made towards syntactic characterizations of classes of equational theories or rewrite systems in which these problems are decidable. The class of *shallow* theories, axiomatized by equations in which variables occur at most at depth one, has been shown by Comon, Haberstrau & Jouannaud (1994) to have a decidable unification problem. They exploit a transformation of the system into an equivalent cycle-syntactic presentation (Kirchner 1986). By a termination analyses under basic superposition Nieuwenhuis (1996) generalized the result to so-called standard theories.

Furthermore, tree automata and tree grammars have also been used for unification purposes. Limet & Réty (1997) use Tree Tuple Synchronized Grammars to generate solutions to unification problems by a simulation of narrowing. In (Kaji, Toru & Kasami 1997) it is shown that the closure with respect to some kind of term rewriting system of the (recognizable set) of ground instances of a

* This work was partially supported by the CONSOLE project.
** This work was supported by the German science foundation program Deduktion.

linear term is recognizable. Similar techniques based on the completion of tree automata are presented by Comon (1995) and Jacquemard (1996) for linear shallow TRS and a generalization called linear growing TRS. The decidability of the word problem as well as restricted cases of unifiability in the concerned theories can be derived from these results.

In this paper we show the decidability of unification in so called semi-linear equational theories which strictly extend shallow theories. Informally, a semi-linear system contains equations in which non-linear variables only appear in the same subterms. For example, the equation $f(f(x,x),y) \approx g(f(x,x))$ is semi-linear whereas $f(g(x),h(x),h(g(y))) \approx h(g(x))$ is not. Our techniques are influenced by tree automata, sorted unification and saturation-based methods.

Sorted shallow equational theories naturally generalize tree automata with equality constraints (Bogaert & Tison 1992) as well as shallow sort theories (Weidenbach 1998). Throughout the paper, we consider the following example of Nieuwenhuis (1996). The equational theory is given by the equations $f(g(x),y) \approx h(y)$ and $f(x,x) \approx g(x)$. The definition of standard theories does not include this case. The closure of the theory under basic superposition leads to an infinite set of equations $g(h^n(g(x))) \approx h^{n+1}(g(x))$. The infinite expansion can be avoided by abstracting the linear (semi-linear) term $g(x)$ into a sort declaration $S(g(x))$. The theory is then transformed into a sorted shallow equational theory consisting of the Horn clauses $\| \rightarrow S(g(x))$, $S(x) \| \rightarrow f(x,y) = h(y)$, $\| \rightarrow f(x,x) = g(x)$. Our notation for clauses is of the form *Sort Constraint* $\|$ *Antecedent* $\rightarrow$ *Succedent* where the sort constraint atoms are particular, monadic antecedent atoms for which special inference rules are provided by our sorted superposition calculus.

The paper is organized as follows: Section 3 starts with a discussion on tree automata with brother constraints. We prove that they are not sufficient for our purpose. Then sorted shallow equational theories are studied. It is shown that saturation under sorted superposition terminates and that unifiability modulo the saturated theory is decidable. A procedure which transforms a sorted semi-linear equational theory into an equivalent sorted shallow one is given in Section 4. This implies the decidability of unifiability modulo a set of (sorted) semi-linear equations. This result strictly embeds previous ones concerning shallow theories by Comon et al. (1994). We show in Section 5 that with similar techniques, we can treat a generalization of standard theories as proposed by Nieuwenhuis (1996). In the same section, we also consider some other extensions for which our method does not work and discuss some related work on E-unification. For more details consider our technical report (Jacquemard, Meyer & Weidenbach 1998).

2 Preliminaries

We adhere to the usual definitions for variables, terms, substitutions, equations, atoms, (positive and negative) literals, multisets, and clauses, see (Dershowitz &

Jouannaud 1990) for what concerns equational theories. We give just the most important definitions for our purpose.

The algebra of terms over a finite set of function symbols $\mathcal{F}$ and a set $\mathcal{X}$ of variables is denoted $\mathcal{T}(\mathcal{F},\mathcal{X})$ and $\mathcal{T}(\mathcal{F})$ is its subalgebra of ground terms. An *equation* is an unoriented pair of terms of $\mathcal{T}(\mathcal{F},\mathcal{X})$ denoted $s \approx t$. For sake of simplicity, we may apply to equations or other atoms the same following notations as for terms. The function *vars* maps terms, atoms, literals, clauses and sets of such objects to the set of variables occurring in these objects. A *position* p in a term (equation, atom) is a word over the natural numbers. For a term (equation, atom) t we define $t|_p$ of t at position p by $t|_\epsilon = t$ and $t|_{i.p} = t_i|_p$ where $t = f(t_1,\ldots,t_n)$ and $1 \leq i \leq n$. We write $t[s]_p$ to denote that $t|_p = s$ and $t[p/s']$ is the term obtained from t by replacing its subterm at position p by s'.

A term is called *complex* if it is neither a constant nor a variable. A term t is called *shallow* if t is a variable or is of the form $f(x_1,\ldots,x_n)$ where the x_i are not necessarily different. An equation $s \approx t$ is called *shallow* if both s and t are shallow. Note that shallow variables in $s \approx t$ can be arbitrarily shared by s and t. A term t is called *linear* if every variable occurs at most once in t. A term t is called *semi-linear* if it is a variable or of the form $f(t_1,\ldots,t_n)$ such that every t_i is semi-linear and whenever $vars(t_i) \cap vars(t_j) \neq \emptyset$ we have $t_i = t_j$ for all i, j. An equation $s \approx t$ is semi-linear if (i) s and t are variables or (ii) $s = f(s_1,\ldots,s_n)$ and if t is a variable and $t \in vars(s_i)$ then $s_i = t$ for all i or if $t = g(t_1,\ldots,t_m)$ and $vars(s_i) \cap vars(t_j) \neq \emptyset$ then $s_i = t_j$ for all i, j. For instance, the term $f(g(x), g(x), h(y,y))$ and the equations $h(g(x), g(x), y) \approx f(y, g(x), y)$ and $f(g(x), g(x), y) \approx y$, are semi-linear, but $f(g(x), g(x), h(x,y))$, $h(g(x), x)$ and $h(g(x), g(x)) \approx x$ are not.

Atoms formed from unary predicates are called *monadic*. For the purpose of this paper, a Horn clause is written in the form $\Theta \,\|\, \Gamma \rightarrow \Delta$ where the *sort constraint* Θ consists of monadic atoms representing the sort restrictions. Γ and Δ denote the antecedent and succedent atoms of the clause, respectively. In the initial clause set we assume the arguments of all sort constraint atoms to be variables. Furthermore, for the theories we consider here, it is always the case that either the antecedent or the succedent of a clause is empty.

We call a sort constraint Θ *solved* in a Horn clause $\Theta \,\|\, \Gamma \rightarrow \Delta$, if $vars(\Theta) \subseteq vars(\Gamma \cup \Delta)$ and all terms occurring in Θ are variables. A Horn clause $T_1(x_1),\ldots,T_n(x_n) \,\|\, \rightarrow S(t)$ is called a *declaration* if $T_1(x_1),\ldots,T_n(x_n)$ is solved. In case t is a variable, a declaration is called a *subsort* declaration. A declaration $T_1(x_1),\ldots,T_n(x_n) \,\|\, \rightarrow S(t)$ is *shallow* (*linear*, *semi-linear*) if t is shallow (linear, semi-linear). A *sort theory* is a finite set of declarations. It is called *shallow* (*linear*, *semi-linear*) if all declarations are shallow (linear, semi-linear). A *sorted equation* (*sorted disequation*) is a clause $\Theta \,\| \rightarrow l \approx r$ (a clause $\Theta \,\|\, l \approx r \rightarrow$) where Θ is solved. A *sorted equational theory* is a finite set of sorted equations and declarations. It is called *shallow* (*semi-linear*) if all equations and all declarations are shallow (semi-linear).

A *substitution* is a mapping from $\mathcal{X}$ to $\mathcal{T}(\mathcal{F},\mathcal{X})$. As usual, we do not distinguish between a substitution and its homomorphic extension in the free algebra

$\mathcal{T}(\mathcal{F},\mathcal{X})$. Given an *equational theory* E, i.e., a finite set of equations, we write $s \xleftrightarrow[E]{} t$ iff there exists an equation $l \approx r \in E$ and a substitution σ such that $s|_p = l\sigma$ and $t = s[p/r\sigma]$. The reflexive symmetric transitive closure of the binary relation $\xleftrightarrow[E]{}$ is denoted $\xleftrightarrow[E]{*}$. Two terms s and t are called *unifiable* modulo an equational theory E iff there exists a substitution σ such that $s\sigma \xleftrightarrow[E]{*} t\sigma$. Note that this is equivalent to stating that the clause set consisting of the equational theory E and the clause $\| s \approx t \rightarrow$ is unsatisfiable.

For a set of Horn clauses $\mathcal{A}$ and a clause C, $\mathcal{A} \models C$ denotes the usual semantic entailment relation where all variables of $\mathcal{A}$ and C are assumed to be universally quantified.

3 Decidability of Shallow

In this section we show how saturation-based methods for sorted shallow equational theories succeed in the decision of unifiability in (non-linear) shallow theories whereas tree automata (with constraints) techniques fail.

3.1 Tree automata and linear shallow theories

We adopt here a definition of tree automata by means of Horn clauses. This definition, though non-standard, is equivalent to the usual ones, e.g., (Bogaert & Tison 1992). A systematic correspondence between various types of Horn clause sets and known classes of tree automata with constraints has been studied by one of the authors (Weidenbach 1998).

Definition 1. A tree automaton $\mathcal{A}$ is a finite set of linear shallow declarations of the form $S_1(x_1), \ldots, S_n(x_n) \, \| \rightarrow S(f(x_1, \ldots, x_n))$.

Following tree automata terminology, the unary predicates are called *states* and the Horn clauses of $\mathcal{A}$ are *transition rules* or just transitions.

A term $t \in \mathcal{T}(\mathcal{F})$ is *recognized* by $\mathcal{A}$ in some state S if $\mathcal{A} \models S(t)$. If we fix in $\mathcal{A}$ a subset $\mathcal{S}$ of *final states* (final predicates), then $t \in \mathcal{T}(\mathcal{F})$ is recognized by $\mathcal{A}$ (with respect to $\mathcal{S}$) if t is recognized by $\mathcal{A}$ in some final state. A set $L \subseteq \mathcal{T}(\mathcal{F})$ is a recognizable language if L is the set of ground terms which are recognized by a tree automaton $\mathcal{A}$ (with respect to some set of final states).

The class of recognizable languages is closed under Boolean operations. Every recognizable language is recognized by one *deterministic* tree automaton $\mathcal{A}$ such that a ground term cannot be recognized by $\mathcal{A}$ in more than one state. Every recognizable language is recognized by one *completely specified* tree automaton $\mathcal{A}$, such that every ground term is recognized by $\mathcal{A}$ at least in one state. It is decidable in polynomial time whether a given term $t \in \mathcal{T}(\mathcal{F})$ is recognized by a tree automaton $\mathcal{A}$. It is decidable in linear time whether the language recognized by some tree automaton $\mathcal{A}$ is empty or not.

Tree automata and grammars have been used by Kaji et al. (1997) and Limet & Réty (1997) to solve word and unifiability problems. In the first paper as well as in the papers by Comon (1995) and Jacquemard (1996) the recognizability of the closure of some recognizable set L with respect to term rewriting systems of restricted classes is investigated. In the following we denote the closure of a set of terms $L \subseteq \mathcal{T}(\mathcal{F})$ with respect to an equational system E: $(\overset{*}{\underset{E}{\leftrightarrow}})(L) := \{s \in \mathcal{T}(\mathcal{F}) \mid \exists t \in L\ t \overset{*}{\underset{E}{\leftrightarrow}} s\}$. For a given system E we can reduce the word problem $s \overset{*}{\underset{E}{\leftrightarrow}} t$ to the membership problem for $s \in (\overset{*}{\underset{E}{\leftrightarrow}})(\{t\})$ if the closure set and L is recognizable. For a goal $s = t$ where s and t are both linear and $vars(s) \cap vars(t) = \emptyset$, unifiability modulo E is equivalent to $\{s\sigma \mid \sigma \text{ ground}\} \cap (\overset{*}{\underset{R}{\leftrightarrow}})(\{t\sigma \mid \sigma \text{ ground}\}) = \emptyset$. Since the set of ground instances of s and t are both recognizable unifiability in this case can be reduced to an emptiness decision problem for tree automata.

Theorem 2. *(Comon 1995). Let E be a linear shallow equational system and L be a recognizable language. Then $(\overset{*}{\underset{E}{\leftrightarrow}})(L)$ is a recognizable language.*

The principle of the construction for linear shallow equational systems is the following. We start with a tree automaton $\mathcal{A}_0$ which recognizes L and contains one state S_{l_i} for each direct ground subterm l_i in equations $f(l_1, \ldots, l_n) \approx r$ in E such that l_i (and only l_i) is recognized by $\mathcal{A}_0$ in S_{l_i}. In some sense these subterms are abstracted by $\mathcal{A}_0$. Then $\mathcal{A}_0$ is completed with respect to inference rules like the one below. Note that the construction of $\mathcal{A}_0$ is similar to our transformation process in Section 4.

$$\text{Inf}\ \frac{\begin{array}{ll} \| \to f(l_1, \ldots, l_n) \approx g(r_1, \ldots, r_m) & \in E \\ S_1(x_1), \ldots, S_n(x_n) \| \to S(f(x_1, \ldots, x_n)) & \end{array}}{T_1(x_1), \ldots, T_m(x_m) \| \to S(g(x_1, \ldots, x_m))}$$

This inference rule is applied providing that for each $i \leq m$ such that r_i is a ground term, $T_i = S_{r_i}$, and for each $i \leq m, j \leq n$ such that l_j and r_i are the same variables, then $T_i = S_j$. If we apply paramodulation to the premises of the above inference rule we obtain the clause $S_1(l_1), \ldots, S_n(l_n) \| \to S(g(r_1, \ldots, r_n))$. With the two above conditions (abstraction of r_i by T_i and equalities between states according to variables), and since E is linear shallow, this clause is equivalent to $T_1(y_1), \ldots, T_m(y_m) \| \to S(g(y_1, \ldots, y_m))$. This relates the automata theoretic approach and its generalization presented in Section 3.3. Unfortunately, the above recognizability result of Theorem 2 cannot be extended to non-linear systems, as the following example shows. Assume f is a binary function symbol, s is unary and a is a constant, and let $L = \{a\}$, $E = \{f(x, x) \approx a\}$. Then it is well known that languages of the form $\{f(t, t)\}$ are not recognizable by tree automata.

3.2 Brothers automata and the non-linearities

Bogaert & Tison (1992) introduce tree automata with constraints which define a strict superclass of recognizable languages to deal with non-linear rewrite systems.

Definition 3. A tree automaton with equality constraints between brother subterms is a finite set of shallow declarations of the form $S_1(x_1), \ldots, S_n(x_n) \| \to S(f(x_1, \ldots, x_n))$ where the x_i are not necessarily different.

We call $Rec_=$ this class of recognizers as well as the class of recognized languages; the notion of recognized terms and languages is the same for tree automata with equality constraints between brother subterms as for (standard) tree automata.

The class $Rec_{\neq}$ (Bogaert & Tison 1992) is strictly larger than $Rec_=$ because (syntactic) disequations between variables $x_i \neq x_j$ are also allowed in the antecedent of clauses. The nice closure properties of tree automata still apply here, namely closure under Boolean operations, under determinism and complete specification. The emptiness problem is also decidable for $Rec_{\neq}$ though EXPTIME-hard. However, disequalities are not necessary for our purpose (see the conclusion for a discussion about this extension), but we can show that neither $Rec_=$ nor $Rec_{\neq}$ suffice to generalize Theorem 2 to the case of non-linear shallow systems.

Lemma 4. *There exists some recognizable set L and (non-linear) shallow equational system E such that the set $(\overset{*}{\underset{E}{\leftrightarrow}})(L)$ is not in $Rec_{\neq}$.*

Proof. Let f, g be two binary function symbols, a be some constant and consider the system $E := \{f(x,x) \to g(x,x)\}$ and language $L := \{g(s_1,s_2) \mid s_1, s_2 \in \mathcal{T}(\mathcal{F})\}$. Assume that $L' := (\overset{*}{\underset{E}{\leftrightarrow}})(L)$ is recognized by some $\mathcal{A} \in Rec_{\neq}$ with respect to the distinguished set of final states $\mathcal{S}$. We may assume without loss of generality that $\mathcal{A}$ is deterministic and completely specified. Let N be the number of states of $\mathcal{A}$. Let us define a sequence of well-balanced ground terms of $\mathcal{T}(\mathcal{F})$ by $t_1 := f(a,a)$ and for all $i \geq 1$, $t_{i+1} := f(t_i, t_i)$. It is easy to check that for all $i \geq 1$, the cardinal of the equivalence class of t_i modulo $\overset{*}{\underset{E}{\leftrightarrow}}$ is $2^{2^i - 1}$. Thus there exists an integer $i_0 = \lceil \log(\log(|Q|+2)) \rceil$ such that for all $i \geq i_0$, we have two distinct ground terms s_i, s'_i both equivalent to t_i modulo $\overset{*}{\underset{E}{\leftrightarrow}}$ and both recognized by $\mathcal{A}$ in the same state called S_i. Moreover, by construction, we have that $f(s_i, s'_i) \in L'$. Thus this term is recognized by $\mathcal{A}$ in some final state $S^{\mathrm{f}}_i \in \mathcal{S}$. By determinism of $\mathcal{A}$, there exists a clause $C_i = S_i(x_1), S_i(x_2) \parallel \Gamma_i \to S^{\mathrm{f}}_i(f(x_1,x_2)) \in \mathcal{A}$, ($\Gamma_i$ is a set of syntactic disequations between variables) such that $\mathcal{A} \models C_i\sigma$ with $\sigma = \{x_1 \mapsto s_j, x_2 \mapsto s_{j'}\}$. Note that for all $i \geq i_0$, the variables x_1 and x_2 are distinct because $s_i \neq s'_i$. On the other hand, there exist two distinct integers $j, j' \geq i_0$ such that $S_j = S_{j'}$. Thus, $\mathcal{A} \models C_j\sigma$ where $\sigma = \{x_1 \mapsto s_j, x_2 \mapsto s_{j'}\}$, because $x_1 \neq x_2$ in C_j and thus $f(s_j, s_{j'})$ is recognized by $\mathcal{A}$ in the final state S^{f}_j. This is a contradiction because this term is not in L'.

We can conclude from Lemma 4 that the syntactic equality constraints of the automata in $Rec_=$ are too rough for our purpose. The sorted shallow equational theories studied in the following section are a strict generalization of $Rec_=$. An important achievement of this approach is that semantic equality tests are possible.

3.3 Saturation

The following inference rules form a sound and refutationally complete calculus for Horn clause sets consisting of declarations and sorted (dis)equations.

They are mainly an adaption of basic superposition with selection (Bachmair, Ganzinger, Lynch & Snyder 1995, Nieuwenhuis & Rubio 1995) to the particular form of the Horn clauses considered here, where the sort constraints are subject to the basic restriction and are solved by a particular selection strategy. This strategy is expressed by the rule Sort Constraint Resolution, see below. As usual, we assume a reduction ordering $\succ$ that is total on ground terms. We call the calculus consisting of the inference rules Sort Constraint Resolution, Superposition Right, Superposition Left and Equality Resolution plus the usual reduction rules subsumption and condensing the *sorted* superposition calculus. Note that the basic restriction does not interfere with subsumption or condensing, because sort constraint atoms are solely subsumed (condensed) by other sort constraint atoms and in considered clause sets no non-variable terms occur in the sort constraint.

Definition 5 (Sort Constraint Resolution). The inference

$$\text{Inf}\ \frac{\begin{array}{r l} T_1(t),\ldots,T_n(t),\Psi \,\|\, & \Gamma \rightarrow \Delta \\ \Theta_1 \,\| & \rightarrow T_1(t_1) \\ \vdots & \\ \Theta_n \,\| & \rightarrow T_n(t_n) \end{array}}{\bigcup_i \Theta_i\sigma, \Psi\sigma \,\|\, \Gamma\sigma \rightarrow \Delta\sigma}$$

where t is either a non-variable term or t is a variable $t = x$ with $x \notin vars(\Gamma \cup \Delta)$ and no non-variable term occurs in Ψ; no further atom $S(t)$ occurs in Ψ, σ is the simultaneous mgu of t, t_1, ..., t_n and all Θ_i are solved is called a *Sort Constraint Resolution* inference.

Definition 6 (Superposition Right). The inference

$$\text{Inf}\ \frac{\begin{array}{r l} \Psi \,\| & \rightarrow s \approx t \\ \Theta \,\| & \rightarrow A[s']_{1.p} \end{array}}{\Psi\sigma, \Theta\sigma \,\| \rightarrow A[1.p/t]\sigma}$$

where σ is the mgu of s and s', $t\sigma \not\succ s\sigma$, s' is not a variable, if A is an equation $l \approx r$ with $l|_p = s'$ then $r\sigma \not\succ l\sigma$ and the sort constraints Ψ, Θ are solved is called a *Superposition Right* inference.

Definition 7 (Superposition Left). The inference

$$\text{Inf}\ \frac{\begin{array}{r l l} \Psi \,\| & & \rightarrow s \approx t \\ \Theta \,\| & l[s']_p \approx r & \rightarrow \end{array}}{\Psi\sigma, \Theta\sigma \,\|\, l[p/t]\sigma \approx r\sigma \rightarrow}$$

where σ is the mgu of s and s', $t\sigma \not\succ s\sigma$, s' is not a variable, $r\sigma \not\succ l\sigma$ and the sort constraints Ψ, Θ are solved is called a *Superposition Left* inference.

Definition 8 (Equality Resolution). The inference

$$\text{Inf}\ \frac{\Theta \,\|\, s \approx t \rightarrow}{\Theta\sigma \,\| \quad\quad \rightarrow}$$

where σ is the mgu of s, t and Θ is solved is called a *Equality Resolution* inference.

Lemma 9. *Sorted shallow equational theories can be finitely saturated by sorted superposition.*

Proof. We shall show that the saturation process results in clauses of the form

$$T_1(t), \ldots, T_n(t), S_1(x_1), \ldots, S_m(x_m) \,\| \rightarrow A$$

where n, m are possibly zero, A is either a monadic atom $T(s)$ or an equation $l \approx r$ and t, s, l and r are always shallow terms. If the saturation process produces only clauses of this form, then it will terminate, because the depth of all these clauses as well as the length of variable chains between their literals are bound. Hence, there are only finitely many different clauses of this form modulo subsumption and condensing.

It remains to prove that all clauses generated by the saturation process have the above form. Obviously, shallow declaration clauses and sorted shallow equations are of the above form, where t as well as the x_i are variables occurring in A. For symmetry reasons it is sufficient to consider three cases of possible inferences: (i) The term t is a non-variable shallow term and we perform a sort constraint resolution inference. (ii) The term t is a variable that does not occur in A and we perform a sort constraint resolution inference. (iii) The sort constraint $T_1(t), \ldots, T_n(t), S_1(x_1), \ldots, S_m(x_m)$ is solved and we perform a superposition right inference. We separately consider these cases:

(i) The other clauses involved in the inference are all of the form $Q_1(y_1), \ldots, Q_{k_i}(y_{k_i}) \,\| \rightarrow T_i(t_i)$ where the y_j occur in t_i and t_i is a shallow term. The unifier σ only maps a variable to a non-variable shallow term if the variable is some t_i. Hence, the result of the inference is a clause of the desired form.

(ii) Again all other clauses involved in the inference are of the form $Q_1(y_1), \ldots, Q_{k_i}(y_{k_i}) \,\| \rightarrow T_i(t_i)$ where the y_j occur in t_i and t_i is a shallow term. The unifier σ possibly maps the variable t to a non-variable shallow term, but since t does not occur in A the result of the inference is again a clause of the desired form.

(iii) Since we do not superpose into variables and for any equation of the form $f(x_1, \ldots, x_n) \approx y$ either $y = x_i$ for some i and hence $f(x_1, \ldots, x_n) \succ x_i$ or y does not occur in $f(x_1, \ldots, x_n)$, a case analysis over the different combinations of the form of A and the involved sorted equation shows that the result is always of the desired form. Note that in the case of a Superposition Right inference, the involved clauses have a solved sort constraint.

For example, we apply the saturation process to the sorted shallow equational theory presented in Section 1:

$$\begin{array}{lll} (1)\ S(x) & \| \rightarrow f(x,y) \approx h(y) \\ (2) & \| \rightarrow f(x,x) \approx g(x) \\ (3) & \| \rightarrow S(g(x)) \end{array}$$

where we assume $f(x,y) \succ g(x) \succ h(x)$. Then the saturation process generates the additional clauses (4) and (5) by Superposition Right inferences.

$$\begin{array}{ll}(4)\ S(x) & \|\rightarrow g(x) \approx h(x) \\ (5)\ S(x) & \|\rightarrow S(h(x))\end{array}$$

The clauses (1)–(5) are saturated by sorted superposition.

Lemma 10. *Unifiability with respect to finitely saturated sorted shallow equational theories is decidable.*

Proof. Two arbitrary terms t, s are unifiable iff we can derive the empty clause from the saturated theory and the goal clause $\| t \approx s \rightarrow$. Since the sorted shallow equational theory is saturated, no inferences inside the theory need to be considered. Furthermore, the goal is purely negative, so we can delete all clauses with an unsolved sort constraint from the saturated theory. We show that the sorted superposition calculus terminates on the goal clause. All generated clauses are of the form:

$$S_1(t_1), \ldots, S_m(t_m) \,\|\, t' \approx s' \rightarrow$$

where t', s' are terms resulting from inference rule applications to clauses inferred from the goal clause, m is possibly zero and $t' \approx s'$ does possibly not exist (after the application of an Equality Resolution inference). All inference rule applications to clauses of the above form, except Sort Constraint Resolution to a variable, are monotone in the well-founded ordering composed of the lexicographic combination of the number of non-linear variables occurring at different depth in some $t_1, \ldots, t_m, t' \approx s'$ and the maximal term depth of $t_1, \ldots, t_m, t' \approx s'$. The rule Sort Constraint Resolution is only applicable to a variable if the variable does not occur in $t' \approx s'$ and all other t_i are variables. Then an application generates at most one new shallow term in the sort constraint and following the argumentation in the proof of Lemma 9 the process of solving the generated sort constraints will eventually terminate. In summary, the term depth in all generated clauses is bound and solving sort constraints with variables that do not occur in the antecedent equation terminates. It remains to show that the length of variable chains in the generated clauses is bound. Obviously, such crucial chains can only occur in the sort constraint between sort constraint atoms that have a complex term as its argument. But this cannot happen, since Sort Constraint Resolution applied to a non-variable term is strictly monotone in the above ordering if the involved declarations have a succedent atom with a non-variable argument. With respect to subsort declarations, the saturation using Sort Constraint Resolution terminates anyway.

We evaluate two example queries with respect to the above saturated sorted shallow equational theory. First, we want to unify $f(x,y)$ and $h(y)$ starting with the goal clause

$$\| f(x,y) \approx h(y) \qquad \rightarrow$$

We apply Superposition Left with (1) giving $S(x) \,\|\, h(y) \approx h(y) \rightarrow$. Next we apply Sort Constraint Resolution with (3) yielding $\| h(y) \approx h(y) \rightarrow$ and finally an application of Equality Resolution yields the empty clause. Therefore, $f(x,y)$ and $h(y)$ are unifiable in the considered shallow equational theory.

Second, consider the unification problem of $f(a,x)$ and $h(x)$ where a is some constant. The problem has no solution justified by the saturated clause set consisting of the clauses (1)–(5) and the clauses below:

$$\begin{array}{rll} & \| f(a,x) \approx h(x) & \rightarrow \\ S(a) & \| & \rightarrow \\ & \| g(a) \approx h(a) & \rightarrow \end{array}$$

4 Semi-Linear Sorted Equational Theories

In this section we prove that unification in semi-linear equational theories is decidable, too. We do so by transforming a semi-linear equational theory into a sorted shallow equational theory, preserving satisfiability. Then we apply Lemma 9 and Lemma 10 to obtain the decidability result. The following rule transforms sorted semi-linear equational theories into sorted shallow equational theories.

Definition 11. The transformation

$$\text{Red}\ \frac{\Psi, S_1(x_1),\ldots,S_m(x_m) \,\|\, \rightarrow A[t]_{p_1}}{\begin{array}{l} S_1(x_1),\ldots,S_m(x_m) \,\|\, \rightarrow T(t) \\ \qquad\qquad\quad T(y),\Psi \,\|\, \rightarrow A[p_1,\ldots,p_n/y] \end{array}}$$

provided t is a non-variable subterm, $x_i \in vars(t)$ for all i, $vars(\Psi) \cap vars(t) = \emptyset$, $|p_i| = 2$ for all i, the positions $p_1,\ldots,p_n$ refer to all positions q of t in A with $|q| = 2$, T is a new monadic predicate and y is new to the replaced clause is called *flattening.*

Lemma 12. *Exhaustive application of flattening to a (sorted) semi-linear equational theory terminates, results in a sorted shallow equational theory and preserves satisfiability.*

Proof. Termination follows from the fact that the transformation replaces a clause by two clauses with fewer function symbols. No transformation is applicable to a clause that is a shallow declaration or a sorted shallow equation, since all terms at depth two of such atoms are always variables (if they exist). On the other hand, if the direct subterm of an atom is not shallow, it has a subterm at depth two which is not a variable and therefore the transformation applies. Hence, the transformation terminates in a sorted shallow equational theory.

By an induction argument it is sufficient to show that a single step of the transformation preserves satisfiability and results in a sorted semi-linear equational theory. The crucial property is that $vars(t) \cap vars(A[p_1,\ldots,p_n/y]) = \emptyset$. We show this by contradiction. Assume that after an application of the transformation there is a variable z occurring in t and $A[p_1,\ldots,p_n/y]$. By construction this can only be the case if z has an occurrence in A that is not inside an occurrence of t in A. So z occurs in some term $s \neq t$ with $A|_q = s$, $|q| = 2$, contradicting that the clause is semi-linear to which the transformation is applied. For the same reason, the result of an application of the transformation is again a sorted semi-linear equational theory and $x_j \notin vars(A[p_1,\ldots,p_n/y])$ for all j.

Theorem 13. *Unifiability in semi-linear equational theories is decidable.*

Proof. By Lemma 12 we can effectively translate semi-linear equational theories into sorted shallow equational theories preserving satisfiability. By Lemma 9 these theories can be effectively saturated by sorted superposition and by Lemma 10 unifiability is decidable with respect to saturated sorted shallow equational theories.

Application of the transformation to the example presented in the introduction yields the sorted shallow equational theory considered in the previous section.

4.1 Applications

Any equational theory E can be transformed into a semi-linear equational theory E' by replacing non-linear variable occurrences with fresh variables. Then E' is an upper approximation for E in the sense that $\overset{*}{\underset{E}{\leftrightarrow}} \subseteq \overset{*}{\underset{E'}{\leftrightarrow}}$, i.e., non-unifiability in E' implies non-unifiability in E. Furthermore, by Theorem 13, non-unifiability in E' is decidable. Ganzinger, Meyer & Weidenbach (1997) showed that in this case non-unifiability in E' can be used to effectively direct the search of a theorem prover in finding proofs with respect to E. One of our future goals is to improve the performance of SPASS (Weidenbach 1997) using this technology. Note that flattening applied to an arbitrary equational theory where we keep some $S_i(x_i)$ in the transformed clause if $x_i \in vars(A[p_1, \ldots, p_n/y])$ is already a transformation that generates an appropriate approximation.

5 Extensions, Limitations and Related Work

5.1 Extensions

A possible extension is to apply our method to compute the (eventual) solution of a unification problem in a semi-linear theory. Sort Constraint Resolution simulates sorted unification. Unification in shallow sort theories is known to be NP-complete and of unification type finitary. This implies that unification in sorted shallow equational theories is NP-hard and also of unification type finitary, if we consider well-sorted unifiers. The results of Theorem 13 in Section 4 obviously extend to sorted semi-linear equational theories.

The standard equations in (Nieuwenhuis 1996) include one form which is not embedded by the semi-linear case: the form $f(\ldots, g(x), \ldots) \approx x$ where g has to be a unary function symbol, assuming additional restrictions on the positions of linear terms and non-linear shallow variables in other equations. Obviously, the subterm $g(x)$ cannot be transformed into a sort declaration. However, we can show that unification in those theories can still be decided by sorted superposition using basic strategies on so-called *semi-standard* equations. We call an equation $f(t_1, \ldots, t_n) \approx x$ *semi-standard* if $f(t_1, \ldots, t_n)$ is semi-linear and moreover, there is one unary symbol g such that for all t_i with $x \in vars(t_i)$ we have that $t_i = g(x)$. An equational theory E is called *semi-standard* if E only contains semi-linear equations or semi-standard equations of the form $f(t_1, \ldots, t_n) \approx x$ where only one t_i can be of the form $g(x)$.

Theorem 14. *Unifiability in semi-standard equational theories is decidable.*

The procedure in the proof of Lemma 12 which transforms a semi-linear theory into a sorted shallow theory can be extended to work for a semi-standard theory. The resulting system may contain clauses of the form $\Theta \,\|\, \rightarrow f(t_1, \ldots, t_n) \approx x$ where the equation is a so-called *semi-shallow* equation. The generalization of shallow equations to semi-shallow equations is similar to the extention of semi-linear equations to semi-standard equations. In the transformation procedure occurrences of subterms of the form $g(x)$ are not abstracted into sorts. Moreover, equations $s \approx t$ where $vars(s) \cap vars(t) = \emptyset$ are transformed into sorted clauses of the form $\Theta \,\|\, \rightarrow x \approx y$ to ensure that non-collapsing equations share at least one variable between both sides. The saturation of semi-shallow theories still terminates by imposing basic restrictions on subterms of the form $g(x)$ which have been introduced by unifiers into the equational part of a clause and which cannot be moved to the sort constraint.

E-unification remains decidable in the according saturated set shown by an analogous of Lemma 10. The problem is that the maximal term depth can be increased by one while the number of variables does not change. However, the according termination ordering can be generalized in a way that basic and non-basic regions of a clause are distinguished. Further detailes can be found in the technical report (Jacquemard et al. 1998).

5.2 Limitations

We present a generalization of semi-linear equational systems which cannot be treated with the methods of Sections 3.3 and 4.

The combination of associativity for one function symbol and a linear (!) shallow sort theory already yields an undecidable unification problem. This can be seen by a reduction of the emptiness of the intersection of context free languages to this problem.

Pseudo-linear theories generalize sorted semi-linear equational theories in a way that multiple occurrences of a variable in an equation are allowed, provided that they occur at the same depth. For instance $f(h(x), g(x)) \approx g(g(x))$ is pseudo-linear (though not semi-linear) and $f(h(x), g(x)) \approx g(x)$ is not. However, emptiness of some sort with respect to the combination of a linear (!) shallow sort theory and a pseudo-linear equational theory is already undecidable as shown in the following proposition.

Proposition 15 (Jacquemard et al. 1998). *The blank accepting problem for a non-deterministic Turing machine can be reduced to the emptiness of the intersection* $(\xleftrightarrow[E]{*})(L_1) \cap L_2$ *where* L_1 *and* L_2 *are two recognizable word languages and* E *is an equational word system with equations of the form* $aa' \approx bb'$.

Note that this kind of theory is an even simple case of a pseudo-linear equational system. The word problem in pseudo-linear word systems is decidable since pseudo-linear word equations are length preserving.

5.3 Related Work

Oyamaguchi (1990) shows that the word problem for right-ground TRS is undecidable whereas the word problem in left-linear and right-ground TRS is decidable in polynomial time. In the undecidability proof for right-ground systems Oyamaguchi used rewrite rules with non-linear variable occurrences at different depth.

Fassbender & Maneth (1996) investigate decidability of E-unification in theories induced by TRS called top-down tree transducers. Syntactic restrictions based on separated function and constructor alphabets are assumed. E-unification in top-down tree transducers with only one function symbol in the alphabet is shown to be decidable. Due to the constructor-based restrictions the results are difficult to compare to semi-linear theories. Otto, Narendran & Dougherty (1995) show that E-unification is decidable in equational theories axiomatized by monadic, confluent string-rewriting systems.

Kaji et al. (1997) show the recognizability of the right-closure of a certain class of right-linear, confluent TRS applied to a linear term. The variables occurring both in the left and right hand side of a rule $l \to r$ are assumed to be linear in l and, moreover, l and the subterms of r are related under additional restrictions which can be effectively computed. The techniques presented in Subsection 3.1 provide a decision method for some restricted unifiability problems modulo the above systems. Actually, the problem addressed in (Kaji et al. 1997) is more general because they deal with "constrained substitutions" which range in some recursively defined (recognizable) set of terms.

Comon et al. (1994) investigate the properties of non-linear shallow theories which are an instance of semi-linear equational theories. Shallow presentations can be transformed into equivalent cycle-syntactic presentations for which decidability of unification has been shown. The first-order theory of the quotient algebra $T(F)/_{=_E}$ is also shown to be decidable where F is finite and E is shallow. However, the proof techniques are entirely different to our approach.

Nieuwenhuis (1996) generalizes the result of Comon et al. (1994) to so-called standard theories. Standard theories extend non-linear shallow theories in a way that non-ground terms containing linear (non-significant) variables are allowed in certain restricted positions in both sides of the equations. An equation $f(s_1, \ldots, s_n) = g(t_1, \ldots, t_m)$ may contain linear terms s_i, respectively t_i, where all other equations with top symbol f, respectively g, must have linear terms in position i. Non-linear variable occurrences are limited to shallow positions [1]. The saturation-based methods are closely related to our work. The decidability results are also obtained by termination analyses of saturation under basic superposition.

[1] Another extension included in standard theories is discussed in Section 5.

Limet & Réty (1997) show the decidability of E-unification in theories represented by a particular class of confluent, constructor-based TRS. The set of possibly infinite solutions is represented by Tree Tuple Synchronized Grammars. A TRS is transformed into such a grammar which then simulates narrowing. The additional restrictions on the TRS are purely syntactic. However, semi-linear systems are difficult to compare to the constructor-based systems in this approach.

6 Conclusions and Future Work

We have shown that unifiability modulo a sorted shallow equational theory is decidable by means of saturation methods under sorted superposition. With the help of a transformation procedure this result extends to (sorted) semi-linear equational theories. Our result strictly embeds previous work concerning shallow theories by Comon et al. (1994). It can be obviously extended into sorted equational theories and also into a generalization of Nieuwenhuis (1996). However, we currently do not have any complexity results concerning the decision procedure or the number of generated mgus. The presented theory is already included in the first-order theorem prover SPASS (Weidenbach 1997) that can therefore be used for experiments with respect to the presented results.

Let us conclude with another possible improvement of this work. Sorted shallow equational theories generalize $Rec_{=}$ tree automata. To subsume the whole class $Rec_{\neq}$ (Bogaert & Tison 1992), it is necessary to add syntactic disequations to clauses while preserving decidability results concerning membership and emptiness problems. This may have interesting applications in call-by-need normalization strategies for TRS. Durand & Middeldorp (1997) use tree automata techniques both to apply a call-by-need strategy based on the detection of needed redexes and to characterize the class of rewrite systems for which it is effective. The key idea is, given a rewrite system $\mathcal{R}$, to recognize the closure $(\xrightarrow[\mathcal{S}]{*})(\mathrm{NF}_{\mathcal{S}})$ by $\mathcal{S}$ of the set of ground $\mathcal{S}$-normal-forms, where $\mathcal{S}$ is a certain approximation of $\mathcal{R}$. If we approximate $\mathcal{R}$ into a non-linear shallow system $\mathcal{S}$, the above set could be a recognized sorted shallow equational theory with syntactic disequalities, generalizing $Rec_{\neq}$ automata. Thus, with the appropriate extension of the theory of needed-redexes, more general call-by-need normalization strategies for some classes of non-linear rewrite systems could be obtained.

References

Bachmair, L., Ganzinger, H., Lynch, C. & Snyder, W. (1995), 'Basic paramodulation', *Information and Computation* **121**(2), 172–192.

Bogaert, B. & Tison, S. (1992), Equality and disequality constraints on direct subterms in tree automata, *in* A. Finkel & M. Jantzen, eds, 'Proceedings of 9th Annual Symposium on Theoretical Aspects of Computer Science, STACS92', Vol. 577 of *LNCS*, Springer, pp. 161–171.

Comon, H. (1995), Sequentiality, second order monadic logic and tree automata, *in* 'Proceedings 10th IEEE Symposium on Logic in Computer Science, LICS'95', IEEE Computer Society Press, pp. 508–517.

Comon, H., Haberstrau, M. & Jouannaud, J.-P. (1994), 'Syntacticness, cycle-syntacticness and shallow theories', *Information and Computation* **111**(1), 154–191.

Dershowitz, N. & Jouannaud, J.-P. (1990), Rewrite systems, *in* J. van Leeuwen, ed., 'Handbook of Theoretical Computer Science', Vol. B, Elsevier Science Publishers, chapter 6, pp. 243–320.

Durand, I. & Middeldorp, A. (1997), Decidable call by need computations in term rewriting (extended abstract), *in* W. McCune, ed., 'Automated Deduction – CADE-14, 14th International Conference on Automated Deduction', LNCS 1249, Springer-Verlag, Townsville, North Queensland, Australia, pp. 4–18.

Fassbender, H. & Maneth, S. (1996), A strict border for the decidability of e-unification for recursive functions, *in* M. Hanus & M. Rodríguez-Artalejo, eds, 'Algebraic and Logic Programming (ALP-5) : 5th international conference, Aachen, Germany, September 25-27, 1996', Vol. 1139 of *Lecture notes in computer science*, Springer, Berlin.

Ganzinger, H., Meyer, C. & Weidenbach, C. (1997), Soft typing for ordered resolution, *in* 'Proceedings of the 14th International Conference on Automated Deduction, CADE-14', Vol. 1249 of *LNAI*, Springer, Townsville, Australia, pp. 321–335.

Jacquemard, F. (1996), Decidable approximations of term rewriting systems, *in* H. Ganzinger, ed., 'Rewriting Techniques and Applications, 7th International Conference, RTA-96', Vol. 1103 of *LNCS*, Springer, pp. 362–376.

Jacquemard, F., Meyer, C. & Weidenbach, C. (1998), Unification in extensions of shallow equational theories, MPI-Report MPI-I-98-2-002, Max-Planck-Institut für Informatik, Saarbrücken, Germany.

Kaji, Y., Toru, F. & Kasami, T. (1997), 'Solving a unification problem under constrained substitutions using tree automata', *Journal of Symbolic Computation* **23**, 79–117.

Kirchner, C. (1986), Computing unification algorithms, *in* 'Proceedings of the First Symposium on Logic in Computer Science', Cambridge, Massachusetts, pp. 206–216.

Limet, S. & Réty, P. (1997), E-unification by means of tree tuple synchronized grammars, *in* M. Bidoit & M. Dauchet, eds, 'TAPSOFT '97: Proceedings of the Seventh Joint Conference on Theory and Practice of Software Development, 7th International Joint Conference CAAP/FASE, Lille, France', Vol. 1214, Springer-Verlag, Lille, France.

Nieuwenhuis, R. (1996), Basic paramodulation and decidable theories (extended abstract), *in* 'Proceedings 11th IEEE Symposium on Logic in Computer Science, LICS'96', IEEE Computer Society Press, pp. 473–482.

Nieuwenhuis, R. & Rubio, A. (1995), 'Theorem proving with ordering and equality constrained clauses', *Journal of Symbolic Computation* **19**, 321–351.

Otto, F., Narendran, P. & Dougherty, D. J. (1995), Some independence results for equational unification, *in* J. Hsiang, ed., 'Rewriting Techniques and Applications, 6th International Conference, RTA-95', LNCS 914, Springer-Verlag, Kaiserslautern, Germany, pp. 367–381.

Oyamaguchi, M. (1990), 'On the word problem for right-ground term-rewriting systems', *The Transactions of the IEICE* **73**(5), 718–723.

Weidenbach, C. (1997), 'Spass version 0.49', *Journal of Automated Reasoning* **18**(2), 247–252.

Weidenbach, C. (1998), Sorted unification and tree automata, *in* W. Bibel & P. H. Schmitt, eds, 'Automated Deduction, A Basis for Applications', Vol. 1, Kluwer, chapter 2.

Unification and Matching in Process Algebras

Qing Guo, Paliath Narendran, and Sandeep K. Shukla*

Department of Computer Science
University at Albany – State University of New York
Albany, NY 12222
Email:{guo,dran,sandeep}@cs.albany.edu**

Abstract. We consider the *compatibility checking* problem for a simple fragment of CCS, called BCCSP[12] , using equational unification techniques. Two high-level specifications given as two process algebraic terms with *free* variables are said to be compatible modulo some equivalence relation if a substitution on the free variables can make the resulting terms equivalent modulo that relation. We formulate this compatibility (modulo an equivalence relation) checking problems as *unification problems* in the equational theory of the the corresponding equivalence relation. We use van Glabbeek's equational axiomatizations [12] for some interesting process algebraic relations. Specifically, we consider equational axiomatizations for *bisimulation* equivalence and *trace* equivalence and establish complexity lower bounds and upper bounds for the corresponding unification and matching problems. We also show some special cases for which efficient algorithmic solutions exist.

1 Introduction

Often we are interested in processes which terminate after a finite number of steps. Such finitistic processes are useful in modeling circuits which engage in a transformational computation. *Sorting circuits* are examples of these. We consider *hierarchical specification* of such processes. In a hierarchical specification such processes are built from modules, where each module is itself a finitistic process. However, at an abstract specification level, some of the modules may be incompletely specified. For example, in a sorting circuit, there could be modules corresponding to the *comparators* as well as modules corresponding to particular *stages* of computation. The main circuit is then an interconnection of these modules. A process term describing this circuit at a high level of abstraction might contain variables representing underspecified modules. If such a specification is written in a process algebraic syntax, like CCS, usually the set of modules is described as a set of equations like

$$X = f(a, b, c, W, Y, Z)$$

* P. Narendran was supported in part by the NSF grant CCR-9404930 and S. Shukla was partially supported by the NSF grant CCR-94-06611.

** Q. Guo and S. Shukla's present addresses: {Qing Guo, Sandeep K. Shukla} GTE Laboratories Inc. 40 Sylvan Road, Waltham, MA 02254.

$$Y = g(a, b, c, W)$$
$$Z = h(a, W)$$

where f, g, h are process expressions involving the variables and action symbols shown in the brackets. Notice that W is not constrained by any equation and hence it stands for an underspecified module.

For simplicity, in this abstract we consider a very simple subset of Milner's CCS [8], called BCCSP in [12], for our syntax. Since we are investigating only finitistic processes, we will only need this fragment of CCS which does not allow recursion.

As explained above, the specification of a process is given in terms of equations. For example, a primitive module which does three (3) a actions and stops is specified as $X = a.a.a.0$. If a higher level module uses this module after a b action, that module is represented by a variable Y where $Y = b.X$. In our framework, underspecified modules are expressed as variables which are not constrained by such equations. Thus we obtain process expressions with free variables.

Given an implementation of such a specification, a standard method of checking the correctness of the implementation with respect to the specification is by checking whether the implementation and the specification are in the same equivalence class for a suitable equivalence relation on processes. Examples of such equivalence relations are bisimulation equivalence, trace equivalence, and many other equivalence relations in the linear time/branching time spectrum [12]. However, usually the implementation has full details of the behavior of the process and hence will be a ground expression denoting a fully specified process. To check if the implementation expression I and the specification expression S are bisimilar, we may use the sound and complete equational axiomatization of bisimulation for BCCSP given in [12], and solve an equational *matching* problem.

However, a more interesting (and often harder) problem is the following: Given two hierarchical specifications of the same system, one might be interested in their *compatibility*. We call two high-level specifications S_1 and S_2 as *ρ-compatible* if they can be instantiated into the same process *modulo* the relation ρ, by the same substitutions of the free variables (in other words, if they have a common implementation). Since we are using variables to stand for modules which are unspecified, two high level designers might come up with distinct expressions, involving these variables, such that both can be claimed to be a specification for the same system. Now checking ρ-compatibility can be formulated as the *unification problem* in the equational theory of the relation ρ in the fragment of the process algebra we are working with.

So far the equivalences we have considered in this framework are the two widely used equivalences, namely, *bisimulation* equivalence and *trace* equivalence. Complete equational axiomatizations of these equivalences for BCCSP appear in [12] and we consider matching and unification problems in these axiomatizations. The two axiomatizations give rise to two different algebras which we call *bisimulation algebra* and *trace algebra.* (In [7], there are sound and complete axiomatizations of similar algebras which are complete with respect to ground expressions. Our approach is applicable there too.) The problems we consider are *equational matching* (E-matching) and *equational unification* modulo these theories.

In the present work, we consider the issue of designing efficient algorithms for matching and unification and also derive complexity bounds. In Section 2 we show that unification modulo bisimulation algebra is an *NP*-complete problem. The problem is decidable in exponential time for trace algebra. We have obtained a *co-NP*-hard lower bound for unification modulo trace algebra. We also address the issue of restricting the problems to special cases for which more tractable algorithms exist.

1.1 Main Results

We show that unification modulo bisimulation is an *NP*-complete problem, even in the simple case when the signature of the algebra contains only one action symbol. For the trace equivalence case, we demonstrate a *co-NP*-hard lower bound and an exponential time upper bound. This upper bound result is obtained via a reduction to the emptiness checking of a particular kind of tree automaton which accepts the solutions of set constraints generated from the trace unification problem. However, in contrast to the bisimulation case, when the trace unification problems are limited to one action signature, we obtain a polynomial time algorithm.

1.2 Preliminary Definitions

In this subsection we define some of the terms that are used in the later sections of this paper. First we define the syntax of BCCSP terms and then we explain the the trace semantics of such terms. This helps in the intuitive understanding of our reduction of the unification problem for trace equivalence to a special class of *set constraints*.

BCCSP Terms: Let $\Sigma = \{0, +\} \cup A$ be a set of special symbols called the *signature*. Let X be a finite set of variables. The set of *open* terms over signature Σ and variable set X, denoted by $T(\Sigma, X)$ is a minimal set satisfying the following conditions:

1. $0 \in T(\Sigma, X)$
2. if $x \in X$, then $x \in T(\Sigma, X)$
3. if $t \in T(\Sigma, X)$ and $a \in A$, then $a.t \in T(\Sigma, X)$
4. if $t_1, t_2 \in T(\Sigma, X)$, then $t_1 + t_2 \in T(\Sigma, X)$

If $X = \phi$, then the terms are *ground* terms and the set of BCCSP ground terms over signature Σ is denoted by $T(\Sigma)$.

Trace: The traces of a ground BCCSP term is a set of strings over A. This set contains all the possible sequences of actions that the process represented by the term can engage in. Given a ground term $t \in T(\Sigma)$, the set $traces(t)$ is a minimal set satisfying the following conditions:

1. $traces(0) = \{\epsilon\}$
2. if $t = a.t'$, then $traces(t) = a.traces(t') \cup \{\epsilon\}$
3. if $t = t_1 + t_2$, then $traces(t) = traces(t_1) \cup traces(t_2)$

Lemma 1. *For any term t in $T(\Sigma)$, $traces(t)$ is a finite nonempty prefix-closed*[3] *set of strings over A.*

Trace Equivalence: Two terms t_1 and t_2 in $T(\Sigma)$ are said to be *trace equivalent* if and only if $traces(t_1) = traces(t_2)$.

In the following sections we overload the symbol = for the different semantics equivalences that we discuss. In section 2, we write $t_1 = t_2$, when the two ground terms t_1 and t_2 are bisimulation equivalent. In section 3, we write $t_1 = t_2$, when t_1 and t_2 are trace equivalent.

Definition 2. The *size* of a term $t \in T(\Sigma, X)$, denoted by $size(t)$ can defined inductively as follows:

1. $size(0) = 1$
2. $size(x) = 1$ for $x \in X$
3. $size(a.t) = 1 + size(t)$
4. $size(t_1 + t_2) = size(t_1) + size(t_2)$

The *size* of an equation over $T(\Sigma, X)$, is the sum of the sizes of both sides of the equation. The *size* of a set of equations is the sum of the sizes of the equations in the set.

2 Unification and Matching in Bisimulation Algebra

The following equational theory is a sound and complete axiomatization of bisimulation equivalence for BCCSP [12].

$$(x + y) + z = x + (y + z)$$
$$x + y = y + x$$
$$x + x = x$$
$$x + 0 = x$$

We call this equational theory BiS. It is the same as the $ACI1$ theory in the unification literature, because '+' satisfies associativity, commutativity, idempotence and identity. It has been shown that the elementary $ACI1$-unification problem can be solved in polynomial time [9].

However, in the bisimulation algebra, we also have an additional set of unary functions: corresponding to each action in the the action set $Act = \{a_i \mid 0 \leq i \leq n\}$, we have a function $a_i\cdot$ which prefixes a process with action a_i. We denote this set of prefixing functions as Act by overloading the notation.

[3] A set of strings is prefix-closed if whenever a string w is in the set, all its prefixes including ϵ (empty string) are also in the set.

Definition 3. *BiS-UNIFIABILITY*, the E-unification problem over the bisimulation algebra is defined as follows:

INSTANCE: A set of pairs $S = \{\langle s_1, t_1\rangle, \langle s_2, t_2\rangle, \ldots, \langle s_m, t_m\rangle\}$, where s_i, t_i are first-order terms over the signature $\{0, Act, +\}$.
QUESTION: Does there exist a unifier σ such that
$\sigma(s_1) =_{BiS} \sigma(t_1),\ \sigma(s_2) =_{BiS} \sigma(t_2),\ \ldots, \sigma(s_m) =_{BiS} \sigma(t_m)$?

If all t_is are ground in S, then the problem is called a *BiS-Matching* problem.

Theorem 4. *BiS-UNIFIABILITY is NP-complete.*

Proof: The elementary *ACI1*-unification problem with linear constant restrictions can be solved in polynomial time [9]. So the *BiS-UNIFIABILITY* problem can be shown to be in *NP* by Theorem 2.1 from [2].

To show *NP*-hardness, we present a polynomial time and linear size reduction of the *NP*-complete problem *GRAPH 3-COLORABILITY* to *BiS*-unification.

Let graph $G = (V, E)$ and color set $C = \{c_1, c_2, c_3\}$ be any instance of *GRAPH 3-COLORABILITY*, where $V = \{v_1, v_2, \ldots, v_n\}$ and $n \geq 3$. We construct an instance of *BiS*-unification as follows.

For every edge $e_k = \{v_i, v_j\} \in E$, we construct the pair

$$\langle a.v_i + a.v_j + a.z_k,\ a.0 + a.a.0 + a.a.a.0\rangle$$

where node names v_i, v_j are used as variables, $Z = \{z_1, z_2, \ldots, z_m\}$ is a set of extra variables, where $m = |E|$, and a is an action. The construction results in an instance of *BiS*-unification. Let $B(t)$ be the equivalence class of BCCSP terms which contains the term t. Then, the colors c_1, c_2 and c_3 are represented by the equivalence classes $B(0)$, $B(a.0)$ and $B(a.a.0)$ respectively. Note that if the above unification pair is *BiS*-unifiable then the variables v_i, v_j, z_k can only be substituted for by terms in $B(0) \cup B(a.0) \cup B(a.a.0)$. Moreover, these terms must be from distinct equivalence classes.

If the input graph is 3-COLORABLE, then there is a color assignment assigning every v_i and v_j two distinct colors, where $\{v_i, v_j\} \in E$. There is one color left out of c_1, c_2, c_3. By assigning to z_k a term from the equivalence class corresponding to the remaining color, the color assignment can be extended to a unifier for the instance of *BiS*-unification.

If the instance of *BiS*-unification problem is solvable then there is a simultaneous unifier θ for all pairs $(a.v_i + a.v_j + a.z_k,\ a.0 + a.a.0 + a.a.a.0)$. As mentioned before, $\theta(v_i)$ and $\theta(v_j)$ can only be ground terms from $B(0) \cup B(a.0) \cup B(a.a.0)$ such that they are from distinct equivalence classes. Since these three equivalence classes $B(0)$, $B(a.0)$ and $B(a.a.0)$ represent the 3 distinct colors, a coloring assignment can be obtained from θ which solves the given instance of *3-COLORABILITY*. □

Note that only one action symbol was needed for the reduction used in the above proof. Furthermore, because the right side of every pair above is ground, the proof also shows that *BiS*-matching is *NP*-hard.

A special case of *BiS*-unification, where the terms are *deterministic* (defined below), can be solved in polynomial time. The algorithm is similar to the one for standard (i.e., empty theory) unification. We omit the details.

Definition 5. *Deterministic* terms:

1. 0 is deterministic.
2. $a_k.0$ is deterministic, $a_k \in Act$.
3. X is deterministic, if X is a variable.
4. $a_{k_1}.E_1 + a_{k_2}.E_2 + \cdots + a_{k_m}.E_m$ is deterministic, where the $E_i's$ are deterministic, $a_{k_i} \in Act$ and $a_{k_i} \neq a_{k_j}$ if $i \neq j$.

3 Unification in Trace Algebra

The equational axiomatization of trace algebra consists of all the axioms of *BiS*, plus an axiom to capture the distributivity of the *prefixing* with actions a over $+$. In other words, the equational theory is

$$(x+y)+z = x+(y+z)$$
$$x+y = y+x$$
$$x+x = x$$
$$x+0 = x$$
$$a.(x+y) = a.x + a.y \qquad \forall\, a \in Act$$

Note that $a.0 \neq 0$. We show that the elementary unification problem modulo trace algebra is *co-NP*-hard. We can prove that it is decidable; however, there is a gap between the lower bound and the upper bound which is *DEXPTIME*.

Theorem 6. *The Unification problem of terms in BCCSP modulo trace algebra is co-NP-hard.*

Proof: The DNF non-tautology problem is known to be *NP*-complete: given a DNF $c = c_1 \vee c_2 \vee \ldots \vee c_m$, where $c_i = x_{i1} \wedge x_{i2} \wedge \ldots \wedge x_{ij}$, it is *NP*-complete to determine if c is NOT a tautology. Hence it is *co-NP*-complete to determine if c is a tautology. We can reduce the tautology problem to the trace unification problem as follows.

Let n be the number of boolean variables in the DNF instance. We create the following set of simultaneous equations:

$$X_1 = a.X_2 + b.X_2$$
$$X_2 = a.X_3 + b.X_3$$
$$\vdots$$
$$X_n = a.0 + b.0$$
$$X_1 = B_1^1 + B_1^2 + \ldots + B_1^m$$

where 0 stands for the zero process and the B_1^i's and X_j's are variables. We now introduce equations for each B_1^i as follows:

if x_1 appears in c_i unnegated then $B_1^i = b.B_2^i$
if x_1 appears in c_i negated then $B_1^i = a.B_2^i$
if x_1 does not appear in c_i then $B_1^i = a.B_2^i + b.B_2^i$.

Note that B_j^i are new variables. We also add equations for each B_j^i as follows:

$$B_j^i = \begin{cases} b.B_{j+1}^i & \text{if } x_j \text{ appears in } c_i \text{ as a positive literal} \\ a.B_{j+1}^i & \text{if } x_j \text{ appears in } c_i \text{ as a negative literal} \\ a.B_{j+1}^i + b.B_{j+1}^i & \text{if } x_j \text{ does not appear in } c_j \end{cases}$$

and $B_{n+1}^i = 0$.

This system of equations has a (unique) solution and hence a (unique) unifier iff c is a tautology. The reason is that $traces(X_1)$ = the prefix closure[4] of $\{a,b\}^n$. However, by setting $X_1 = B_1^1 + ... + B_1^m$ and coding all possible assignments that make c true in the B_1^i's, we get the desired effect. Also, note that the size of this set of equations is $O(mn)$ and the reduction is polynomial time.□

Now we show how the trace unification problem can be solved in exponential time using tree automata working on finite trees [4].

Definition 7. A set of equations over $T(\Sigma, X)$ is said to be in *reduced form* if all the equations in the set are of one of the following three forms:

1. $x = 0$ for some $x \in X$
2. $x = a.y$ for some $x, y \in X$ and $a \in A$
3. $x = y + z$ for some $x, y, z \in X$

It can easily seen that any set of equations over $T(\Sigma, X)$ can be reduced (preserving the solutions) to a *reduced form* with at most a linear increase in the size of the set of the equations.

Lemma 8. *Any set of equations over $T(\Sigma, X)$ can be reduced (preserving the solutions) to a* reduced form *with at most a linear increase in the size of the set.*

By virtue of this lemma, from now on, we only consider equations in reduced form.

Given an equation of the form $x = 0$, from the definition,

$$traces(x) = \{\epsilon\}$$

Similarly, given an equation of the form $x = a.y$, from the definition of traces

$$traces(x) = a.traces(y) \cup \{\epsilon\}$$

[4] The prefix closure of a string is the set of all prefixes of the string, including the empty string and the string itself. The prefix closure of a set of strings is the union of the prefix closures of the elements of the set.

Also, for an equation of the form $x = y + z$, it must be that

$$traces(x) = traces(y) \cup traces(z)$$

By denoting each $traces(x)$ by set variables T_x which can take up values over nonempty, prefix-closed sets of strings over A, we can reduce the unification problem over $T(\Sigma, X)$ to a set of set constraints *with equality only*. For example, the following set of equations

$$x = y + z, y = a.w, z = 0$$

can be reduced to

$$T_x = T_y \cup T_z, T_y = a.T_w \cup \{\epsilon\}, T_z = \{\epsilon\}$$

Let Ψ be the above transformation which takes a unification problem modulo trace algebra and reduces it to a system of set equations as shown. It is not hard to show that if these set equations have a solution, say θ, over finite nonempty sets of strings over A, then they also have a solution over finite, nonempty, *prefix-closed* sets of strings over A—to get the new solution we merely have to take the prefix-closures of each $\theta(x)$ for variables x.[5] In short, if P is an instance of unification modulo trace algebra, then P is unifiable if and only if $\Psi(P)$ has a solution over finite nonempty sets of strings over A. (Thus the requirement that the sets be prefix-closed is removed.)

We can transform this set of equations further, with at most a linear increase in size, to a similar reduced form as defined in this section. In the above example, we can introduce new set variables T_u and T_w^a and change the equations (preserving solutions) as follows:

$$T_x = T_y \cup T_z, T_y = T_w^a \cup T_u, T_u = \{\epsilon\}, T_w^a = a.T_w, T_z = \{\epsilon\}$$

In other words,

Theorem 9. *Every elementary unification problem modulo trace algebra can be reduced to a system of set equations, in which every equation is either of the form $X = \{\epsilon\}$, $X = a.Y$, or $X = Y \cup Z$ where X, Y, Z are variables over finite, nonempty sets of strings over A and $a \in A$.*

We want to point out that these set constraints are **not** a subcase of those considered by Aiken et. al. [1] since they allow finite as well as infinite sets in the solutions.

Given a set of such equations, one can design, in exponential time, a tree automaton that accepts their solutions (encoded in a tree form). Since the emptiness problem for tree automata can be solved in polynomial time, we get our main decidability result. However, before we construct the tree automata, one further step is needed. Let *rev* be the string reverse function defined on a set of strings. $rev(X)$ *reverses* every string in the set X. For instance

[5] Note that this technique does not work *in general*; it works for equations that Ψ produces. For instance, the equation $X = a.Y$ does not have a prefix-closed solution.

$rev(\{c.b.a,\ a.b,\ c\}) = \{a.b.c,\ b.a,\ c\}$. It is clear that $X = a.Y$ is solvable if and only if $rev(X) = rev(Y).a$ is solvable. Recall that by Theorem 9, an elementary *Trace*-unification problem can be reduced to a system of set equations, in which every equation is of the form $X = \{\epsilon\}$, $X = a_i.Y$, or $X = Y \cup Z$. By applying the reverse function this set of equations can be transformed into one where every equation is either of the form $X = \{\epsilon\}$, $X = Y.a$, or $X = Y \cup Z$ where X, Y, Z are variables over finite, nonempty sets of strings over A and $a \in A$.

Now, we discuss the method of solving such set equations using Doner's tree acceptor [4]. First we discuss how to encode sets of strings as a Σ_n-tree.

Let Σ_n be the set of n-tuples over $\{0, 1\}$; 0^n denotes the n-tuple consisting entirely of 0's.

Definition 10. A Σ_n-tree of order p is a function $\tau : A \to \Sigma_n$, where A is a finite prefix-closed subset of $\{a_0, a_1, \ldots, a_{p-1}\}^*$, i.e., if $uv \in A$ then $u \in A$.

Intuitively, a Σ_n-tree τ of order p is a tree such that every node has at most p children. The root node represents the empty string, and if the root node has p children, then the children of the root node from left to right represent the strings a_0,...,a_{p-1}. Generally, if a node represents a string u, and has degree p then the p children of that node from left to right will represent the strings $u.a_0$, ...$u.a_{p-1}$ respectively. The mapping τ labels each node by n-tuples from Σ_n. Given n finite sets $X_1, \ldots, X_n$ of strings over A, they can be coded in the form of a Σ_n-tree τ. If the ith component of $\tau(v)$ at the node v is 1 then v belongs to the set X_i, otherwise it does not belong to X_i.[6]

The class of all Σ_n-trees of order p is denoted by $\Sigma_n^\#$. For notational convenience, we restrict our examples in this paper to trees of order 2, i.e., binary trees. The domain of trees of order 2 is denoted by $\{a, b\}^*$. Let τ be a binary tree. The subtree of τ beginning at a is denoted by $\tau \uparrow_a$ and the subtree of τ beginning at b is denoted by $\tau \uparrow_b$. We use λ to denote the empty tree. For all $\sigma \in \Sigma_n$ and $\tau, \tau' \in \Sigma_n^\#$, $\sigma[\tau,\ \tau']$ is the unique tree π such that $\pi(w) = \sigma$, $\pi \uparrow_a = \tau$ and $\pi \uparrow_b = \tau'$.

Definition 11 (Doner [4])**.** A Σ_n-tree acceptor is a 4-tuple $\mathcal{A} = \langle Q,\ t,\ q_0,\ F\rangle$ where

1. Q is a finite set of states;
2. $t : Q \times Q \times \Sigma_n \to Q$ is the transition function;
3. $q_0 \in Q$ is the initial state;
4. $F \subseteq Q$ is the set of accept states.

Function $\bar{t} : \Sigma_n^\# \to Q$ is defined by

$$\bar{t}(\lambda) = q_0$$
$$\bar{t}(\sigma[\tau, \tau']) = t(\bar{t}(\tau), \bar{t}(\tau'), \sigma)$$

$\mathcal{A}$ *accepts* a tree $\tau \in \Sigma_n^\#$ if $\bar{t}(\tau) \in F$. $T(\mathcal{A})$ denotes the set of Σ_n-trees accepted by $\mathcal{A}$.

[6] This coding technique was first used by Rabin [10].

Theorem 12. [11] *If $\mathcal{A}$ is any tree acceptor then the emptiness checking of $T(\mathcal{A})$ can be done in time polynomial in the size of $\mathcal{A}$.*

Proof: One way to show this is by reducing the problem (in polynomial time and space) to the satisfiability problem for Horn formulae (HORNSAT). Since HORNSAT is decidable in linear time [5], the result follows. See [6], for instance, for a detailed proof. $\square$

Theorem 13 (Doner [4]). *Let $\mathcal{A}$ and $\mathcal{B}$ be two tree acceptors. A tree acceptor $\mathcal{C}$ can be effectively constructed (in polynomial time) such that $T(\mathcal{C}) = T(\mathcal{A}) \cap T(\mathcal{B})$.*

Given an equation on set variables $X_1, \ldots, X_n$, the variables can be coded in the form of n-tuples. Then we can construct a Σ_n-tree acceptor $\mathcal{A}$ such that if $\tau \in T(\mathcal{A})$ then the labels of the tree τ give a solution of the equation (under the interpretation given above). Hence $T(\mathcal{A}) \neq \emptyset$ if and only if the equation is solvable.

Given two equations e_1, e_2 over set variables $X_1, \ldots, X_n$, we can construct two Σ_n-tree acceptors $\mathcal{A}_1$ and $\mathcal{A}_2$ for e_1 and e_2 respectively. By Theorem 13 we can construct another Σ_n-tree acceptor $\mathcal{A}$ such that $T(\mathcal{A}) = T(\mathcal{A}_1) \cap T(\mathcal{A}_2)$. Then e_1 and e_2 are solvable simultaneously if and only if $T(\mathcal{A}) \neq \emptyset$. This is the basic idea of solving set equations using Σ_n-tree acceptors.

Example 1. Let $X = Y.a$ be an equation, where X, Y are set variables. We construct a Σ_2-tree acceptor (of order 2) $\mathcal{A}$ such that $T(\mathcal{A})$ contains trees which satisfy the above equation.
$\mathcal{A} = \langle \{q_0, q_x, q_a, q_r\}, t, q_0, \{q_a\}\rangle$ where

- q_0: the initial state.
- q_x: current node is in X.
- q_a: the tree below this node is acceptable.
- q_r: reject state.

The transition function t is defined as follows:

$$t(q_0, q_0, 00) = q_0 \qquad t(q_0, q_a, 00) = q_a$$
$$t(q_0, q_0, 10) = q_x \qquad t(q_0, q_a, 10) = q_x$$
$$t(q_0, q_0, \_1) = q_r \qquad t(q_0, q_a, \_1) = q_r$$

$$t(q_x, q_0, \_0) = q_r \qquad t(q_x, q_a, \_0) = q_r$$
$$t(q_x, q_0, 01) = q_a \qquad t(q_x, q_a, 01) = q_a$$
$$t(q_x, q_0, 11) = q_x \qquad t(q_x, q_a, 11) = q_x$$

$$t(q_a, q_0, 00) = q_a \qquad t(q_a, q_a, 00) = q_a$$
$$t(q_a, q_0, 10) = q_x \qquad t(q_a, q_a, 10) = q_x$$
$$t(q_a, q_0, \_1) = q_r \qquad t(q_a, q_a, \_1) = q_r$$

$$t(\_\_, q_x, \_\_) = q_r \qquad t(\_\_, q_r, \_\_) = q_r$$
$$t(q_r, \_\_, \_\_) = q_r$$

Equations $X = Y \cup Z$ and $X = \{\epsilon\}$ can be handled similarly. It is straightforward to construct a Σ_3-tree acceptor for the former and a Σ_1-tree acceptor for the latter.

Given a system P of m equations on set variables $X_1, X_2, \ldots, X_n$, in which each equation is either in the form $X = \{\epsilon\}$, or $X_i = X_j.a$, or $X_i = X_j \cup X_k$, a tree acceptor $\mathcal{A}$ can be constructed such that P is solvable if and only if $T(\mathcal{A}) \neq \emptyset$. (If the number of actions is κ then we have to construct Σ_n-tree acceptors of order κ for all equations before we perform the intersection.) It can be shown that the time complexity of constructing acceptor $\mathcal{A}$ for the equation system is $O(2^{mn})$. Hence, by Theorem 12, the time complexity of the emptiness checking of $T(\mathcal{A})$ is in polynomial of 2^{mn}. Thus the equation system P can be solved in exponential time.

Theorem 14. *Elementary unifiability problem modulo trace algebra can be solved in exponential time.*

Surprisingly, the problem turns out to be solvable in polynomial time when there is only one action symbol. This is in marked contrast to the bisimulation case, where the problem is *NP*-hard even when there is only one action.

4 Single-action Trace Algebra

We consider an interesting subclass of the trace algebra, called the *single-action trace algebra*, where only one action symbol is allowed, i.e., *Act* $= \{a\}$. We show that with this restriction the unifiability problem can be solved in polynomial time. The proof is by reduction to HORNSAT.

Lemma 15. *Let S be a set of elementary unification pairs in the single-action trace algebra. S can be reduced to a set of equations W, in which each equation is in one of the following forms:*

$$x_m = max(x_i, x_j)$$
$$x_i = x_j + 1$$
$$x_i = 0$$

where x_i, x_j, x_m are variables over non-negative integers. S is unifiable in the single-action trace algebra iff W is solvable.

Proof: Straightforward by Theorem 9 and since there is only one action symbol a, every ground term is equivalent to a term of the form $a^i.0$. □

Now, we study the solvability of equation systems which are in the form of the above equation system W.

Lemma 16. *Let $X = \{x_1, x_2, \ldots, x_n\}$ be a set of variables over non-negative integers, and W be a set of equations, in which each equation is in one of the* *following forms:*

$$x_l = max(x_i, x_j)$$
$$x_i = x_j + 1$$
$$x_i = 0 \qquad\qquad x_i,\ x_j,\ x_l \in X$$

If W is solvable then there is a solution in which $\forall\, 1 \le i \le n : x_i < n$.

Proof: If W only contains equations of the form $x_l = max(x_i, x_j)$ then W is solvable and this lemma is trivially true. Without loss of generality, we assume that there is at least one equation in W, which is in the form of $x_i = x_j + 1$ or $x_i = 0$.

A solution of W assigns a nonnegative integer to each variable. If there is a solution for W then based on the value of each variable, variables in X can be arranged in the following order:

$$x_{i_1} \le x_{i_2} \le \ldots \le x_{i_n}$$

Furthermore, all variables which have the same value can be renamed to be the same. For instance, if $x_i = 3, x_j = 3, x_k = 3$ then x_j and x_k can be renamed by x_i. After renaming, we have a strict ordering on variables:

$$x_{i_1} < x_{i_2} < \ldots < x_{i_m}$$

where $m \le n$ and x_{i_m} has the largest value in the solution of W.

After renaming, equations in the form $x_l = max(x_i, x_j)$ become redundant and hence can be removed from W.

Thus we are left with equations of the form

$$x_k = x_j + 1, \text{ or}$$
$$x_k = 0$$

where $j, k \in \{i_1, \ldots, i_m\}$.

Now we prove that this system of inequalities, comprising the ordering chain and W, has a solution $\{x_{i_j} \leftarrow j - 1\}$ by induction on m, the number of variables in the strict ordering chain.

Claim: Let I_W be a system of inequalities over variables $\{x_1, \ldots, x_m\}$ consisting of

(i) a total order $x_1 < x_2 < \ldots < x_m$

(ii) equations of the form $x_i = x_j + 1$ and $x_k = 0$.

Then if I_W is solvable, then $\theta = \{x_i \leftarrow i - 1\}$ is a solution.

<u>Base case</u>: $m = 1$. There is only one variable in the chain. The only possible equation in I_W is $x_1 = 0$. (An equation of the form $x_1 = x_1 + 1$ is not solvable.) Hence $\{x_1 \leftarrow 0\}$ is a solution.

<u>Induction step</u>: Assume the solution holds for $m = p$. Consider $m = p + 1$. Because x_{p+1} is the largest element in the variable chain, there can be at most

one equation of the form $x_{p+1} = x_p + 1$ in I_W. Remove that equation from I_W to get I'_W. Applying the induction hypothesis on I'_W, there must be a solution where $x_{i_q} = q - 1$ for all $q \leq p$, and clearly this solution can be extended to a solution of I_W where $x_{p+1} = p$. □

Given this bound, we can reduce the solvability problem for W to the satisfiability problem for a set of Horn clauses. We introduce propositional variables $Q_{x \leq i}$ for every variable x and $0 \leq i \leq n-1$. The Horn clauses are constructed as follows:

(A) For each variable x, we add a unit clause $Q_{x \leq n-1}$ as well as clauses

$$\forall\, 0 \leq i < (n-1): \qquad Q_{x \leq i} \Rightarrow Q_{x \leq i+1}$$

(B) For each equation of the form $z = max(x, y)$, we form the clauses

$$\forall\, 0 \leq i \leq (n-1): \qquad \begin{array}{r} Q_{x \leq i} \wedge Q_{y \leq i} \Rightarrow Q_{z \leq i} \\ Q_{z \leq i} \Rightarrow Q_{x \leq i} \\ Q_{z \leq i} \Rightarrow Q_{y \leq i} \end{array}$$

(C) For each equation of the form $y = x + 1$, we form the clauses

$$\begin{array}{rr} & \neg Q_{y \leq 0} \\ \forall\, 0 \leq i \leq n-2: & Q_{x \leq i} \Rightarrow Q_{y \leq i+1} \\ & Q_{y \leq i+1} \Rightarrow Q_{x \leq i} \end{array}$$

(D) For each equation of the form $x = 0$, we form the clause $Q_{x \leq 0}$.

It can now be shown that this set of clauses is satisfiable if and only if the set of equations W we started with has a solution. It should not be hard to see that given a bounded solution of W we can construct a satisfying assignment for the Horn clauses. On the other hand, let ϕ be a satisfying assignment for the set of Horn clauses. For each variable x pick the smallest i such that $Q_{x \leq i}$ is set to true by ϕ. It can be shown that an assignment constructed like this is a solution for W. We informally outline the argument below.

(i) Let $z = max(x, y)$ be an equation in W. Let $i = i_x$ be the smallest i such that $Q_{x \leq i_x}$ is set to true by ϕ, $i = i_y$ be the smallest i such that $Q_{y \leq i_y}$ is set to true by ϕ, and $i = i_z$ be the smallest i such that $Q_{z \leq i_z}$ is set to true by ϕ. We need to show that $i_z = max(i_x, i_y)$. This can be proved by contradiction as follows. Assume $i_z \neq max(i_x, i_y)$. Without loss of generality, we can assume that $i_x \geq i_y$. There are two cases:

1. $i_z > max(i_x, i_y)$, i.e., $i_z > i_x$.
   Because i_z is the smallest i such that $Q_{z \leq i_z}$ is set to true by ϕ. $Q_{z \leq i_x}$ must be set to false by ϕ. Therefore the clause $Q_{x \leq i_x} \wedge Q_{y \leq i_x} \Rightarrow Q_{z \leq i_x}$ is not satisfiable by ϕ, which is a contradiction.

2. $i_z < max(i_x, i_y)$, i.e., $i_z < i_x$.
   Because i_x is the smallest i such that $Q_{x \le i_x}$ is set to true by ϕ, $Q_{x \le i_z}$ must be set to false by ϕ. Therefore the clause $Q_{z \le i_z} \Rightarrow Q_{x \le i_z}$ is not satisfiable by ϕ, which is a contradiction.

(ii) Let $y = x + 1$ be an equation in W. Let $i = i_x$ be the smallest i such that $Q_{x \le i_x}$ is set to true by ϕ, and $i = i_y$ be the smallest i such that $Q_{y \le i_y}$ is set to true by ϕ. We need to show that $i_y = i_x + 1$. This can be proved by contradiction as follows. Assume $i_y \neq i_x + 1$. There are two cases:

1. $i_y > i_x + 1$.
   $Q_{y \le i_x + 1}$ must be set to false by ϕ. Therefore the clause $Q_{x \le i_x} \Rightarrow Q_{y \le i_x + 1}$ is not satisfiable by ϕ, which is a contradiction.
2. $i_y < i_x + 1$.
   Because $i_y > 0$ ($\neg Q_{y \le 0}$ is set to true in ϕ), it must be that $i_x > 0$. $Q_{x \le i_x - 1}$ must be set to false by ϕ. Therefore the clause $Q_{y \le i_x} \Rightarrow Q_{x \le i_x - 1}$ is not satisfiable by ϕ, which is a contradiction.

(iii) Let $x = 0$ is an equation in W. By construction of the Horn clauses, $Q_{x \le 0}$ is set to true in ϕ.

If m is the number of equations in W, then the construction above yields a set of Horn clauses with cardinality at most $3mn + n^2$ (over n^2 variables). Since the satisfiability problem of a set of Horn clauses is solvable in polynomial time [5], we get

Lemma 17. *Let $X = \{x_1, x_2, \ldots, x_n\}$ be a set of variables over non-negative integers, and W be a set of equations, in which each equation is in one of the following forms:*

$$\begin{array}{ll} x_l = max(x_i, x_j) & \\ x_i = x_j + 1 & \qquad x_i,\ x_j,\ x_l \in X, \\ x_i = 0 & \end{array}$$

Solvability of W can be checked in polynomial time.

Theorem 18. *The unification problem in the single-action trace algebra can be solved in polynomial time.*

5 Future Work

The astute reader would have observed that there is a gap in the complexity bounds for unification modulo trace algebra — the lower bound is co-NP-hard and the upper bound is DEXPTIME. Future work, from the complexity point of view, should concentrate on eliminating this gap. From the practical point of view, there are several directions we plan to explore:

1. Allow recursion in the specification of processes. This could turn out to be difficult, but even partial results would be immensely useful.

2. Consider cases where the actions satisfy additional constraints.
3. Consider other equivalences such as *branching bisimulation* [13].

Acknowledgments: We thank Siva Anantharaman, Franz Baader, Rob van Glabbeek, Ralf Treinen and the referees for their comments and suggestions.

References

1. A. Aiken and D. Kozen and M. Vardi and E. Wimmers. The Complexity of Set Constraints. *Proceedings of Computer Science Logic,* (E. Börger, Y. Gurevich and K. Meinke, eds.) 1993, 1–17.
2. F. Baader and K.U. Schulz. Unification in the Union of Disjoint Equational Theories: Combining Decision Procedures. *J. of Symbolic Computation,* 21:211–243, 1996.
3. J.C.M. Baeten and W.P. Weijland. *Process Algebra.* Cambridge Tracts in Theoretical Computer Science, Vol. 18, Cambridge University Press, Cambridge, England.
4. J. Doner. Tree acceptors and some of their applications. *Journal of Computer and System Sciences* 4(3):406–451, 1970.
5. W. Dowling and J. Gallier. Linear-time algorithms for testing the satisfiability of propositional Horn formulae. *Journal of Logic Programming,* 1(3):267–284, 1984.
6. Q. Guo. *Nilpotence, Bisimulation and the Unification Workbench.* PhD thesis, State University of New York at Albany, 1997.
7. M. Hennessy. *Algebraic Theory of Processes.* Foundations of Computing Series, The MIT Press, 1988.
8. R. Milner. *Communication and Concurrency.* International Series in Computer Science. Prentice Hall, 1989. SU Fisher Research 511/24.
9. P. Narendran. Unification modulo ACI + 1 + 0. *Fundamenta Informaticae* 25 (1):49–57, 1996.
10. M.O. Rabin. Weakly definable relations and special automata. In: *Mathematical Logic and Foundations of Set Theory* (Y. Bar-Hillel, ed.), North-Holland, 1970, 1–23.
11. W. Thomas. Automata on infinite objects. *Handbook of Theoretical Computer Science* (J. van Leeuwen, ed.), Elsevier Science Publishers, 1990, 133–191.
12. R.J. van Glabbeek. The linear time—branching time spectrum. Technical Report CS-R9029, Computer Science Department, CWI, Centre for Mathematics and Computer Science, Netherlands, 1990.
13. R.J. van Glabbeek and Peter Weijland. Branching Time and Abstraction in Bisimulation Semantics. *Journal of the ACM* 43 (3), May 1996, 555–600.

E-Unification for Subsystems of *S4*

Renate A. Schmidt*

Department of Computing, Manchester Metropolitan University,
Chester Street, Manchester M1 5GD, United Kingdom
E-mail: R.A.Schmidt@doc.mmu.ac.uk

Abstract. This paper is concerned with the unification problem in the path logics associated by the optimised functional translation method with the propositional modal logics *K*, *KD*, *KT*, *KD4*, *S4* and *S5*. It presents improved unification algorithms for certain forms of the right identity and associativity laws. The algorithms employ mutation rules, which have the advantage that terms are worked off from the outside inward, making paramodulating into terms superfluous.

1 Introduction

An area of application for unification theory which has not been explored much is modal logic. Modal inference can be facilitated by theory resolution via the so-called functional translation or its variation for propositional modal logics, the optimised functional translation approach. The functional translation method was proposed independently in the late eighties by a number of groups. Fariñas del Cerro and Herzig (1989, 1995) describe a transformation of quantified modal logics into so-called deterministic logics and use a modal resolution calculus. Ohlbach (1988, 1991) and Auffray and Enjalbert (1992) embed quantified modal logics into fragments of first-order logic and employ first-order resolution theorem proving. Zamov (1989) describes a lock decision procedure for the translation of *S4*. All procedures involve theory unification.

The *optimised functional translation* method (Herzig 1989, Ohlbach and Schmidt 1997) applies to propositional normal modal logics and gives rise to a class of *path logics*, which this paper considers. Very much like modal logics, path logics form a lattice with the weakest being the *basic path logic* associated with the basic modal logic *K* and also *KD*. Different path logics are distinguished by different theories involving equations. This paper focuses on a subclass of path logics with theories consisting exclusively of equations. Path logics with equational theories are associated with serial modal logics, which are modal logics stronger than *KD*. Clauses in path logics satisfy two important properties. One, they satisfy prefix stability which determines a certain ordering of the variables, and two, all Skolem functions in input clauses are nullary.

* I thank H. Ganzinger, A. Herzig, U. Hustadt, A. Nonnengart, H. J. Ohlbach and especially the anonymous referees for their comments. This research was conducted while I was employed at the Max-Planck-Institut für Informatik in Saarbrücken, Germany, and was funded by the MPG and the DFG through the TRALOS-Project.

The purpose of this paper is to give a formal treatment of unification and normalisation for equational path theories explaining the core issues exemplified for the equations corresponding to the modal schemas T and 4. Due to the characteristic properties of clauses the unification problems are easier than in semi-groups or monoids, for example. Related unification algorithms and resolution calculi found in the literature differ from ours in at least three respects. One, they are all designed for the non-optimised translations which require extended (strong) forms of Skolemisation in order that a particular ordering within terms is preserved. Accordingly, our unification algorithms are more elegant and the proofs are considerably simpler, though remaining technical. Two, most of the systems are incomplete. Three, our exposition pays special attention to normalisation.

The paper is organised as follows. Section 2 defines the class of path logics and the equational schemas we consider. Section 3 considers unification for the basic path logic, recalling some essential definitions and facts of syntactic unification. In Sect. 4 we discuss E-unification for the schemas T and 4 reviewing what is known from the literature. The main parts are Sects. 5 and 6. Section 5 presents a mutation algorithm for the combination of T and 4 illustrating the computational gain and outlining new proofs of termination, soundness and completeness. Section 6 proves prefix stability is an invariance property under binary E-resolution. The conclusion mentions some open problems. Due to space restrictions most proofs are omitted, but can be found in Schmidt (1998).

2 Path logics

The *basic path logic* is a clausal logic defined over a language with two principal sorts: the sort W and the sort AF. Expressions of the sort W and AF are called 'world' and 'functional expressions'. The vocabulary includes finitely many unary predicates $P, Q, \ldots$, variables $\alpha, \beta, \ldots$ of sort AF, finitely many constants $\underline{\alpha}, \underline{\beta}, \ldots$ of sort AF, a special constant $[]$ of sort W, and one binary function $[\cdot, \cdot] : W \times AF \longrightarrow W$. Terms in the basic path logic have the form $t = [[[[[]u_1]u_2]\ldots]u_m]$, or in shorthand notation $t = [u_1 u_2 \ldots u_m]$, with any u_i denoting a functional term (by which we mean a term of sort AF). A term defines a set of paths in the functional semantics of modal logics. $[]$ is the initial world and the term $[\alpha\underline{\beta}]$, for example, defines the set of worlds reached from $[]$ via any α-step followed by some $\underline{\beta}$-step. We also refer to such terms as *paths.*

By definition, the *prefix* of a variable or constant u_i in t is $[]$ if $i = 1$ and $[u_1 \ldots u_{i-1}]$, otherwise. A set T of terms is said to be *prefix stable* if for any variable α all its occurrences in T have one common prefix. A clause is said to be prefix stable if the set of its terms has this property. By definition, a formula ψ is a *path formula* iff it is a conjunction of prefix stable clauses.

Stronger path logics which we consider are obtained by extending the basic path logic with (finite equational) presentations given by a subset of the following two schemas.

$$\textit{right identity:}\quad [x\underline{e}] = x \tag{1}$$

$$\textit{associativity:}\quad [x(\alpha \circ \beta)] = [x\alpha\beta] \tag{A}$$

The symbol x denotes a variable of sort world and $[xu_1u_2]$ is an abbreviation for $[[xu_1]u_2]$. The symbol $\underline{e}$ (the identity constant) is a functional constant and $\circ$ (composition) is an operation of the kind $AF \times AF \longrightarrow AF$. The universal closures of the two equations are the global versions of the functional correspondence properties of the modal schemas T and 4, respectively. Recall, the relational correspondence properties of T and 4 are reflexivity and transitivity, respectively. *S4* is the logic *KT4* (or *KDT4*).

This completes the syntactic definition of basic path logic and some of its extensions. Their semantics is defined as usual by Herbrand models.

Later we will refer to the following alternative characterisation of prefix stability. A set T of terms is prefix stable iff for any two terms $[u_1 \ldots u_m]$ and $[v_1 \ldots v_n]$ in T these conditions hold for variables:

T1 If some variable u_i and some variable v_j are identical then $i = j$, and
T2 the terms of each pair u_k and v_k preceding u_i and v_i, respectively, are also identical.

T1 implies paths are linear terms, and it also implies every variable that occurs at position i in some term of the set T occurs at position i in every term, when it does occur in that term.

A note on our notation is in order. The symbols $u, u_1, u_2, \ldots$ and $v, v_1, v_2, \ldots$ are reserved for terms of sort AF. The symbols $s, t, \ldots$ are reserved for any world terms. Strictly, the term $[s\,t]$ is malformed since $[\cdot,\cdot]$ is assumed to be left-associative, but when we write $[s\,t]$ we mean the term $[s\,u_1 \ldots u_i]$ provided $t = [u_1 \ldots u_i]$. Given a term $s = [u_1 \ldots u_m]$, define $s|_i$ by $s|_0 = []$ and $s|_i = u_i$ for any $0 < i \leq m$. If each subterm u_i of s is either a variable or a constant then s is called a *basic path*.

3 Unification for the basic path logic

Because the basic path language has no compound functional terms, any non-empty substitution σ defined over sets of basic paths consists of bindings that have one of two forms, namely $\alpha \mapsto \beta$ or $\alpha \mapsto \underline{\gamma}$. A substitution is said to be *admissible* for the basic path logic iff its bindings have this form. It is immediate that admissible substitutions or unifiers do not change the depths (or lengths) of paths, and only terms of equal depth are unifiable.

The general transformation rules of syntactic tree based unification (from Jouannaud and Kirchner (1991), for example) adapt to those of Fig. 1 for the basic path logic. P denotes a problem set of pairs $s =^? t$, of world terms, or $u =^? v$, of functional terms. The equality relation $=^?$ is assumed to be symmetric. The symbol $\rightsquigarrow$ denotes the derivability relation in a unification calculus.

Delete	$P \cup \{s =^? s\}$	$\rightsquigarrow P$	(for world terms)
	$P \cup \{u =^? u\}$	$\rightsquigarrow P$	(for terms of sort AF)
Decompose	$P \cup \{[su] =^? [tv]\}$	$\rightsquigarrow P \cup \{u =^? v,\ s =^? t\}$	
Conflict	$P \cup \{[su] =^? []\}$	$\rightsquigarrow \bot$	
	$P \cup \{\underline{\alpha} =^? \underline{\beta}\}$	$\rightsquigarrow \bot$	provided $\underline{\alpha} \neq \underline{\beta}$
Coalesce	$P \cup \{\alpha =^? \beta\}$	$\rightsquigarrow P\{\alpha \mapsto \beta\} \cup \{\alpha =^? \beta\}$	
	provided $\alpha \neq \beta$ are variables both occurring in P.		
Eliminate	$P \cup \{\alpha =^? \underline{\beta}\}$	$\rightsquigarrow P\{\alpha \mapsto \underline{\beta}\} \cup \{\alpha =^? \underline{\beta}\}$	
	provided α is a variable occurring in P and $\underline{\beta}$ is constant.		

Fig. 1. Syntactic unification rules for the basic path logic

$\bot$ indicates failure of the unification problem. All other symbols have the usual interpretation, and s and t may be empty paths. It is important that we keep in mind any term $[u_1 \ldots u_m]$ is an abbreviation for a nested term $[[[[[]u_1]u_2] \ldots]u_m]$. This determines how paths are decomposed, namely from right to left. Evidently, any most general unifier of a unification problem over basic paths obtained by the rules of Fig. 1 is an admissible substitution. As no world variables occur in basic path clauses, the occurs check rule is superfluous. Soundness and completeness is immediate by soundness and completeness of the general rules for syntactic unification.

For singleton problem sets two rules are redundant:

Theorem 1. *Let P be a singleton set $\{s =^? t\}$ with s and t terms for which T1 for variables holds. Then the rules Coalesce and Eliminate are redundant.*

The situation is pleasantly simple for the logic *S5* (which coincides with *KDB4*, *KTB4* and *KT5*). In *S5* any sequence of modal operators can be replaced by the first one in the sequence, and *S5* corresponds to the fragment of monadic first-order logic in one variable (via the relational translation). This is reflected in the corresponding path logic by the fact that any singleton unification problem $[u_1 \ldots u_m] =^? [v_1 \ldots v_n]$ can be seen to reduce to the unification problem of $[u_1] =^? [v_1]$. Such problems can be solved modulo (a subset of) the rules of Fig. 1.

4 Unification for (1) and (A)

We turn to unification of paths under the right identity law (1) and the associativity law (A).

Unification of paths under (1) is finitary and decidable. This can be seen easily by considering a unification problem in n variables and forming 2^n syntactic unification problems by replacing some of the variables by $\underline{e}$. Each of the problems is decidable by syntactic unification in linear time. Therefore, the decision problem of unification under (1) is in NP, and by a result of Arnborg and Tidén (1985) for standard right identity it is at least NP-complete.

Unification under (A) is related to unification under standard associativity. Plotkin (1972) shows unification in free semi-groups is infinitary and he gives a unification algorithm that is sound and complete, but it is not guaranteed to terminate. There are decision procedures by Makanin (1977) and Jaffar (1990), for example, but these are far too complex for our purposes. Fortunately, though Plotkin's algorithm is non-terminating in the general case, it decides unification problems of one linear equation, or one equation in which no variable occurs more than twice (Schulz 1992). This implies, unification of one pair of paths under the form of associativity we consider is also finitary and decidable. (By exploiting the correspondence to regular expressions and using methods from automata theory, we expect that unifiability under (1) and/or (A) can be decided in polynomial time.)

Delete	$P \cup \{s =^? s\}$	$\rightsquigarrow P$
Decompose	$P \cup \{s * s' =^? t * t'\}$	$\rightsquigarrow P \cup \{s =^? t,\ s' =^? t'\}$
	when s and t are non-empty strings	
Identity	$P \cup \{s * \alpha * s' =^? t\}$	$\rightsquigarrow P \cup \{\alpha =^? \underline{e},\ s * s' =^? t\}$
Path-separat.	$P \cup \{\alpha * s =^? t * t'\}$	$\rightsquigarrow P \cup \{\alpha =^? t,\ s =^? t'\}$
Splitting	$P \cup \{\alpha * s'' * s =^? t * t'' * \beta * t'\}$	
	$\rightsquigarrow P \cup \{\alpha =^? t * t'' * \beta_1,\ \beta =^? \beta_1 * \beta_2,\ s'' * s =^? \beta_2 * t'\}$	
	when s'' and t'' are non-empty and β_1, β_2 are new variables.	

Fig. 2. Ohlbach's unification rules for (1) and (A)

Unification algorithms are described in Ohlbach (1988, 1991), Fariñas del Cerro and Herzig (1995) and Auffray and Enjalbert (1992) for the non-optimised translation of quantified modal logics and in Zamov (1989) for the non-optimised translation of propositional *S4*. The first three algorithms are not complete. Problems of the form $\{s =^? s * \underline{e}\}$, $\{s =^? s\,;\textsc{id}\}$ and $\{s\,!\,\alpha\,!\,f(s\,!\,\alpha) =^? s\,!\,f(s)\}$ (using in essence the notation of the respective authors) are not treated properly, which can be rectified by adding a rule for deleting the identity constant. The standard deletion rule suffices for solving singleton problems of basic paths though, so that under this condition the rules from Ohlbach (1991) relevant for propositional *S4*, listed in Fig. 2, form a complete system. The language is a variation from ours. Terms are strings built from variables and constants of the sort AF with an associative operation $*$, making the additional operation $\circ$ superfluous. (The correspondence to world terms of basic path logic is given by $h([[]u]) = u$ and $h([su]) = h(s) * u$ when $s \neq []$.) The symbols s, s', t and t' in the figure denote (possibly empty) strings.

By way of an example we will demonstrate the system can be improved. Fig. 3 sketches a derivation of A1-unifiers for $\{\alpha * \underline{\beta} * \underline{\gamma} * \delta =^? \underline{\alpha} * \beta * \underline{\gamma}\}$.[1] Terms are decomposed from left to right. Failure branches are those that do not produce

[1] An *A1-unifier* is a unifier modulo the equations (1) and (A).

$$\alpha * \underline{\beta} * \underline{\gamma} * \delta =^? \underline{\alpha} * \beta * \underline{\gamma}$$
$$\overset{\text{Dec}}{\rightsquigarrow}\ \alpha =^? \underline{\alpha},\ \underline{\beta} * \underline{\gamma} * \delta =^? \beta * \underline{\gamma}$$
$$\begin{array}{l} \qquad \overset{\text{Dec}}{\rightsquigarrow}\ \beta =^? \underline{\beta},\ \underline{\gamma} * \delta =^? *\underline{\gamma}\ \overset{\text{Id}}{\rightsquigarrow}\ \delta =^? \underline{e} \rightsquigarrow 1. \\ \qquad\qquad\qquad\qquad\qquad\qquad\qquad \overset{\text{Dec}}{\rightsquigarrow}\ \delta =^? \underline{e} \quad \text{redundant} \\ \qquad \overset{\text{Id}}{\rightsquigarrow}\ \beta =^? \underline{e},\ \underline{\beta} * \underline{\gamma} * \delta =^? \underline{\gamma} \rightsquigarrow \ldots \quad \text{not solved} \\ \qquad \overset{\text{Sep}}{\rightsquigarrow}\ \beta =^? \underline{\beta} * \underline{\gamma},\ \delta =^? \underline{\gamma} \rightsquigarrow 2. \\ \qquad \overset{\text{Sep}}{\rightsquigarrow}\ \beta =^? \underline{\beta},\ \underline{\gamma} * \delta =^? \underline{\gamma} \rightsquigarrow \ldots \quad \text{not solved} \\ \qquad \overset{\text{Sep}}{\rightsquigarrow}\ \beta =^? \underline{e},\ \underline{\beta} * \underline{\gamma} * \delta =^? \underline{\gamma} \quad \text{redundant} \\ \qquad \overset{\text{Id}}{\rightsquigarrow}\ \delta =^? \underline{e},\ \underline{\beta} * \underline{\gamma} =^? \beta * \underline{\gamma} \\ \qquad\qquad\qquad \overset{\text{Dec,Del}}{\rightsquigarrow}\ \beta =^? \underline{\beta} \rightsquigarrow 3. \\ \qquad\qquad\qquad \overset{\text{Id}}{\rightsquigarrow}\ \beta =^? \underline{e},\ \underline{\beta} * \underline{\gamma} =^? \underline{\gamma} \quad \text{not solved} \\ \qquad\qquad\qquad \overset{\text{Sep}}{\rightsquigarrow}\ \beta =^? \underline{e},\ \underline{\beta} * \underline{\gamma} =^? \underline{\gamma} \quad \text{redundant} \\ \qquad \overset{\text{Split}}{\rightsquigarrow}\ \beta =^? \underline{\beta} * \underline{\gamma} * \delta_1,\ \delta =^? \delta_1 * \delta_2,\ \delta_2 =^? \underline{\gamma} \rightsquigarrow 4. \end{array} \left.\vphantom{\begin{array}{l}a\\a\\a\\a\\a\\a\\a\\a\\a\\a\\a\end{array}}\right\} (')$$
$$\overset{\text{Split}}{\rightsquigarrow}\ \alpha =^? \underline{\alpha} * \beta_1,\ \beta =^? \beta_1 * \beta_2,\ \underline{\beta} * \underline{\gamma} * \delta =^? \beta_2 * \underline{\gamma} \rightsquigarrow \ldots \text{ like } (')$$
$$\overset{\text{Id}}{\rightsquigarrow}\ \alpha =^? \underline{e},\ \underline{\beta} * \underline{\gamma} * \delta =^? \underline{\alpha} * \beta * \underline{\gamma} \rightsquigarrow \ldots$$
$$\overset{\text{Id}}{\rightsquigarrow}\ \delta =^? \underline{e},\ \alpha * \underline{\beta} * \underline{\gamma} =^? \underline{\alpha} * \beta * \underline{\gamma} \rightsquigarrow \ldots$$
$$\overset{\text{Id}}{\rightsquigarrow}\ \beta =^? \underline{e},\ \alpha * \underline{\beta} * \underline{\gamma} * \delta =^? \underline{\alpha} * \underline{\gamma} \rightsquigarrow \ldots$$

Fig. 3. A sample derivation of A1-unifiers according to Ohlbach's algorithm

solved forms, that is, sets of the form $\{\alpha_1 =^? u_1, \ldots, \alpha_n =^? u_n\}$ with each α_i occurring exactly once in the set. The successful branches in the derivation tree are those marked with numbers, whose solved forms yield the following unifiers:

1. $\{\alpha \mapsto \underline{\alpha},\ \beta \mapsto \underline{\beta},\ \delta \mapsto \underline{e}\}$
2. $\{\alpha \mapsto \underline{\alpha},\ \beta \mapsto \underline{\beta} * \underline{\gamma},\ \delta \mapsto \underline{\gamma}\}$
3. $\{\alpha \mapsto \underline{\alpha},\ \beta \mapsto \underline{\beta},\ \delta \mapsto \underline{e}\}$ (a duplicate of 1.)
4. $\{\alpha \mapsto \underline{\alpha},\ \beta \mapsto \underline{\beta} * \underline{\gamma} * \delta_1,\ \delta \mapsto \delta_1 * \underline{\gamma}\}$

etc.

In the next section we will give a set of rules applying paramodulation only at the top symbols of the terms of an equation $s =^? t$ bearing a more efficient unification algorithm. These restricted forms of paramodulation rules are known as *mutation rules* and are sound and complete only for very particular E. For instance, they may be applied where E (is a finite resolvent set of equations and) defines a syntactic theory. Two results from the literature are of relevant. Kirchner and Klay (1990) prove mutation rules are complete for syntactic collapse-free theories. Comon, Haberstrau and Jouannaud (1994) consider mutation with (and without) collapsing equations for shallow theories and prove any shallow theory is syntactic. This is relevant for the identity law which is collapsing and shallow. The result of Kirchner and Klay is relevant for our associativity law which can be shown to be syntactic by an analogous argument as for ordinary associativity. However, it is not clear from the literature whether the combination of mutation rules for shallow and collapsing axioms and those for syntactic axioms automatically bear a complete procedure. The next section outlines a proof for the completeness of the combination of right identity and associativity, without relying on the notion of syntactic-ness.

5 Mutation and normalisation for (1) and (A)

In this section we present a unification system with mutation rules for (1) and (A). This system and its appropriate subsystems are to be used in resolution calculi, denoted by $\mathcal{R}^E_{N_E}$ and $\mathcal{R}^E_{\text{COND} \circ N_E}$, defined by binary E-resolution, syntactic factoring and normalisation modulo E and possibly condensing. Normalisation and condensing are applied eagerly.

The normalising functions N_1 and N_A rearrange terms according to the rewrite rules $[x\underline{e}] \to x$ and $[x(\alpha \circ \beta)] \to [x\alpha\beta]$. Inductive specifications of N_1 and N_A are: $N_1([]) = []$ and $N_1([s\underline{e}]) = N_1(s)$, and $N_A([]) = []$,

$$\begin{aligned} N_A([su]) &= [N_A(s)u] \quad \text{provided } u \text{ is a variable or constant, and} \\ N_A([s(v \circ v')] &= N_A([N_A([sv])v']). \end{aligned}$$

Normalisation under both (1) and (A) is by $N_{A1}(s) = N_1(N_A(s))$, which first eliminates the operation $\circ$ and then the identity constant $\underline{e}$. Clearly, all three functions are recursive. Any term $N_E(s)$ is said to be in *E-normal form.*

As we employ a resolution calculus requiring unification under a non-empty theory E in the resolution rule only, and not the factoring rule, we make the following assumption.

Assumption: Any unification problem has the form $P = \{s =^? t\}$ where s and t are variable disjoint basic paths, (i) $\{s, t\}$ is prefix stable, (ii) s and t do not contain world variables, and (iii) are normalised by N_{A1}.

(ii) is ensured for the negation of the translation of any modal formula and it is preserved since no world variables will be introduced during unification. Thus, *admissible* substitutions have the form $\alpha \mapsto u$ with u a functional term.

Delete	$P \cup \{s =^? s\}$	$\rightsquigarrow$	P
Variable Eliminate	$P \cup \{\alpha =^? u\}$	$\rightsquigarrow$	$P\{\alpha \mapsto u\}$
	when α is an introduced variable and does not occur in u.		
Decompose	$P \cup \{[su] =^? [tv]\}$	$\rightsquigarrow$	$P \cup \{s =^? t,\ u =^? v\}$
Mutate-1	$P \cup \{[s\alpha] =^? t\}$	$\rightsquigarrow$	$P \cup \{s =^? t, \alpha =^? \underline{e}\}$
Mutate-A	$P \cup \{[s\alpha] =^? [tv]\}$	$\rightsquigarrow$	$P \cup \{[s\alpha'] =^? t,\ \alpha =^? \alpha' \circ v\}$
	when α' is a new variable and not both s and t are empty.		

Fig. 4. Unification rules for the path logics closed under (1) and (A)

Our unification rules for path logics closed under (1) or (A), or both, are those listed in Fig. 4. s and t may denote empty paths, except where stated otherwise.

Observe, the variable elimination rule applies only to introduced variables, by which we mean the variables not present in the original problem set. The system does not decompose or mutate functional terms involving $\circ$, and no normalisation is done in the unification algorithm. The rules are (in essence) instances of the mutation rules of Comon et al. (1994).[2]

Mutate-1 binds a variable in a right-most position with the identity constant $\underline{e}$ and deletes the variable from the original term. For example, the only minimal (most general) 1-unifier for $\{[\alpha\underline{\beta}] =^? [\gamma]\}$ is $\{\alpha \mapsto \underline{e}, \gamma \mapsto \underline{\beta}\}$. The unification problem $\{[\alpha\beta] =^? [\gamma]\}$ has two minimal 1-unifiers: $\{\alpha \mapsto \gamma,\ \beta \mapsto \underline{e}\}$ and $\{\alpha \mapsto \underline{e},\ \beta \mapsto \gamma\}$. The algorithm computes a third unifier, namely $\{\alpha \mapsto \underline{e},\ \beta \mapsto \underline{e},\ \gamma \mapsto \underline{e}\}$, which is not most general.

Mutate-A applies to terms $s = [u_1 \ldots u_{m+1}] =^? [v_1 \ldots v_{n+1}] = t$ with the pair (u_{m+1}, v_{n+1}) being either a variable-constant pair, a constant-variable pair or a variable-variable pair. For the first two constellations there is one transformation by Mutate-A and for the last constellation there are two.

$$\{[u_1 \ldots u_m\alpha] =^? [v_1 \ldots v_n\underline{\beta}]\} \overset{\text{A}}{\rightsquigarrow} \{\alpha =^? \alpha' \circ \underline{\beta},\ [u_1 \ldots u_m\alpha'] =^? [v_1 \ldots v_n]\}$$
$$\{[u_1 \ldots u_m\alpha] =^? [v_1 \ldots v_n\gamma]\} \overset{\text{A}}{\rightsquigarrow} \{\alpha =^? \alpha' \circ \gamma,\ [u_1 \ldots u_m\alpha'] =^? [v_1 \ldots v_n]\}$$
$$\text{or } \{\gamma =^? \gamma' \circ \gamma,\ [u_1 \ldots u_m] =^? [v_1 \ldots v_n\gamma']\}.$$

This illustrates that the search tree for transformations with Mutate-A can be seen to be an instance of the search tree of Plotkin's (1972) algorithm for semi-groups (applied to paths and employing right-to-left as opposed to left-to-right decomposition).

Compare the derivation in Fig. 3 according to Ohlbach's method with the derivation in Fig. 5 according to the mutation system. The successful branches in the derivation tree yield the following unifiers:

1. $\{\delta \mapsto \underline{\gamma},\ \beta \mapsto \underline{\beta} \circ \underline{\gamma},\ \alpha \mapsto \underline{\alpha}\}$
2. $\{\delta \mapsto \underline{\gamma},\ \beta \mapsto (\beta'' \circ \underline{\beta}) \circ \underline{\gamma},\ \alpha \mapsto \underline{\alpha} \circ \beta''\}$
3. $\{\delta \mapsto \underline{e},\ \beta \mapsto \underline{\beta},\ \alpha \mapsto \underline{\alpha}\}$
4. $\{\delta \mapsto \underline{e},\ \beta \mapsto \beta' \circ \underline{\beta},\ \alpha \mapsto \underline{\alpha} \circ \beta'\}$
5. $\{\delta \mapsto \delta' \circ \underline{\gamma},\ \beta \mapsto (\underline{\beta} \circ \underline{\gamma}) \circ \delta',\ \alpha \mapsto \underline{\alpha}\}$
6. $\{\delta \mapsto \delta' \circ \underline{\gamma},\ \beta \mapsto ((\beta'' \circ \underline{\beta}) \circ \underline{\gamma}) \circ \delta',\ \alpha \mapsto \underline{\alpha} \circ \beta''\}$.

Clearly, the search tree is considerably smaller and there are no repetitions in the solution set. The solution set is not minimal though.

Now, we prove our system is sound and complete. By definition, a set of transformation rules is *sound and complete* in a theory E if the following two conditions hold:

[2] For readers familiar with this paper we note, in our context their cycle breaking rule can be easily seen to be superfluous, since there are no world variables in the original problem, and for the functional variables Cycle applies to equations of the form $\alpha \circ u =^? \alpha$ or $u \circ \alpha =^? \alpha$, which our algorithm does not produce as we will see.

$$[\alpha\underline{\beta}\gamma\delta] =^? [\underline{\alpha}\beta\underline{\gamma}]$$
$$\overset{\text{Dec}}{\rightsquigarrow} \delta =^? \underline{\gamma},\ [\alpha\underline{\beta}\underline{\gamma}] =^? [\underline{\alpha}\beta]$$
$$\left.\begin{array}{l} \overset{\text{Dec}}{\rightsquigarrow} \beta =^? \underline{\gamma},\ [\alpha\underline{\beta}] =^? [\underline{\alpha}] \quad \text{not solved} \\ \overset{1}{\rightsquigarrow} \beta =^? \underline{e},\ [\alpha\underline{\beta}\gamma] =^? [\underline{\alpha}] \quad \text{not solved} \\ \overset{\text{A}}{\rightsquigarrow} \beta =^? \beta' \circ \underline{\gamma},\ [\alpha\underline{\beta}] =^? [\underline{\alpha}\beta'] \\ \qquad \overset{2\times\text{Dec,Del}}{\rightsquigarrow} \beta' =^? \underline{\beta},\ \alpha =^? \underline{\alpha} \rightsquigarrow 1. \\ \qquad \overset{1}{\rightsquigarrow} \beta' =^? \underline{e},\ [\alpha\underline{\beta}] =^? [\underline{\alpha}] \quad \text{not solved} \\ \qquad \overset{2\times A\text{,Dec,Elim}}{\rightsquigarrow} \beta' =^? \beta'' \circ \underline{\beta},\ \alpha =^? \underline{\alpha} \circ \beta'' \rightsquigarrow 2. \end{array}\right\} ('')$$
$$\overset{1}{\rightsquigarrow} \delta =^? \underline{e},\ [\alpha\underline{\beta}\gamma] =^? [\underline{\alpha}\beta\underline{\gamma}] \overset{\text{Dec,Del}}{\rightsquigarrow} [\alpha\underline{\beta}] =^? [\underline{\alpha}\beta]$$
$$\overset{2\times\text{Dec}}{\rightsquigarrow} \beta =^? \underline{\beta},\ \alpha =^? \underline{\alpha} \rightsquigarrow 3.$$
$$\overset{1}{\rightsquigarrow} \beta =^? \underline{e},\ [\alpha\underline{\beta}] =^? [\underline{\alpha}] \quad \text{not solved}$$
$$\overset{2\times A\text{,Dec,Elim}}{\rightsquigarrow} \beta =^? \beta' \circ \underline{\beta},\ \alpha =^? \underline{\alpha} \circ \beta' \rightsquigarrow 4.$$
$$\overset{\text{A}}{\rightsquigarrow} \delta =^? \delta' \circ \underline{\gamma},\ [\alpha\underline{\beta}\underline{\gamma}\delta'] =^? [\underline{\alpha}\beta]$$
$$\overset{\text{Dec}}{\rightsquigarrow} \beta =^? \delta',\ [\alpha\underline{\beta}\gamma] =^? [\underline{\alpha}] \quad \text{not solved}$$
$$\overset{1}{\rightsquigarrow} \delta' =^? \underline{e},\ [\alpha\underline{\beta}\underline{\gamma} =^? [\underline{\alpha}\beta] \rightsquigarrow \ldots \quad \text{not solved}$$
$$\overset{1}{\rightsquigarrow} \beta =^? \underline{e}, \ldots \quad \text{not solved}$$
$$\overset{\text{A}}{\rightsquigarrow} \beta =^? \beta' \circ \delta',\ [\alpha\underline{\beta}\underline{\gamma}] =^? [\underline{\alpha}\beta] \rightsquigarrow \ldots \text{ like } ('') \rightsquigarrow \text{5. and 6.}$$

Fig. 5. A sample derivation of A1-unifiers

Soundness: If P transforms to P' by the application of any of the transformation rules, written $P \overset{*}{\rightsquigarrow} P'$, then every E-unifier of P' is an E-unifier of P.

Completeness: For any E-unifier θ of P, there is some P' in solved form such that $P \overset{*}{\rightsquigarrow} P'$ and the idempotent unifier σ associated with P' is more general than θ with respect to the variables occurring in P, written $\sigma \leq_E \theta[\mathit{Var}(P)]$.

Formally, a *solved form* is either the empty set or a finite set of the form $\{\alpha_1 =^? u_1, \ldots, \alpha_n =^? u_n\}$ and $\alpha_1, \ldots, \alpha_n$ are distinct variables occurring in no u_i. A variable α is *solved* in a set P if P includes a pair $\alpha =^? u$ (or $u =^? \alpha$) and α occurs exactly once in P. A variable that is not solved is an *unsolved variable.* By definition, $\sigma =_E \theta[V]$ iff for any variable x in V, $x\sigma$ and $x\theta$ are E-equivalent, and $\sigma \leq_E \theta[V]$ iff there is a substitution such that $\sigma\sigma' =_E \theta[V]$.

Equivalence (inequivalence and inclusion) modulo right identity and associativity will be denoted by $=_{\mathrm{A1}}$ ($\neq_{\mathrm{A1}}$ and $\leq_{\mathrm{A1}}$). The following is also easy to verify by inspecting the rules.

Theorem 2. *The system of Fig. 4 is sound.*

In the remainder of the section, P denotes a singleton unification problem of variable disjoint basic paths satisfying (i), (ii) and (iii) of the assumption, and P' denotes a set obtained from P by any sequence of transformations in Fig. 4. For the next lemmas it is important that the initial pair in P is variable disjoint.

Lemma 1. *For any identity $\alpha =^? u$ in P', the variable α does not occur in u.*

Lemma 2. *Each P' irreducible by the rules of Fig. 4 is in solved form or it is unsatisfiable in $=_{A1}$.*

We now sketch the proof of completeness.

Theorem 3. *The system of Fig. 4 is complete.*

The core structure of the proof is standard. We let

$$P = \{[su] =^? [tv]\}$$

with $s = [s|_1 \ldots s|_m]$ and $t = [t|_1 \ldots t|_n]$, each non-empty, that is, $1 \leq m, n$. We let θ be any A1-unifier of P, that is,

$$[su]\theta =_{A1} [tv]\theta.$$

The aim is to show there is a sequence of transformations of P to a solved P' such that the associated unifier σ is more general than θ. In parallel to transforming P we extend the unifier θ by adding bindings of new variables to θ obtaining θ'. Below, in Lemmas 6 and 7, we will define θ' in such a way that if θ unifies P and $P \rightsquigarrow P'$, that is, if P transforms to P' in one step, then θ' unifies P'. The resulting procedure starts with the pair (P, θ) and computes at least one pair (P', θ'), such that 1. $P \overset{*}{\rightsquigarrow} P'$, 2. P' is in solved form, 3. $\theta \subseteq \theta'$ and the restriction of θ' to the variables of θ coincides with θ. By assumption θ is a unifier of P, and consequently, by induction on the proof length, θ' of the final pair is a unifier of P'. This establishes completeness, when every derivation is finite. The next lemmas supply the technical details.

Lemma 3. *The unifier σ associated with the solved P' is more general than θ' with respect to the variables of P'.*

Lemma 4. *The unifier σ associated with the solved P' is more general than θ with respect to the variables of P' and P.*

Lemma 5. *Any fair implementation of a unification algorithm for the transformation system of Fig. 4 terminates for paths.*

Proof. Let $\tau(s)$ denote the functional depth of a term s. Define a measure μ of any unification problem P by $\mu(P) = (d, v)$, where v denotes the number of unsolved variables in P, and $d = \tau(s) + \tau(t)$ for $s =^? t \in P$ and both s and t are of type world. Examine each transformation rule in turn to see that $\mu(P')$ is smaller than $\mu(P)$ under the lexicographical ordering. The rules do not convert the status of any variable from solved to unsolved. □

The following two lemmas are concerned with the one step conversions of any pair (P, θ) to a suitable pair (P', θ').

Lemma 6. *Consider $P \cup \{[su] =^? [tv]\}$ with $s = [s|_1 \ldots s|_m]$ and $t = [t|_1 \ldots t|_n]$ for $1 \leq m, n$. The terms $[su]$ and $[tv]$ are assumed to be in A1-normal form. Let θ be any A1-unifier of $P \cup \{[su] =^? [tv]\}$, in particular, $[su]\theta =_{A1} [tv]\theta$. θ', as defined in the following, is in each case an A1-unifier of P'.*

1. *If $u\theta =_{A1} v\theta$, then let $\theta' = \theta$ and apply Decompose to P, yielding*

$$P' = P \cup \{s =^? t,\ u =^? v\}.$$

2. *If $u\theta =_{A1} \underline{e}$, then let $\theta' = \theta$ and apply Mutate-1 to P, yielding*

$$P' = P \cup \{s =^? [tv],\ u =^? \underline{e}\}.$$

3. *If u is a variable α, say, and $u\theta =_{A1} [t|_k \dots t|_n v]\theta$ for some $1 \le k \le n$, then let $\theta' = \theta\theta_0$ with $\theta_0 = \{\alpha' \mapsto [t|_k \dots t|_n]\}$ and apply Mutate-A to P, yielding*

$$P' = P \cup \{[s\alpha'] =^? t,\ \alpha =^? \alpha' \circ v\},$$

for α' a new variable not occurring in P or θ.

The lemma also covers the cases that $v\theta =_{A1} \underline{e}$ and $v\theta =_{A1} [s|_k \dots s|_n]\theta$ for some $1 \le k \le m$ and v a variable. Observe that when $u\theta =_{A1} [t|_k \dots t|_n v]\theta$ but both u and v are constants, the conditions of either 1. or 2. hold. If u and v are both constants then either (a) $u = v$ or (b) $u = \underline{\alpha}$, say, and $v = \underline{e}$. (a) implies $u\theta = v\theta$, and (b) implies $v\theta = \underline{e}$.

It remains to clarify whether there are cases that the lemma does not cover. The answer is, yes, as in this example

$$P = \{[\underline{\alpha}\underline{\alpha}'\beta] =^? [\underline{\alpha}\delta\underline{\gamma}]\} \quad \text{and} \quad \theta = \{\beta \mapsto \underline{\epsilon} \circ \underline{\gamma},\ \delta \mapsto \underline{\alpha}' \circ \underline{\epsilon}\} \tag{$*$}$$

when in the general case $[s|_k \dots s|_m u]\theta =_{A1} [t|_l \dots t|_n v]\theta$ is true, for some $1 \le k \le m$ and $1 \le l \le n$. If u and v are both constants then, as above, either $u\theta = v\theta$ or $u\theta = \underline{e}$ or $v\theta = \underline{e}$. The following result deals with the case that one of u or v is a variable. (It implies 3. of the previous lemma.)

Lemma 7. *Let θ be an A1-unifier of $P \cup \{[s\alpha] =^? [tv]\}$ with $s = [s|_1 \dots s|_m]$ and $t = [t|_1 \dots t|_n]$ for $1 \le m, n$, and both $[s\alpha]$ and $[tv]$ are in A1-normal form. Let $[s|_k \dots s|_m \alpha]\theta =_{A1} [t|_l \dots t|_n v]\theta$ for some $1 \le k \le m$ and $1 \le l \le n$. If θ includes a binding of α to u, that is, $\alpha \mapsto u \in \theta$, then let*

$$\theta' = \theta\theta_0 \quad \text{with } \theta_0 = \{\alpha' \mapsto u'\}$$

where u' is given by $u =_A u' \circ v'$ and $v' = v\theta$, and apply Mutate-A to P, yielding

$$P' = P \cup \{[s\alpha'] =^? t,\ \alpha =^? \alpha' \circ v\},$$

for α' a new variable not occurring in P or θ. Then θ' unifies P'.

For example, the pair $(*)$ is converted to

$$P' = \{[\underline{\alpha}\underline{\alpha}'\beta'] =^? [\underline{\alpha}\delta],\ \beta =^? \beta' \circ \underline{\gamma}\} \text{ and } \theta' = \{\beta \mapsto \underline{\epsilon} \circ \underline{\gamma},\ \delta \mapsto \underline{\alpha}' \circ \underline{\epsilon},\ \beta' \mapsto \underline{\epsilon}\}.$$

The lemma makes assumptions, which are not met in the following two situations. First, if no u' exists such that $u =_A u' \circ (v\theta)$ then $v\theta$ is equivalent to $\underline{e}$. This case is dealt with in 2. of the previous lemma. Second, the situation that neither α nor v are in the domain of θ and $\alpha\theta \neq_{A1} v\theta$ is impossible (for otherwise $[s\alpha]$ and $[tv]$ are not unifiable).

6 Preservation of prefix stability

Now, we verify that the application of A1-unifiers followed immediately by normalisation under N_{A1} preserves prefix stability. This justifies the assumptions made in the previous section, namely, that the terms in the initial problem set are basic paths and the world terms on the left hand sides of the transformation rules of Fig. 4 are also basic paths. We also prove a preservation result for forming $\mathcal{R}^{\mathrm{A1}}_{N_{\mathrm{A1}}}$ and $\mathcal{R}^{\mathrm{A1}}_{\mathrm{COND}\,\circ\,N_{\mathrm{A1}}}$-resolvents. The proofs are very similar to those for applying syntactic unifiers and forming standard resolvents. We start by considering the preservation of prefix stability under syntactic bindings.

Theorem 4. *Let T be a set of terms in the vocabulary of the basic path logic. Let $s = [u_1 \dots u_m]$ and $t = [v_1 \dots v_n]$ be two terms in T such that for some $k > 0$,*

$$u_1 = v_1, \dots, u_{k-1} = v_{k-1} \quad \textit{and} \quad u_k \neq v_k$$

and u_k is a variable. Let σ be the substitution $\{u_k \mapsto v_k\}$. Then $T\sigma$ satisfies T1 and T2, provided T does.

Based on this result it is not difficult to prove that prefix stability is preserved under syntactic factoring and ordinary resolution. Also, as prefix stability remains invariant under the formation of subsets, it is immediate that prefix stability is preserved by subsumption deletion and condensing. Thus, the basic path logic is closed under ordinary resolution, syntactic factoring, subsumption deletion and condensing.

Now, we address closure of the extensions of the basic path logic under the fundamental operations in our resolution calculus for $E = \{\mathrm{A}, 1\}$. We let T be a set of terms in the vocabulary of basic path logic, because remember, every theory resolvent is immediately normalised by N_{A1}. The analogue of Theorem 4 is not true in its full generality for bindings of A1-unifiers. It is true when suffixes are variable disjoint, and when more restrictions (to be made precise below) hold for instantiations with $\circ$ terms. For bindings of the form $\alpha \mapsto \underline{e}$ the following is immediate by Theorem 4.

Corollary 1. *Let T be a set of terms in the vocabulary of basic path logic. Let $s = [u_1 \dots u_m]$ and $t = [v_1 \dots v_n]$ be two terms in T such that for some $k > 0$,*

$$u_1 = v_1, \dots, u_{k-1} = v_{k-1} \quad \textit{and} \quad u_k \neq v_k,$$

u_k is a variable, and the suffixes $[u_{k+1} \dots u_m]$ and $[v_{k+1} \dots v_n]$ are variable disjoint. Let σ be a substitution $\{u_k \mapsto \underline{e}\}$. Then $N_1(T\sigma)$ satisfies T1 and T2, provided T does.

For bindings of the form $u_k \mapsto v \circ v'$ which cause the term depth to increase we need the concept of (k,l)-equality. Two basic paths s and t are (k,l)-*equal* if t is like s except possibly the term $s|_k$ in the k-th position is replaced by a string $w_1 \dots w_l$ of length l. In other words, s and t are (k,l)-equal provided $s = t$, or when $s = [s|_1 \dots s|_m]$ then $t = [s|_1 \dots s|_{k-1} w_1 \dots w_l s|_{k+1} \dots s|_m]$, or the other way around.

Theorem 5. *Let s and t be two terms in T defined as in the previous result. Let σ be a substitution $\{u_k \mapsto w\}$ where*

$$N_{A1}([w]) = [w_1 \dots w_l] \quad \textit{and} \quad w_1 = v_k,$$

and s and w are variable disjoint.[3] *Then $N_{A1}(T\sigma)$ satisfies T1 and T2, provided T, $N_{A1}([w])$ and the set $\{[w], [v_k \dots v_n]\}$ do.*

Proof. Let s and t be any terms in T that satisfy the conditions *T1* and *T2* and consider $N_{A1}(s\sigma)$ and $N_{A1}(t\sigma)$ in $N_{A1}(T\sigma)$. It is not difficult to verify that the pairs $N_{A1}(s)$ and $N_{A1}(s\sigma)$, and also $N_{A1}(t)$ and $N_{A1}(t\sigma)$, are (k,l)-equal. We assume without loss of generality that $s|_k = u_k$ (for otherwise if u_k does not occur in either of s or t then the result is trivially true). Then

$$N_{A1}(s\sigma) = [s|_1 \dots s|_{k-1} w_1 \dots w_l s|_{k+1} \dots s|_m].$$

Since s and $[w_1 \dots w_l]$ have no common variables and w satisfies *T1* and *T2*, so does $s\sigma$. Now, consider two cases: 1. σ leaves t unchanged so that $t\sigma = t$ and 2. it does not. In the either case we need to prove *T1* and *T2* hold for i and j strictly below $k + l$. As this is tedious we omit the details. □

Lemma 8. *Let s and t be two variable disjoint basic paths. Let σ be any A1-unifier computed by the system of Fig. 4. Then*

1. *σ is an idempotent unifier.*
2. *$\sigma = \sigma_1 \dots \sigma_l$, where the σ_i are of the form $\{\alpha_i \mapsto w\}$ such that for any pair σ_i and σ_j with $1 \le i < j \le n$, if α_i and α_j occur at positions k_i and k_j in s or t, then $k_i \le k_j$.*
3. *If α_1 of σ_1 is a variable occurring in s then the following are equivalent.*
   (a) *$s = [u_1 \dots u_m]$ and $t = [v_1 \dots v_n]$ have a common prefix $[u_1 \dots u_{k+1}]$, $u_k \neq v_k$ and u_k is a variable.*
   (b) *Either $\sigma_1 = \{u_k \mapsto \underline{e}\}$ or $\sigma_1 = \{u_k \mapsto w\}$ where $N_A([w]) = [w_1 \dots w_l]$ and $w_1 = v_k$.*
4. *$N_A([w])$ satisfies T1 and T2 provided s and t do.*
5. *$\{N_A([w]), [v_k \dots v_n]\}$ satisfies T1 and T2 provided s and t do.*

Theorem 6. *Let σ be an A1-unifier of two variable disjoint terms s and t in T. If T satisfies properties T1 and T2 then so does $N_{A1}(T\sigma)$.*

Proof. The proof is by an induction argument over the decomposition into bindings of idempotent unifiers. Let σ be $\sigma_1 \dots \sigma_l$ as in 2. of the previous lemma. Iteratively, consider the triples s, t and σ_1, then $N_{A1}(s\sigma_1)$, $N_{A1}(t\sigma_1)$ and σ_2, etcetera, and apply Corollary 1 and Theorem 5. By 3., 4. and 5. of the previous lemma, in any iteration the conditions *T1* and *T2* are satisfied by any $N_{A1}(s\sigma_1 \dots \sigma_i)$, $N_{A1}(t\sigma_1 \dots \sigma_i)$ and σ_{i+1}. □

[3] More accurately, $N_{A1}([w])$ coincides with $N_{A1}([[], w]) = [[] w_1 \dots w_l]$.

Consequently, as the union of two variable disjoint sets of prefix stable terms is prefix stable, the preservation result for binary $\mathcal{R}^{\mathrm{A1}}_{N_{\mathrm{A1}}}$-resolvents follows. More generally, specialisation to just (1) or (A) renders:

Theorem 7. *For $E \subseteq \{A, 1\}$, the binary $\mathcal{R}^{E}_{N_E}$-resolvent of two variable disjoint clauses satisfying T1 and T2 also satisfies T1 and T2.*

The main preservation theorem follows:

Theorem 8. *Let S be a finite set of basic path clauses. Then $(\mathcal{R}^{E}_{N_E})^n(S)$ and $(\mathcal{R}^{E}_{\mathrm{COND}\,\circ\, N_E})^n(S)$, for any n, are well-formed in the basic path logic, when $E \subseteq \{A, 1\}$.*

7 Conclusion

In summary, we have discussed issues concerning unification and normalisation of E-resolution for certain path logics, namely, those closed under right identity and associativity, or both. We have defined complete (and terminating) unification algorithms employing mutation rules. We have shown the search spaces are considerably smaller than those of Ohlbach's procedure. And, we have proved syntactic unification can be simplified for singleton problems.

We conclude with some remarks concerning further work.

Due to the assumption we make in Sect. 5, in particular, that the input set consists of one variable disjoint pair of terms, our resolution calculi are defined by binary E-resolution and syntactic factoring. For semantic factoring we need general E-unification for which our algorithm is not sufficient (this would require a deletion rule of the identity constant and a more general form of the variable elimination rule). Given a set of terms (literals), computing the syntactic most general unifier (when it exists) is easier than computing the set of minimal E-unifiers. Semantic factoring can produce an exponential number of factors causing a significant overhead. The price we pay for using syntactic factoring is incompatibility with strategies like tautology deletion. So, evidently there is a tradeoff which should be kept in mind and deserves further investigation.

Unification for other path theories has not been examined. Ohlbach (1988, 1991) considers unification for the modal schema B in the non-optimised context. In our context using the global form of the correspondence property of B is not sound and we are forced to use the local form, namely $[x\alpha\, i(x,\alpha)] = x$. Unification by mutation rules will not do in this case. For example, the solution $\{\gamma =^? i([s\alpha],\beta),\ \delta =^? i([s],\alpha)\}$ of the problem $\{[s\alpha\beta\gamma\delta] =^? s\}$ can only be derived by paramodulating into the left term, at a position not at the top. As many other path theories (not considered here) are collapse-free, the results of Kirchner and Klay (1990) and also Doggaz and Kirchner (1991), which are about collapse-free syntactic theories, may be of value for developing terminating (mutation) unification algorithms. The latter paper presents a completion algorithm for automatically converting a presentation of linear and collapse-free equations to a finite resolvent set of equations.

References

Arnborg, S. and Tidén, E. (1985), Unification problems with one-sided distributivity, *in* J.-P. Jouannaud (ed.), *Proc. Intern. Conf. on Rewriting Techniques and Applications*, Vol. 202 of *LNCS*, Springer, pp. 398–406.

Auffray, Y. and Enjalbert, P. (1992), Modal theorem proving: An equational viewpoint, *Journal of Logic and Computation* **2**(3), 247–297.

Comon, H., Haberstrau, M. and Jouannaud, J.-P. (1994), Syntacticness, cycle-syntacticness and shallow theories, *Information and Computation* **111**(1), 154–191.

Doggaz, N. and Kirchner, C. (1991), Completion for unification, *Theoretical Computer Science* **85**, 231–251.

Fariñas del Cerro, L. and Herzig, A. (1989), Automated quantified modal logic, *in* P. B. Brazdil and K. Konolige (eds), *Machine Learning, Meta-Reasoning and Logics*, Kluwer, pp. 301–317.

Fariñas del Cerro, L. and Herzig, A. (1995), Modal deduction with applications in epistemic and temporal logics, *in* D. M. Gabbay, C. J. Hogger and J. A. Robinson (eds), *Handbook of Logic in Artificial Intelligence and Logic Programming*, Vol. 4, Clarendon Press, Oxford, pp. 499–594.

Herzig, A. (1989), *Raisonnement automatique en logique modale et algorithmes d'unification.*, PhD thesis, Univ. Paul-Sabatier, Toulouse.

Jaffar, J. (1990), Minimal and complete word unification, *J. ACM* **37**(1), 47–85.

Jouannaud, J.-P. and Kirchner, C. (1991), Solving equations in abstract algebras: A rule-based survey of unification, *in* J.-L. Lassez and G. Plotkin (eds), *Computational Logic: Essays in Honor of Alan Robinson*, MIT-Press, pp. 257–321.

Kirchner, C. and Klay, F. (1990), Syntactic theories and unification, *in* J. C. Mitchell (ed.), *Proc. LICS'90*, IEEE Computer Society Press, Philadelphia, pp. 270–277.

Makanin, G. S. (1977), The problem of solvability of equations in a free semigroup, *Math. USSR Sbornik* **32**(2), 129–198.

Ohlbach, H. J. (1988), *A Resolution Calculus for Modal Logics*, PhD thesis, Univ. Kaiserslautern, Germany.

Ohlbach, H. J. (1991), Semantics based translation methods for modal logics, *Journal of Logic and Computation* **1**(5), 691–746.

Ohlbach, H. J. and Schmidt, R. A. (1997), Functional translation and second-order frame properties of modal logics, *Journal of Logic and Computation* **7**(5), 581–603.

Plotkin, G. (1972), Building-in equational theories, *in* B. Meltzer and D. Michie (eds), *Machine Intelligence 7*, American Elsevier, New York, pp. 73–90.

Schmidt, R. A. (1998), *E*-unification for subsystems of S4, *Research Report MPI-I-98-2-003*, Max-Planck-Institut für Informatik, Saarbrücken, Germany.

Schulz, K. U. (1992), Makanin's algorithm for word equations: Two improvements and a generalization, *in* K. U. Schulz (ed.), *Word Equations and Related Topics*, Vol. 572 of *LNCS*, Springer, pp. 85–150.

Zamov, N. K. (1989), Modal resolutions, *Soviet Mathematics* **33**(9), 22–29. Translated from *Izv. Vyssh. Uchebn. Zaved. Mat.* **9** (328) (1989) 22–29.

Solving Disequations Modulo Some Class of Rewrite Systems

Sébastien Limet and Pierre Réty

LIFO - Université d'Orléans
B.P. 6759, 45067 Orléans cedex 2, France
e-mail : {limet, rety}@lifo.univ-orleans.fr

Abstract. This paper gives a procedure for solving disequations modulo equational theories, and to decide existence of solutions. For this, we assume that the equational theory is specified by a confluent and constructor-based rewrite system, and that four additional restrictions are satisfied. The procedure represents the possibly infinite set of solutions thanks to a grammar, and decides existence of solutions thanks to an emptiness test. As a consequence, checking whether a linear equality is an inductive theorem is decidable, if assuming moreover sufficient completeness.

1 Introduction

The problem that consists in solving symbolic equations modulo a theory is called equational *unification*. A lot of work has already studied this subject in a theoretical way to know when the problem can be decided, as well as in a practical way to find efficient algorithms that solve the problem. Another interesting problem consists in solving the negation of equations, called disequations, i.e. in finding the solutions of $s \overset{?}{\neq} t$ modulo a theory. This problem is a particular case of the equational *disunification* (see [2] for a survey), when formulas to be solved are just disequations, instead of general first order formulas. Disunification has a large interest in the domain of automated deduction. In symbolic constraint solving, disunification problems appear as soon as negation is handled. From functional logic programming point of view, disunification is used to treat negative information too. Handling negation improves the expressiveness of the programming language. For example, it becomes natural to define notions like lists of integers that are not multiple of z as follows:

$$list\_not\_z(c(x,y),z) \rightarrow list\_not\_z(y,z) : -remainder(x,z) \neq 0.$$
$$list\_not\_z(nil,z) \quad \rightarrow true.$$

$remainder(x,y)$ is a function defined by some other rewrite rules, that gives the remainder of $x \div y$. When using the first clause, a disequation has to be solved modulo a theory.

Few authors have studied equational disunification. Decidability results have been established for quasi-free theories and compact theories [3], and for shallow theories [4]. Other work gives algorithms that enumerate the solutions, as in syntactic theories [2], or modulo theories presented by a convergent set of rewrite rules [6]. These enumerations do not terminate in general, therefore these works give semi-decision procedures of existence of solutions.

In this paper our goal is both to decide existence of solutions of disequations and to give a finite representation of them, modulo equational theories given as term rewrite systems. Of course, to get decidability, we need stronger restrictions than [6]. In [9], we described a new equational unification algorithm based on tree grammars. This algorithm computes a tree grammar of a new kind called Tree Tuple Synchronized Grammar (TTSG for short), that fulfills both aims. In this paper, it is shown that TTSG's can also be used to deal with disunification.

In a few words, our idea is the following. We assume that the equational theory is defined by a confluent and constructor-based rewrite system, and that four additional restrictions are satisfied (see Section 3).

Example 1. Consider the TRS that defines functions f and g on natural integers.

$$f(s(x)) \xrightarrow{1} s(f(x)),\ f(0) \xrightarrow{2} 0,$$
$$g(s(x)) \xrightarrow{3} s(s(g(x))),\ g(0) \xrightarrow{4} 0$$

f is the identity whereas g is the multiplication per two. We want to solve $f(y) \stackrel{?}{\neq} g(x)$. A solution is found whenever the instance of y is not the double of that of x.

Thanks to a TTSG, we can express the set (or language) $\mathcal{L}_1$ of pairs of terms (t'_1, t_1) where t'_1 is a ground instance of a term obtained from g(x) by narrowing, that contains only constructors, while t_1 is the corresponding instance of x (see Section 4). In the same way, we can express for $f(y)$ a set $\mathcal{L}_2$ of pairs of terms (t'_2, t_2).

From $\mathcal{L}_1$ and $\mathcal{L}_2$, we are able to compute the language $\mathcal{L}_3$ of solutions. Indeed from lifting lemma [8], $(t'_1, t_1) \in \mathcal{L}_1$ and $(t'_2, t_2) \in \mathcal{L}_2$ is equivalent to that $g(t_1)$ rewrites into t'_1 and $f(t_2)$ rewrites into t'_2. t'_1, t'_2 contain only constructors, and due to the constructor discipline, there is no relation between constructors. Thus t'_1 is the normal form of $g(t_1)$, and t'_2 is the normal form of $f(t_2)$. Moreover normal forms are unique thanks to confluence. Thus, according to the semantics of functional-logic languages (see Section 3), the substitution $\{x \mapsto t_1,\ y \mapsto t_2\}$ is a solution of $f(y) \stackrel{?}{\neq} g(x)$ iff $t'_1 \neq t'_2$. Therefore to compute solutions, we have to achieved a kind of dis-intersection (also called anti-join) between the first components of $\mathcal{L}_1$ and $\mathcal{L}_2$ (see Section 5). A TTSG that recognizes $\mathcal{L}_3$ is thus computed, and thanks to the emptiness test of [9] we can decide existence of solutions.

2 Preliminaries

We assume that the reader is familiar with standard definitions of one-sorted terms, substitutions, equations, rewrite systems (see [5]), and we just recall here the main definitions and notations used in the paper.

Let Σ be a finite set of symbols and V be an infinite set of variables, $T_{\Sigma \cup V}$ is the term algebra over Σ and V. Σ is partitioned in two parts: the set $\mathcal{F}$ of **function symbols**, and the set $\mathcal{C}$ of **constructors**. The terms of $T_{\mathcal{C} \cup V}$ are called

data-terms. A term is said **linear** if it does not contain several occurrences of the same variable. A **data-substitution** σ is a substitution such that for each variable x, σx is a data-term.

In the following x, y, z denote variables, s, t, l, r denote terms, f, g, h function symbols, c a constructor symbol, and u, v, w occurrences. Let t be a term, $\boldsymbol{O(t)}$ is the set of occurrences of t, $\boldsymbol{t|_u}$ is the subterm of t at occurrence u and $\boldsymbol{t(u)}$ is the symbol that labels the occurrence u of t. $\boldsymbol{t[u \leftarrow s]}$ is the term obtained by replacing in t the subterm at occurrence u by s. We generalize the occurrences (as well as the above notations) to tuples. Moreover we define the **concatenation** of two tuples by $(t_1, \ldots, t_n) * (t'_1, \ldots, t'_{n'}) = (t_1, \ldots, t_n, t'_1, \ldots, t'_{n'})$ and the **component elimination** by $(t_1, \ldots, t_i, \ldots, t_n)\backslash_i = (t_1, \ldots, t_{i-1}, t_{i+1}, \ldots, t_n)$

A term rewrite system (**TRS**) is a finite set of oriented equations called rewrite rules or rules. **lhs** means left-hand-side and **rhs** means right-hand-side. For a TRS R, the rewrite relation is denoted by $\rightarrow_R$ and is defined by $t \rightarrow_R s$ if there exists a rule $l \rightarrow r$ in R and a non-variable occurrence u in t such that $t|_u = \sigma l$ and $s = t[u \leftarrow \sigma r]$. The reflexive transitive closure of $\rightarrow_R$ is denoted by $\rightarrow_R^*$, and the symmetric closure of $\rightarrow_R^*$ is denoted by $=_R$. $t \neq_R t'$ means $not(t =_R t')$. An **R-disunifier** of t and t' is a substitution σ such that $\sigma t \neq_R \sigma t'$.

A TRS is said **confluent** if $t \rightarrow_R^* t_1$ and $t \rightarrow_R^* t_2$ implies $t_1 \rightarrow_R^* t_3$ and $t_2 \rightarrow_R^* t_3$ for some t_3. If the lhs (resp. rhs) of every rule is linear the TRS is said **left-**(resp. **right-**)**linear**. If it is both left and right-linear the TRS is said **linear**. A TRS is **constructor based** if every rule is of the form $f(t_1, \ldots, t_n) \rightarrow r$ where the t_i's are data-terms. t **narrows** into s, if there exists a rule $l \rightarrow r$ in R, a non-variable occurrence u of t, such that $\sigma(t|_u) = \sigma l$ where $\sigma = mgu(t|_u, l)$ and $s = (\sigma t)[u \leftarrow \sigma r]$. We write $\boldsymbol{t \rightsquigarrow_{[u, l \rightarrow r, \sigma]} s}$.

3 Restrictions and semantics

3.1 Restrictions

The considered rewrite systems are supposed to be constructor-based and confluent[1]. Our four additional restrictions are:

1. *Linearity of rewrite rules*: every rewrite rule side is linear.
2. *No σ_{in}*: if a subterm r of some rhs unifies with some lhs l (after variable renaming to avoid conflicts) then the mgu σ does not modify the variables of l[2].
3. *No nested functions in rhs's*: the function symbols in the rhs's may not appear at comparable occurrences. For example f and g are nested in $f(g(x))$ but not in $c(f(x), g(y))$.
4. *Linearity of the disequation*: the disequation does not contain several occurrences of the same variable.

[1] Actually, ground confluence suffices.

[2] In other words, if σ is split into $\sigma = \sigma_{in} \cup \sigma_{out}$ where $\sigma_{in} = \sigma\,[var(l)]$ and $\sigma_{out} = \sigma\,[var(r)]$ then σ_{in} must be a variable renaming.

Note that these four restrictions together allow non-finitary theories, and they define a class of rewrite systems incomparable with those for which decidability results on disunification are known. Theories like addition in the positive and negative integers can be encoded with a TRS respecting restrictions 1. 2. 3. We have proved in [9] that if only one of these restrictions is not respected then the unifiability problem is undecidable.

3.2 Semantics

Since the signatures we deal with follow a constructor discipline, the validity of an equation is defined as a strict equality on terms, in the spirit of functional logic languages like K-LEAF [7] and BABEL [10]. So, as in [1], σ is a solution of the equation $s \stackrel{?}{=} t$ iff σ is a data substitution and σs and σt rewrite into the same data-term. The validity of a disequation is defined in a similar way: the ground data-substitution σ is a **solution** of the disequation $t \stackrel{?}{\neq} t'$ iff $\sigma t \rightarrow_R^* s$ and $\sigma t' \rightarrow_R^* s'$ where s, s' are non equal data-terms[3]. For example, the disequation $1 \div x \stackrel{?}{\neq} 1 \div 0$ where $\div$ is the euclidean division defined by some rewrite rules, is not equivalent to the disequation $x \stackrel{?}{\neq} 0$, since the former is actually not defined.

In Section 5, we decide existence of solutions (as defined above) assuming the previous restrictions. This is equivalent to the existence of classical ground R-disunifiers, if assuming moreover the sufficient completeness.

Lemma 1. *If the TRS is sufficiently complete*[4]*, we have: there exists a solution for* $t \stackrel{?}{\neq} t'$ $\Longleftrightarrow$ *there exists a ground R-disunifier of* t *and* t'.

Proof: $\Longrightarrow$. A solution is trivially a ground R-disunifier.

$\Longleftarrow$. Let θ be a ground R-disunifier, we have $\theta t \neq_R \theta t'$. Thanks to sufficient completeness, there exists a ground data-substitution σ such that $\theta \rightarrow_R^* \sigma$, and we get $\theta t \rightarrow_R^* \sigma t \rightarrow_R^* s$ and $\theta t' \rightarrow_R^* \sigma t' \rightarrow_R^* s'$ where s, s' are data-terms. Moreover we have $s \neq_R s'$, then $s \neq s'$. Therefore σ is a solution of $t \stackrel{?}{\neq} t'$. □

Assuming still sufficient completeness, this has an unexpected consequence, as the following reasoning shows:

There exists no solution for $t \stackrel{?}{\neq} t'$
$\Longleftrightarrow$ not(there exists a solution of $t \stackrel{?}{\neq} t'$)
$\Longleftrightarrow$ not(there exists a ground R-disunifier of t and t')
$\Longleftrightarrow$ not(there exists a ground substitution σ s.t. $\sigma t \neq_R \sigma t'$)
$\Longleftrightarrow$ for all ground substitution σ, $\sigma t =_R \sigma t'$
$\Longleftrightarrow$ $t \stackrel{?}{=} t'$ is an inductive theorem.

---

[3] Note that such data-terms s and s' are irreducible because there is no relation between constructors, and they are then unique because of confluence.

[4] This means that every function is completely defined on ground data-terms, and as a consequence every ground term rewrites into a data-term.

Therefore checking whether $t \stackrel{?}{=} t'$ is an inductive theorem is decidable. However, because of our four restrictions, $t \stackrel{?}{=} t'$ is assumed to be linear. So we can decide only for linear equalities.

4 Encoding narrowing by TTSG's

4.1 Principle

We explain here the principles of our narrowing encoding method, already used in [9] for solving unification. This is why we just recall the ideas here, for more details see [9]. Consider again the TRS of Example 1. Let us narrow for instance $f(g(x))$.

Step 1. The term $f(g(x))$ is decomposed into two parts, $f(y)$ and $g(x)$, where y is a new variable. From lifting lemma [8], the ground instances of the data-terms obtained by narrowing from $g(x)$, and the corresponding instances of x, can be viewed as an infinite set of pairs of terms defined by $\{(t'_1, t_1) | g(t_1) \rightarrow^* t'_1\}$. This set is considered as a language (says $\mathcal{L}_1$) of pairs of trees where the two components are not independent. In the same way, for $f(y)$ we consider the language (says $\mathcal{L}_2$) of pairs of trees that describes the set $\{(t'_2, t_2) | f(t_2) \rightarrow^* t'_2\}$. These languages can be described by TTSG's. The grammars are computed from the rewrite system and the term to be narrowed.

Step 2. Once these two TTSG's are constructed, the initial term $f(g(x))$ is re-composed. The languages $\mathcal{L}_1$ and $\mathcal{L}_2$ are combined to get the language $\mathcal{L}_0$ of the ground instances of the data-terms issued by narrowing from $f(g(x))$, as well as the corresponding instances of x. This is done by computing a special kind of intersection between two TTSG's that corresponds to the join operation in relational data-bases. The result is a TTSG that describes the language of triples of trees defined by $\{(t_1, t_2, t_3) | (t_1, t_2) \in \mathcal{L}_2 \text{ and } (t_2, t_3) \in \mathcal{L}_1\}$. In other words, t_2 is the result of $g(x)$ when instantiating x by t_3, moreover t_2 belongs to the definition domain of the function f, and t_1 is the result of $f(t_2)$, i.e. of $f(g(t_3))$.

Let us now explain in a few words the way the two steps are actually made. The first step consists in building some TTSG's from the TRS and the term to be narrowed. In this context the constructor symbols are the terminals and the occurrences of the terms are the non-terminals.

The meaning of the non-terminals is the following: the non-terminal L^i_u represents the subterm at occurrence u of the lhs of the ith rewrite rule. (R^i_u is similar for the rhs). The only exception is R^i_ϵ that represents both the lhs and the rhs of the ith rewrite rule in order to simulate the rewrite relation. X^i represents the variable x of the ith rewrite rule. A similar numbering of non-terminals is defined for the term to be narrowed (or for the equation or the disequation to be solved). For our example let us suppose that G^l_ϵ represents $f(y)$, A^l_1 represents y, G^l_1 for $g(x)$ and X^l for x.

Then, two kinds of productions are deduced. The **free productions** that are similar to the productions of regular tree grammars. These productions generate constructor symbols and are deduced from subterm relations. In our example L^1_1 represents the subterm at occurrence 1 of the lhs of the first rewrite rule i.e. $s(x)$. This term is headed by s and has x (represented by the non-terminal X^1) as argument, so the production $L^1_1 \Rightarrow s(X^1)$ is deduced.

The second kind of productions, the **synchronized productions**, come from syntactically unifiable terms. When productions are synchronized, they have to be applied at the same time. For example we have the synchronized productions $\{R^1_1 \Rightarrow R^2_\epsilon, X^1 \Rightarrow L^2_1\}$ which means that the subterm at occurrence 1 of the rhs of rewrite rule 1 i.e. $f(x)$ unifies with the lhs of the rewrite rule 2 i.e. $f(0)$ and this unification sets the variable x represented by X^1 to 0 represented by L^2_1.

One grammar is defined per function call in the term to be narrowed (or the disequation). All the grammars have the same terminals (the constructors), the same non-terminals, and the same productions (those defined before). Just the axioms (tuples of non-terminals) are different. In our example the two axioms are (G^l_ϵ, A^l_1) (which represents $(f(y), y)$) for language $\mathcal{L}_2$ and (G^l_1, X^l) (which represents $(g(x), x)$) for language $\mathcal{L}_1$. The two grammars of our example are:

- Gr^l_ϵ which generates the language $\mathcal{L}_2$ is defined by the axiom (G^l_ϵ, A^l_1), and its usable[5] productions are $\{G^l_\epsilon \Rightarrow R^1_\epsilon, A^l_1 \Rightarrow L^1_1\}$, $\{G^l_\epsilon \Rightarrow R^2_\epsilon, A^l_1 \Rightarrow L^2_1\}$
  $\{R^1_1 \Rightarrow R^1_\epsilon, X^1 \Rightarrow L^1_1\}$, $\{R^1_1 \Rightarrow R^2_\epsilon, X^1 \Rightarrow L^2_1\}$
  $L^1_1 \Rightarrow s(X^1)$, $R^1_\epsilon \Rightarrow s(R^1_1)$, $L^2_1 \Rightarrow 0$, $R^2_\epsilon \Rightarrow 0$
- Gr^l_1 which generates the language $\mathcal{L}_1$ is defined by the axiom (G^l_1, X^l), and its usable productions are $\{G^l_1 \Rightarrow R^3_\epsilon, X^l \Rightarrow L^3_1\}$, $\{G^l_1 \Rightarrow R^4_\epsilon, X^l \Rightarrow L^4_1\}$,
  $\{R^3_{1.1} \Rightarrow R^3_\epsilon, X^3 \Rightarrow L^3_1\}$, $\{R^3_{1.1} \Rightarrow R^4_\epsilon, X^3 \Rightarrow L^4_1\}$
  $L^3_1 \Rightarrow s(X^3)$, $R^3_\epsilon \Rightarrow s(R^3_1)$, $R^3_1 \Rightarrow s(R^3_{1.1})$, $L^4_1 \Rightarrow 0$, $R^4_\epsilon \Rightarrow 0$

One derivation of Gr^l_ϵ is:
$(G^l_\epsilon, A^l_1) \Rightarrow (R^1_\epsilon, L^1_1) \Rightarrow^* (s(R^1_1), s(X^1)) \Rightarrow (s(R^2_\epsilon), s(L^2_1)) \Rightarrow^* (s(0), s(0))$. It simulates narrowing in the following way. The first pack of productions that is applied is $\{G^l_\epsilon \Rightarrow R^1_\epsilon, A^l_1 \Rightarrow L^1_1\}$ which means that $f(y)$ unifies with $f(s(x))$ instantiating y by $s(x)$. Thus $f(y)$ can be narrowed using the first rewrite rule. We obtain the narrowing step $f(y) \rightsquigarrow_{[\epsilon,1,y\mapsto s(x)]} s(f(x))$. The two free productions applied next instantiate y on one hand and computes the term obtain by the narrowing step on the other hand. $s(R^1_1)$ actually represents $s(f(x))$ and $s(X^1)$ represents $s(x)$. Then a second pack of synchronized productions is applied i.e. $\{R^1_1 \Rightarrow R^2_\epsilon, X^1 \Rightarrow L^2_1\}$ which means that $f(x)$ unifies with $f(0)$ instantiating x by 0. So $s(f(x))$ can be narrowed by the second rewrite rule. The narrowing step is then $s(f(x)) \rightsquigarrow_{[1,2,x\mapsto 0]} s(0)$. So this derivation of Gr^l_ϵ which recognizes the tuple $(s(0), s(0))$ simulates the narrowing derivation $f(y) \rightsquigarrow_{[\epsilon,1,y\mapsto s(x)]} s(f(x)) \rightsquigarrow_{[1,2,x\mapsto 0]} s(0)$ which maps y to $s(0)$ and transforms $f(y)$ to $s(0)$.

The problem of the grammars we have just defined is that they do not take into account variable renamings. The solution of this problem consists in using an integer number, called **control**, to encode variable renamings. In a grammar computation, each non-terminal is coupled with an integer of control, which is

[5] The grammars may contain some productions that cannot be applied starting from the axiom.

incremented into a not yet used value when a pack of synchronized productions is applied on it. When a free production is applied, the control number is preserved.

The second step of our unification algorithm consists in computing a kind of intersection on TTSG's called the n, m intersection over one component. This operation is complex, and need the use of a tuples of integers instead of single integers as control. It is explained precisely and proved in [9].

4.2 Formal definitions of TTSG's

In the following, NT is a finite set of non-terminal symbols and recall that C is the set of constructor symbols. Upper-case letters denote elements of NT.

Definition 2. A **production** is a rule of the form $X \Rightarrow t$ where $X \in NT$ and $t \in T_{C \cup NT}$. A **pack of productions** is a set of productions coupled with a non negative integer and denoted $\{X_1 \Rightarrow t_1, \ldots, X_n \Rightarrow t_n\}_k$ (k is the rank of field (also called **level**) in the control tuple that is incremented when applying the pack of productions).

When $k = 0$ the pack is a singleton and it is of the form $\{X_1 \Rightarrow c(Y_1, \ldots, Y_n)\}_0$ where c is a constructor and $Y_1, \ldots, Y_n$ non-terminals. The production is said **free**, and is written more simply $X_1 \Rightarrow c(Y_1, \ldots, Y_n)$.

When $k > 0$ the pack is of the form $\{X_1 \Rightarrow Y_1, \ldots, X_n \Rightarrow Y_n\}_k$ where $Y_1, \ldots, Y_n$ are non-terminals. The productions of the pack are said **synchronized**.

Definition 3. A **TTSG** is defined by a 5-tuple (Sz, C, NT, PP, TI) where Sz is a positive integer that defines the size of the tuple of control, C is the set of constructors (terminals in the terminology of grammars), NT is the finite set of non-terminals, PP is a finite set of packs of productions, TI is the axiom of the TTSG. It is a tuple $((I_1, ct_1), \ldots, (I_n, ct_n))$ where every I_i is a non-terminal, and every ct_i is a Sz-tuple of control containing 0's and $\perp$'s.

Intuitively a free production $X \Rightarrow c(Y_1, \ldots, Y_n)$ can be applied as soon as X appears in a computation of the grammar, and then $Y_1, \ldots, Y_n$ preserves the same control as X. On the other hand a pack of productions $\{X_1 \Rightarrow Y_1, \ldots, X_n \Rightarrow Y_n\}_k$ can be applied iff $X_1, \ldots X_n$ occur in the same computation and the kth components of their controls are identical (and are not $\perp$). The X_i's are then replaced by the Y_i's and the kth component of control is set to a new fresh value.

Definition 4. The set of **computations** of a TTSG $Gr = (Sz, C, NT, PP, TI)$ denoted $\boldsymbol{Comp(Gr)}$ is the smallest set defined by:
- TI is in $Comp(Gr)$,
- if tp is in $Comp(Gr)$ and $tp|_u = (X, ct)$ and the free production $X \Rightarrow c(Y_1, \ldots, Y_n)$ is in PP then $tp[u \leftarrow c((Y_1, ct), \ldots, (Y_n, ct))]$ is in $Comp(Gr)$,
- if tp is in $Comp(Gr)$ and there exists $a \in \mathbb{N}$ as well as n pairwise different occurrences $u_1, \ldots, u_n$ of tp such that $\forall i \in [1, n]$ $tp|_{u_i} = (X_i, ct_i)$ and $ct_i|_k = a$ and the pack of productions $\{X_1 \Rightarrow Y_1, \ldots, X_n \Rightarrow Y_n\}_k \in PP$, then

$tp[u_1 \leftarrow (Y_1, ct_1[k \leftarrow b])] \ldots [u_n \leftarrow (Y_n, ct_n[k \leftarrow b])]$ (where b is a new integer[6]) is in $Comp(Gr)$.

The symbol $\Rightarrow$ denoting also the above two deduction steps, a **derivation** of Gr is a sequence of computations $TI \Rightarrow tp_1 \Rightarrow \ldots \Rightarrow tp_n$.

The language **recognized** by a TTSG Gr denoted $\boldsymbol{Rec(Gr)}$ is the set of tuples of ground data-terms $Comp(Gr) \cap T_C^n$.

5 Disunification algorithm

Warning : to get simple explanations, we consider as an example the simple disequation $f(y) \stackrel{?}{\neq} g(x)$. So the language $\mathcal{L}_0$ defined in Section 4 will not be used in this section. It is used when solving for instance $f(g(x)) \stackrel{?}{\neq} 0$.

5.1 The anti-join operation

In Section 4, it is explained how to transform the TRS R into some TTSG's that recognize the languages $\mathcal{L}_2 = \{(t_2', t_2)\}$ and $\mathcal{L}_1 = \{(t_1', t_1)\}$ where $g(t_1) \rightarrow^* t_1'$ and $f(t_2) \rightarrow^* t_2'$. This section defines the operation on $\mathcal{L}_2$ and $\mathcal{L}_1$ that computes the language $\mathcal{L}_3$ of the solutions of the disequation $f(y) \stackrel{?}{\neq} g(x)$.

The solutions of the disequation $f(y) \stackrel{?}{\neq} g(x)$ is the set of ground data substitutions $\sigma = \{x \mapsto t_1, y \mapsto t_2\}$ such that $\sigma(f(y)) \rightarrow^* t_2'$ and $\sigma(g(x)) \rightarrow^* t_1'$. t_1', t_2' are both ground data terms and $t_1' \neq t_2'$. In other words, we search the language $\mathcal{L}_3 = \{(t_1, t_2) | (t_1', t_1) \in \mathcal{L}_1$ and $(t_2', t_2) \in \mathcal{L}_2$ and $t_1' \neq t_2'\}$. Since $t_1' = 2 * t_1$ and $t_2' = t_2$, the solution of $f(y) \stackrel{?}{\neq} g(x)$ are the pairs (t_1, t_2) such that t_2 *is not* the double of t_1. By analogy with relational data-bases, the operation achieved to compute $\mathcal{L}_3$ is called **anti-join**.

Definition 5. Let E_1 and E_2 be two sets of tuples. The k_1, k_2 **anti-join** operation of E_1 and E_2, denoted $E_1 \overline{\bowtie}_{k_1,k_2} E_2$ is the set of tuples defined by
$E_1 \overline{\bowtie}_{k_1,k_2} E_2 = \{(tp_1 \backslash_{k_1}) * (tp_2 \backslash_{k_2}) \,|\, tp_1 \in E_1 \text{ and } tp_2 \in E_2 \text{ and } tp_1|_{k_1} \neq tp_2|_{k_2}\}$

For example, if $E_1 = \{(1, a), (0, b)\}$ and $E_2 = \{(0, a), (1, c)\}$ then
$E_1 \overline{\bowtie}_{1,1} E_2 = \{(a, a), (b, c)\}$.

5.2 Algorithm : informal presentation

We have just defined the operation we want to compute, so we have now to describe how to compute it. In order to have an intuitive idea of the problem, let us consider again Example 1. We saw that solving $f(y) \stackrel{?}{\neq} g(x)$ amounts to compute $\mathcal{L}_1 \overline{\bowtie}_{1,1} \mathcal{L}_2$ i.e. the set $\{ (t_1, t_2) | (t_1', t_1) \in \mathcal{L}_1, (t_2', t_2) \in \mathcal{L}_2$ and $t_1' \neq t_2' \}$.

So from the TTSG's Gr_ϵ^l and Gr_1^l, that recognize respectively $\mathcal{L}_2$ and $\mathcal{L}_1$, we want to compute the TTSG Gr_{dis} that recognizes $\mathcal{L}_1 \overline{\bowtie}_{1,1} \mathcal{L}_2$. To be able

[6] In practice, the smallest integer not yet used.

to test disequality of 2 terms in grammar formalism, there is not any other way but generating them at least partially. This is why Gr_{dis} computes an additional field that enables to perform this test. So, the language recognized by Gr_{dis} is roughly defined as the set $\{(t_{test}, t_1, t_2) | (t'_1, t_1) \in \mathcal{L}_1$ and $(t'_2, t_2) \in \mathcal{L}_2$ and t_{test} is computed from t'_1 and t'_2 and ensures that $t'_1 \neq t'_2\}$.

In fact, t_{test} is a kind of hybrid made of subterms of t'_1 and t'_2. The test it performs, is the following:
two terms t'_1 and t'_2 are different iff $\exists u \in O(t'_1) \cap O(t'_2)$ such that $t'_1(u) \neq t'_2(u)$
So the algorithm searches one occurrence where the two terms are different thanks to t_{test}. For that, it uses a new set of non-terminals which is the cartesian product of the sets of non-terminals of Gr^l_ϵ and Gr^l_1. In a computation (s_{test}, s_1, s_2) of Gr_{dis}, there is at most one occurrence u of s_{test} such that $s_{test}(u) = XY$ and XY belongs to this new set of non-terminals.

- if there exists such an occurrence u of s_{test} then this means that $g(s_1) \rightarrow^* s'_1$ and $f(s_2) \rightarrow^* s'_2$ are such that for each occurrence v over u $s'_1(v) = s'_2(v)$, in other words no difference between the two terms has been detected yet along this branch (note that the other branches of s_{test} are meaningless),
- otherwise this means that $s'_1 \neq s'_2$.

The field t_{test} plays a second part. It enables to preserve synchronization. In other words, it ensures that if (t_{test}, t_1, t_2) is recognized by Gr_{dis} then there do exist (t'_1, t_1) in $\mathcal{L}_1$ and (t'_2, t_2) in $\mathcal{L}_2$. This is why t_{test} is made with subterms of t'_1 and t'_2. Let us detail all this on a very simple example, that contains only free productions.

Example 2. Let us consider two TTSG's

- $G_4 = (1, \mathcal{C}, NT_4, PP_4, TI_4)$ whose set of productions is $PP_4 = \{X_4 \Rightarrow 0,\ Y_4 \Rightarrow c(Z_4, T_4),\ Z_4 \Rightarrow s(Z'_4),\ Z'_4 \Rightarrow 0,\ T_4 \Rightarrow a\}$ and whose axiom is $TI_4 = (Y_4, X_4)$,
- $G_5 = (1, \mathcal{C}, NT_5, PP_5, TI_5)$ with $PP_5 = \{X_5 \Rightarrow 0,\ Y_5 \Rightarrow c(Z_5, T_5),\ Z_5 \Rightarrow s(Z'_5),\ Z'_5 \Rightarrow 0,\ T_5 \Rightarrow b\}$ and $TI_5 = (Y_5, X_5)$

Clearly the languages recognized by G_4 and G_5 are respectively

$$Rec(G_4) = (c(s(0), a), 0) \text{ and } Rec(G_5) = (c(s(0), b), 0)$$

and then $Rec(G_4) \bowtie_{1,1} Rec(G_5) = (0, 0)$. The TTSG G_6 that computes this language is constructed as follows. Its axiom is $(Y_4Y_5,\ X_4,\ X_5)$. The first field performs the disequality test. The set NT_6 of non-terminals of G_6 is the union of NT_4, NT_5, $NT_4 \times NT_5$ and a new set called NT_Δ. The set of productions PP_6 of G_6 is the union of PP_4 and PP_5 to which the following sets of productions are added.

- $Cons = \{Y_4Y_5 \Rightarrow c(Z_4Z_5, \Delta_{T_4,T_5}),\ Y_4Y_5 \Rightarrow c(\Delta_{Z_4,Z_5}, T_4T_5),$ $Z_4Z_5 \Rightarrow s(Z'_4Z'_5)\}$ Intuitively $Y_4Y_5 \Rightarrow c(Z_4Z_5, \Delta_{T_4,T_5})$ means that the constructor c can be generated both from Y_4 in G_4 and Y_5 in G_5, so generating a c does not ensure that the two terms are different: we generate a c and search a difference on the first argument of c and do not care of the second argument. The second production is the symmetric of the first one. Note that $Z'_4Z'_5 \Rightarrow 0$ has not been included in this set because this production would replace Z_4Z_5 by a terminal without detecting a difference.

- *Diff* $= \{T_4T_5 \Rightarrow \delta\}$ (δ and δ_2 are new terminals)
  This production means that since $T_4 \Rightarrow a$ and $T_5 \Rightarrow b$, a difference has been detected, therefore there is no more need of a non-terminal of $NT_4 \times NT_5$.
- *Diff*$' = \{\Delta_{Z_4,Z_5} \Rightarrow \delta_2(Z_4, Z_5),\ \Delta_{T_4,T_5} \Rightarrow \delta_2(T_4, T_5)\}$
  The first production means that the terms generated from Z_4 by G_4, and from Z_5 by G_5 need not be different. So they are generated (to preserve synchronization constraints, when any) in two independent branches.

Here are two possible derivations of G_6

$$(Y_4Y_5,\ X_4,\ X_5) \Rightarrow^* (Y_4Y_5,\ 0,\ 0) \Rightarrow (c(Z_4Z_5, \Delta_{T_4,T_5}), 0, 0)$$
$$\Rightarrow (c(s(Z'_4Z'_5), \Delta_{T_4,T_5}), 0, 0) \Rightarrow^* (c(s(Z'_4Z'_5), \delta_2(a,b)), 0, 0)$$

and this derivation fails because no more productions can be applied.

$$(Y_4Y_5,\ X_4,\ X_5) \Rightarrow^* (Y_4Y_5,\ 0,\ 0) \Rightarrow (c(\Delta_{Z_4,Z_5}, T_4T_5),\ 0,\ 0)$$
$$\Rightarrow (c(\delta_2(Z_4, Z_5), T_4T_5),\ 0,\ 0) \Rightarrow^* (c(\delta_2(s(0), s(0)), T_4T_5),\ 0,\ 0)$$
$$\Rightarrow (c(\delta_2(s(0), s(0)), \delta),\ 0,\ 0)$$

which means that the pair $(0,0)$ belongs to $Rec(G_4) \overline{\bowtie}_{1,1} Rec(G_5)$ because a difference between $c(s(0), a)$ and $c(s(0), b)$ have been found on the second argument.

The actual algorithm has to take into account two additional problems. The first one concerns the control (note that to simplify the example, no control has been introduced). Indeed when computing the anti-join, controls must not be merged together. For that we need to introduce a tuple as control (in the same way as [9] for unification). The second one is that synchronized productions do not produce any constructor, therefore no production of *Diff* must not be generated using both the premise of a free production and the premise of a synchronized production.

5.3 Algorithm : formal presentation

Theorem 6. (anti-join algorithm) *Let* $G_1 = (Sz_1, \mathcal{C}_1, NT_1, PP_1, TI_1)$ *and* $G_2 = (Sz_2,\ \mathcal{C}_2,\ NT_2,\ PP_2,\ TI_2)$ *be two TTSG's deduced from a TRS. The language recognized by the TTSG* $G_3 = (Sz_3, \mathcal{C}_3,\ NT_3, PP_3, TI_3)$ *as defined below is the* i_1, i_2 *anti-join of* G_1 *and* G_2.

(I_{i_1}, ct_{i_1}) (resp. (I_{i_2}, ct_{i_2})) denotes the i_1^{th} (resp. i_2^{th}) component of the axiom TI_1 (resp. TI_2) of G_1 (resp. G_2).

- $Sz_3 = Sz_1 + Sz_2$
- $\mathcal{C}_3 = \mathcal{C}_1 \cup \mathcal{C}_2 \cup \mathcal{C}_\delta$ *with* $\mathcal{C}_\delta = \{\delta_0, \ldots, \delta_n\}$ *where* n *is the addition of the maximal arity of* $\mathcal{C}_1$ *and the one of* $\mathcal{C}_2$,
- $NT_3 = NT_1 \cup NT_2 \cup (NT_1 \times NT_2) \cup NT_\Delta$[7] *where*
  $NT_\Delta = \{\Delta_{X_1,X_2} | X_1 \in NT_1$ *and* $X_2 \in NT_2\}$
- $PP_3 = PP_1 \cup PP_2 \cup Cons \cup \mathrm{Diff} \cup \mathrm{Diff}' \cup Sync_1 \cup Sync_2$ *where*

[7] NT_Δ is introduced in order to respect the property that a free production introduces one and only one constructor.

- $Cons = \{X_1X_2 \Rightarrow c(\Delta_{Y_{1,1},Y_{2,1}}, \ldots, Y_{1,i}Y_{2,i}, \ldots, \Delta_{Y_{1,n},Y_{2,n}})$ such that c is not a constant, $i \in [1,n]$, $X_1 \Rightarrow c(Y_{1,1}, \ldots, Y_{1,n}) \in PP_1$ and $X_2 \Rightarrow c(Y_{2,1}, \ldots, Y_{2,n}) \in PP_2\}$
- $Diff = \{X_1X_2 \Rightarrow \delta_{n+n'}(Y_{1,1}, \ldots, Y_{1,n}, Y_{2,1}, \ldots, Y_{2,n'})$ such that $X_1 \Rightarrow c(Y_{1,1}, \ldots, Y_{1,n}) \in PP_1$, $X_2 \Rightarrow c'(Y_{2,1}, \ldots, Y_{2,n'}) \in PP_2$ and $c \neq c'\}$
- $Diff' = \{\Delta_{X_1,X_2} \Rightarrow \delta_2(X_1, X_2) | \Delta_{X_1,X_2} \in NT_\Delta\}$
- $Sync_1 = \{\{X_1X_2 \Rightarrow Y_1X_2, Z_1 \Rightarrow Z'_1, \ldots, Z_n \Rightarrow Z'_n\}_k$ such that $\{X_1 \Rightarrow Y_1, Z_1 \Rightarrow Z'_1, \ldots, Z_n \Rightarrow Z'_n\}_k \in PP_1$ and $X_2 \in PP_2\}$
- $Sync_2 = \{\{X_1X_2 \Rightarrow X_1Y_2, Z_1 \Rightarrow Z'_1, \ldots, Z_n \Rightarrow Z'_n\}_{Sz_1+k}$ such that $\{X_2 \Rightarrow Y_2, Z_1 \Rightarrow Z'_1, \ldots, Z_n \Rightarrow Z'_n\}_k \in PP_2$ and $X_1 \in PP_1\}$

– $TI_3 = (TI'_1[i_1 \leftarrow (I_{i_1}I_{i_2}, ct_{i_1} * ct_{i_2})]) * (TI'_2 \backslash_{i_2})$ *where* TI'_1 *(resp.* TI'_2*) is built from* TI_1 *(resp.* TI_2*) by replacing each control tuple* ct_j *by* $ct_j * (\bot_1, \ldots, \bot_{Sz_2})$ *(resp.* $(\bot_1, \ldots, \bot_{Sz_1}) * ct_j$*).*

The technical definition below is just used to prove Theorem 6.

Definition 7. Let tp and tp' be two computations of (eventually different) TT-SG's. Let $t = tp|_i$ and $t' = tp'|_{i'}$. We write $t \approx_l t'$ (resp $t \approx_r t'$) if

1. $\forall u \in O(t)$, $t(u)$ is a constructor symbol implies $t(u) = t'(u)$,
2. and $\forall u \in O(t)$, $t(u) = (X, ct)$ implies $t'(u) = (X', ct')$ and $ct' = ct * (\bot, \ldots, \bot)$ (resp $ct' = (\bot, \ldots, \bot) * ct$).

On the other hand, we write $t =_{cons} t'$ if 1. is verified and $\forall u \in O(t)$, $t(u) = (X, ct)$ implies $t'(u) = (X', ct')$ and $t \neq_{cons} t'$ if $\exists u \in O(t) \cap O(t')$ such that $t(u)$ and $t'(u)$ are constructor symbols and $t(u) \neq t'(u)$.

The set of non-terminal leaves of a computation tp is denoted by $NT(tp)$ and is defined by $NT(tp) = \{(X, ct) | \exists u \in O(tp)$ such that $tp(u) = (X, ct)\}$

Proof: (of Theorem 6)
First, we consider that n_1 is the number of components of the tuples of G_1 and n_2 is the number of components of G_2. In other words, if $tp_1 \in Rec(G_1)$, $tp1 = (t_1, \ldots, t_{n_1})$ and if $tp_2 \in Rec(G_2)$, $tp_2 = (t'_1, \ldots, t'_{n_2})$.
We want to prove that $Rec(G_3)\backslash_{i_1} = Rec(G_1) \bowtie_{i_1,i_2} Rec(G_2)$.
$\triangleright$ First we prove that $Rec(G_3)\backslash_{i_1} \subseteq Rec(G_1) \bowtie_{i_1,i_2} Rec(G_2)$. For that we demonstrate that $\forall tp_3 \in Rec(G_3)$, $\exists tp_1 \in Rec(G_1)$ and $tp_2 \in Rec(G_2)$ such that $tp_3\backslash_{i_1} = tp_1 \backslash_{i_1} * tp_2\backslash_{i_2}$ and $tp_1|_{i_1} \neq tp_2|_{i_2}$.

Therefore, it suffices to prove that $\forall tp_3 \in Comp(G_3)$ that does not contain any non-terminal of NT_Δ, $\exists tp_1 \in Comp(G_1)$ and $tp_2 \in Comp(G_2)$ such that

(a) $\forall j \in [1, n_1]$ and $j \neq i_1$ we have $tp_3|_j \approx_l tp_1|_j$

(b) $\forall j \in [n_1+1, n_1+i_2[$ we have $tp_3|_j \approx_r tp_1|_{j-n_1}$
$\forall j \in]n_1+i_2, n_1+n_2]$ we have $tp_3|_j \approx_r tp_2|_{j+1-n_1}$

(c) if $\exists u \in O(tp_3|_{i_1})$ s.t. $tp_3(i_1.u) = (X_1X_2, ct_1 * ct_2)$ with $X_1X_2 \in NT_1 \times NT_2$ then
$tp_1(i_1.u) = (X_1, ct_1)$ and $tp_2(i_2.u) = (X_2, ct_2)$ and
$(NT(tp_3|_{i_1}) \setminus \{(X_1X_2, ct_1 * ct_2)\}) \cup \{(X_1, ct_1), (X_2, ct_2)\}) = NT(tp_1|_{i_1}) \cup NT(tp_2|_{i_2})$

(d) if $\nexists u \in O(tp_3|_{i_1})$ s.t. $tp_3(i_1.u) = (X_1X_2, ct_1 * ct_2)$ with $X_1X_2 \in NT_1 \times NT_2$ then
$tp_1|_{i_1} \neq tp_2|_{i_2}$ and $NT(tp_3|_{i_1}) = NT(tp_1|_{i_1}) \cup NT(tp_2|_{i_2})$

(a) and (b) ensure that $tp_3\backslash_{i_1} = tp_1 \backslash_{i_1} * tp_2\backslash_{i_2}$ and (d) ensures that if tp_3 does not contains any non-terminals of $NT_1 \times NT_2$ then $tp_1|_{i_1}$ is different from $tp_2|_{i_2}$. (c) and (d) ensure that any production that can be applied on tp_1 or on tp_2 can be applied on tp_3 too because tp_3 contains the same non-terminals (but at most one) with the same control.

We prove it by induction on the length n of the derivation $TI_3 \Rightarrow^* tp_3$

If $n = 0$ then $tp_3 = TI_3$ it is easy to see that TI_1, TI_2 and TI_3 verify (a), (b), (c) and (d).

For $n > 0$, we suppose for all $tp_3 \in Comp(G_3)$ such that tp_3 does not contain any non-terminals of NT_Δ and $TI_3 \Rightarrow^* tp_3$ with a derivation of length less than n that there exists $tp_1 \in Comp(G_1)$ and $tp_2 \in Comp(G_2)$ that verify (a), (b), (c) and (d). Suppose that next pack of productions applied on tp_3 is pp and $tp_3 \Rightarrow tp'_3$.

If pp does not affect a non-terminal of $NT_1 \times NT_2$ or if it is a synchronized pack then we can easily find tp'_1 and tp'_2 that verify (a), (b), (c) and (d) by applying the corresponding pack of PP_1 or PP_2.

If $\exists u \in O(tp_3|_{i_1})$ such that $tp_3(i_1.u) = (X_1X_2, ct_1 \times ct_2)$ with $X_1X_2 \in NT_1 \times NT_2$ and pp is a free production that applies on u then

- if $pp \in Cons$ then $pp = X_1X_2 \Rightarrow c(\Delta_{Y_{1,1},Y_{2,1}}, \ldots, Y_{1,i}Y_{2,i}, \ldots, \Delta_{Y_{1,n},Y_{2,n}})$. We know that there exist $X_1 \Rightarrow c(Y_{1,1}, \ldots, Y_{1,n}) \in PP_1$ and $X_2 \Rightarrow c(Y_{2,1}, \ldots, Y_{2,n}) \in PP_2$. Moreover from (c) we know that $tp_1(i_1.u) = (X_1, ct_1)$ and $tp_2(i_2.u) = (X_2, ct_2)$ therefore the two productions can be applied respectively on tp_1 and tp_2 at occurrence u giving tp'_1 and tp'_2. On the other hand all the non-terminals of NT_Δ can be derived by their corresponding production in $Diff\,'$ so $tp'_3 \Rightarrow^* tp''_3$ and tp''_3 does not contain any non-terminals of NT_Δ. As no other field of tp_3 but i_1 has been modified in tp''_3, it is easy to see that tp'_1 verifies (a) and tp'_2 verifies (b). Moreover $tp''_3(i_1.u.i) = (Y_{1,i}Y_{2,i}, ct_1 * ct_2)$ and $tp'_1(i_1.u.i) = (Y_{1,i}, ct_1)$ and $tp'_2(i_2.u.i) = (Y_{2,i}, ct_2)$. Endly the new non-terminals brought in tp'_1 and tp'_2 are the same as those brought in tp''_3 but $Y_{1,i}Y_{2,i}$ so we can verify the last part of property (c).
- if $pp \in Diff$ then $pp = X_1X_2 \Rightarrow \delta_{n+n'}(Y_{1,1}, \ldots, Y_{1,n}, Y_{2,1}, \ldots, Y_{2,n'})$ and we know that there exist $X_1 \Rightarrow c(Y_{1,1}, \ldots, Y_{1,n}) \in PP_1$, $X_2 \Rightarrow c'(Y_{2,1}, \ldots, Y_{2,n'}) \in PP_2$ and $c \neq c'$. So from property (c) we have $tp_1(i_1.u) = (X_1, ct_1)$ and $tp_2(i_2.u) = (X_2, ct_2)$ therefore two productions can be applied respectively on tp_1 and tp_2 at occurrence u giving tp'_1 and tp'_2. tp'_3 does not contain any more a non-terminal of NT_Δ, moreover the new non-terminals introduced in tp'_3 are the union of those introduced in tp'_1 and tp'_2 so we can verify (d). Note that (a) and (b) are easily verified because no fields but i_1 of tp_3 have been modified.

From (d), we can deduce that if $tp_3 \in Rec(G_3)$ then $tp_1|_{i_1} \neq tp_2|_{i_2}$ because if $tp_3 \in Rec(G_3)$, $\nexists u \in O(tp_3|_{i_1})$ s.t. $tp_3(i_1.u) = (X_1X_2, ct_1 * ct_2)$ with $X_1X_2 \in NT_1 \times NT_2$, therefore $tp_1|_{i_1} \neq tp_2|_{i_2}$.

▷ Now we want to prove that $\forall tp_1 \in Rec(G_1)$ and $tp_2 \in Rec(G_2)$ such that $tp_1|_{i_1} \neq tp_2|_{i_2}$ there exists $tp_3 \in Rec(G_3)$ such that $tp_3\backslash_{i_1} = tp_1 \backslash_{i_1} * tp_2\backslash_{i_2}$ (In other words we want to prove that $Rec(G_1) \bowtie_{i_1,i_2} Rec(G_2) \subseteq Rec(G_3)\backslash_{i_1})$.

For that we prove that $\forall tp_1 \in Comp(G_1) \setminus Rec(G_1)$ and $tp_2 \in Comp(G_2) \setminus Rec(G_2)$ there exists $tp_3 \in Comp(G_3)$ such that (a) (b) and

(e) if $tp_1|_{i_1} =_{cons} tp_2|_{i_2}$ then in one hand, $\exists u \in O(tp_3|_{i_1})$ such that $tp_3(i_1.u) = (X_1X_2, ct_1 * ct_2)$ when $tp_1(i_1.u) = (X_1, ct_1)$ and $tp_2(i_2.u) = (X_2, ct_2)$.
In the other hand $(NT(tp_3|_{i_1}) \backslash \{(X_1X_2, ct_1 * ct_2)\}) \cup \{(X_1, ct_1), (X_2, ct_2)\}) = NT(tp_1|_{i_1}) \cup NT(tp_2|_{i_2})$.

(f) If $tp_1|_{i_1} \neq_{cons} tp_2|_{i_2}$ then $tp_3|_{i_1}$ does not contain any non-terminal from $NT_1 \times NT_2$ and $NT(tp_3|_{i_1}) = NT(tp_1|_{i_1}) \cup NT(tp_2|_{i_2})$.

We prove it on induction on the sum n of the length of $TI_1 \Rightarrow^* tp_1$ and $TI_2 \Rightarrow^* tp_2$.

- for $n = 0$, TI_1, TI_2 and TI_3 verify (a), (b), (e) and (f).
- for $n > 0$, we suppose that $TI_1 \Rightarrow^* tp_1$ and $TI_2 \Rightarrow^* tp_2$ and there exists $tp_3 \in Comp(G_3)$ that verifies (a),(b), (e) and (f). Suppose that a new pack of productions pp is applied on tp_1 (or on tp_2).
If pp is a synchronized pack of productions or if $tp_1 \neq_{cons} tp_2$ then we can find tp'_3 such that (a), (b), (e) and (f) are verified by applying the corresponding pack of PP_3 on tp_3.
If $tp_1|_{i_1} =_{cons} tp_2|_{i_2}$ and $tp_3|_{i_1}(u) = (X_1X_2, (ct_1 * ct_2))$ and pp_1 is a free production $X_1 \Rightarrow c(Y_{1,n}, \ldots, Y_{1,n})$ of PP_1 applied at occurrence u, (we obtain tp'_1) then three cases are possible:
  1. X_2 is the premise of a synchronized production, then we have to apply this production[8] on tp_2 and the corresponding one of $Sync_2$ on tp_3 and restart the reasoning.
  2. X_2 is not the premise of a synchronized production and $X_2 \Rightarrow c(Y_{2,n}, \ldots, Y_{2,n})$. In this case there exists $X_1X_2 \Rightarrow c(\Delta_{Y_{1,1},Y_{2,1}}, \ldots, Y_{1,i}Y_{2,i}, \ldots, \Delta_{Y_{1,n},Y_{2,n}}) \in Cons$. Then if we apply these two productions respectively on tp_2 and tp_3 we get tp'_2 and tp'_3. $tp'_3 \Rightarrow^* tp''_3$ with the production of $Diff$ so tp''_3 does not contain any non-terminals of NT_Δ. We can verify that tp'_1, tp'_2 and tp''_3 verify (a), (b) and (e) because $tp''_3(i_1.u.i) = (Y_{1,i}Y_{2,i}, ct_1 * ct_2)$ and $tp'_1(i_1.u.i) = (Y_{1,i}, ct_1)$ and $tp'_2(i_2.u.i) = (Y_{2,i}, ct_2)$ in one hand and in the other hand tp''_3 contains $Y_{1,1}, \ldots, Y_{1,n}, Y_{2,1}, \ldots, Y_{2,n}$.
  3. X_2 is not the premise of a synchronized production and $X_2 \Rightarrow c'(Y_{2,n}, \ldots, Y_{2,n'})$ with $c \neq c'$. In this case there exists $X_1X_2 \Rightarrow \delta_{n+n'}(Y_{1,1}, \ldots, Y_{1,n}, Y_{2,1}, \ldots, Y_{2,n}) \in Diff$ that can be applied on tp_3 giving tp'_3. Applying $X_2 \Rightarrow c'(Y_{2,n}, \ldots, Y_{2,n'})$ on tp_2 gives tp'_2. We can verify then that tp'_1, tp'_2 and tp'_3 verifies (a) (b) and (f) because $tp'_1|_{i_1} \neq_{cons} tp'_2|_{i_2}$ and tp'_3 does not contains any non-terminal of $NT_1 \times NT_2$. Moreover tp'_3 contains $Y_{1,1}, \ldots, Y_{1,n}, Y_{2,1}, \ldots, Y_{2,n}$.

So we have proved that $\forall tp_1 \in Comp(G_1) \setminus Rec(G_1)$ and $tp_2 \in Comp(G_2) \setminus Rec(G_2)$ there exists $tp_3 \in Comp(G_3)$ such that (a), (b), (e) and (f). This means that if $tp_1 \in Rec(G_1)$ and $tp_2 \in Rec(G_2)$ and $tp_1|_{i_1} = tp_2|_{i_2}$ then there is no

[8] It is always possible because G_1 and G_2 are deduced from a TRS.

$tp_3 \in Rec(G_3)$ that corresponds to these tuples because (e) is verified and there is no production in NT_3 such that $X_1X_2 \Rightarrow c$ where c is a constant and X_1X_2 belongs to $NT_1 \times NT_2$. On the other hand if $tp_1 \in Rec(G_1)$ and $tp_2 \in Rec(G_2)$ and $tp_1|_{i_1} \neq tp_2|_{i_2}$ then there is a corresponding $tp_3 \in Rec(G_3)$ because all the non-terminals of $NT_1 \times NT_2$ are eliminated. □

Example 3. For the TTSG's Gr^l_ϵ and Gr^l_1 computed in Section 4 for Example 1, we obtain the TTSG Gr_{dis} whose productions are those of Gr^l_ϵ and Gr^l_1 to which the following ones are added[9] (_ replaces any non-terminal).

- $Sync_1 = \{\{G^l_\epsilon\_ \Rightarrow R^1_\epsilon\_, A^l_1 \Rightarrow L^1_1\}_1, \{G^l_\epsilon\_ \Rightarrow R^2_\epsilon\_, A^l_1 \Rightarrow L^2_1\}_1,$
  $\{R^1_1\_ \Rightarrow R^1_\epsilon\_, X^1 \Rightarrow L^1_1\}_1, \{R^1_1\_ \Rightarrow R^2_\epsilon\_, X^1 \Rightarrow L^2_1\}_1\}$
- $Sync_2 = \{\{\_G^l_1 \Rightarrow \_R^3_\epsilon, X^l \Rightarrow L^3_1\}_2, \{\_G^l_1 \Rightarrow \_R^4_\epsilon, X^l \Rightarrow L^4_1\}_2,$
  $\{\_R^3_{1.1} \Rightarrow \_R^3_\epsilon, X^3 \Rightarrow L^3_1\}_2, \{\_R^3_{1.1} \Rightarrow \_R^4_\epsilon, X^3 \Rightarrow L^4_1\}_2\}$
- $Cons = \{R^1_\epsilon R^3_\epsilon \Rightarrow s(R^1_1 R^3_1), R^1_\epsilon R^3_1 \Rightarrow s(R^1_1 R^3_{1.1})\}$
- $Diff = \{R^1_\epsilon R^4_\epsilon \Rightarrow \delta_1(R^1_1), R^2_\epsilon R^3_\epsilon \Rightarrow \delta_1(R^3_1), R^2_\epsilon R^3_1 \Rightarrow \delta_1(R^3_{1.1}), \}$

The axiom is $((G^l_\epsilon G^l_1, (0,0)), (A^l_1, (0,\bot)), (X^l, (\bot, 0)))$.

One possible derivation is

$$((G^l_\epsilon G^l_1, (0,0)), (A^l_1, (0,\bot)), (X^l, (\bot,0)))$$
$$\Rightarrow ((R^1_\epsilon G^l_1, (1,0)), (L^1_1, (1,\bot)), (X^l, (\bot,0)))$$
$$\Rightarrow ((R^1_\epsilon R^4_\epsilon, (1,1)), (L^1_1, (1,\bot)), (L^4_1, (\bot,1)))$$
$$\Rightarrow ((R^1_\epsilon R^4_\epsilon, (1,1)), (L^1_1, (1,\bot)), 0) \Rightarrow (\delta_1(R^1_1, (1,1)), (L^1_1, (1,\bot)), 0)$$
$$\Rightarrow (\delta_1(R^1_1, (1,1)), (s(X^1, (1,\bot))), 0) \Rightarrow (\delta_1(R^2_1, (2,1)), (s(L^2_1, (2,\bot))), 0)$$
$$\Rightarrow (\delta_1(0), (s(L^2_1, (2,\bot))), 0) \Rightarrow (\delta_1(0), s(0), 0)$$

This means that $\{x \mapsto 0, y \mapsto s(0)\}$ is a solution of the disequation $f(y) \overset{?}{\neq} g(x)$. In this derivation we can remark that it is useful to preserve the non-terminal R^1_1 in the production $R^1_\epsilon R^4_\epsilon \Rightarrow \delta(R^1_1)$ of *Diff* to be able to compute the second field of the tuple.

5.4 Decidability result

It is easy to prove that the TTSG obtained by anti-join is of the same kind as those computed when solving a unification problem. So by using the results of [9] we get :

Lemma 8. *Emptiness of languages recognized by TTSG's built from disunification problems is decidable.*

Thus we get the decidability result :

Theorem 9. *The satisfiability of linear disequations in theories given as confluent, constructor based, linear, without σ_{in}, without nested functions in rhs's, rewrite systems is decidable. Moreover the set of solutions can be expressed by a tree tuple synchronized grammar.*

---

[9] Only the useful productions are given, i.e. those that can be applied in at least one derivation of Gr_{dis}.

$$
\begin{array}{ll}
r_1 : f(c(x,x'),c(y,y')) & \rightarrow c(f(x,y'),f(x',y)) \\
r_2 : f(0,0) & \rightarrow 0 \\
r_3 : g(c(x,y)) & \rightarrow c(g(x),g(y)) \\
r_4 : g(0) & \rightarrow 0
\end{array}
$$

Example 4. This system provides an idea of the expressiveness of TTSG's because $f(x,y)$ is defined only if the instance of x is the symmetric of that of y, and returns the instance of x. g is the identity function on binary trees. Then solving $f(x,y) \stackrel{?}{\neq} g(z)$ amounts to look for z such that z is not equal to x and so z is not the symmetric tree of y.

6 Conclusion

We have presented an original approach using Tree Tuple Synchronized Grammars (TTSG) to decide the existence of solutions for linear disequations modulo theories given as rewrite systems satisfying some restrictions. Consequently checking whether a linear equality is an inductive theorem is decidable, if assuming moreover that the theory is sufficiently complete. Moreover, our method gives a way to represent finitely the set of solutions.

References

1. A. Antoy, R. Echahed, and M. Hanus. A Needed Narrowing Strategy. In *Proceedings 21st ACM Symposium on Principle of Programming Languages, Portland*, pages 268–279, 1994.
2. H. Comon. Disunification: a Survey. In J.-L. Lassez and G. Plotkin, editors, *Computational Logic: Essays in Honor of Alan Robinson.* MIT Press, 1991.
3. H. Comon. Complete Axiomatizations of some Quotient Term Algebras. *Theoretical Computer Science*, 118(2), 1993.
4. H. Comon, M. Haberstrau, and J.-P. Jouannaud. Syntacticness, Cycle-Syntacticness and Shallow Theories. *Information and Computation*, 111(1):154–191, 1994.
5. N. Dershowitz and J.-P. Jouannaud. Rewrite Systems. In J. Van Leuven, editor, *Handbook of Theoretical Computer Science.* Elsevier Science Publishers, 1990.
6. M. Fernández. Narrowing Based Procedures for Equational Disunification. *Applicable Algebra in Engineering Communication and Computing*, 3:1–26, 1992.
7. E. Giovannetti, G. Levi, C. Moiso, and C. Palamidessi. Kernel LEAF: a Logic plus Functional Language. *The Journal of Computer and System Sciences*, 42(3):139–185, 1991.
8. J.-M. Hullot. Canonical Forms and Unification. In W. Bibel and R. Kowalski, editors, *Proceedings 5th International Conference on Automated Deduction, Les Arcs (France)*, volume 87 of *LNCS*, pages 318–334. Springer-Verlag, 1980.
9. S. Limet and P. Réty. E-Unification by Means of Tree Tuple Synchronized Grammars. *Discrete Mathematics and Theoritical Computer Science (http://www.chapmanhall.com/dm)*, 1:69–98, 1997.
10. J.J. Moreno-Navarro and M. Rodriguez-Artalejo. Logic Programming with Functions and Predicates : The Language BABEL. *Journal of Logic Programming*, 12(3):191–223, 1992.

About Proofs by Consistency
(Abstract)

Hubert Comon
Laboratoire Spécification et Vérification, URA 2236 du CNRS
École Normale Supérieure de Cachan
61 avenue du président Wilson
94235 Cachan cedex, France
comon@lsv.ens-cachan.fr

The *proof by consistency* method has been introduced in 1980 by several authors including D. Musser [19], J. Goguen [7], D. Lankford [18] and G. Huet and J.-M. Hullot [8]. The name, which is more appropriate than "inductionless induction", as we will see, was introduced by D. Kapur and D. Musser [12]. It was a success of term rewriting in the area of inductive proofs: completion techniques were able to prove fully automatically some inductive theorems which seemed at that time impossible to prove without any user interaction (lemmas). The most famous example, given in [8], concerns the equation $rev(rev(x)) = x$ which usually requires a lemma stating that *append* is associative. There are several works explaining or extending this approach; let us cite for instance [12, 11, 2].

In this talk, we want to contribute here to this field, trying to give again an explanation of the method and suggest some extensions. In particular, we emphasize the role of some axiomatizations of the initial model(s), which was already pointed out in [5] and give some hints on how to compute them. For example, we would like to use associative and commutative symbols in the specification, without having to check for AC-ground reducibility (which is undecidable). More generally, we would like to relax the condition of the initial set of axioms to form a ground convergent rewrite system. We give some examples of inductive completion techniques (following [6, 16, 1]) which do not require terminationn nor ground confluence of the initial set of axioms.

The proof by consistency approach is very elegant, refutationally complete and requires less user interaction than the classical explicit induction technique. So, one may ask the question: why has it been very popular in the 80s and almost abandoned since ? The main problem is that the proof by consistency approach does not work in practice. It works on toy examples but in most cases, the completion runs forever without any answer. What can we learn from this failure ? Actually, many ideas borrowed from the work on proofs by consistency can be reused in an other way. Let us cite for instance [15, 20, 9]. The work on proofs by consistency also brought to light the key notion of *ground reducibility* [10] which was studied by many authors among which [17, 14, 13, 3, 4]. We will also see how this notion can be generalized and possibly yield new approaches.

References

1. L. Bachmair. Proof by consistency in equational theories. In *Proc. 3rd IEEE Symp. Logic in Computer Science, Edinburgh*, July 1988.
2. L. Bachmair. *Canonical Equational Proofs*. Birkhäuser, Boston, 1991.
3. A.-C. Caron, J.-L. Coquidé, and M. Dauchet. Encompassment properties and automata with constraints. In C. Kirchner, editor, *5th International Conference on Rewriting Techniques and Applications*, volume 690 of *Lecture Notes in Computer Science*, Montreal, Canada, June 1993. Springer-Verlag.
4. H. Comon and F. Jacquemard. Ground reducibility is exptime-complete. In *Proc. IEEE Symp. on Logic in Computer Science*, Varsaw, June 1997. IEEE Comp. Soc. Press.
5. L. Fribourg. A narrowing procedure for theories with constructors. In R. Shostak, editor, *Proc. 7th Int. Conf. on Automated Deduction*, volume 170 of *Lecture Notes in Computer Science*, pages 259–281, Napa, CA., 1984. Springer-Verlag.
6. L. Fribourg. A strong restriction of the inductive completion procedure. *Journal of Symbolic Computation*, 8:253–276, 1989.
7. J. A. Goguen. How to prove inductive hypothesis without induction. In *Proc. 5th Conf. on Automated Deduction, Les Arcs, France, LNCS 87*, July 1980.
8. G. Huet and J.-M. Hullot. Proofs by induction in equational theories with constructors. *J. Comput. Syst. Sci.*, 25(2), 1982.
9. J.-P. Jouannaud and A. Bouhoula. Automata-driven automated induction. In *Twelfth Annual IEEE Symposium on Logic in Computer Science*, Warsaw,Poland, June 1997. IEEE Comp. Soc. Press.
10. J.-P. Jouannaud and E. Kounalis. Automatic proofs by induction in equational theories without constructors. In *Proc. 1st IEEE Symp. Logic in Computer Science, Cambridge, Mass.*, June 1986.
11. J.-P. Jouannaud and E. Kounalis. Automatic proofs by induction in theories without constructors. *Information and Computation*, 82(1), July 1989.
12. D. Kapur and D. Musser. Proof by consistency. *Artificial Intelligence*, 31(2), Feb. 1987.
13. D. Kapur, P. Narendran, D. Rosenkrantz, and H. Zhang. Sufficient completeness, ground reducibility and their complexity. *Acta Inf.*, 28:311–350, 1991.
14. D. Kapur, P. Narendran, and H. Zhang. On sufficient completeness and related properties of term rewriting systems. *Acta Inf.*, 24(4):395–415, 1987.
15. D. Kapur and H. Zhang. An overview of the rewrite rule laboratory. *Journal of Mathematics of Computation*, 1995.
16. W. Küchlin. Inductive completion by ground proof transformation. In H. Ait-Kaci and M. Nivat, editors, *Proc. Coll. on the Resolution of Equations in Algebraic Structures, Lakeway*. Academic Press, May 1987.
17. E. Kounalis. Testing for the ground (co)-reducibility in term rewriting systems. *Theoretical Computer Science*, 106(1):87–117, 1992.
18. D. S. Lankford. A simple explanation of inductionless induction. Technical Report MTP-14, Mathematics Department, Louisiana Tech. Univ., 1981.
19. D. Musser. Proving inductive properties of abstract data types. In *Proc. 7th ACM Symp. on Principles of Programming Languages, Las Vegas*, 1980.
20. U. Reddy. Term rewriting induction. In *Proc. 10th Int. Conf. on Automated Deduction, Kaiserslautern, LNCS 449*, 1990.

Normalization of S-Terms is Decidable

Johannes Waldmann

Institut für Informatik, Friedrich-Schiller-Universität, D-07740 Jena, Germany
joe@informatik.uni-jena.de, http://www5.informatik.uni-jena.de/~joe/

Abstract. The combinator **S** has the reduction rule **S** $x\ y\ z \to x\ z\ (y\ z)$. We investigate properties of ground terms built from **S** alone. We show that it is decidable whether such an **S**-term has a normal form. The decision procedure makes use of rational tree languages. We also exemplify and summarize other properties of **S**-terms and hint at open questions.

1 Introduction

It is well known that the combinators **S** and **K** with their reduction rules **S** $x\ y\ z \to x\ z\ (y\ z)$ and **K** $x\ y \to x$ form a complete basis for combinatory logic. Therefore, most of the interesting properties of an (**S**, **K**)-term are necessarily undecidable.

Now **CL**(**K**) is strongly normalizing. One is lead to assume that the difficulty of **CL** (**S**, **K**) comes from **S**. That's the reason for studying the system **CL**(**S**). Somewhat surprisingly it turns out that normalization *is* decidable for **S**-terms. (Since **CL**(**S**) is orthogonal and non-erasing, an **S**-term is normalizing (has a normal form) iff it is terminating (has no infinite reduction)).

We will start by giving two examples of infinite reductions in **CL**(**S**).

Example 1. Let $\mathsf{T} = \mathbf{S}\ \mathbf{S}; X_0 = \mathbf{S}\ \mathsf{T}\ \mathsf{T}$. The expression $X_0\ X_0$ has an infinite reduction. (It had already been described by E. Zachos [Zac78].)

Define $X_{n+1} = \mathsf{T}\ X_n$. All redexes that occur during the reduction of $X_0\ X_0$ have the form $X_n\ X_m$ for some n and m. There are two types of reduction steps: one goes *downward*: $X_{n+1}\ X_m = \mathbf{S}\ \mathbf{S}\ X_n\ X_m \to \mathbf{S}\ X_m\ \underline{(X_n\ X_m)}$ and the other goes *upward*: $X_0\ X_m = \mathbf{S}\ \mathsf{T}\ \mathsf{T}\ X_m \to \mathsf{T}\ X_m\ (\mathsf{T}\ X_m) = \overline{X_{m+1}\ X_{m+1}}$.

It is interesting to note that the reduction graph of $X_0\ X_0$ is a *pure line* (each expression contains exactly one redex).

Example 2. The expression $\mathsf{A}\ \mathsf{A}\ \mathsf{A}$ where $\mathsf{A} = \mathbf{S}\ \mathbf{S}\ \mathbf{S}$ has an infinite reduction. (It had been found by H. Barendregt [Bar84].) Again the reduction can be described by a pattern: $Y_0 = \mathbf{S}\ \mathsf{A}\ (\mathbf{S}\ \mathsf{A}\ \mathsf{A})\,; Y_{n+1} = \mathsf{A}\ Y_n$.

First we show that the pattern is reached: $\mathsf{A}\ \mathsf{A}\ \mathsf{A} \to \mathbf{S}\ \mathsf{A}\ (\mathbf{S}\ \mathsf{A})\ \mathsf{A} \to \mathsf{A}\ \mathsf{A}\ (\mathbf{S}\ \mathsf{A}\ \mathsf{A}) \to \mathbf{S}\ \mathsf{A}\ (\mathbf{S}\ \mathsf{A})\ (\mathbf{S}\ \mathsf{A}\ \mathsf{A}) \to \mathsf{A}\ (\mathbf{S}\ \mathsf{A}\ \mathsf{A})\ Y_0 \to \mathbf{S}\ (\mathbf{S}\ \mathsf{A}\ \mathsf{A})\ (\mathbf{S}\ (\mathbf{S}\ \mathsf{A}\ \mathsf{A}))\ Y_0 \to \mathbf{S}\ \mathsf{A}\ \mathsf{A}\ Y_0\ * \to Y_1\ Y_1\ *$. (Here, $*$ denotes a subexpression whose value is irrelevant because no later reduction touches it.)

Then we show how the pattern reduces. There is a *downward* move that in general goes like $Y_{k+1}\ Z = \mathsf{A}\ Y_k\ Z \to \mathbf{S}\ Y_k\ (\mathbf{S}\ Y_k)\ Z \to Y_k\ Z\ *$ and an *upward* move $Y_0\ Z = \mathbf{S}\ \mathsf{A}\ (\mathbf{S}\ \mathsf{A}\ \mathsf{A})\ Z \to \mathsf{A}\ Z\ (\mathbf{S}\ \mathsf{A}\ \mathsf{A}\ Z) \to \mathbf{S}\ Z\ (\mathbf{S}\ Z)\ (\mathbf{S}\ \mathsf{A}\ \mathsf{A}\ Z) \to Z\ (\mathbf{S}\ \mathsf{A}\ \mathsf{A}\ Z)\ *$.
The placeholder Z is instantiated with Y_k to give $Y_0\ Y_k \twoheadrightarrow Y_k\ (\mathbf{S}\ \mathsf{A}\ \mathsf{A}\ Y_k)\ *$ and now $Z = \mathbf{S}\ \mathsf{A}\ \mathsf{A}\ Y_k$ gives $Y_k\ (\mathbf{S}\ \mathsf{A}\ \mathsf{A}\ Y_k) \twoheadrightarrow Y_0\ (\mathbf{S}\ \mathsf{A}\ \mathsf{A}\ Y_k) \ldots \twoheadrightarrow \mathbf{S}\ \mathsf{A}\ \mathsf{A}\ Y_k\ * \ldots \to Y_{k+1}\ Y_{k+1}\ \ldots$ This is an infinite head reduction.

Note that in both examples, the reduction sequence could be called *doubly periodic*. It is quite easy to see [BK79] that **CL(S)** has no *cycles*, i. e. reductions $X \to^+ X$. See [Klo80] for an investigation of cycles in Combinatory Logic (with various base sets). It has been conjectured that **CL(S)** even does not admit *loops*, i. e. reductions $X \to^+ C[X\sigma]$, where $C[]$ is a context and σ is a substitution. The absence of *ground* loops (where σ is the identity) has been shown recently [Wal97]. For loops in term rewriting in general, see [ZG96].
The rest of the paper is organized as follows. First we fix some notation, and recall basic facts on **CL(S)**. We will introduce *directors*, in analogy to [KS88]. Then we show examples of infinite reductions in **CL(S)**. They cover all possible cases that have to be considered in the decision procedure for normaliziation, which is given next. Finally, we mention open questions and related work.

2 Notations and Preliminaries

We consider the term rewriting system **CL(S)**. Its signature consists of one nullary symbol (**S**) and one binary symbol $\circ$ (application). **CL(S)** has only one rule $(((\mathbf{S} \circ x) \circ y) \circ z) \to ((x \circ z) \circ (y \circ z))$.
We follow the usual conventions about suppressing the application symbol and omitting parentheses. So the rule just given is written as $\mathbf{S}\ x\ y\ z \to x\ z\ (y\ z)$.
The system is left-linear and non-overlapping, thus confluent. It is moreover non-erasing and therefore weak and strong normalization coincide (if one reduction leads to normal form, then all do).

Symbols appear in different fonts. We use bold face, upper case, for combinators $(\mathbf{S}, \mathbf{K}, \ldots)$, italic face, lower case, for variables in the system $(x, y, \ldots)$, italic face, upper case, for meta-variables (that denote terms or sets of terms) $(X, Y, \ldots)$, calligraphic face, upper case, for fixed sets of terms $(\mathcal{P}, \mathcal{Q}, \ldots)$, sans serif face, upper case, for certain fixed terms (T, A).
Unless stated otherwise explicitly, all terms are ground.

$X \trianglelefteq Y$ $(X \triangleleft Y)$ means X is a (strict) subterm of Y.
$X \to Y$ means X reduces to Y in one step.
$X \hookrightarrow Y$ means when reducing X, Y occurs as a subterm: $\exists\ Z : X \to Z$ and $Z \trianglerighteq Y$.
$\twoheadrightarrow$ denotes the reflexive and transitive closure of $\to$, likewise for $\hookrightarrow\!\!\!\!\rightarrow$.

We write $\downarrow(X)$ if X normalizes, and $X \twoheadrightarrow \infty$ if X does not. Remember that weak and strong normalization coincide in **CL(S)**.

We often write a term when meaning a set containing exactly this term.

Operations are understood to be extended from terms to sets of terms. If E and F are sets of terms, then EF denotes the set of terms $X\ Y$ with $X \in E$ and $Y \in F$.

If E and F are sets, $E \to F$ denotes $\forall X \in E : \exists Y \in F : X \to Y$, while $E \not\to F$ stands for $\forall X \in E : \neg\exists Y \in F : X \to Y$.

Also, $\downarrow(E)$ denotes $\forall X \in E : \downarrow(X)$, and $E \twoheadrightarrow \infty$ denotes $\forall X \in E : X \twoheadrightarrow \infty$.

Now we will define some sets of ground **S**-terms that will be used frequently.

Definition 3. $\mathsf{T} = \mathbf{S}\ \mathbf{S}$, $\mathsf{A} = \mathsf{T}\ \mathbf{S} = \mathbf{S}\ \mathbf{S}\ \mathbf{S}$.

Definition 4. $\mathcal{M}$ denotes the set of *all* terms. $\mathcal{N}$ denotes the set of all terms in *normal* form.

(These sets are rational: we have $\mathcal{N} = \mathbf{S} \cup \mathbf{S}\,\mathcal{N} \cup \mathbf{S}\,\mathcal{N}\,\mathcal{N}$ and $\mathcal{M} = \mathbf{S} \cup \mathcal{M}\,\mathcal{M}$.) Inside an expression, we often just write $*$ for $\mathcal{M}$.

Definition 5. $\mathcal{M}_0 = \mathcal{M}; \mathcal{M}_{k+1} = \mathcal{M}_k\ \mathcal{M}$.

The set $\mathcal{M}_k$ consists of all terms whose left spine has length $\geq k$. For example, $\mathcal{M}_0$ are all terms, and $\mathcal{M}_1$ are all terms except **S**.

Definition 6. $\mathsf{S}_1 = \mathbf{S}$, for $k \geq 1 : \mathsf{S}_{k+1} = \mathbf{S}\,\mathsf{S}_k$.
For $k \geq 0 : \mathcal{P}_k = \mathsf{S}_1 \cup \ldots \cup \mathsf{S}_k$. For $k \geq 0 : \mathcal{Q}_k = \mathcal{M} \setminus \mathcal{P}_k$.

The term S_k has a right spine of length $k-1$ and all its left children are **S**. The set $\mathcal{P}_k$ is the collection of all S_i up to k (while $\mathcal{P}_0 = \emptyset$). The complement of $\mathcal{P}_k$ is denoted by $\mathcal{Q}_k$. These are exactly the terms whose right spine has a length $\geq k$ or who have some subterm with left spine longer than 1. For example, $\mathcal{Q}_0$ are all terms, $\mathcal{Q}_2$ are all terms except **S** and T, and $\mathsf{A} = \mathbf{S}\ \mathbf{S}\ \mathbf{S} \in \mathcal{Q}_k$ for any k.
From the above description we immediately derive

Lemma 7. For all $k \geq 0$ we have $\mathcal{P}_k \subseteq \mathcal{P}_{k+1}$ and $\mathcal{Q}_k \supseteq \mathcal{Q}_{k+1}$. □

Lemma 8. If $X \in \mathcal{Q}_k$, and $X \lhd Y$, then $Y \in \mathcal{Q}_{k+1}$ □

Definition 9. For a set of terms Y, the set of Y*-directors*, denoted $\langle Y \rangle$, is the least set D such that $D = Y \cup \mathbf{S}\,D\,\mathcal{M} \cup \mathbf{S}\,\mathcal{M}\,D$

The name has been chosen in analogy to Kennaway's *director strings* [KS88], because an argument Z of an Y-director can be *directed* towards Y:

Proposition 10. $\langle Y\rangle\ Z \hookrightarrow\!\!\!\!\rightarrow Y\ Z$

Proof. The claim is proved by structural induction. Let $Y' \in \langle Y\rangle$.

1. $Y' = Y$. Then the claim is vacuously true.
2. $Y' \in \mathbf{S}\ \langle Y\rangle\ *$. Then $Y'Z \in \mathbf{S}\ \langle Y\rangle\ *\ Z \to \underline{(\langle Y\rangle\ Z)}\ (*\ Z) \rhd \langle Y\rangle\ Z$.
3. $Y' \in \mathbf{S} * \langle Y\rangle$. Then $Y'Z \in \mathbf{S} * \langle Y\rangle\ Z \to * Z\ \underline{(\langle Y\rangle\ Z)} \rhd \langle Y\rangle\ Z$. □

Definition 11. For sets X and Y, the set $\langle {}^{X}/\,{}_{Y}\rangle$ is the least set D such that $D = Y \cup X\ D$

In the literature this is also denoted by $X^{*}Y$ but we chose this notation because of the similarity to $\langle Y\rangle$:

Lemma 12. $\langle {}^{\mathbf{S}\ X}/\,{}_{Y}\rangle \subseteq \langle Y\rangle$. □

The following will be needed later. It is easily proved by structural induction.

Proposition 13. $\langle {}^{\mathbf{S}\ X}/\,{}_{Y}\rangle\ Z \twoheadrightarrow \langle {}^{X\ Z}/\,{}_{Y\ Z}\rangle$ □

3 Examples of Infinite Reductions

We now give three classes of expressions not having a normal form. They are all of the shape $\langle * \rangle\ *$. (The next section shows that the examples presented here are essentially complete.)
We start by generalizing the reduction shown in example 1. We note that each redex has the form $\langle \mathbf{S}\ \mathsf{T}\ \mathsf{T}\rangle\ \langle \mathbf{S}\ \mathsf{T}\ \mathsf{T}\rangle$ and that $\mathbf{S}\ \mathsf{T}\ \mathsf{T}$ is the smallest term in the set $\mathcal{Q}_2\ \mathcal{Q}_1$.

Definition 14. $\mathcal{F} = \langle \mathcal{Q}_2\ \mathcal{Q}_1\rangle$

We are going to show that $\mathcal{F}\ \mathcal{F} \twoheadrightarrow \infty$. We need

Lemma 15. $\mathcal{Q}_1\ \mathcal{F} \subseteq \mathcal{F}$

Proof. Let $X\ Y \in \mathcal{Q}_1\ \mathcal{F}$. We use the fact that $\mathcal{Q}_1 = \mathsf{T} \cup \mathcal{Q}_2$.

1. $X = \mathsf{T}$. Then $X\ Y \in \mathbf{S} * \mathcal{F} \subseteq \langle \mathcal{F}\rangle \subseteq \mathcal{F}$.
2. $X \in \mathcal{Q}_2$. As $Y \in \mathcal{F} \subseteq \mathcal{Q}_1$, we have $X\ Y \in \mathcal{Q}_2\ \mathcal{Q}_1 \subseteq \mathcal{F}$. □

Proposition 16. $\langle \mathcal{Q}_2\ \mathcal{Q}_1\rangle\ \langle \mathcal{Q}_2\ \mathcal{Q}_1\rangle \twoheadrightarrow \infty$

Proof. We have $\langle \mathcal{Q}_2\ \mathcal{Q}_1\rangle\ \langle \mathcal{Q}_2\ \mathcal{Q}_1\rangle \hookrightarrow\!\!\!\!\rightarrow \mathcal{Q}_2\ \mathcal{Q}_1\ \mathcal{F}$. Let $X\ Y\ Z \in \mathcal{Q}_2\ \mathcal{Q}_1\ \mathcal{F}$.
If $X \twoheadrightarrow \infty$, then the $X\ Y\ Z \twoheadrightarrow \infty$. So assume X has a normal form. It could have shape $\mathbf{S}\ *$ or $\mathbf{S} * *$.

1. $X \twoheadrightarrow \mathbf{S}\ \mathcal{Q}_1$.
   $\mathbf{S}\ \mathcal{Q}_1\ Y\ \mathcal{F} \to \mathcal{Q}_1\ \mathcal{F}\ (Y\ \mathcal{F}) \subseteq \mathcal{Q}_1\ \mathcal{F}\ (\mathcal{Q}_1\ \mathcal{F}) \subseteq \mathcal{F}\ \mathcal{F} \twoheadrightarrow \infty$
2. $X \twoheadrightarrow \mathbf{S} * *$.
   $\mathbf{S} * *\ \mathcal{Q}_1\ \mathcal{F} \to *\ \mathcal{Q}_1\ (*\ \mathcal{Q}_1)\ \mathcal{F}$. By lemmata 8 and 7 we have $*\ \mathcal{Q}_1 \subseteq \mathcal{Q}_2 \subseteq \mathcal{Q}_1$, so $*\ \mathcal{Q}_1\ (*\ \mathcal{Q}_1)\ \mathcal{F} \subseteq \mathcal{Q}_2\ \mathcal{Q}_1\ \mathcal{F} \subseteq \mathcal{F}\ \mathcal{F} \twoheadrightarrow \infty$.

In either case we find an infinite reduction because we've done at least one reduction step and the result again has the form $\mathcal{F}\ \mathcal{F}$. □

The following pattern turns out to be a generalization of $\mathsf{A\ A\ A} \twoheadrightarrow \infty$ (example 2). We recall that $\mathsf{A} \in \mathcal{Q}_k$ for any k.

Proposition 17. $\langle \mathcal{Q}_3\ \mathcal{Q}_2 \rangle\ \mathcal{Q}_1 \twoheadrightarrow \infty$

Proof. Let $\langle \mathcal{Q}_3\ \mathcal{Q}_2 \rangle\ \mathcal{Q}_1 \hookrightarrow\!\!\!\!\twoheadrightarrow \mathcal{Q}_3\ \mathcal{Q}_2\ \mathcal{Q}_1$, and $X\ Y\ Z \in \mathcal{Q}_3\ \mathcal{Q}_2\ \mathcal{Q}_1$.
If $X \twoheadrightarrow \infty$, then the $X\ Y\ Z \twoheadrightarrow \infty$. So assume X has a normal form. It could be $\mathbf{S}\ \mathcal{Q}_2$ or $\mathbf{S} * *$.

1. $X \twoheadrightarrow \mathbf{S}\ \mathcal{Q}_2$.
   $\mathbf{S}\ \mathcal{Q}_2\ \mathcal{Q}_2\ \mathcal{Q}_1 \to \mathcal{Q}_2\ \mathcal{Q}_1\ (\mathcal{Q}_2\ \mathcal{Q}_1) \subseteq \mathcal{F}\ \mathcal{F} \twoheadrightarrow \infty$ by proposition 16
2. $X \twoheadrightarrow \mathbf{S} * *$.
   $\mathbf{S} * *\ \mathcal{Q}_2\ \mathcal{Q}_1 \to *\ \mathcal{Q}_2\ (*\ \mathcal{Q}_2)\ \mathcal{Q}_1$ By lemmata 8 and 7 we have $*\ \mathcal{Q}_2 \subseteq \mathcal{Q}_3 \subseteq \mathcal{Q}_2$, so $*\ \mathcal{Q}_2\ (*\ \mathcal{Q}_2)\ \mathcal{Q}_1 \subseteq \mathcal{Q}_3\ \mathcal{Q}_2\ \mathcal{Q}_1 \twoheadrightarrow \infty$ by induction. □

Finally we mention a variant of the $\mathcal{Q}_3\ \mathcal{Q}_2\ \mathcal{Q}_1$ pattern:

Proposition 18. $\langle \mathbf{S}\ \mathcal{Q}_3\ \mathcal{Q}_0 \rangle\ \mathcal{Q}_2 \twoheadrightarrow \infty$

Proof. $\langle \mathbf{S}\ \mathcal{Q}_3\ \mathcal{Q}_0 \rangle\ \mathcal{Q}_2 \hookrightarrow\!\!\!\!\twoheadrightarrow \mathbf{S}\ \mathcal{Q}_3\ \mathcal{Q}_0\ \mathcal{Q}_2 \to \mathcal{Q}_3\ \mathcal{Q}_2\ (\mathcal{Q}_0\ \mathcal{Q}_2) \subseteq \mathcal{Q}_3\ \mathcal{Q}_2\ \mathcal{Q}_1$. Then apply proposition 17. □

Example 19. An interesting term is $\mathbf{S}\ \mathsf{A}\ \mathbf{S}\ (\mathbf{S}\ \mathsf{A}\ \mathbf{S}) \in \langle \mathbf{S}\ \mathcal{Q}_3\ \mathcal{Q}_0 \rangle\ \mathcal{Q}_2$. Its reduction graph is a pure line, and all reductions are head reductions.

4 The Decision Procedure

4.1 Outline

To decide whether an arbitrary ground **S**-term normalizes, it is enough to have a procedure that decides this for the case $\mathcal{N}\ \mathcal{N}$, i. e. it may assume that both children of the root already are in normal form. If they were not, we could first apply the procedure to these subterms recursively and compute their normal forms if the procedure says they exist.

We will construct subsets of $\mathcal{N}\ \mathcal{N}$ that contain terms with similar reduction properties. The construction begins by partitioning the set $\mathcal{N}$ itself into classes

$\mathcal{N}_0$, $\mathcal{N}_1$ and $\mathcal{N}_2$, and then combines pairs of these. It turns out that $\mathcal{N}_0$ is responsible for finite reductions, and $\mathcal{N}_2$ for infinite ones.

During the process, the case $\mathcal{N}_1\ \mathcal{N}_1$ needs special attention. Here, a further partition of $\mathcal{N}_1$ into sets $\mathcal{L}_0$, $\mathcal{L}_1$ and $\mathcal{L}_2$ is needed to completely ananlyze this case.

Finally we consider the remaining case $\mathcal{N}_2\ \mathcal{P}_2$ by showing that after some reductions it is transformed into one of the cases that have already been dealt with.

Most of these constructions have been found empirically, by looking at lots of examples. So the presentation here necessarily has some ad-hoc feeling, whereas a straight top-down approach might be more desirable. But that seems to transcend today's knowledge of **CL(S)**.

At a few places during the presentation, we rely on inclusion/exclusion relations between certain rational tree languages.

We exploit that $\mathcal{M}$ (all terms) and $\mathcal{N}$ (all normal forms) are rational, as are director languages $\langle Y \rangle$ and $\langle {}^{X}/\, Y \rangle$ (definitions 9, 11) for rational X and Y, and rational languages are closed under boolean operations. For a general introduction to tree automata and languages, see [GS84].

The relations claimed in the present paper (in the proofs of propositions 31, 40, 41) can be verified by pencil and paper, which is a rather tedious exercise. Alternatively, they can be verified by the program `RX`, which handles tree automata and languages, and which is freely available from the author's `WWW` page.

4.2 The case $\mathcal{N}\ \mathcal{N}$

We will classify the set $\mathcal{N}$ of normal forms according to the existence of certain subterms that might or might not occur during the reduction of $X\ a$ where $X \in \mathcal{N}$ and a is a variable.

The most restricted class is

Definition 20. $\mathcal{N}_0 = \left\langle {}^{\mathsf{T}}/\ \mathbf{S} \cup \mathbf{S}\,\mathcal{N} \right\rangle$

Example 21. $\mathsf{A} = \mathbf{S\ S\ S} = \mathsf{T\ S} \in \left\langle {}^{\mathsf{T}}/\ \mathbf{S} \right\rangle \subseteq \mathcal{N}_0$.

A term from $\mathcal{N}_0$ never moves its argument into functional position (it never *activates* it):

Example 22. $\mathsf{A}\ a = \mathbf{S\ S\ S}\ a \to \mathbf{S}\ a\ (\mathbf{S}\ a)$

Proposition 23. $\downarrow (\mathcal{N}_0\ a)$, and $\mathcal{N}_0\ a \not\twoheadrightarrow a\ *$

Proof. By proposition 13, $\mathcal{N}_0\ a = \left\langle {}^{\mathsf{T}}/\ \mathbf{S} \cup \mathbf{S}\,\mathcal{N} \right\rangle\ a \twoheadrightarrow \langle {}^{\mathbf{S}\ a}/\ \mathbf{S}\ a \cup \mathbf{S}\,\mathcal{N}\ a \rangle$ and no further reductions can happen. We see that a always is the right child of its parent. □

Since $\mathcal{N}_0$ effectively ignores its argument, we have

Proposition 24. $\downarrow(\mathcal{N}_0\ \mathcal{N})$

Proof. The normal form of $\mathcal{N}_0\ \mathcal{N}$ is that of $\mathcal{N}_0\ a$ with a replaced by $\mathcal{N}$. No reductions can happen inside $\mathcal{N}$, because it already is a normal form. No new redexes can be created. □

A slightly more extended class is

Definition 25. $\mathcal{N}_{0,1} = \left\langle \mathsf{S}\ \mathcal{P}_2 \middle/ \mathsf{S} \cup \mathsf{S}\,\mathcal{N} \cup \mathsf{S}\ (\mathsf{S}\ \mathsf{T})\ \mathcal{P}_2 \right\rangle$

Example 26. $\mathsf{S}\ \mathsf{T}\ \mathsf{T} \in \mathsf{S}\ \mathcal{P}_2\ (\mathsf{S}\ \mathcal{N}) \subseteq \langle \mathsf{S}\ \mathcal{P}_2 / \mathsf{S}\ \mathcal{N} \rangle \subseteq \mathcal{N}_{0,1}$

A term from $\mathcal{N}_{0,1}$ might activate its argument, but there at most one argument is supplied.

Example 27. $\mathsf{S}\ \mathsf{T}\ \mathsf{T}\ a \to \mathsf{T}\ a\ (\mathsf{T}\ a) \to \mathsf{S}\ (\mathsf{T}\ a)\ \underline{(a\ (\mathsf{T}\ a))}$

Proposition 28. $\downarrow(\mathcal{N}_{0,1}\ a)$, and $\mathcal{N}_{0,1}\ a \not\hookrightarrow\!\!\!\!\twoheadrightarrow a * *$.
Moreover, if $\mathcal{N}_{0,1}\ a \hookrightarrow\!\!\!\!\twoheadrightarrow a\ X$, then $a \lhd X$.

Proof. The base cases are

1. $\mathsf{S}\ a$ is in normal form.
2. $\mathsf{S}\ \mathcal{N}\ a$ is in normal form.
3. $\mathsf{S}\ (\mathsf{S}\ \mathsf{T})\ \mathsf{S}\ a \to \mathsf{S}\ \mathsf{T}\ a\ (\mathsf{S}\ a) \to \mathsf{T}\ (\mathsf{S}\ a)\ (a\ (\mathsf{S}\ a))$
   $\to \mathsf{S}\ \underline{(a\ (\mathsf{S}\ a))}\ \left(\mathsf{S}\ a\ \underline{(a\ (\mathsf{S}\ a))}\right)$
4. $\mathsf{S}\ (\mathsf{S}\ \mathsf{T})\ \mathsf{T}\ a \to \mathsf{S}\ \mathsf{T}\ a\ (\mathsf{T}\ a) \to \mathsf{T}\ (\mathsf{T}\ a)\ (a\ (\mathsf{T}\ a))$
   $\to \mathsf{S}\ (a\ (\mathsf{T}\ a))\ (\mathsf{T}\ a\ (a\ (\mathsf{T}\ a))) \to \mathsf{S}\ \underline{(a\ (\mathsf{T}\ a))}\ \left(\mathsf{S}\ \underline{(a\ (\mathsf{T}\ a))}\ \underline{\left(a\ (a\ (\mathsf{T}\ a))\right)}\right)$

where the claim can be verified at the underlined subterms.
The inductive steps are

1. $\mathsf{S}\ \mathsf{S}\ \mathcal{N}_{0,1}\ a \to \mathsf{S}\ a\ \underline{(\mathcal{N}_{0,1}\ a)}$
2. $\mathsf{S}\ \mathsf{T}\ \mathcal{N}_{0,1}\ a \to \mathsf{T}\ a\ (\mathcal{N}_{0,1}\ a) \to \mathsf{S}\ \underline{(\mathcal{N}_{0,1}\ a)}\ \underline{\left(a\ \underline{(\mathcal{N}_{0,1}\ a)}\right)}$ □

Proposition 28 will now be used to exhibit another set of normalizing terms:

Proposition 29. $\downarrow(\mathcal{N}_{0,1}\ \mathcal{N}_0)$

Proof. First reduce $\mathcal{N}_{0,1}\ \mathcal{N}_0$ as if $\mathcal{N}_0$ were a variable. Using the same reasoning as in the proof for proposition 28, the only redexes that might arise are of the form $\mathcal{N}_0\ *$. They might be nested, but all of them can be reduced to normal form by proposition 24. □

By definition we have $\mathcal{N}_0 \subseteq \mathcal{N}_{0,1}$. We introduce some more names:

Definition 30. $\mathcal{N}_1 = \mathcal{N}_{0,1} \setminus \mathcal{N}_0; \mathcal{N}_2 = \mathcal{N} \setminus \mathcal{N}_{0,1}$.

So $\mathcal{N}$ is the disjoint union of $\mathcal{N}_0$, $\mathcal{N}_1$, and $\mathcal{N}_2$. Recall that $\mathcal{N}_0$ *ignores* its argument, while $\mathcal{N}_1$ *activates* it *once*.

It would have been possible to consider propositions 23 resp. 28 as definitions, and then to derive the definitions 20 resp. 25. But the given presentation is more compact (although less intuitive).

Essentially, $\mathcal{N}_0$ is responsible for normalization, and $\mathcal{N}_2$ for non-normalization, as will be highlighted in the next two statements.

Proposition 31. $\mathcal{N}_2\ \mathcal{Q}_2 \twoheadrightarrow \infty$.

Proof. It can be shown that $\mathcal{N}_2 \subseteq \langle \mathcal{Q}_3\ \mathcal{Q}_2 \rangle \cup \langle \mathbf{S}\ \mathcal{Q}_3\ \mathcal{Q}_0 \rangle$. (See the introductory remark on verification of such relations.) This is exactly the right form to apply propositions 17 and 18. □

Proposition 32. $\mathcal{N}_1\ \mathcal{N}_2 \twoheadrightarrow \infty$.

Proof. Let $X\ Y \in \mathcal{N}_1\ \mathcal{N}_2$. There is a reduction from $X\ Y$ that activates Y, i. e. that produces a subterm $Y\ Z$. By the second part of proposition 28, we have $Y \lhd Z$. Since $Y \in \mathcal{N}_2 \subseteq \mathcal{Q}_1$, clearly $Z \in \mathcal{Q}_2$. So we have $Y\ Z \in \mathcal{N}_2\ \mathcal{Q}_2$ and can apply proposition 31. □

Combining propositions 24, 29 on one hand and 31, 32 on the other, the cases still missing in the analysis of $\mathcal{N}\ \mathcal{N}$ are $\mathcal{N}_2\ \mathcal{P}_2$ and $\mathcal{N}_1\ \mathcal{N}_1$.

4.3 The sub-case $\mathcal{N}_1\ \mathcal{N}_1$

We look deeper into the set $\mathcal{N}_1$. We single out the subset of terms that move their argument a into functional position but which only supply them with the argument $\mathbf{S}\ a$.

Definition 33. $\mathcal{L}_0 = \left\langle \mathsf{T} \middle/ \mathbf{S}\ (\mathbf{S}\ \mathcal{P}_2)\ \mathbf{S} \right\rangle$

Proposition 34. $\downarrow (\mathcal{L}_0\ a)$, and $\{Z : \mathcal{L}_0\ a \hookrightarrow\!\!\!\!\rightarrow a\ Z\} = \{\mathbf{S}\ a\}$.

Proof. The base cases are

1. $\mathbf{S}\ \mathsf{T}\ \mathbf{S}\ a \to \mathsf{T}\ a\ (\mathbf{S}\ a) \to \mathbf{S}\ (\mathbf{S}\ a)\ \underline{(a\ (\mathbf{S}\ a))}$
2. $\mathbf{S}\ (\mathbf{S}\ \mathsf{T})\ \mathbf{S}\ a \to \mathbf{S}\ \mathsf{T}\ a\ (\mathbf{S}\ a) \to \mathsf{T}\ (\mathbf{S}\ a)\ (a\ (\mathbf{S}\ a))$
   $\to \mathbf{S}\ \underline{(a\ (\mathbf{S}\ a))}\ \left(\mathbf{S}\ a\ \underline{(a\ (\mathbf{S}\ a))}\right)$

Then inductive step reads $\mathsf{T}\ \mathcal{L}_0\ a \to \mathbf{S}\ a\ \underline{(\mathcal{L}_0\ a)}$. □

Using $\mathcal{L}_0$, we find another class of normalizing terms.

Proposition 35. $\downarrow(\mathcal{L}_0\ \mathcal{N}_1)$.

Proof. Reduce $\mathcal{L}_0\ \mathcal{N}_1$ as if $\mathcal{N}_1$ were a variable. By proposition 34, the only redexes that may arise have the form $\mathcal{N}_1\ (\mathbf{S}\ \mathcal{N}_1)$. But $\mathbf{S}\ \mathcal{N}_1 \subseteq \mathbf{S}\ \mathcal{N} \subseteq \mathcal{N}_0$, so $\mathcal{N}_1\ (\mathbf{S}\ \mathcal{N}_1) \subseteq \mathcal{N}_1\ \mathcal{N}_0$ and therefore they normalize by proposition 29. □

Another useful subset of $\mathcal{N}_1$ is

Definition 36. $\mathcal{L}_1 = \left\langle \mathsf{T} \big/ \mathbf{S}\ \mathsf{T}\ (\mathbf{S}\ \mathcal{P}_2) \right\rangle$.

It has the following property:

Proposition 37. $\downarrow(\mathcal{L}_1\ a)$, and $\{Z : \mathcal{L}_1\ a \hookrightarrow\!\!\!\rightarrow a\ Z\} = \{\mathbf{S}\ \mathcal{P}_2\ a\}$.

Proof. The base case is $\mathbf{S}\ \mathsf{T}\ (\mathbf{S}\ \mathcal{P}_2)\ a \rightarrow \mathsf{T}\ a\ (\mathbf{S}\ \mathcal{P}_2\ a) \rightarrow \mathbf{S}\ (\mathbf{S}\ \mathcal{P}_2\ a)\ \underline{(a\ (\mathbf{S}\ \mathcal{P}_2\ a))}$. The inductive step is $\mathsf{T}\ \mathcal{L}_1\ a \rightarrow \mathbf{S}\ a\ \underline{(\mathcal{L}_1\ a)}$. □

This leads to yet another set of normalizing **S**-terms.

Proposition 38. $\downarrow(\mathcal{L}_1\ \mathcal{L}_0)$

Proof. Reduce $\mathcal{L}_1\ \mathcal{L}_0$, treating $\mathcal{L}_0$ as a variable. By proposition 37, the only new redexes have the form $\mathcal{L}_0\ (\mathbf{S}\ \mathcal{P}_2\ \mathcal{L}_0)$. We have $\mathcal{L}_0 \subseteq \mathcal{N}_1$, therefore $\mathbf{S}\ \mathcal{P}_2\ \mathcal{L}_0 \subseteq \mathbf{S}\ \mathcal{P}_2\ \mathcal{N}_1 \subseteq \mathcal{N}_1$ by definition. But then $\mathcal{L}_0\ (\mathbf{S}\ \mathcal{P}_2\ \mathcal{L}_0) \subseteq \mathcal{L}_0\ \mathcal{N}_1$ which normalizes by proposition 35. □

Up to here we found some normalizing subsets of $\mathcal{N}_1\ \mathcal{N}_1$. Now we will show that all other elements of $\mathcal{N}_1\ \mathcal{N}_1$ have an infinite reduction.
We introduce even more names:

Definition 39. $\mathcal{L}_{1,2} = \mathcal{N}_1 \setminus \mathcal{L}_0; \mathcal{L}_2 = \mathcal{L}_{1,2} \setminus \mathcal{L}_1$.

So $\mathcal{N}_1$ is the disjoint union of $\mathcal{L}_0$, $\mathcal{L}_1$, and $\mathcal{L}_2$.
From $\mathcal{L}_{1,2}$ we build more non-normalizing terms.

Proposition 40. $\mathcal{L}_{1,2}\ \mathcal{L}_{1,2} \twoheadrightarrow \infty$

Proof. We have $\mathcal{L}_{1,2} \subseteq \langle \mathcal{Q}_2\ \mathcal{Q}_1 \rangle$. (See introductory remark on verification of such relations.) So proposition 16 is applicable. □

Combining propositions 35, 38, and 40, there is exactly one case remaining in the analysis of $\mathcal{N}_1\ \mathcal{N}_1$, and that's $\mathcal{L}_2\ \mathcal{L}_0$. It is covered by the following

Proposition 41. $\mathcal{L}_2\ \mathcal{N}_1 \twoheadrightarrow \infty$

Proof. We use the fact that $\mathcal{L}_2 \subseteq \left\langle \mathsf{T} \middle/ \mathbf{S}\ (\mathbf{S}\ \mathsf{T})\ \mathsf{T} \cup \mathbf{S}\ \mathsf{T}\ \mathcal{Q}_3 \right\rangle$. (See introductory remark on verification.) Also, for any k, we have $\mathcal{N}_1 \subseteq \mathcal{Q}_k$, because $\mathcal{N}_1 \cap \mathcal{P}_k = \emptyset$ since $\mathcal{P}_k \subseteq \mathcal{N}_0$. Then we have two base cases:

1. $\mathbf{S}\ \mathsf{T}\ \mathcal{Q}_3\ \mathcal{N}_1 \rightarrow \mathsf{T}\ \mathcal{N}_1\ (\mathcal{Q}_3\ \mathcal{N}_1) \rightarrow \mathbf{S}\ (\mathcal{Q}_3\ \mathcal{N}_1)\ \underline{(\mathcal{N}_1\ (\mathcal{Q}_3\ \mathcal{N}_1))}$. By proposition 28, applied to the underlined expression, there is a reduction $\mathcal{N}_1\ (\mathcal{Q}_3\ \mathcal{N}_1) \hookrightarrow\!\!\!\!\rightarrow \mathcal{Q}_3\ \mathcal{N}_1\ X$ with $\mathcal{Q}_3\ \mathcal{N}_1 \lhd X$, so that $X \subseteq \mathcal{Q}_2$. Moreover we have $\mathcal{N}_1 \subseteq \mathcal{Q}_2$, therefore the $\mathcal{Q}_3\ \mathcal{N}_1\ X \subseteq \mathcal{Q}_3\ \mathcal{Q}_2\ \mathcal{Q}_1 \twoheadrightarrow \infty$ by proposition 17.
2. $\mathbf{S}\ (\mathbf{S}\ \mathsf{T})\ \mathsf{T}\ \mathcal{N}_1 \rightarrow \mathbf{S}\ \mathsf{T}\ \mathcal{N}_1\ (\mathsf{T}\ \mathcal{N}_1) \subseteq \mathbf{S}\ \mathsf{T}\ \mathcal{Q}_3\ \mathcal{N}_1$ because $\mathcal{N}_1 \subseteq \mathcal{Q}_3$. This gets us back to the previous case.

The inductive step is just $\mathsf{T}\ \mathcal{L}_2\ \mathcal{N}_1 \rightarrow \mathbf{S}\ \mathcal{N}_1\ \underline{(\mathcal{L}_2\ \mathcal{N}_1)}$. □

4.4 The sub-case $\mathcal{N}_2\ \mathcal{P}_2$

Combining all of the previous, the analysis for $\mathcal{N}_2\ \mathcal{Q}_2$ is complete, but $\mathcal{N}_2\ \mathcal{P}_2 = \mathcal{N}_2\ (\mathbf{S} \cup \mathsf{T})$ is missing. We will show that after doing some reductions, the decision procedure for $\mathcal{N}_2\ \mathcal{Q}_2$ can be used.

We define a notion of *complete reduction*. After one complete reduction step of an $X\ \mathbf{S}$ redex, with X in normal form, no further redexes $X'\ \mathbf{S}$ can occur. And after a complete reduction step of an $X\ \mathsf{T}$ redex, no further redexes $X'\ \mathcal{P}_2$ are possible. That's why we then may use the procedure that decides normalization for $\mathcal{N}\ \mathcal{Q}_2$.

Definition 42. The complete reduction relation $X \overset{!}{\rightarrow} W$ is defined for all $X \in \mathcal{N}\ \mathcal{M}$ and $W \in \mathcal{M}$ by

$$\mathbf{S}\ Z \overset{!}{\rightarrow} \mathbf{S}\ Z; \mathbf{S}\ X\ Z \overset{!}{\rightarrow} \mathbf{S}\ X\ Z$$

$$\mathbf{S}\ X\ Y\ Z \overset{!}{\rightarrow} X'\ Y' \quad \text{where} \quad X\ Z \overset{!}{\rightarrow} X' \text{ and } Y\ Z \overset{!}{\rightarrow} Y'$$

Lemma 43. Obviously $X \overset{!}{\rightarrow} W$ implies $X \twoheadrightarrow W$. For each $X \in \mathcal{N}\ \mathcal{M}$, there is exactly one W with $X \overset{!}{\rightarrow} W$, so we can view this relation as a function, defined by structural recursion on the normal forms. □

In a redex $\mathbf{S}\ X\ Y\ Z$, call Z the *argument* of the redex. We are also interested in subterms that might become arguments later. In a term, call a subterm (position) an *argument candidate* if it occurs as the right child of a subterm that has a left spine longer than two (i. e. whose left child is in $\mathcal{M}_2$). If $X\ Y \overset{!}{\rightarrow} Z$, then all argument candidates in terms reachable from Z are larger than Y. This is made precise by

Proposition 44. If $X \in \mathcal{N}$, and $X\ \mathsf{S}_k \overset{!}{\rightarrow} Z_0 \rightarrow Z_1 \rightarrow \ldots$, then none of the Z_i contains a subterm $\mathcal{M}_2\ \mathcal{P}_k$.

Proof. We prove that the property holds for Z_0. Then we proceed by induction.

We are going to show that $X\ \mathsf{S}_k \xrightarrow{!} Z_0$ implies that no $\mathcal{M}_2\ \mathcal{P}_k \trianglelefteq Z_0$. We use structural induction on X.

1. $X \in \mathbf{S} \cup \mathbf{S}\ X'$.
   Then $X\ \mathsf{S}_k \xrightarrow{!} X\ \mathsf{S}_k$ and does not contain $\mathcal{M}_2\ \mathcal{P}_k$ because it's a normal form and $\mathcal{M}_2\ \mathcal{P}_k$ is not.
2. $X = \mathbf{S}\ X_1\ X_2$.
   Then $X\ \mathsf{S}_k \xrightarrow{!} X_1'\ X_2'$ where $X_1\ \mathsf{S}_k \xrightarrow{!} X_1'$ and $X_2\ \mathsf{S}_k \xrightarrow{!} X_2'$. By induction, the claim is true for X_1' and X_2'.
   (a) If $X_2 = \mathbf{S}$, then $X_2' = \mathbf{S}\ \mathsf{S}_k = \mathsf{S}_{k+1} \in \mathcal{Q}_k$.
   (b) If $X_2 \neq \mathbf{S}$, then $X_2\ \mathsf{S}_k \in \mathcal{M}_1 * \subseteq \mathcal{M}_2 \subseteq \mathcal{Q}_k$. Now $\mathcal{Q}_k$ is forward closed under reduction.
   In either case we have $X_2' \in \mathcal{Q}_k$ and therefore $X_1'\ X_2' \notin \mathcal{M}_2\ \mathcal{P}_k$.

Now chose the smallest i such that Z_i contains $\mathcal{M}_2\ \mathcal{P}_k$. (We already know $i > 0$.) We list the possible relative positions of the redex $\mathbf{S}\ C\ D\ E$ that is reduced in Z_{i-1} (resp. its contractum $C\ E\ (D\ E)$ in Z_i) and the pattern $\mathcal{M}_2\ \mathcal{P}_k$. Pattern and redex position must overlap, since i should be minimal.

1. The pattern occurs inside C, D, or E.
   Then it would already have occured in Z_{i-1}, so i is not minimal.
2. $C\ E \in \mathcal{M}_2\ \mathcal{P}_k$ or $D\ E \in \mathcal{M}_2\ \mathcal{P}_k$
   Then $E \in \mathcal{P}_k$ and in Z_{i-1} the redex $\mathbf{S}\ C\ D\ E \in \mathbf{S} * * \mathcal{P}_k \subseteq \mathcal{M}_2\ \mathcal{P}_k$ and i is not minimal.
3. $C\ E\ (D\ E) \in \mathcal{M}_2\ \mathcal{P}_k$.
   In Z_{i-1}, we had $\mathbf{S}\ C\ D\ E \in \mathbf{S} * * E \subseteq \mathcal{M}_2\ E$. Because i is minimal, we have $E \in \mathcal{Q}_k$ leading to $D\ E \in \mathcal{Q}_k$ and the contradiction $C\ E\ (D\ E) \in \mathcal{M}_2\ \mathcal{Q}_k$.
4. $C\ E\ (D\ E)\ \ldots\ G \in \mathcal{M}_2\ \mathcal{P}_k$.
   That is, $G \in \mathcal{P}_k$. But then $\mathbf{S}\ C\ D\ E\ \ldots\ G \in \mathcal{M}_2\ \mathcal{P}_k$ in Z_{i-1} as well, contradicting the minimality of i.

So an i with the required property does not exist, which proves the claim. □

4.5 Deciding Normalization

Theorem 45. *There is a procedure that decides whether a ground term in* $\mathbf{CL}(\mathbf{S})$ *has a normal form.*

Proof. $\mathbf{S}$ has a normal form. Therefore assume the expression is $X\ Y$.

Recursively apply the procedure to X and Y and compute their normal forms if they exist. This recursion is terminating because it's structural. If one (or both) of the normal forms don't exist, halt and answer *not normalizing.* Otherwise call them X' and Y', respectively.

If $Y' \in \mathcal{Q}_2$, answer *normalizing* if $X'\,Y' \in \mathcal{N}_0\,\mathcal{N} \cup \mathcal{L}_0\,\mathcal{N}_1 \cup \mathcal{L}_1\,\mathcal{L}_0 \cup \mathcal{N}_1\mathcal{N}_0$, (justified by propositions 24, 35, 38, 29) and *not normalizing* else (justified by propositions 31, 32, 40, 41).

If $Y' \in \mathcal{P}_2$, then case T or **S** applies:

Case T: If $Y' = \mathsf{T}$, determine Z from $X'\,Y' \overset{!}{\to} Z$. Then apply the procedure to Z. By proposition 44, no redex $\mathcal{M}_2$ T will occur in reductions from Z. Therefore case T will not happen again, so recursion terminates.

Case **S**: If $Y' = \mathbf{S}$, determine Z from $X'\,Y' \overset{!}{\to} Z$. Then apply the procedure to Z. By proposition 44, no redex $\mathcal{M}_2$ **S** will occur in reductions from Z. Therefore case **S** will not happen again, so recursion terminates. □

It is possible to omit cases **S** and T from the decision procedure, by explicitly giving rational sets X_S and X_T such that X **S** (for $X \in \mathcal{N}$) normalizes iff $X \in X_S$, and similar for T.

In the procedure given above, of course one would even like to remove the restriction that X and Y be already in normal form, to save all those recursive calls. Indeed this seems possible.

Conjecture 46. The set of normalizing terms in **CL(S)** is rational.

5 Related Work

Apparently, a seminar held by Barendregt, Bergstra, Klop, and Volken in 1975 started the detailed study of **S**-terms. They placed a bet on terms without normal form, and Barendregt collected 25 guilders from the other three by producing A A A (and later **S** A A (**S** A A)). Later Duboué found **S** T **S S S S** (and its reduct A T **S S**) as the smallest **S**-terms that don't normalize.

The decidability of normalization had already been conjectured by E. Zachos in 1978. He completely analyzes the normalization of **S**-terms of size up to 9, by either finding the normal form or exhibiting patterns (similar to those given in the examples 1 and 2) that lead to an infinite reduction.

Still a proof that each non-normalizing term admits such a pattern would be desirable. (It does not follow directly from the results presented here. The main problem is that the case distinctions in the proofs of propositions 16 and 17 are not "constructive".)

The combinator **L** with reduction rule **L** $x\ y \to x\ (y\ y)$ is called the *lark* in Smullyan's book [Smu85]. It has been investigated in depth by R. Statman ([Sta89]). He showed that convertibility is decidable for **CL(L)**. I conjecture that this should also be decidable for **S** (the *starling*) but it might be substantially more difficult than for the lark: If deciding normalization is a step towards deciding convertibility, than this first step is trivial for the lark, but, as we've seen, quite hard for the starling.

I proved [Wal97] that **CL(S)** is *top-terminating*, meaning that, even in infinite reductions, the root of a term is rewritten only finitely often. Top-termination guarantees the existence of limits (of fair infinite reductions), and perhaps these could be used to decide convertibility of non-normalizing terms.

In all, the seemingly small **S** rule is far from being completely understood. It is a challenging test case for the application of known term rewriting results, and a motivation to develop new techniques that would better explain the results obtained so far, by putting them in a more general framework.

6 Acknowledgments

I would like to thank Henk Barendregt, Thomas Genet, Alfons Geser, Dieter Hofbauer, Jan Willem Klop, Vincent van Oostrom, and Rick Statman, who have provided many useful comments as well as general remarks and historical anecdotes on combinators. I also would like to thank the anonymous referees for valuable suggestions for improving the presentation of this paper.

7 References

[Bar84] H. P. Barendregt. *The Lambda Calculus, Its Syntax and Sematics.* Elsevier Science Publishers, 1984.

[BK79] Jan Bergstra and Jan Willem Klop. Church–Rosser strategies in the Lambda calculus. *Theoretical Computer Science*, 9:27–38, 1979.

[GS84] F. Gécseg and M. Steinby. *Tree Automata.* Akadémiai Kiadó, Budapest, 1984.

[Klo80] Jan Willem Klop. *Reduction Cycles in Combinatory Logic*, pages 193–214. In Seldin and Hindley [SH80], 1980.

[KS88] Richard Kennaway and Ronan Sleep. Director strings as combinators. *ACM Transactions on Programming Languages and Systems*, 10(4):602–626, October 1988.

[SH80] J. P. Seldin and J. R. Hindley, editors. *To H. B. Curry: Essays on Combinatory Logic, Lambda Calculus and Formalism.* Academic Press, 1980.

[Smu85] Raymond Smullyan. *To Mock a Mockingbird: and other logic puzzles including an amazing adventure in combinatory logic.* Knopf, New York, 1985.

[Sta89] Richard Statmann. The word problem for Smullyan's lark combinator is decidable. *J. Symbolic Comput.*, 7:103–112, 1989.

[Wal97] J. Waldmann. Nimm Zwei. Internal Report IR-432, Vrije Universiteit Amsterdam, Research Group Theoretical Computer Science, September 1997.

[Zac78] Efstathios Zachos. Kombinatorische Logik und S-Terme. Berichte des Instituts für Informatik 26, Eidgenössische Technische Hochschule Zürich, 1978.

[ZG96] H. Zantema and A. Geser. Non-looping rewriting. Technical Report UU-CS-1996-03, Utrecht University, January 1996.

Decidable Approximations of Sets of Descendants and Sets of Normal Forms

Thomas Genet

INRIA Lorraine & CRIN CNRS - BP 101
54602 Villers-lès-Nancy Cedex FRANCE
Phone: (+33) 3-83-59-30-18 - Fax: (+33) 3-83-27-83-19
E-mail: Thomas.Genet@loria.fr
http://www.loria.fr/equipe/protheo.html/

Abstract. We present here decidable approximations of sets of descendants and sets of normal forms of Term Rewriting Systems, based on specific tree automata techniques. In the context of rewriting logic, a Term Rewriting System is a program, and a normal form is a result of the program. Thus, approximations of sets of descendants and sets of normal forms provide tools for analysing a few properties of programs: we show how to compute a superset of results, to prove the sufficient completeness property, or to find a criterion for proving termination under a specific strategy, the sequential reduction strategy. The main technical contribution of the paper is the construction of an approximation automaton which recognises a superset of the set of normal forms of terms in a set E, w.r.t. a Term Rewriting System $\mathcal{R}$.

Introduction

In the context of programming languages such as ELAN [18], a Term Rewriting System (TRS for short) is a program. We propose here to use tree automata techniques for proving various properties on TRSs and thus on programs. For a given TRS $\mathcal{R}$ and a set of terms E, these proofs are based on the computation of approximations of the set of $\mathcal{R}$-descendants of E and the set of $\mathcal{R}$-normal forms of E. For that, we build an approximation automaton which recognises a superset of the set of $\mathcal{R}$-descendants and $\mathcal{R}$-normal forms of terms in E. Considering $\mathcal{R}$ as a program and E as the set of possible inputs of the program, the set of $\mathcal{R}$-descendants of E represents all intermediate results of the program at every step of its execution on the given set of possible inputs. The set of $\mathcal{R}$-normal forms of E represents the set of all possible results obtained by executing the program $\mathcal{R}$ on the set of possible given inputs E, when the program stops. Thanks to those two sets, we show how to prove sufficient completeness of a program on a set of possible initial inputs, how to achieve some reachability testing on a program, and how to prove termination of a program represented by a TRS and a strategy of application of rewrite rules called sequential reduction strategy.

In Section 1, we recall basic definitions of terms, term rewriting systems, and tree automata. In Section 2, we briefly present sufficient completeness, reachability testing and termination proof under the sequential reduction strategy. Then, in Section 3, we recall some undecidability results on the set of descendants and the set of normal forms motivating our approach by approximation. We also detail the approximation construction which is based on specific matching and rewriting techniques on tree automata, schematising matching and rewriting on sets of terms. Feasibility of the approximation construction and its appropriateness for our purpose is shown in Section 4 on some examples. Some automatic proofs achieved by our prototype are also presented in Section 4. Finally we conclude on this work in Section 5.

1 Preliminaries

We now introduce some notations and basic definitions. Comprehensive surveys can be found in [9] for term rewriting systems, in [11,4] for tree automata and tree language theory, and in [15] for connections between regular tree languages and term rewriting systems.

Terms, Substitutions, Rewriting systems

Let $\mathcal{F}$ be a finite set of symbols associated with an arity function denoted by $ar : \mathcal{F} \mapsto \mathbb{N}$, $\mathcal{X}$ be a countable set of variables, $\mathcal{T}(\mathcal{F}, \mathcal{X})$ the set of terms, and $\mathcal{T}(\mathcal{F})$ the set of ground terms (terms without variables). Positions in a term are represented as sequences of integers. The set of positions in a term t, denoted by $\mathcal{P}os(t)$, is ordered by lexicographic ordering $\prec$. The empty sequence ϵ denotes the top-most position. $\mathcal{R}oot(t)$ denotes the symbol at position ϵ in t. For any term $s \in \mathcal{T}(\mathcal{F}, \mathcal{X})$, we denote by $\mathcal{P}os_{\mathcal{F}}(s)$ the set of functional positions in s, i.e. $\{p \in \mathcal{P}os(s) \mid p \neq \epsilon \text{ and } \mathcal{R}oot(s|_p) \in \mathcal{F}\}$. If $p \in \mathcal{P}os(t)$, then $t|_p$ denotes the subterm of t at position p and $t[s]_p$ denotes the term obtained by replacement of the subterm $t|_p$ at position p by the term s. A *ground context* is a term of $\mathcal{T}(\mathcal{F} \cup \{\Box\})$ with only one occurrence of $\Box$, where $\Box$ is a special constant not occurring in $\mathcal{F}$. For any term $t \in \mathcal{T}(\mathcal{F})$, $C[t]$ denotes the term obtained after replacement of $\Box$ by t in the ground context $C[\,]$. The set of variables of a term t is denoted by $\mathcal{V}ar(t)$. A term is linear if any variable of $\mathcal{V}ar(t)$ has exactly one occurrence in t. A substitution is a mapping σ from $\mathcal{X}$ into $\mathcal{T}(\mathcal{F}, \mathcal{X})$, which can uniquely be extended to an endomorphism of $\mathcal{T}(\mathcal{F}, \mathcal{X})$. Its domain $\mathcal{D}om(\sigma)$ is $\{x \in X \mid x\sigma \neq x\}$.

A term rewriting system $\mathcal{R}$ is a set of *rewrite rules* $l \to r$, where $l, r \in \mathcal{T}(\mathcal{F}, \mathcal{X})$, $l \notin \mathcal{X}$, and $\mathcal{V}ar(l) \supseteq \mathcal{V}ar(r)$. A rewrite rule $l \to r$ is *left-linear* (resp. *right-linear*) if the left-hand side (resp. right-hand side) of the rule is linear. A rule is linear if it is both left and right-linear. A TRS $\mathcal{R}$ is linear (resp. left-linear, right-linear) if every rewrite rule $l \to r$ of $\mathcal{R}$ is linear (resp. left-linear, right-linear).

The relation $\to_{\mathcal{R}}$ induced by $\mathcal{R}$ is defined as follows: for any $s, t \in \mathcal{T}(\mathcal{F}, \mathcal{X})$, $s \to_{\mathcal{R}} t$ if there exist a rule $l \to r$ in $\mathcal{R}$, a position $p \in \mathcal{P}os(s)$ and a substitution σ such that $l\sigma = s|_p$ and $t = s[r\sigma]_p$. The transitive (resp. reflexive transitive) closure of $\to_{\mathcal{R}}$ is denoted by $\to_{\mathcal{R}}^{+}$ (resp. $\to_{\mathcal{R}}^{*}$). A term s is *reducible* by $\mathcal{R}$ if there exists t s.t. $s \to_{\mathcal{R}} t$.

A term s is in *$\mathcal{R}$-normal form* (or is *$\mathcal{R}$-irreducible*) if s is not reducible by $\mathcal{R}$. A term s *has a normal form* if there exists a term t in $\mathcal{R}$-normal form s.t. $s \to_{\mathcal{R}}^* t$. The set of all ground terms in $\mathcal{R}$-normal form is denoted by $IRR(\mathcal{R})$, and $s \to_{\mathcal{R}}^* t$ with $t \in IRR(\mathcal{R})$ is denoted by $s \to_{\mathcal{R}}^! t$. The set of $\mathcal{R}$-descendants of a set of ground terms E is denoted by $\mathcal{R}^*(E)$ and $\mathcal{R}^*(E) = \{t \in \mathcal{T}(\mathcal{F}) \mid \exists s \in E \text{ s.t. } s \to_{\mathcal{R}}^* t\}$. The set of ground $\mathcal{R}$-normal forms of E is denoted by $\mathcal{R}^!(E)$ and $\mathcal{R}^!(E) = \{t \in \mathcal{T}(\mathcal{F}) \mid \exists s \in E \text{ s.t. } s \to_{\mathcal{R}}^! t\}$. Moreover, $\mathcal{R}^!(E) = \mathcal{R}^*(E) \cap IRR(\mathcal{R})$.

A rewriting system $\mathcal{R}$ is (1) *terminating* or *strongly normalising* if there exists no infinite derivation $s_1 \to_{\mathcal{R}} s_2 \to_{\mathcal{R}} \dots$ where $s_1, s_2, \dots \in \mathcal{T}(\mathcal{F}, \mathcal{X})$, (2) *weakly normalising* (WN for short) if every s of $\mathcal{T}(\mathcal{F}, \mathcal{X})$ has a normal form, (3) *weakly normalising on* $E \subseteq \mathcal{T}(\mathcal{F}, \mathcal{X})$ (WN on E) if every $s \in E$ has a normal form.

The set of function symbols $\mathcal{F}$ occurring in a TRS $\mathcal{R}$ can be partitioned into the set of *defined symbols* $\mathcal{D} = \{Root(l) \mid l \to r \in \mathcal{R}\}$ and the set of *constructors* $\mathcal{C} = \mathcal{F} \backslash \mathcal{D}$. A *constructor term*, is a ground term with no defined symbol. The set of constructor terms is denoted by $\mathcal{T}(\mathcal{C})$. Let $\mathcal{R}_1$ and $\mathcal{R}_2$ be TRSs with respective sets of symbols $\mathcal{F}_1$ and $\mathcal{F}_2$, respective sets of defined symbols $\mathcal{D}_1$ and $\mathcal{D}_2$, and respective sets of constructors $\mathcal{C}_1$ and $\mathcal{C}_2$. TRSs $\mathcal{R}_1$ and $\mathcal{R}_2$ are *hierarchical* if $\mathcal{F}_2 \cap \mathcal{D}_1 = \emptyset$ and $\mathcal{R}_1 \subset \mathcal{T}(\mathcal{F}_1 \setminus \mathcal{D}_2, \mathcal{X}) \times \mathcal{T}(\mathcal{F}_1, \mathcal{X})$.

Automata, Regular Tree Languages

Let $\mathcal{Q}$ be a finite set of symbols, with arity 0, called *states*. $\mathcal{T}(\mathcal{F} \cup \mathcal{Q})$ is called the set of *configurations*. A *transition* is a rewrite rule $c \to q$, where $c \in \mathcal{T}(\mathcal{F} \cup \mathcal{Q})$ and $q \in \mathcal{Q}$. A *normalised transition* is a transition $c \to q$ where $c = f(q_1, \dots, q_n)$, $f \in \mathcal{F}$, $ar(f) = n$, and $q_1, \dots, q_n \in \mathcal{Q}$. A bottom-up finite tree automaton (tree automaton for short) is a quadruple $A = \langle \mathcal{F}, \mathcal{Q}, \mathcal{Q}_f, \Delta \rangle$, where $\mathcal{Q}_f \subseteq \mathcal{Q}$ and Δ is a set of normalised transitions. The rewriting relation induced by Δ is denoted by $\to_\Delta$. The tree language recognised by A is $\mathcal{L}(A) = \{t \in \mathcal{T}(\mathcal{F}) \mid \exists q \in \mathcal{Q}_f \text{ s.t. } t \to_\Delta^* q\}$. For a given $q \in \mathcal{Q}$, the tree language recognised by A and q is $\mathcal{L}(A, q) = \{t \in \mathcal{T}(\mathcal{F}) \mid t \to_\Delta^* q\}$. A tree language (or a set of terms) E is *regular* if there exists a bottom-up tree automaton A such that $\mathcal{L}(A) = E$. The class of regular tree language is closed under boolean operations $\cup, \cap, \backslash$, and inclusion is decidable. A $\mathcal{Q}$-substitution is a substitution σ s.t. $\forall x \in \mathcal{D}om(\sigma)$, $x\sigma \in \mathcal{Q}$. Let $\Sigma(\mathcal{Q}, \mathcal{X})$ be the set of $\mathcal{Q}$-substitutions. For every transition, there exists an equivalent set of normalised transitions. Normalisation consists in decomposing a transition $s \to q$, into a set $Norm(s \to q)$ of flat transitions $f(u_1, \dots, u_n) \to q'$ where $u_1, \dots, u_n$, and q' are states, by abstracting subterms $s' \notin \mathcal{Q}$ of s by states of $\mathcal{Q}$. We first define the abstraction function as follows:

Definition 1. Let $\mathcal{F}$ be a set of symbols, and $\mathcal{Q}$ a set of states. For a given configuration $s \in \mathcal{T}(\mathcal{F} \cup \mathcal{Q}) \setminus \mathcal{Q}$, an abstraction of s is a surjective mapping α:

$$\alpha : \{s|_p \mid p \in \mathcal{P}os_{\mathcal{F}}(s)\} \mapsto \mathcal{Q}$$

The mapping α is extended on $\mathcal{T}(\mathcal{F} \cup \mathcal{Q})$ by defining α as identity on $\mathcal{Q}$.

Definition 2. Let $\mathcal{F}$ be a set of symbols, $\mathcal{Q}$ a set of states, $s \to q$ a transition s.t. $s \in \mathcal{T}(\mathcal{F} \cup \mathcal{Q})$ and $q \in \mathcal{Q}$, and α an abstraction of s. The set $Norm_\alpha(s \to q)$ of normalised transitions is inductively defined by:

1. if $s = q$, then $Norm_\alpha(s \to q) = \emptyset$, and
2. if $s \in \mathcal{Q}$ and $s \neq q$, then $Norm_\alpha(s \to q) = \{s \to q\}$, and
3. if $s = f(t_1, \ldots, t_n)$, then $Norm_\alpha(s \to q) = \{f(\alpha(t_1), \ldots, \alpha(t_n)) \to q\} \cup \bigcup_{i=1}^{n} Norm_\alpha(t_i \to \alpha(t_i))$.

Example 3. Let $\mathcal{F} = \{f, g, a\}$ and $A = \langle \mathcal{F}, \mathcal{Q}, \mathcal{Q}_f, \Delta \rangle$, where $\mathcal{Q} = \{q_0, q_1, q_2, q_3, q_4\}$, $\mathcal{Q}_f = \{q_0\}$, and $\Delta = \{f(q_1) \to q_0, g(q_1, q_1) \to q_1, a \to q_1\}$.

• The languages recognised by q_1 and q_0 are the following: $\mathcal{L}(A, q_1) = \mathcal{T}(\{g, a\})$, and $\mathcal{L}(A, q_0) = \mathcal{L}(A) = \{f(x) \mid x \in \mathcal{L}(A, q_1)\}$.

• Let $s = f(g(q_1, f(a)))$, and α_1 be an abstraction of s, mapping any subterm $s|_p$ with $p \in \mathcal{P}os_{\mathcal{F}}(s)$, to distinct states in $\{q_2, q_3, q_4\}$. A possible normalisation of transition $f(g(q_1, f(a))) \to q_0$ with abstraction α_1 is the following: $Norm_{\alpha_1}(f(g(q_1, f(a))) \to q_0) = \{f(q_2) \to q_0, g(q_1, q_3) \to q_2, f(q_4) \to q_3, a \to q_4\}$.

2 Applications of $\mathcal{R}^*(E)$ and $\mathcal{R}^!(E)$

In this section we present three applications of the set of descendants and the set of normal forms to program verification.

2.1 Sufficient Completeness

This property has already been much investigated [3,19,23,17], in the context of algebraic specifications. We give here a definition of sufficient completeness of a TRS on a subset of the set of ground terms $E \subseteq \mathcal{T}(\mathcal{F})$.

Definition 4. A TRS $\mathcal{R}$ is sufficiently complete on $E \subseteq \mathcal{T}(\mathcal{F})$ if $\forall s \in E$, $\exists t \in \mathcal{T}(\mathcal{C})$ s.t. $s \to_{\mathcal{R}}^* t$, where $\mathcal{C}$ is the set of constructors in $\mathcal{F}$.

Usual methods for checking this property on algebraic specifications are either based on enumeration and testing techniques [19,23,17] or on disunification [3]. We propose, here, to check this property thanks to the set $\mathcal{R}^!(E)$.

Proposition 5. *If the TRS $\mathcal{R}$ is WN on $E \subseteq \mathcal{T}(\mathcal{F})$, and $\mathcal{R}^!(E) \subseteq \mathcal{T}(\mathcal{C})$, then $\mathcal{R}$ is sufficiently complete on E.*

This comes from the fact that since $\mathcal{R}$ is WN on E, for all terms $s \in E$, $\exists t \in IRR(\mathcal{R})$ s.t. $s \to_{\mathcal{R}}^* t$. Moreover, $t \in \mathcal{R}^!(E)$. Since $\mathcal{R}^!(E) \subseteq \mathcal{T}(\mathcal{C})$, we have $t \in \mathcal{T}(\mathcal{C})$.

Example 6. Let $\mathcal{R} = \{app(nil, x) \to x, app(cons(x, y), z) \to cons(x, app(y, z))\}$, $\mathcal{F} = \mathcal{D} \cup \mathcal{C}$, where $\mathcal{D} = \{app\}$ and $\mathcal{C} = \{cons, nil, a\}$, and $E = \{app(nil, nil), app(cons(a, nil), nil), \ldots, app(nil, cons(a, nil)), \ldots\}$. In this context, $\mathcal{R}^!(E) = \{nil, cons(a, nil), cons(a, cons(a, nil)), \ldots\}$. Since $\mathcal{R}$ is terminating, $\mathcal{R}$ is WN on E and since $\mathcal{R}^!(E) \subseteq \mathcal{T}(\mathcal{C})$, then $\mathcal{R}$ is sufficiently complete on E.

On the other hand, sufficient completeness on E does not necessarily imply that $\mathcal{R}^!(E) \subseteq \mathcal{T}(\mathcal{C})$. For example, let $\mathcal{R} = \{f(a) \to a, f(a) \to f(b)\}$, $\mathcal{C} = \{a, b\}$ and let $E = \{f(a)\}$. Then $\mathcal{R}$ is sufficiently complete on E, since $f(a) \to_{\mathcal{R}} a$, but $\mathcal{R}^!(E) = \{a, f(b)\} \not\subseteq \mathcal{T}(\mathcal{C})$.

2.2 Reachability Testing

Reachability testing consists in verifying if a term, or a term containing a pattern, can be reached by rewriting from an initial set E.

Definition 7. Let $\mathcal{R}$ be a TRS, $E \subseteq \mathcal{T}(\mathcal{F})$ and $t \in \mathcal{T}(\mathcal{F}, \mathcal{X})$. The *pattern* t *is* $\mathcal{R}$*-reachable from* E if there exists a ground context $C[\,]$, a term $s \in E$, and a substitution σ s.t. $s \rightarrow^*_{\mathcal{R}} C[t\sigma]$.

It is clear that

Proposition 8. *A pattern* t *is* $\mathcal{R}$*-reachable from* E *if an instance of* t *is a subterm of an element of* $\mathcal{R}^*(E)$.

Let us now show what can be the use of reachability testing on a simple example.

Example 9. Assume that we want to compute $A^p_n = \frac{n!}{(n-p)!}$ with the following TRS:

$$\mathcal{R} = \left\{ \begin{array}{ll} A(n,p) \rightarrow fact(n)/fact(n-p) & 0 * x \rightarrow x \\ fact(0) \rightarrow s(0) & s(x) * y \rightarrow (x * y) + y \\ fact(s(x)) \rightarrow s(x) * fact(x) & 0/s(y) \rightarrow 0 \\ x - 0 \rightarrow x & s(x)/s(y) \rightarrow s((x-y)/s(y)) \\ 0 - x \rightarrow 0 & x + 0 \rightarrow x \\ s(x) - s(y) \rightarrow x - y & x + s(y) \rightarrow s(x+y) \end{array} \right\}$$

on the domain $E = \{A(n,p) \mid n, p \in Nat\}$, where $Nat = \{0, s(0), \dots\}$. Verifying if a division by 0 can occur is equivalent to check whether the pattern $div(x, 0)$ is $\mathcal{R}$-reachable from E, i.e. whether $\exists C[\,]$, $\exists \sigma$, s.t. $C[div(x,0)\sigma] \in \mathcal{R}^*(E)$.

2.3 Termination under Sequential Reduction Strategy

Some works are devoted to automatising termination proofs of TRSs [1,14]. On the other hand, it is interesting to study weaker forms of termination, since for many purposes weak normalisation is enough [2]. In theorem provers and programming languages, rules are always applied under a specific strategy, and it is enough to ensure termination under this strategy. In addition, proving termination or WN on $\mathcal{T}(\mathcal{F}, \mathcal{X})$ or on $\mathcal{T}(\mathcal{F})$ is not always needed. In practice, a TRS is often designed to rewrite terms from a subset $E \subseteq \mathcal{T}(\mathcal{F})$, for example logical formulas in disjunctive normal form, flattened lists, or well-typed terms. Moreover, some TRSs are WN on $E \subset \mathcal{T}(\mathcal{F})$, but not on $\mathcal{T}(\mathcal{F})$ [13].

The strategy studied here is called the *Sequential Reduction Strategy* (SRS for short) and consists in separating a TRS $\mathcal{R}$ into several TRSs $\mathcal{R}_1, \dots, \mathcal{R}_n$ s.t. $\mathcal{R} = \mathcal{R}_1 \cup \ldots \cup \mathcal{R}_n$ and in normalising terms successively w.r.t. $\mathcal{R}_1, \dots, \mathcal{R}_n$. This rewriting relation under SRS is denoted by $\rightarrow_{\mathcal{R}_1;\dots;\mathcal{R}_n}$, and is based on *modular reduction relation* [20].

Definition 10. Let $\mathcal{R} = \mathcal{R}_1 \cup \ldots \cup \mathcal{R}_n$ be TRSs. For $s, t \in \mathcal{T}(\mathcal{F}, \mathcal{X})$, $s \rightarrow_{\mathcal{R}_1;\dots;\mathcal{R}_n} t$ if s is reducible by $\mathcal{R}$ and $\exists s_1, \dots, s_{n-1} \in \mathcal{T}(\mathcal{F}, \mathcal{X})$ s.t. $s \rightarrow^!_{\mathcal{R}_1} s_1$ and $s_1 \rightarrow^!_{\mathcal{R}_2} s_2$ and ... and $s_{n-1} \rightarrow^!_{\mathcal{R}_n} t$.

This kind of strategy is of great interest when normalising terms w.r.t. a TRS splitted into several hierarchical TRSs (or modules) $\mathcal{R}_1, \ldots, \mathcal{R}_n$. In this situation, interleaving of rewriting steps w.r.t. to $\mathcal{R}_1, \ldots, \mathcal{R}_n$ is often not needed. If modules $\mathcal{R}_1, \ldots, \mathcal{R}_n$ are WN and share only constructors, then $\twoheadrightarrow_{\mathcal{R}_1;\ldots;\mathcal{R}_n}$ is terminating [13], as a corollary of results of [21,24]. Now let us give an intuition on how to prove termination of $\twoheadrightarrow_{\mathcal{R}_1;\ldots;\mathcal{R}_n}$ for WN TRSs $\mathcal{R}_1, \ldots, \mathcal{R}_n$ sharing function symbols. Let $\mathcal{R} = \mathcal{R}_1 \cup \mathcal{R}_2$, for example. For proving termination of $\twoheadrightarrow_{\mathcal{R}_1;\mathcal{R}_2}$ on a set of initial terms $E \subseteq \mathcal{T}(\mathcal{F})$, we need to prove that for any term $s \in E$, there is no possible infinite derivation $s \rightarrow^!_{\mathcal{R}_1} s'_1 \rightarrow^!_{\mathcal{R}_2} s_1 \rightarrow^!_{\mathcal{R}_1} s'_2 \rightarrow^!_{\mathcal{R}_2} s_2 \rightarrow^!_{\mathcal{R}_1} \ldots$ In that case, a criterion for proving termination of $\twoheadrightarrow_{\mathcal{R}_1;\mathcal{R}_2}$ on E is the following: construct the sets $G_1 = \mathcal{R}^!_2(\mathcal{R}^!_1(E))$, $G_2 = \mathcal{R}^!_2(\mathcal{R}^!_1(G_1))$, ... until we get a fixpoint G_m s.t. $G_m = \mathcal{R}^!_2(\mathcal{R}^!_1(G_m))$. Then if $\mathcal{R}_1$ WN on $E, G_1, G_2, \ldots$, and $\mathcal{R}_2$ WN on $\mathcal{R}^!_1(E), \mathcal{R}^!_1(G_1), \ldots$ and if $G_m \subseteq IRR(\mathcal{R}_1 \cup \mathcal{R}_2)$, then $\mathcal{R}$ is WN on E and $\mathcal{R}$ is terminating on E under SRS.

Proposition 11. *If $\mathcal{R}_1, \ldots, \mathcal{R}_n$ are WN resp. on subsets $E_1, \ldots, E_n$ of $\mathcal{T}(\mathcal{F})$, the rewriting relation under SRS $\twoheadrightarrow_{\mathcal{R}_1;\ldots;\mathcal{R}_n}$ is terminating on E if the iterated sequence of sets $G_{k+1} = \mathcal{R}^!_n(\ldots \mathcal{R}^!_1(G_k)\ldots)$, starting from $G_0 = E$, has a fixpoint which is a subset of $IRR(\mathcal{R}_1 \cup \ldots \cup \mathcal{R}_n)$, and for all $k \geq 0$, $G_k \subseteq E_1, \mathcal{R}^!_1(G_k) \subseteq E_2, \ldots$, and $\mathcal{R}^!_{n-1}(\ldots \mathcal{R}^!_1(G_k)\ldots) \subseteq E_n$.*

3 Approximating $\mathcal{R}^*(E)$ and $\mathcal{R}^!(E)$

First, recall that $\mathcal{R}^!(E) = \mathcal{R}^*(E) \cap IRR(\mathcal{R})$. $IRR(\mathcal{R})$ is a regular tree language if $\mathcal{R}$ is left-linear [10], and a procedure for building a regular tree grammar (resp. a tree automaton) producing (resp. recognising) $IRR(\mathcal{R})$ can be found in [5]. However, $\mathcal{R}^*(E)$ is not necessarily a regular tree language, even if E is. The language $\mathcal{R}^*(E)$ is regular if E is regular and if $\mathcal{R}$ is either a ground TRS [7], a right-linear and monadic TRS [25], a linear and semi-monadic TRS [6] or an "inversely-growing" TRS [16], where "inversely-growing" means that every right-hand side is either a variable, or a term $f(t_1, \ldots, t_n)$ where $f \in \mathcal{F}$, $ar(f) = n$, and $\forall i = 1, \ldots, n$, t_i is a variable, a ground term, or a term whose variables do not occur in the left-hand side. On the other hand, for a given regular language E, $\mathcal{R}^*(E)$ is not necessarily regular, even if $\mathcal{R}$ is a confluent and terminating linear TRS [15]. If $\mathcal{R}$ is not "inversely-growing", then $\mathcal{R}^*(E)$ is not necessarily regular [16].

Since our purpose is to deal with TRSs representing programs, we cannot stick to the decidable class of "inversely-growing" TRSs which is not expressive enough. Our goal here is to define, an *approximation of* $\mathcal{R}^*(E)$ i.e. a regular superset of $\mathcal{R}^*(E)$ for left-linear TRSs and regular sets E. Then, since regular langages are closed by intersection, the intersection between the regular superset of $\mathcal{R}^*(E)$ and $IRR(\mathcal{R})$ gives a regular superset of $\mathcal{R}^!(E)$. Before going into details of the construction of the approximation itself, let us first show why an approximation is sufficient for proving properties addressed in Section 2. Let $\mathcal{R}$ be a TRS, and $E \subseteq \mathcal{T}(\mathcal{F})$. For any set G, let *super*(G) be a regular superset of G. For sufficient completeness: if *super*$(\mathcal{R}^!(E)) \subseteq \mathcal{T}(\mathcal{C})$ then $\mathcal{R}^!(E) \subseteq \mathcal{T}(\mathcal{C})$.

For reachability testing: if $C[t\sigma] \notin super(\mathcal{R}^*(E))$ then $C[t\sigma] \notin \mathcal{R}^*(E)$. For termination under SRS: if $super(\mathcal{R}_n^!(\dots super(\mathcal{R}_1^!(G_k))\dots\dots))$ is a subset of $IRR(\mathcal{R}_1 \cup \dots \cup \mathcal{R}_n)$ then so is $\mathcal{R}_n^!(\dots \mathcal{R}_1^!(G_k)\dots)$.

Now, starting from a tree automaton A s.t. $\mathcal{L}(A) = E$ and a left-linear TRS $\mathcal{R}$, we show how to build a tree automaton $\mathcal{T}_{\mathcal{R}}\uparrow(A)$ s.t. $\mathcal{L}(\mathcal{T}_{\mathcal{R}}\uparrow(A)) \supseteq \mathcal{R}^*(\mathcal{L}(A))$. The next proposition gives a sufficient condition for an automaton B to have such a property.

Proposition 12. *Let $\mathcal{R}$ be a left-linear TRS, $A = \langle \mathcal{F}, \mathcal{Q}, \mathcal{Q}_f, \Delta\rangle$, and $B = \langle \mathcal{F}, \mathcal{Q}', \mathcal{Q}_f, \Delta'\rangle$ two tree automata. $\mathcal{R}^*(\mathcal{L}(A)) \subseteq \mathcal{L}(B)$ if*

1. *$\Delta \subseteq \Delta'$, and*
2. *$\forall l \to r \in \mathcal{R}$, $\forall q \in \mathcal{Q}'$, $\forall \sigma \in \Sigma(\mathcal{Q}', \mathcal{X})$, $l\sigma \to_{\Delta'}^* q$ implies $r\sigma \to_{\Delta'}^* q$.*

Proof. (sketch) By definition, any term t of $\mathcal{R}^*(\mathcal{L}(A))$ is such that $\exists s \in \mathcal{L}(A)$ s.t. $s \to_{\mathcal{R}}^* t$. By induction on the size of the derivation $s \to_{\mathcal{R}}^* t$, we prove that if $s \to_{\mathcal{R}}^* t$ and $s \to_{\Delta'}^* q$ with $q \in \mathcal{Q}_f$ then $t \to_{\Delta'}^* q$, which implies that $t \in \mathcal{L}(B)$. See [12] for a detailed proof.

For building $\mathcal{T}_{\mathcal{R}}\uparrow(A)$, the algorithm we propose starts from the tree automaton A and incrementally adds to Δ the transitions necessary to ensure Condition 2 of Proposition 12, by computing critical peaks between rules of $\mathcal{R}$ and rules of Δ: $r\sigma\ {}_{\mathcal{R}}\!\!\leftarrow l\sigma \to_{\Delta}^* q$. If $r\sigma \not\to_{\Delta}^* q$, then it is necessary to add the transition $r\sigma \to q$ to Δ. If the transition $r\sigma \to q$ is not normalised, then it has to be normalised according to Definition 2. The choice of new states used to normalise $r\sigma \to q$ is guided by the approximation function γ defined below:

Definition 13. Let $\mathcal{Q}$ be a set of states, $\mathcal{Q}_{new}$ be a set of new states s.t. $\mathcal{Q} \cap \mathcal{Q}_{new} = \emptyset$, and $\mathcal{Q}_{new}^*$ the set of sequences $q_1 \cdots q_k$ of states in $\mathcal{Q}_{new}$. An approximation function is a mapping $\gamma : \mathcal{R} \times (\mathcal{Q} \cup \mathcal{Q}_{new}) \times \Sigma(\mathcal{Q} \cup \mathcal{Q}_{new}, \mathcal{X}) \mapsto \mathcal{Q}_{new}^*$, such that $\gamma(l \to r, q, \sigma) = q_1 \cdots q_k$, where $k = Card(\mathcal{P}os_{\mathcal{F}}(r))$.

In the following, for any sequence $S = q_1 \cdots q_k \in \mathcal{Q}_{new}^*$, and for all i s.t. $1 \leq i \leq k$, $\pi_i(S)$ denotes the i-th element of the sequence S, i.e. q_i.

Definition 14. (Approximation Automaton) Let $A = \langle \mathcal{F}, \mathcal{Q}, \mathcal{Q}_f, \Delta\rangle$ be a tree automaton, $\mathcal{R}$ a left-linear TRS, $\mathcal{Q}_{new}$ a set of new states s.t. $\mathcal{Q} \cap \mathcal{Q}_{new} = \emptyset$, and γ an approximation function. An approximation automaton $\mathcal{T}_{\mathcal{R}}\uparrow(A)$ is a tree automaton $\langle \mathcal{F}, \mathcal{Q}', \mathcal{Q}_f, \Delta'\rangle$ s.t.

(1) $\mathcal{Q}' = \mathcal{Q} \cup \mathcal{Q}_{new}$, and
(2) $\Delta \subseteq \Delta'$, and
(3) $\forall l \to r \in \mathcal{R}$, $\forall q \in \mathcal{Q}'$, $\forall \sigma \in \Sigma(\mathcal{Q}', \mathcal{X})$, $l\sigma \to_{\Delta'}^* q$ implies

$$Norm_{\alpha}(r\sigma \to q) \subseteq \Delta'$$

where α is the abstraction of $r\sigma$ defined by: $\alpha(r\sigma|_{p_i}) = \pi_i(\gamma(l \to r, q, \sigma))$, for all $p_i \in \mathcal{P}os_{\mathcal{F}}(r) = \{p_1, \dots, p_k\}$, s.t. $p_i \prec p_{i+1}$ for $i = 1 \dots k-1$ (where $\prec$ is the lexicographic ordering).

By choosing specific approximation functions γ, we obtain specific approximations.

Theorem 15. *Given a tree automaton A and a left-linear TRS $\mathcal{R}$, every approximation automaton satisfies: for any approximation function γ,*

$$\mathcal{L}(\mathcal{T}_{\mathcal{R}}\!\uparrow(A)) \supseteq \mathcal{R}^*(\mathcal{L}(A))$$

Proof. (sketch) For proving $\mathcal{L}(\mathcal{T}_{\mathcal{R}}\!\uparrow(A)) \supseteq \mathcal{R}^*\mathcal{L}(A)$, it is enough to prove that the approximation automata verifies Conditions 1 and 2 of Proposition 12, for all approximation functions γ. By Definition 14, $\mathcal{T}_{\mathcal{R}}\!\uparrow(A)$ trivially verifies Condition 1. Then, to prove that $\mathcal{T}_{\mathcal{R}}\!\uparrow(A)$ also verifies Condition 2 of Proposition 12, it is enough to prove that $Norm_\alpha(r\sigma \to q) \subseteq \Delta'$ implies $r\sigma \to^*_{\Delta'} q$. See [12] for a detailed proof.

For any rule $l \to r \in \mathcal{R}$, in order to find a $\mathcal{Q}$-substitution σ and a state $q \in \mathcal{Q}$ s.t. $l\sigma \to^*_\Delta q$, it is possible to enumerate every possible combination of σ and q and check whether $l\sigma \to^*_\Delta q$. However, this solution is not usable in practice, especially when $\mathcal{Q}$ is a large set, due to the huge number of possible σ and q to consider. In [13,12], we detail a *matching algorithm* which starts from a *matching problem* $l \trianglelefteq q$ and a set of transitions Δ, and gives every solution $\sigma : \mathcal{X} \mapsto \mathcal{Q}$ s.t. $l\sigma \to^*_\Delta q$. This algorithm is used in our implementation.

However, adding transitions to Δ may not terminate, depending on the approximation function γ used, as in the following example.

Example 16. Let A be an automaton s.t. $\Delta = \{app(q_0, q_0) \to q_1, cons(q_2, q_1) \to q_0, nil \to q_0, nil \to q_1, a \to q_2\}$, $rl = app(cons(x, y), z) \to cons(x, app(y, z))$, $\mathcal{R} = \{rl\}$, and let γ be the approximation function mapping every tuple (rl, q, σ) to one new state (since $Card(\mathcal{P}os_{\mathcal{F}}(cons(x, app(y, z)))) = 1$).

Step 1 If we apply the matching algorithm on $app(cons(x, y), z) \trianglelefteq q_1$, we obtain a solution $\sigma = \{x \mapsto q_2, y \mapsto q_1, z \to q_0\}$, corresponding to the following critical peak: $cons(q_2, app(q_1, q_0))\ {}_{\mathcal{R}}\!\leftarrow app(cons(q_2, q_1), q_0) \to^*_\Delta q_1$. Thus, the transition to be added is $cons(q_2, app(q_1, q_0)) \to q_1$. Let q_3 be the new state s.t. $\gamma(rl, q_1, \sigma) = q_3$. Then, since $\mathcal{P}os_{\mathcal{F}}(cons(q_2, app(q_1, q_0)) = \{p_1\} = \{2\}$, we have $\alpha(app(q_1, q_0)) = \pi_1(\gamma(rl, q_1, \sigma)) = q_3$, and the set of normalised transitions to be added to Δ is:
$Norm_\alpha(cons(q_2, app(q_1, q_0)) \to q_1) = \{cons(q_2, q_3) \to q_1, app(q_1, q_0) \to q_3\}$.

Step 2 Applying the matching algorithm on $app(cons(x, y), z) \trianglelefteq q_3$ gives a solution $\sigma' = \{x \mapsto q_2, y \mapsto q_3, z \mapsto q_0\}$, corresponding to the following critical peak: $cons(q_2, app(q_3, q_0))\ {}_{\mathcal{R}}\!\leftarrow app(cons(q_2, q_3), q_0) \to^*_\Delta q_3$. Thus, the transition to be added is $cons(q_2, app(q_3, q_0)) \to q_3$. Let q_4 be the new state s.t. $\gamma(rl, q_3, \sigma') = q_4$. Then, $\alpha(app(q_3, q_0)) = q_4$, and the set of normalised transitions to be added to Δ is:
$Norm_\alpha(cons(q_2, app(q_3, q_0)) \to q_3) = \{cons(q_2, q_4) \to q_1, app(q_3, q_0) \to q_4\}$.

This process can go on forever and add infinitely many new states. This is due to the fact that we can apply recursively the rule $app(cons(x, y), z) \to cons(x, app(y, z))$ onto infinitely growing terms recognised by the automaton A (with transitions Δ), as shown on the following figure.

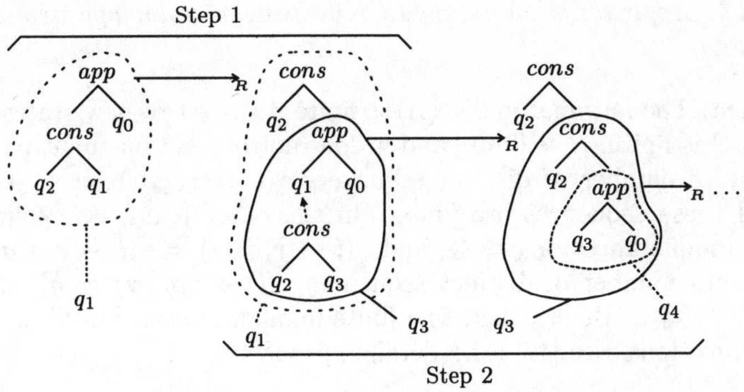

In order to have a finite automaton approximating the set $\mathcal{R}^*(\mathcal{L}(A))$, the intuition is to *fold recursive calls* into a unique new state. In the previous example, during Step 1, by applying the rule of $app(cons(x,y),z) \to cons(x, app(y,z))$ on $app(cons(q_2,q_1),q_0)$, we have obtained the configuration $cons(q_2, app(q_1,q_0))$, and we have created a new state q_3 recognising the subterm $app(q_1,q_0)$. During Step 2 we have applied *the same rule* on the subterm $app(q_1,q_0)$ recognised by q_3. In order to fold this recursive call in Step 2, we simply *re-use* the state q_3, instead of creating a new state q_4 for normalising the transition $cons(q_2, app(q_3,q_0)) \to q_3$ obtained in Step 2. Thus we obtain the set of normalised transitions $\{cons(q_2,q_3) \to q_3, app(q_3,q_0) \to q_3\}$ to be added to Δ. No more state nor transition needs to be further added and this automaton recognises a superset of $\mathcal{R}^*(\mathcal{L}(A))$. This is one of the basic idea of the ancestor approximation, which is formalised below.

Informally, every state $q \in \mathcal{Q}' = \mathcal{Q} \cup \mathcal{Q}_{new}$ has a unique ancestor $q_a \in \mathcal{Q}$. The ancestor of any state $q \in \mathcal{Q}$ is q itself, and the ancestor of every new state $q' \in \mathcal{Q}_{new}$ occuring in the sequence $\gamma(l \to r, q, \sigma)$ (used to normalise a new transition $r\sigma \to q$), is the ancestor of q. In the ancestor approximation, (1) the γ function does not depend on the σ parameter and, (2) for any new state $q' \in \mathcal{Q}_{new}$, $\gamma(l \to r, q', \sigma) = \gamma(l \to r, q, \sigma)$, where $q \in \mathcal{Q}$ is the ancestor of q'.

Definition 17. An approximation function γ is called ancestor approximation if

1. $\forall l \to r \in \mathcal{R}, \forall q \in \mathcal{Q}', \forall \sigma_1, \sigma_2 \in \Sigma(\mathcal{Q}', \mathcal{X})$,

$$\gamma(l \to r, q, \sigma_1) = \gamma(l \to r, q, \sigma_2), \text{ and}$$

2. $\forall l_1 \to r_1, l_2 \to r_2 \in \mathcal{R}, \forall q \in \mathcal{Q}', \forall q_1, \ldots, q_k \in \mathcal{Q}_{new}, \sigma_1, \sigma_2 \in \Sigma(\mathcal{Q}', \mathcal{X})$,

$$\gamma(l_1 \to r_1, q, \sigma_1) = q_1 \ldots q_k \Rightarrow \forall i = 1 \ldots k, \gamma(l_2 \to r_2, q_i, \sigma_2) = \gamma(l_2 \to r_2, q, \sigma_2).$$

Note that in the particular case of Example 16, using the ancestor approximation, we have $\gamma(rl, q_1, \sigma) = q_3$, and by case 2 of Definition 17 we get $\gamma(rl, q_3, \sigma') = \gamma(rl, q_1, \sigma')$, by case 1 we get that $\gamma(rl, q_1, \sigma') = \gamma(rl, q_1, \sigma) = q_3$, thus we have $\gamma(rl, q_3, \sigma') = q_3$, and the construction of $\mathcal{T}_{\mathcal{R}}\!\uparrow\!(A)$ becomes finite.

Theorem 18. *Approximation automata built using ancestor approximation are finite automata.*

Proof. (sketch) The automaton $\mathcal{T}_{\mathcal{R}}\!\uparrow(A)$ is finite if the set of new states $\mathcal{Q}_{new}$ is finite. Since $\mathcal{Q}$ is finite, $\mathcal{R}$ is finite, and γ does not depend on the σ parameter, there is a finite number of distinct sequences $\gamma(l \to r, q, \sigma)$ for $l \to r \in \mathcal{R}$, $q \in \mathcal{Q}$, and these sequences are finite. On the other hand, every state $q' \in \mathcal{Q}_{new}$ has a unique ancestor $q \in \mathcal{Q}$, and $\gamma(l \to r, q', \sigma) = \gamma(l \to r, q, \sigma)$. Thus, there is a finite number of distinct sequences $\gamma(l \to r, q', \sigma) = q'_1 \dots q'_n$ with $q', q'_1, \dots, q'_n \in \mathcal{Q}_{new}$. Hence, there is a finite number of states in $\mathcal{Q}_{new}$, used to normalise transitions. See [12] for a detailed proof.

4 Experiments

Working on tree automaton by hand is always a heavy task. In order to experiment and check feasibility of the method, we have implemented in ELAN [18] a library of usual algorithms on tree automaton: union, intersection, cleaning, inclusion test, as well as algorithms for building the tree automata $\mathcal{T}_{\mathcal{R}}\!\uparrow(A)$, and $A_{IRR(\mathcal{R})}$ (the automaton recognising the set $IRR(\mathcal{R})$) for a given automaton A and a given left-linear TRS $\mathcal{R}$. In all the following examples, we use the same ancestor approximation method. We have experimented with several other approximations: if the γ function does not depend on the rule $l \to r$, on the state q or on the position p, then the approximation automaton is smaller, and faster to compute. However, the recognised language is bigger and sometimes not precise enough for our purpose. On the other hand, if for every σ, the γ function have distinct values, then the construction of the automaton is not necessarily terminating.

4.1 Reachability Testing

Let $\mathcal{R}_1$ be a TRS computing the function $A_n^p = \frac{n!}{(n-p)!}$, and $Aut(0)$ a tree automaton recognising the set $\mathcal{L}(Aut(0)) = \mathcal{L}(Aut(0), q_0) = \{A(n,p) \mid n, p \in \mathcal{L}(Aut(0), q_1)\}$ where $\mathcal{L}(Aut(0), q_1) = Nat = \{0, s(0), \dots\}$. The TRS $\mathcal{R}_1$ and the automaton $Aut(0)$ are given as input to our prototype in the following syntax:

```
specification Anp
Vars    x y n p
Ops
    A:2 minus:2 div:2 o:0 s:1 fact:1 plus:2 mult:2
R1
    A(n, p) -> div(fact(n), fact(minus(n, p)))
    fact(s(x)) -> mult(s(x), fact(x))
    fact(o) -> s(o)
    mult(o, x) -> o
    mult(s(x), y) -> plus(mult(x, y), y)
    div(o, s(y)) -> o
    div(s(x), s(y)) -> s(div(minus(x, y), s(y)))
    plus(x, o) -> x
    plus(x, s(y)) -> s(plus(x, y))
    minus(x, o) -> x
    minus(o, x) -> o
    minus(s(x), s(y)) -> minus(x, y)
    nil
```

```
Automata
    Description of Aut(0)
    states q|0.q|1.nil
    final states q|0.nil
    transitions A(q|1, q|1) -> q|0
            o-> q|1
            s(q|1) -> q|1
            nil
    End of Description
    nil
```

Computing the automaton $\mathcal{T}_{\mathcal{R}_1}\!\uparrow(Aut(0))$ s.t. $\mathcal{L}(\mathcal{T}_{\mathcal{R}_1}\!\uparrow(Aut(0))) \supseteq \mathcal{R}_1^*(\mathcal{L}(Aut(0)))$, can be achieved by evaluating the following query: `T_up(R1) on (Aut(0))`, and the result is the automaton $Aut(1)$:

```
[] result term:
   Description of Aut(1) states
   q|12.q|13.q|11.q|10.q|9.q|8.q|6.q|7.q|2.q|5.q|3.q|0.q|1.nil final states q|0.nil
   transitions s(q|11)->q|11.o->q|11.minus(q|10,q|10)->q|12.minus(q|8,q|10)->q|12. s(q|10)->
   q|13.s(q|10)->q|3.minus(q|10,q|8)->q|12.minus(q|8,q|8)->q|12.s(q|8)->q|13.div(q|12,q|13)
   ->q|11.s(q|11)->q|0.plus(q|9,q|7)->q|3.s(q|8)->q|3.mult(q|6,q|7)->q|3.plus(q|9,q|10)->
   q|10.plus(q|9,q|10)->q|12.s(q|10)->q|7.s(q|10)->q|10.s(q|10)->q|12.s(q|10)->q|9.
   plus(q|9,q|8)->q|10. plus(q|9,q|8)->q|12.s(q|10)->q|2.plus(q|9,q|7)->q|2.plus(q|9,q|7)->
   q|9.plus(q|9,q|7)->q|10.plus(q|9,q|7)->q|12.o->q|9.o->q|10.o->q|12.mult(q|1,q|7)->q|9.
   mult(q|1,q|7)->q|10.mult(q|1,q|7)->q|12.plus(q|9,q|7)->q|7.s(q|8)->q|2.o->q|8.s(q|8)->q|7.
   mult(q|6,q|7)->q|7.s(q|1)->q|6.fact(q|1)->q|7.mult(q|6,q|7)->q|2.fact(q|1)->q|2.
   minus(q|1,q|1)->q|5.fact(q|5)->q|3.div(q|2,q|3)->q|0.A(q|1,q|1)->q|0.o->q|1.o->q|5.s(q|1)
   ->q|1.s(q|1)->q|5.nil End of Description
```

The pattern $div(x,0)$ is not $\mathcal{R}_1$-reachable from $\mathcal{L}(Aut(0))$ if for all ground contexts $C[\,]$ and all substitutions σ, $C[dix(x,0)\sigma] \notin \mathcal{L}(\mathcal{T}_{\mathcal{R}_1}\!\uparrow(Aut(0)))$. This can be checked evaluating the query: `(div(x, o) ?= states) with (Aut(1))`. The result is

```
[] result term:
   nil
```

meaning that there exists no substitution σ and no state $q \in \mathcal{Q}$ s.t. $div(x,0)\sigma \to_{\Delta}^{*} q$, where $\mathcal{Q}$ and Δ are respectively the set of states and the set of transitions of $Aut(1)$. An interesting aspect of this method is that the automaton $\mathcal{T}_{\mathcal{R}}\!\uparrow(A)$ is computed once for all, and the check itself is a simple and low cost operation. Another advantage is that for computing $\mathcal{T}_{\mathcal{R}}\!\uparrow(A)$, the TRS $\mathcal{R}$ is not supposed to be terminating nor even weakly normalising. This is of great interest when using TRS to encode non-terminating systems, like systems of communicating processes, for example. Note that such non-terminating TRS cannot be handled by induction proof techniques that need a well-founded ordering for proving termination of the TRS.

4.2 Sufficient Completeness

In order to prove sufficient completeness of $A(n,p)$ with $n, p \in Nat$, we compute the intersection automaton between $Aut(1)$, computed previously, and the automaton recognising the set $IRR(\mathcal{R}_1)$, computed by the function `build_nf(R1)`. The query is `simplify(Aut(1) inter build_nf(R1))`, and the result is:

```
[] result term:
    Description of Aut(2) states q|0.q|1.nil final states q|1.nil
    transitions s(q|0)->q|1.s(q|0)->q|0.o->q|0.nil End of Description
```

Thus, the superset of $\mathcal{R}_1^!(\mathcal{L}(Aut(0)))$ is $\mathcal{L}(Aut(2)) = \mathcal{L}(Aut(2), q_1) = \{s(x) \mid x \in \mathcal{L}(Aut(2), q_0)\}$, and $\mathcal{L}(Aut(2), q_0) = \{0, s(0), \ldots\}$. Thus $\mathcal{L}(Aut(2), q_0) = Nat$, and $\mathcal{L}(Aut(2)) = Nat^*$. Therefore, we trivially have $\mathcal{R}_1^!(\mathcal{L}(Aut(0))) \subseteq Nat^* \subseteq \mathcal{T}(\mathcal{C})$ and if $\mathcal{R}_1$ is weakly normalising on terms $A(n,p)$ with $n, p \in Nat$, then $\mathcal{R}_1$ is also sufficiently complete on those terms. Note that, if $Aut(2)$ is more complex, inclusion between automaton $Aut(2)$ and an automaton recognising exactly $\mathcal{T}(\mathcal{C})$ can also be verified automatically by our prototype.

4.3 Sequential Reduction Strategy

In this third example, we show that sequential reduction strategy is interesting for proving termination of programs combining different methods of termination proof. The following specification defines a function $fact\_list(i, j)$, that constructs a list of naturals $(i!, (i+1)!, \dots, (j-1)!, j!)$. The module $\mathcal{R}_1$ constructs the list and the module $\mathcal{R}_2$ achieves the computation of the factorial function.

```
specification fact_list
Vars     x y z

Ops
  o:0 p:1 s:1 fact:1 plus:2 mult:2 cons:2 int:2
  intlist:1 null:0 fact_list:2 appfact:1

R1
  fact_list(x, y) -> appfact(int(x, y))
  appfact(null) -> null
  appfact(cons(x,y)) -> cons(fact(x),appfact(y))
  intlist(null) -> null
  intlist(cons(x, y)) -> cons(s(x), intlist(y))
  int(o,o) -> cons(o,null)
  int(o,s(y)) -> cons(o, int(s(o), s(y)))
  int(s(x),o) -> null
  int(s(x), s(y)) -> intlist(int(x, y))
  nil
```

```
R2
  p(s(x)) -> x
  mult(o, x) -> x
  mult(s(x), y) -> plus(mult(x, y), y)
  plus(x, o) -> x
  plus(x, s(y)) -> s(plus(x, y))
  fact(s(x)) -> mult(s(x), fact(p(s(x))))
  fact(o) -> s(o)
  nil

Automata
  Description of Aut(0)
  states q|0.q|1.nil
  final states q|0.nil
  transitions
    fact_list(q|1,q|1) -> q|0
    o -> q|1 s(q|1) -> q|1
    nil
  End of Description
  nil
```

Note that neither termination of $\mathcal{R}_1$ nor termination of $\mathcal{R}_2$ can be proven by a simplification ordering. However, termination of $\mathcal{R}_1$ can be proved by the dependency pair method [2], and on the other hand, termination of $\mathcal{R}_2$ can be proved by GPO [8]. Instead of reconsidering the termination of the whole TRS $\mathcal{R}_1 \cup \mathcal{R}_2$, we can automatically verify that the (hierarchical) combination of those two systems is terminating under the sequential reduction strategy, for every initial term from the regular set $\mathcal{L}(Aut(0)) = \mathcal{L}(Aut(0), q_0) = \{fact\_list(n,p) \mid n, p \in \mathcal{L}(Aut(0), q_1)\}$ where $\mathcal{L}(Aut(0), q_1) = \{0, s(0), \dots\} = Nat$. The query `start(Aut(0))` iterates the process described in Section 2.3, implemented with the `T_up` and `build_nf` operations, until we get a fixpoint. The result of this proof is the following:

```
[] result term:
   [true,Description of nil states q|0.q|1.q|2.q|3.q|4.nil final
   states q|4.nil transitions cons(q|2,q|3)->q|3.null->q|4.null->q|3.cons(q|2,q|3)->q|4.
   s(q|0)->q|1.o->q|0.s(q|1)->q|1.s(q|1)->q|2.s(q|0)->q|2.nil End of Description]
```

where the first field is `true` — the combination is terminating under the sequential reduction strategy — and the second field contains the automaton recognising the superset of the normal forms: lists (possibly empty) of strictly positive natural numbers, which is what was expected by definition of function $fact\_list$, and which also proves sufficient completeness of $\mathcal{R}_1 \cup \mathcal{R}_2$ under sequential reduction strategy on $\mathcal{L}(Aut(0))$.

4.4 Testing co-domains of functions

This is a last example showing that computing a superset of the set of normal forms may be of great help also in debugging a functional program. Assume that you have the following program defining a function which reverses a list of elements.

```
specification reverse
Vars    x y z
Ops
    a:0 b:0 rev:1 cons:2 append:2 null:0
R1
    rev(null) -> null
    rev(cons(x, y)) -> append(rev(y), cons(x, null))
    append(null, x) -> null
    append(cons(x, y), z) -> cons(x, append(y, z))
    nil
```

```
Automata
    Description of Aut(0)
    states q|0.q|1.q|2.nil
    final states q|0.nil
    transitions rev(q|1) -> q|0.
                cons(q|2, q|1) -> q|1.
                null -> q|1.
                a -> q|2.
                b -> q|2.
                nil
    End of Description
```

where $\mathcal{L}(Aut(0)) = \mathcal{L}(Aut(0), q_0) = \{rev(l) \mid l \in \mathcal{L}(Aut(0), q_1)\}$, $\mathcal{L}(Aut(0), q_1) = \{null, cons(x, y) \mid x \in \mathcal{L}(Aut(0), q_2), y \in \mathcal{L}(Aut(0), q_1)\}$, and $\mathcal{L}(Aut(0), q_2) = \{a, b\}$. In other words, $\mathcal{L}(Aut(0))$ is of the form $rev(l)$ where l is any flat list of a and b, possibly empty. If we compute the automaton recognising the superset of $\mathcal{R}_1^!(\mathcal{L}(Aut(0)))$, the superset of co-domain, by evaluating the query `simplify(T_up(R1) on(Aut(0)) inter build_nf(R1))`, we obtain:

```
[] result term:
   Description of Aut(1) states q|0.nil final
   states q|0.nil transitions null->q|0.nil End of Description
```

Thus $\mathcal{L}(Aut(1))$, the superset of $\mathcal{R}_1^!(\mathcal{L}(Aut(0)))$, is the singleton $\{null\}$, the empty list. That is clearly not what is expected from the reverse function. If you check TRS $\mathcal{R}_1$ in detail, you will notice that it is wrong: in the third rule of $\mathcal{R}_1$, the right-hand side should be x rather than $null$. The interesting remark here is that $\mathcal{R}_1$ has all usual good properties: it is terminating, confluent, and sufficiently complete on $\mathcal{L}(Aut(0))$. Note also that typing $\mathcal{R}$ would not detect any error. The main interest of the co-domain estimation is to be complementary to usual verification techniques used on TRSs: confluence, termination, sufficient completeness, and typing. After fixing the bug in $\mathcal{R}_1$, we obtain:

```
[] result term:
   Description of Aut(1) states q|0.q|1.q|2.q|3.nil final states q|3.nil
   transitions b->q|1.a->q|1.null->q|3.cons(q|1,q|0)->q|3.cons(q|1,q|0)
   ->q|2.null->q|0.cons(q|1,q|2)->q|2.cons(q|1,q|2)->q|3.nil
   End of Description
```

where $Aut(1)$ recognise any flat list of a and b, possibly empty.

5 Conclusion

We have shown in this work that the computation of regular supersets of $\mathcal{R}$-descendants and $\mathcal{R}$-normal forms using tree automata techniques can provide assistance for checking a few properties of TRSs seen as functional programs.

An important part of this work is devoted to the computation of a regular superset of the set of descendants $\mathcal{R}^*(E)$ for any left-linear TRS $\mathcal{R}$ and any regular set of terms E. The approach proposed here is based on the computation of an approximation automaton recognising a superset of $\mathcal{R}^*(E)$. This approximation seems to be sufficient for our purposes in many practical cases. Approximation of regular language is a notion that was already used in in [16], but in a different way and for a different purpose. In [16], Jacquemard approximates a TRS by another one for which the set of descendants is regular, whereas in our approach, we approximate the set of new states used for normalising transitions, in order to fold recursion when necessary. The set of descendants can be computed exactly thanks to the Tree Tuple Synchronised Grammars (TTSG) approach of non-regular langages proposed in [22]. However, this approach deals with more restricted classes of TRSs; namely linear confluent constructor systems. Moreover, in practice, efficiency of TTSGs for our purposes is not obvious.

A promising application area is the study of non-terminating TRSs encoding the behaviour of systems of communicating processes or systems of parallel processes sharing memory. In this framework, we can prove that there is no deadlock and also some general "reachability" properties: ensure mutual exclusion, ensure that a process never stops, etc. In further research, we intend to compute another regular approximation: a *subset of* $\mathcal{R}^*(E)$ in order to achieve some reachability testing in the other way: for instance to prove that a specific behaviour must occur, we may have to check that a specific pattern *does occur* in the set of $\mathcal{R}$-descendants. We also would like to get rid of the left-linear limitation in order to enlarge the class of programs to be checked, and to compute more precise approximations.

Acknowledgements

I would like to thank Hélène Kirchner, Christophe Ringeissen, Aart Middeldorp, Bernhard Gramlich, Isabelle Gnaedig for comments and discussion on this work, Florent Jacquemard, Marc Tommasi, Gregory Kucherov for discussion on tree automata and regular tree langages, as well as referees for their useful comments.

References

1. T. Arts and J. Giesl. Automatically proving termination where simplification orderings fail. In M. Bidoit and M. Dauchet, editors, *Proc. 22nd CAAP Conf., Lille (France)*, volume 1214 of *LNCS*, pages 261–272. Springer-Verlag, 1997.
2. T. Arts and J. Giesl. Proving innermost termination automatically. In *Proc. 7th RTA Conf., Sitges (Spain)*, volume 1232 of *LNCS*, pages 157–171. Springer-Verlag, 1997.
3. H. Comon. Sufficient completeness, term rewriting system and anti-unification. In J. Siekmann, editor, *Proc. 8th CADE Conf., Oxford (UK)*, volume 230 of *LNCS*, pages 128–140. Springer-Verlag, 1986.
4. H. Comon, M. Dauchet, R. Gilleron, D. Lugiez, S. Tison, and Tommasi. Tree automata techniques and applications. Preliminary Version, `http://l3ux02.univ-lille3.fr/tata/`, 1997.

5. H. Comon and J.-L. Rémy. How to characterize the language of ground normal forms. Technical Report 676, INRIA-Lorraine, 1987.
6. J. Coquidé, M. Dauchet, R. Gilleron, and S. Vágvölgyi. Bottom-up tree pushdown automata and rewrite systems. In R. V. Book, editor, *Proc. 4th RTA Conf., Como (Italy)*, volume 488 of *LNCS*, pages 287–298. Springer-Verlag, 1991.
7. M. Dauchet and S. Tison. The theory of ground rewrite systems is decidable. In *Proc. 5th LICS Symp., Philadelphia (Pa., USA)*, pages 242–248, June 1990.
8. N. Dershowitz and C. Hoot. Natural termination. *TCS*, 142(2):179–207, May 1995.
9. N. Dershowitz and J.-P. Jouannaud. *Handbook of Theoretical Computer Science*, volume B, chapter 6: Rewrite Systems, pages 244–320. Elsevier Science Publishers B. V. (North-Holland), 1990. Also as: Research report 478, LRI.
10. J. H. Gallier and R. V. Book. Reductions in tree replacement systems. *TCS*, 37:123–150, 1985.
11. F. Gécseg and M. Steinby. *Tree automata*. Akadémiai Kiadó, Budapest, Hungary, 1984.
12. T. Genet. Decidable approximations of sets of descendants and sets of normal forms (extended version). Technical Report RR-3325, INRIA, 1997.
13. T. Genet. Proving termination of sequential reduction relation using tree automata. Technical Report 97-R-091, CRIN, 1997.
14. T. Genet and I. Gnaedig. Termination proofs using gpo ordering constraints. In M. Dauchet, editor, *Proc. 22nd CAAP Conf., Lille (France)*, volume 1214 of *LNCS*, pages 249–260. Springer-Verlag, 1997.
15. R. Gilleron and S. Tison. Regular tree languages and rewrite systems. *Fundamenta Informaticae*, 24:157–175, 1995.
16. F. Jacquemard. Decidable approximations of term rewriting systems. In H. Ganzinger, editor, *Proc. 7th RTA Conf., New Brunswick (New Jersey, USA)*, pages 362–376. Springer-Verlag, 1996.
17. D. Kapur, P. Narendran, and H. Zhang. On sufficient completeness and related properties of term rewriting systems. *Acta Informatica*, 24:395–415, 1987.
18. C. Kirchner, H. Kirchner, and M. Vittek. Designing constraint logic programming languages using computational systems. In P. Van Hentenryck and V. Saraswat, editors, *Principles and Practice of Constraint Programming. The Newport Papers.*, chapter 8, pages 131–158. The MIT press, 1995.
19. E. Kounalis. Completeness in data type specifications. In B. Buchberger, editor, *Proceedings EUROCAL Conference, Linz (Austria)*, volume 204 of *LNCS*, pages 348–362. Springer-Verlag, 1985.
20. M. Kurihara and I. Kaji. Modular term rewriting systems and the termination. *IPL*, 34:1–4, Feb. 1990.
21. M. Kurihara and A. Ohuchi. Modular term rewriting systems with shared constructors. *Journal of Information Processing of Japan*, 14(3):357–358, 1991.
22. S. Limet and P. Réty. E-unification by means of tree tuple synchronized grammars. In M. Dauchet, editor, *Proc. 22nd CAAP Conf., Lille (France)*, volume 1214 of *LNCS*, pages 429–440. Springer-Verlag, 1997.
23. T. Nipkow and G. Weikum. A decidability result about sufficient completeness of axiomatically specified abstract data types. In *6th GI Conference*, volume 145 of *LNCS*, pages 257–268. Springer-Verlag, 1983.
24. E. Ohlebusch. *Modular Properties of Composable Term Rewriting Systems.* PhD thesis, Universität Bielefeld, Bielefeld, 1994.
25. K. Salomaa. Deterministic Tree Pushdown Automata and Monadic Tree Rewriting Systems. *J. of Computer and System Sciences*, 37:367–394, 1988.

Algorithms and Reductions for Rewriting Problems

[1]Rakesh M. Verma*, [2]Michael Rusinowitch, [3]Denis Lugiez
1. Computer Science Dept., University of Houston, TX 77204, USA.
2. INRIA Lorraine, BP101-54602 Villers les Nancy, France.
3. CMI, 39 r Juliot-Curie, F-13453 Marseille Cedex 13, France.

Abstract. In this paper we initiate a study of polynomial-time reductions for some basic decision problems of rewrite systems. We then give a polynomial-time algorithm for Unique-normal-form property of ground systems for the first time. Next we prove undecidability of these problems for a fixed string rewriting system using our reductions. Finally, we prove partial decidability results for Confluence of commutative semi-thue systems. The Confluence and Unique-normal-form property are shown Expspace-hard for commutative semi-thue systems. We also show that there is a family of string rewrite systems for which the word problem is trivially decidable but confluence undecidable, and we show a linear equational theory with decidable word problem but undecidable linear equational matching.

1 Introduction

In this paper, we initiate a systematic study of reducibilities among some basic decision problems for rewrite systems. We also develop efficient algorithms and prove decidability/undecidability for some restricted versions of these problems. The importance of these problems is well-established and it is well-known that they are all undecidable in general. Surprisingly, although these problems are well-known, so far – to the best of our knowledge – there is no systematic attempt to establish tight relationships between them. Specifically, we are not aware of any effort to relate them via the well-known complexity-theoretic concept of a polynomial-time reduction or transformation. The benefit of such reductions between problems cannot be overemphasized. Using reducibilities it is possible to economically derive and express many results for these problems.

The problems considered here are: Reachability, Joinability, Word problem, Confluence, Unique normalization, Unique-normal-form property, and Equational matching. We prove many polynomial-time reductions among these problems for ordinary rewriting as well as for rewriting modulo equational theories. In designing these reductions one of the properties we wish to preserve is the groundness of instances. Using these reductions and the undecidability of a restricted version of Reachability we are able to show that all these problems are undecidable for a fixed ground rewriting system modulo associativity. We also show that there is a family of string rewrite systems for which the word problem is trivially decidable but confluence undecidable, and we give a linear equational theory with decidable word problem but undecidable linear equational matching.

* Research supported in part by NSF grant CCR-9303011 and INRIA.

We show that there is a polynomial-time algorithm for the Unique-normal-form property of ground rewrite systems, and for ground rewriting modulo commutativity. This result is interesting since this property may be regarded as being "close" to Confluence and there is no good upper bound for Confluence of ground systems. Using this result and polynomial reductions we are able to show membership of some of these problems in P for ground rewriting and for ground rewriting modulo commutativity. Finally, we show the decidability of confluence for rewrite rules involving one AC operator and constants. Confluence testing for ground rewrite systems modulo a single AC operator is shown to be Expspace-hard via a simple reduction from the word problem for commutative semigroups, which is proved Expspace-complete in [12]. Some proofs are omitted to save space, the full version is available as an INRIA technical report.

We now discuss earlier related work on these problems. Decidability of Confluence and Reachability for ground systems was shown by [3, 14]. Decidability of Reachability, Joinability and Confluence for left-linear, right-ground systems was shown in [3]. Decidability of Reachability and Joinability for right-ground rewrite systems was shown in [15]. In [17] a polynomial time algorithm for reachability of ground systems is given. In [6] the decidability of reachability for ground systems and ground rewriting modulo associativity, commutativity is studied. Decidability of the theory of ground rewriting is studied in [4] and the word problem for right ground systems in [16].

2 Preliminaries

We assume familiarity with basic notions of rewriting (see [5, 9]). Let V be a countable set of *variables* and Σ be a countable set of function symbols with $\Sigma \cap V = \emptyset$. $\mathcal{T}$ is the set of all terms of a first-order language constructed from V and Σ. It is convenient to think of terms as ordered rooted trees. If $s \in \mathcal{T}$, then $Var(s)$ denotes the *set* of variables in s. A *rule* is a pair of terms $l \to r$, such that $l \notin V$ and $Var(r) \subseteq Var(l)$ (the variables in a rule are implicitly universally quantified). A *system* R is a finite set of rules. A rule $l \to r$ is *collapsing* if $r \in V$ and *noncollapsing* otherwise, it is *var-preserving* if $Var(l) = Var(r)$. Similarly, an equation $s = t$ is collapsing if either s or t is a variable, it is var-preserving if $Var(s) = Var(t)$, and it is *weakly var-preserving* if either $Var(s) \subseteq Var(t)$ or vice versa. We extend the concepts noncollapsing and var-preserving to sets of rules and equations in the obvious way. The notion of a *path* or *occurrence* is used to refer to subterms in a term as follows. A path is either the empty string λ that reaches the root or $o.i$ (o is a path and i an integer) which reaches the ith argument of the root of the subterm reached by o. $o \leq q$ whenever $\exists p\ o.p = q$; if $p \neq \lambda$ also, then $o < q$. t/o refers to the subterm of t reached by o. $O(t)$ denotes the set of occurrences of t.

We use $\to$ to denote rewrite relations. Let $\to$ be a relation over $\mathcal{T}$. We say that $\to$ is: *stable* iff $\forall \sigma,\ \forall s, t, s \to t \Rightarrow \sigma(s) \to \sigma(t)$; *compatible* iff $\forall s,\ \forall o \in O(s)$, $\forall t, u, t \to u \Rightarrow s[o \leftarrow t] \to s[o \leftarrow u]$. Clearly the ordinary rewrite relation is the smallest compatible stable relation containing R. We say that relation $\to$

is *confluent* (CR) if and only if $\forall s, t, u\ s \xrightarrow{*} t$ and $s \xrightarrow{*} u$ implies $\exists\ v$ such that $t \xrightarrow{*} v$ and $u \xrightarrow{*} v$.

Notation. We use $=_R$ to represent the least congruence relation containing $\rightarrow_R$. We also say that $a =_R b$ is an equational proof. When the set of rules R is clear from the context, we drop the subscripts from $\rightarrow$. The reflexive-transitive closure of $\rightarrow$ is denoted by $\xrightarrow{*}$, the transitive closure by $\xrightarrow{+}$, and the symmetric closure by $\leftrightarrow$. We use the term *root reduction* to indicate reduction of the entire term, and *nonroot reduction* to indicate reduction at a proper subterm. A polynomial-time reduction between two decision problems is denoted $\leq_P$.

Proposition 1 *(i) $s =_R t$ implies $\sigma(s) =_R \sigma(t)$ for any substitution σ (stability under substitution). (ii) $s =_R t$ implies $C[s] =_R C[t]$ for any context C (stability under contexts). (iii) Let $s =_R t$ and let O be the set of occurrences at which reductions are carried out in this proof. If $o \in O$ is any minimal occurrence, then $s/o =_R t/o$ (projection property).*

Definition 2 *Let R be a rewrite system. A term t is a* normal form *if there is no u such that $t \rightarrow u$. A term t* has a normal form *if there is a normal form u such that $t \xrightarrow{*} u$. R (or $\rightarrow_R$) is uniquely normalizing (is $UN^{\rightarrow}$) if for all s, t, u such that $s \xrightarrow{*} t$ and $s \xrightarrow{*} u$ and t, u are normal forms we have $t = u$. R (or $\rightarrow_R$) has unique normal forms (is $UN^{=}$, also unicity for short) if for all normal forms s, t with $s =_R t$ we have $s = t$.*

Lemma 3 *For every system R: $CR \Rightarrow UN^{=} \Rightarrow UN^{\rightarrow}$, but not the reverse.*

The following eight decision problems are studied in this paper.

Reachability. *Instance:* Rewrite System, R, terms s, t. *Question:* Does $s \xrightarrow{*}_R t$?

Normal form reachability. *Instance:* Rewrite System, R, terms s, t, where t is an R normal form. *Question:* Does $s \xrightarrow{*}_R t$?

Joinability. *Instance:* Rewrite System, R, terms s, t. *Question:* Is there a term u such that $s \xrightarrow{*}_R u$ and $t \xrightarrow{*}_R u$?

For the three problems above, we also define versions in which the input rewrite system satisfies some property P. To indicate these versions, we attach the phrase "for P systems" to the basic problem.

Confluence/UN$^{=}$/UN$^{\rightarrow}$. *Instance:* System R. *Question:* Is R confluent, etc.?

Word problem for Equational theories (WP). *Instance:* Equational theory E, terms s, t. *Question:* Does $s =_E t$?

Linear E-Matching. *Instance:* Linear equational theory E, linear term $s \in T(\Sigma, V)$ and $t \in T(\Sigma)$. *Question:* Is there a substitution σ such that $\sigma(s) =_E t$?

3 Reductions

In the following whenever R is ground, then, without loss of generality, we may assume that all terms appearing in the problem instances are also ground. Note that one of the desired properties of the following reductions is the preservation of groundness of the instances.

Theorem 4 *Normal form reachability for confluent systems $\leq_P$ Joinability for confluent systems.*

Proof: Given R, s, t, where t is a normal form, we construct an instance of Joinability R', s', t' by setting $R' = R$, $s' = s$ and $t' = t$. The rest is easy. □

Remark. The converse is not difficult to show, but in contrast to the reduction above this reduction does not preserve groundness. Also note that the restriction "confluent" can be dropped totally in the above theorem.

The following lemma will prove useful below. It means that an equational proof step remains valid when substituting a free constant by any term. We denote $A[e_1 \parallel f_1, \ldots, e_n \parallel f_n]$ the parallel replacement of all occurrences of ground terms e_i by f_i, for $i = 1, \ldots, n$.

Lemma 5 *Let E be a set of equations, let $c_1, \ldots, c_n$ be a set of constants not occurring in E, and let $A, B, s_1, \ldots, s_n$ be terms. If $A =_E B$ then $A[c_1 \parallel s_1, \ldots, c_n \parallel s_n] =_E B[c_1 \parallel s_1, \ldots, c_n \parallel s_n]$*

It is easily shown that Joinability for confluent systems $\leq_P$ WP. We now show:

Theorem 6 *Joinability for confluent systems $\leq_P$ Not $UN^=$.*

Proof: Given R, s, t with R confluent, we construct below an instance R' of $UN^=$.
Case 1: R is ground. Let $R' = R \cup \{s \to c, t \to d\} \cup T = \{a \to a \mid$ for all constants $a \in R, s, t\}$, where c, d are new constants not in R, s, t. T is a set of rules that reduces every ground term over the signature of R, s, t to itself. We must show that $n =_{R'} N$ for distinct normal forms $n \neq N$ iff $s \downarrow_R t$. The if direction is obvious. For the only if direction, observe that since R is confluent, $R \cup T$ is also confluent and hence also $UN^=$. Also because of the T rules we must have $\{n, N\} = \{c, d\}$. Therefore, by selecting a shortest length equational proof $c =_{R'} d$, we must have $s =_{R \cup T} t$ since R' is ground. Since $R \cup T$ is also confluent, we have $s \downarrow_{R \cup T} t$, which implies $s \downarrow_R t$.
Case 2: R is not ground. Wlog we may assume that both s and t cannot be variables, since in this case they are never joinable. If one of s or t is a variable the corresponding rule is dropped from R' below. We may also assume that $(Var(s) \cup Var(t)) \cap Var(R) = \emptyset$. The variables of s and t become new constants of R'. Let $R' = R \cup \{s \to c, t \to d\} \cup T'$, where $T' = T \cup \{x \to x \mid x \in Var(s) \cup Var(t)\} \cup \{h(z_1, \ldots, z_p) \to h(z_1, \ldots, z_p) \mid$ for every function symbol h of arity $p > 0$ in $R, s, t\}$. We must show that $n =_{R'} N$ for distinct normal forms n, N iff $s \downarrow_R t$. The if direction is obvious. For the only if direction observe that since R is confluent, $R \cup T'$ is also confluent and hence $UN^=$. Therefore, because of the T' rules we conclude that $\{n, N\} = \{c, d\}$. The rest is justified in the Appendix.

Note that the reductions takes only logspace and polynomial time. □

Theorem 7 *Joinability for confluent systems $\leq_P$ Confluence.*

Proof: Given R, s, t with R confluent, we build the system $R' = R \cup T$ where $T = \{a \to s', a \to t'\}$, a is a new constant and s', t' are obtained by replacing every variable x of s, t by a new constant x'. We show that $s \downarrow_R t$ iff R' is confluent. For the if direction, since R' is confluent we have $s' \rightarrow_{R'} u$ and $t' \rightarrow_{R'} u$ for some u. Since no rule in R' can introduce a subterm a there are no occurrence of a in the derivation. Hence we have $s' \rightarrow_R u$ and $t' \rightarrow_R u$ and therefore $s \downarrow_R t$. For the only if direction, we shall prove that $\xrightarrow{*}_T \cdot \xrightarrow{*}_R$ is strongly confluent in Lemma 10 below. Since strong confluence implies confluence [9] we will get that $\xrightarrow{*}_T \cdot \xrightarrow{*}_R$ is confluent and since $\xrightarrow{*}_{R'} = (\xrightarrow{*}_T \cdot \xrightarrow{*}_R)^*$, this will finish the proof. □

Lemma 8 *If $u \xrightarrow{*}_T v$ and $u \xrightarrow{*}_T w$ then there exists v', w', h such that $v \xrightarrow{*}_T v' \xrightarrow{*}_R h$, $w \xrightarrow{*}_T w' \xrightarrow{*}_R h$, and there is no occurrence of a in v', w', h.*

Proof: Let U be the set of positions of a in u. Take $v' = v[a \parallel s']$, $w' = w[a \parallel s']$. Note that v' and w' may differ only below occurrences $\epsilon \in U$, and for all $\epsilon \in U$, $v'/\epsilon \in \{s', t'\}$, $w'/\epsilon \in \{s', t'\}$. By hypothesis there exists β a common descendant of s' and t' by R. If $v'/\epsilon = s'$ and $w'/\epsilon = t'$, then rewriting s' (resp., t') to β in v' (resp., w') allows to derive terms that are identical below ϵ. Using similar reductions at the other occurrences of U (note that they are independent) we obtain h as in the lemma. □

Lemma 9 *If $u \xrightarrow{*}_T v$, $u \xrightarrow{*}_R w$ and v does not contain any occurrence of a then there exists w', h such that $v \xrightarrow{*}_R h$, $w \xrightarrow{*}_T w' \xrightarrow{*}_R h$, and there is no occurrence of a in w'.*

Proof: Let U be the set of positions of a in u. Since v does not contain any a and the positions $u_1, \ldots, u_m$ in U are independent, we can build a derivation $v \xrightarrow{*}_R v' = u[a \parallel \beta]$, by rewriting all the s', t' in v occurring at a position $\epsilon \in U$ to β. By induction on the number n of R steps in $u \xrightarrow{*}_R w$ we can show that $v' \xrightarrow{*}_R w[a \parallel \beta]$. If $n = 0$ it is obvious. Otherwise $u \xrightarrow{n-1}_R p \xrightarrow{l \rightarrow r, \epsilon}_R w$. By induction hypothesis: $u[a \parallel \beta] \xrightarrow{*}_R p[a \parallel \beta]$. Since a behaves as a free constant for R, we have that $p \xrightarrow{l \rightarrow r, \epsilon}_R w$ implies $p[a \parallel \beta] \xrightarrow{l \rightarrow r, \epsilon}_R w[a \parallel \beta]$ and this finishes the induction step. We can take $w[a \parallel \beta]$ for h and $w[a \parallel s'] = w'$ since $w \xrightarrow{*}_T w' \xrightarrow{*}_R w[a \parallel \beta]$. □

Lemma 10 $\xrightarrow{*}_T \cdot \xrightarrow{*}_R$ *is strongly confluent.*

Proof: Follows from Lemma 8, Lemma 9 and confluence of R. □

Theorem 11 *Reachability $\leq_P$ Joinability.*

Proof: Omitted. □

Theorem 12 *Joinability $\leq_P$ Reachability.*

Proof: Given instance R, s, t of Joinability, let $R' = R \cup \{equal(x, x) \rightarrow true\}$, where *equal* and *true* are new function symbols not in the signature of R, s, t. Let $s' = equal(s, t)$ and $t' = true$. It is easy to see that $s \downarrow_R t$ iff $s' \xrightarrow{*}_{R'} t'$. □

Thus, we have the following tight relationship for general systems:

Corollary 13 *Reachability $\equiv_P$ Joinability.*

Remarks. Note that in Theorem 11 the reduction preserves all properties of the given rewrite system, e.g., groundness, left-linearity, etc. However, in Theorem 12 this is not the case. We can change the reduction to preserve left-linearity by allowing variables in the right-hand sides, but whether groundness can be preserved is open.

Theorem 14 *Joinability for non-collapsing, var-preserving systems* $\leq_P$ *Not* $UN^{\rightarrow}$.

Proof: Omitted. □

Undecidable Confluence, Linear E-Matching with decidable WP

Now we show that there is a recursive family of string rewrite systems with a trivially decidable Word Problem but for which Confluence is undecidable.

Theorem 15 *There exists a family of string rewrite systems for which WP is trivially decidable and Confluence is undecidable.*

Proof: Let R be a Thue system on alphabet $\Sigma = \{0, 1\}$. Let R' be the system $T \cup S$, where $S = \{s \rightarrow t \mid s = t \text{ or } t = s \in R\}$ and $T = \{g \rightarrow 0g, g \rightarrow 1g, g \rightarrow \lambda\}$. Whether the monoid generated by R is trivial is known to be undecidable, since *triviality* is a Markov property.

Assume that R is nontrivial. Then there exists two words w_1, w_2 on Σ such that $w_1 \overset{*}{\not\leftrightarrow}_R w_2$. Since g generates Σ^*, we have $g \overset{*}{\rightarrow}_T w_1$ and $g \overset{*}{\rightarrow}_T w_2$. Since w_1 and w_2 do not have any g symbol they can be rewritten only by rules from S. But w_1, w_2 cannot be joinable since they are not congruent in R.

Assume that the monoid associated to R is trivial. Then note that every rewrite derivation $a \overset{*}{\rightarrow}_{R'} b$ can be extended to $b \overset{*}{\rightarrow}_T c$ where c is in Σ^* and $c \overset{*}{\rightarrow}_S \lambda$. Therefore every words rewrite to λ which implies that R' is confluent.

WP for R' is trivial since every word is equal to g modulo R'. □

It is easily seen that $WP \leq_P$ E-matching. We now show:

Theorem 16 *There exists a linear equational theory E for which Linear E-Matching is undecidable and WP is decidable.*

Proof (sketch): Let $P = \{(v_i, w_i) | i = 1, \ldots n\}$ be a fixed set of pairs of strings on Σ. Let us consider the following variant of Modified PCP from [11]. Given two strings $v_0, w_0 \in \Sigma^*$ is there a sequence $i_1, \ldots, i_k$ of numbers with each i_j between 1 and n such that $v_{i_k} v_{i_{k-1}} \ldots v_{i_1} v_0 = w_{i_k} w_{i_{k-1}} \ldots w_{i_1} w_0$.

We consider the following system R_P whose signature contains n unary symbols $1, \ldots, n$, a unary symbol for each element of the alphabet Σ of P, a single constant $\perp$ and functions k, *check* :

$$
\begin{array}{lll}
1\ k(y, a(x), a(u), v, w) & \rightarrow k(y, x, u, a(v), a(w)) \text{ for any } a \in \Sigma & \\
2\ k(y, \perp, \perp, v, w) & \rightarrow \mathit{check}(y, v, w) & \\
3\ \mathit{check}(i(y), v_i(x), w_i(u)) & \rightarrow \mathit{check}(y, x, u) & \text{for any } (v_i, w_i) \in P
\end{array}
$$

Let E be the equational theory associated with R_P. Since R_P terminates and has no critical pairs it is confluent and WP is decidable. Consider linear E-matching with $s = k(y, x, u, \perp, \perp)$ and $t = \mathit{check}(\perp, v_0(\perp), w_0(\perp))$. □

3.1 Reductions for Rewriting modulo E

Note that most of these reductions work for rewriting modulo E for general E. However, for the purpose of this paper it is sufficient to observe that all of the reductions work for rewriting modulo E for var-preserving, noncollapsing equational theories E. Var-preserving, noncollapsing equational theories include associativity and commutativity. The proof of this observation is straightforward since R/E is equivalent to $R \cup \{s \rightarrow t \mid s = t \in E \vee t = s \in E\}$ for var-preserving, noncollapsing E.

4 Rewriting Problems with Polynomial-time Algorithms

4.1 $UN^= \in P$ for ground systems

We now give a polynomial time algorithm for deciding the unicity property of ground rewrite systems. Note that in the proof-by-consistency approach for inductive proofs, consistency is often ensured by requiring the $UN^=$ property. Our algorithm can be extended to give an algorithm for deciding $UN^=$ property of left-linear, right-ground systems using ideas from [16]. This may be used as a decidable sufficient condition ensuring $UN^=$ for left-linear systems using approximation techniques. We remark that decidability of $UN^{\rightarrow}$ and $UN^=$ for ground rewrite systems follows from the work of [4]. However, their work does not address the complexity issue, and polynomial bounds are generally not possible using tree-automata techniques used by [3, 4] because exponential-time algorithms for determinization are usually needed.

The algorithm makes use of the Nelson-Oppen congruence closure algorithm [7, 10, 13] to keep the necessary (potentially infinite) set of equational proofs in a finite compact data structure. The data structure consists of a set of equivalence classes of terms known to be equivalent using the given rules, which are treated as equations. Terms are implicitly represented by signatures. Each class is assigned a number and the signature of a term $f(t_1, \ldots, t_n)$ is the $n+1$-tuple $\langle f, \#[t_1], \ldots, \#[t_n]\rangle$, where $\#[t_i]$ is the number of the equivalence class which contains the signature representing t_i. We say that an *equivalence class represents a term t* if it contains the signature representing t.

Our algorithm works in two phases. In the first phase the algorithm constructs the signatures of all the subterms appearing in the given rewrite system and uses the rules as equations to form the union of the equivalence classes containing signatures representing the lhs and rhs of the rule. This union may cause other signatures to change and this may create further unions. After all the rules have been used the algorithm enters the second phase, which is described below. The polynomial time implementation of the following algorithm is discussed later.
Algorithm: UN(R) /* R is a ground rewrite system */

1. Construct Nelson-Oppen data structure using the rules as equations.
Let C represent the set of all equivalence classes in the final data structure.
2(a). $H = 0$; $NF = T' = T = \emptyset$; $Type1class = Type2class = C$;

2(b). **repeat**

Let T be the set of terms of height H whose arguments are
terms all of which are represented by signatures in *Type2class* ;
Let T' be the members of T that are lhs's of R and let $NF = NF \cup T - T'$;
$Type1class = Type1class \cap \{c \mid c \in C$ and c represents no member of $NF\}$;
$Type2class = C - Type1class$;
If $Type2class = \emptyset$ **return**(true) and **halt** ;
If there is a $c \in Type2class$ representing 2 elements of NF,
return(false) and **halt** ;
$H = H + 1$;

until $H = |R|$;
return(true) and **halt** ;

We now prove the correctness of this algorithm. We say that a symbol appears in R if it appears in the lhs or rhs of some rule in R. The *size* of a term is the number of function and variable symbols in it. The size of a rewrite system R, denoted $|R|$, is the sum of the sizes of all the lhs's and rhs's of R. The *height* of a term is the number of edges in a longest path from the root to a leaf in its tree representation.

Proposition 17 *Let R be any ground system. If there are two distinct normal forms t_1 and t_2 satisfying $t_1 =_R t_2$, then there are two distinct ground normal forms s_1 and s_2 satisfying $s_1 =_R s_2$. Further the s_i's can be chosen to contain only function symbols appearing in the rules of R.*

Proof: Straightforward using the projection property for equational proofs. □

The first lemma below allows us to restrict the search for equational proofs to only those that are present in the output of the congruence closure algorithm on the rewrite system R. However, since even this set can be "very large" (infinite in general) the next lemma allows us to restrict the search space even further to those that are between terms of "minimal" height.

Lemma 18 *Let R be any ground system. If there exist two distinct normal forms t_1 and t_2 such that $t_1 =_R t_2$, then there exist two distinct ground normal forms s_1 and s_2 such that $s_1 =_R s_2$ and both s_1 and s_2 are represented by signatures in the output of the congruence-closure algorithm on R.*

Proof: By Proposition 17, we can restrict ourselves to equational proofs between ground normal forms containing only the function symbols of R. Let s and t be any two ground terms such that $q : s =_R t$. Let q be $s = s_0 \leftrightarrow s_1 \ldots \leftrightarrow s_n = t$ for some $n \geq 0$. First, we show that if any one of the terms in the equational proof q is represented in the output of the congruence closure algorithm on R, then *every* term must be represented in the output of the congruence closure algorithm on R. Suppose that s_i is represented for some i, $0 \leq i \leq n$. We show that both s_{i-1} and s_{i+1} (if they exist) must also be represented.

Consider $s_{i-1} \leftrightarrow s_i$. If $s_i \rightarrow s_{i-1}$, then $s_i = C[S]$ for some context C and term S and $s_{i-1} = C[T]$ and $S \rightarrow T$ is a rule in R. Since we start with all the subterms of lhs's and rhs's represented and we never lose any represented terms by applying congruence closure, both S and T are represented. Since $S \to T$ both S and T are represented by the same class in the output of the congruence-closure algorithm (CCA) on R. Since $C[S] = s_i$ is represented, there are signatures corresponding to each superterm (in s_i) of S in the output of CCA and since S and T are represented by the same class and $s_{i-1} = C[T]$, the *same* signatures represent every superterm (in s_{i-1}) of T in $C[T]$ and hence s_{i-1} is also represented. We argue similarly if $s_{i-1} \to s_i$.

Finally, suppose there are two distinct ground normal forms with an equational proof $q : t_1 =_R t_2$ such that every term in q is not represented in the output of CCA on R. This implies that there is no root reduction in q, since a root reduction implies that the corresponding terms are represented and hence all terms are represented, as we have just now shown. Therefore, we consider the minimal occurrences at which reductions are applied in q (there must be at least one such occurrence; since t_1 and t_2 are distinct q cannot be a null proof) and use the projection property of equational proofs to get two distinct ground normal forms t_1' and t_2' which are proper subterms of t_1 and t_2 such that $q' : t_1' =_R t_2'$ and there is a root reduction in q'. The rest follows. □

Lemma 19 *Let R be any ground system. If there exist two distinct normal forms t_1 and t_2 such that $t_1 =_R t_2$, then there are two distinct ground normal forms s_1 and s_2 such that $s_1 =_R s_2$ with height of the s_i's at most $|C| \leq |R|$.*

Proof: By Proposition 17 and Lemma 18, we can restrict ourselves to equational proofs between distinct ground normal forms containing only the function symbols of R and which are represented in the output of CCA on R. Suppose that the height of any one of the ground normal forms, say t_1, in the equational proof $q : t_1 =_R t_2$ is more than $|C|$, where both t_1 and t_2 are represented in the output of CCA.

Consider a longest path in t_1 from the root of t_1 to a leaf. Since this path has at least $|C|$ edges, there are at least $|C| + 1$ nodes on this path including the root and the leaf. Hence there are at least $|C| + 1$ distinct ground subterms of t_1 on this path. Since t_1 is a represented term, every subterm of t_1 is also represented by some class and since there are only $|C|$ equivalence classes, by the pigeonhole principle there must be two distinct ground subterms s and t on this path that are represented by the same equivalence class. By the property of the congruence-closure algorithm we have $s =_R t$. Since subterms of normal forms are normal forms (by definition), s and t are ground normal forms. Choose s and t of minimal height with these properties. □

Theorem 20 *Let R be any ground rewrite system. R has the unicity property iff Algorithm UN returns true.*

Proof: The proof is by induction and omitted for lack of space. □

Now we analyze the complexity of Algorithm UN and show that it runs in time polynomial in the input size $|R|$. The main difficulty is to construct the set of normal forms represented in C efficiently. A naive procedure that generates all possible normal forms of height h using the function symbols appearing in R

in a bottom-up manner and checks whether they are represented by some class in C may take exponential time. We omit the proof to save space.

Theorem 21 *The running time of Algorithm UN is a polynomial in* $|R|$.

4.2 Other rewriting problems in P

As a consequence of our reductions and Algorithm UN, we painlessly derive:

Corollary 22 *There are polynomial-time algorithms for: Normal form reachability for confluent ground systems and joinability of confluent ground systems.*

Proof: Using the reductions to $UN^=$ and the above algorithm for $UN^=$. □

Note that these problems are known to be decidable for arbitrary ground systems [14, 15, 3, 6] by proofs from scratch. However, no upper bounds are given for the decision procedures in these papers. It can be shown that the algorithm for reachability in [14] takes polynomial time, and a polynomial-time algorithm for reachability of ground systems is explicitly given in [17].

4.3 Polynomial Algorithms for Problems modulo Commutativity

The Nelson-Oppen congruence-closure algorithm and the polynomial-time algorithm for $UN^=$ can be generalized to handle commutative operators. The idea is to sort signatures based on the ordering of class numbers. The signature of a term is constructed in a bottom-up manner and the two arguments of each commutative operator are sorted according to the class representing them. Thus, we also have polynomial-time algorithms for normal form reachability of confluent ground systems modulo commutativity and joinability of confluent ground systems modulo commutativity, since one can check that the reductions of these problems to Not $UN^=$ are valid for ground rewriting modulo commutativity.

5 Rewriting Problems modulo Associativity

We show that all rewriting problems of Section 3 are undecidable for a fixed ground rewriting system modulo associativity by showing that Normal form reachability for a fixed ground confluent system modulo associativity is undecidable. The latter is undecidable by reduction from the halting problem of Turing Machines that halt with blank tape, as detailed below. We use a coding of TM as a rewrite system R similar to [1]. Let $M = (\Sigma_0, Q, q_0, \delta)$ be a fixed deterministic TM whose halting problem is undecidable, where Σ_0 is the tape alphabet, Q the finite set of states, q_0 the initial state, and $\delta : Q \times \Sigma_0 \to Q \times (\Sigma_0 \cup \{L, R\})$ the transition function. The blank symbol is denoted by s_0 and the left (resp. right) move is denoted by L (resp. R). Without loss of generality we may assume that $q_a \in Q$ is the unique accepting state and there are no transitions from this state. We consider now the following signature of constant function symbols for the rewrite system R: $\Sigma_0 \cup \Sigma'_0 \cup Q \cup \{h, h', q\}$ where $\Sigma'_0 = \{s'_i | s_i \in \Sigma_0\}$, $\Sigma_0 \cap \Sigma'_0 = \emptyset$, and h, h', q are new symbols.

$q_i s_p \to q_j s_l$ **if** $(q_i, s_p, q_j, s_l) \in \delta$ $\quad s'_l q_i s_p \to q_j s_l s_p$ **if** $(q_i, s_l, q_j, L) \in \delta$
$q_i h \to q_j s_l h$ **if** $(q_i, s_0, q_j, s_l) \in \delta$ $\quad s'_l q_i h \to q_j s_l h$ **if** $(q_i, s_l, q_j, L) \in \delta$
$q_i s_p \to s'_p q_j$ **if** $(q_i, s_p, q_j, R) \in \delta$ $\quad h' q_i s_p \to h' q_j s_0 s_p$ **if** $(q_i, s_p, q_j, L) \in \delta$
$q_i h \to s'_0 q_j h$ **if** $(q_i, s_0, q_j, R) \in \delta$ $\quad h' q_i h \to h' q_j s_0 h$ **if** $(q_i, s_0, q_j, L) \in \delta$

Reachability for R is undecidable since TM M halts on string w with the tape blank iff $h' q_0 w h \xrightarrow{*} h' q_a h$. Note that $h' q_a h$ is a normal form. We have:

Theorem 1. *The following are undecidable for a fixed ground rewrite system modulo associativity: Normal form reachability for confluent systems, Joinability for confluent systems, $UN^=$, Confluence, Reachability, Joinability, WP, and $UN^{\to}$.*

Proof: Note that if the restricted version of a problem is undecidable, then the general version is also undecidable. Hence we only need to prove the first four undecidable because of the reduction to $UN^{\to}$. Also, because of the reductions among the first four problems we only need to show that the first is undecidable. Therefore, the only point we are left to prove is that R above is confluent. A variant of this system R is proved confluent in [1] (lemma 2.6) under the assumption that the TM is terminating. We now relax this assumption. Note that there no overlaps between left-hand sides. We decompose any word as subwords of type $w'qw$ and $w'w$ where w (resp. w') is a maximal Σ_0 (resp. Σ'_0) subword and $q \in Q$. We can notice that such a decomposition is unique. Let w be a word and $W_1.W_2 \ldots W_k$ its decomposition. If $w \xrightarrow{*} w'$ then w' can be decomposed as $W'_1.W'_2 \ldots W'_k$ with $W_i \xrightarrow{*} W'_i$ for $i = 1 \ldots k$. This is because rewriting can take place only in subwords of type $w'qw$. Now if $w \xrightarrow{*} w'$ and $w \xrightarrow{*} w''$, it implies that $W_i \xrightarrow{*} W'_i$ and $W_i \xrightarrow{*} W''_i$ for $i = 1 \ldots k$. Since the TM is deterministic either $W'_i \xrightarrow{*} W''_i$ or $W''_i \xrightarrow{*} W'_i$. Therefore since rewriting steps occurring in different components commute we have $w' \downarrow w''$. □

Note that in [1] a family of rewrite systems modulo associativity is constructed and confluence is shown undecidable for this family whereas all the undecidability results here are for a fixed rewrite system modulo associativity.

6 Rewrite Problems in Commutative Semigroups

6.1 A lower bound on Confluence, $UN^=$ modulo AC

Theorem 23 *WP for weakly var-preserving theories $\leq_P$ Not $UN^=$.*

Proof: Let $R = \{l \to r \mid l = r \in E, Var(r) \subseteq Var(l)\} \cup \{r \to l \mid l = r \in E, Var(l) \subseteq Var(r)\} \cup \{s \to c, t \to d\} \cup T'$, where c, d are new constants not in E, s, t and T' is as in the proof of Theorem 6. We must show that $s =_E t$ iff $c =_R d$, i.e., R is Not $UN^=$. This follows from lemma 5 and the proposition in the Appendix. □

Theorem 24 *WP for noncollapsing, var-preserving theories $\leq_P$ Confluence.*

Proof: Given E, s, t we construct an instance R of Confluence by taking $R = \{l \to r \mid l = r \in E \vee r = l \in E\} \cup \{a \to s, a \to t\}$, where a is a new constant (variables of s and t are also new constants of R). □

We consider now the case where the signature of the rewrite system R is built solely from constants and a single AC symbol. These rewrite systems are also called *commutative semi-Thue systems*. They present commutative semigroups. It appears that the confluence problem with just one constant and one AC symbol is already harder than the word problem for this signature, since the confluence problem is not studied in [2] whereas the word problem with this restriction is shown decidable there. Note that since the word problem is Expspace hard for commutative semigroups [12], we have:

Corollary 25 *$UN^{=}$ and Confluence for commutative semi-Thue systems are Expspace-hard.*

Terms can be considered here as commutative words on an alphabet: $A = \{a_1, \ldots, a_n\}$. Therefore each term can be written $a_1^{m_1} \ldots a_n^{m_n}$ and identified to a vector $(m_1, \ldots, m_n)$ of $\mathbb{N}^n$ where the m_i are nonnegative integers, which is assumed below. Vector addition is denoted by $+$ and n is called the *dimension* of the semigroup. A word s is greater than a word t, denoted by $s \succ t$ iff each component of s is greater than or equal to the corresponding component of t and at least one of them is strictly greater.

6.2 Decidability of confluence of rewrite systems

We show how to encode the confluence problem into the home space problem for Petri nets which is decidable in our case. Let us recall that a *Petri net* is a tuple $(\Pi, \mathcal{T}, F)$ where $\Pi = \{p_1, \ldots, p_n)$ is a finite set of *places*, $\mathcal{T} = \{t_1, \ldots, t_p\}$ is a finite set of *transitions*, $F : \Pi \times \mathcal{T} \cup \mathcal{T} \times \Pi \to \mathbb{N}$ is the flow function. A marking : $\Pi \to \mathbb{N}$ attachs non-negative integers to places. Given a marking M, the *transition* τ is enabled if $M(p) \geq F(p, \tau)$ for all $p \in \Pi$ and firing τ yields a new marking M' such that $M'(p) = M(p) - F(p, \tau) + F(\tau, p)$. A marking M' is *reachable* from another marking M if it can be obtained by a sequence of transitions. Given an initial marking M_0, the set of marking reachable from M_0 is the reachability set of M_0 called $Reach(M_0)$.

The *home space problem* for Petri net is the following: *Instance:* a Petri net $(\Pi, \mathcal{T}, F)$, an initial marking M_0, E a set of markings. *Question:* from every marking $M \in Reach(M_0)$, can we reach a marking of E? This problem is decidable for E a linear set of $\mathbb{N}^{|\Pi|}$ [8]. The following Petri net P is used to encode the confluence problem for a rewrite system $l_i \to r_i$ for $i = 1, \ldots, p$ and $l_i, r_i \in \mathbb{N}^n$ into a home space problem for P.

- The set of places $\Pi = \{p_1, \ldots, p_n, p'_1, \ldots, p'_n, q_0, q_1\}$ has $2n + 2$ places, the $2n$ first ones are used to represent a pair of terms of $\mathbb{N}^n$ and the two last ones are used to separate different stages of the computation. Actually the two last places can be seen as states since their values will always be 0 or 1, with $q_i = 1$ (resp. 0) indicating that we are (resp. are not) in state q_i.

- The initial marking is $M_0 = (\underbrace{0\ldots,0,}_{2n}1,0)$
- We have several group of transitions. For each group of transition we indicate why they are introduced. If unspecified the value of $F(p,\tau)$ or $F(\tau,p)$ is 0.
  - Compute a marking $(s_1,\ldots,s_n,s_1,\ldots,s_n,1,0)$ from M_0 while staying in state q_0.
  $\mathcal{T}$ contains the transition τ_i^1 for $i=1,\ldots,n$. These transitions add 1 at component i and $i+n$, i.e. $F(\tau_i^1,p_i)=F(\tau_{i+n}^1,p_i')=1$.
  - Stop the previous computation and go to state q_1.
  $\mathcal{T}$ contains the transition τ_{q_0,q_1} with $F(q_0,\tau_{q_0,q_1})=F(\tau_{q_0,q_1},q_1)=1$.
  - Mimic rewrite rule on the first term (components $1,\ldots,n$) and stay in state q_1.
  For each rule $l\to r$ with $l=(\alpha_1,\ldots,\alpha_n)$ and $r=(\beta_1,\ldots,\beta_n)$ add the transition $\tau_{l\to r}^1$ where $F(p_i,\tau_{l\to r}^1)=\alpha_i$ and $F(\tau_{l\to r}^1,p_i)=\beta_i$ for $i=1,\ldots,n$, and $F(\tau_{l\to r}^1,q_1)=F(q_1,\tau_{l\to r}^1)=1$.
  - Mimic rewrite rule on the second term (components $n+1,\ldots,2n$) and stay in state q_1.
  Define transitions $\tau_{l\to r}^2$ similarly as above (replace p_i by p_i').

We shall identify a marking M and the vector $(M(p_1),\ldots,M(p_n),M(p_1'),\ldots,M(p_n'),M(q_0),M(q_1))$. In the same way if s is an n-tuple, we write $(s,s,\ldots)$ for $(s_1,\ldots,s_n,s_1,\ldots,s_n,\ldots)$.

Proposition 26 *The marking $(t_1,t_2,0,1)$ is reachable from M_0 iff there is some s such that $s\overset{*}{\to}t_1$ and $s\overset{*}{\to}t_2$.*

Proof: It is sufficient to realize that the following facts are true.

Fact 1. From $(0,\ldots,0,1,0)$ we necessarily reach markings of the form $(s,s,1,0)$. Conversely any marking of this form can be reached from $(0,\ldots,0,1,0)$. This is proved by an easy induction on the number of transitions for the first direction, and on the size of the marking for the other direction.

Fact 2. Any marking $(s,s,0,1)$ can be reached from $(0,\ldots,0,1,0)$. Use the previous point and use the second transition.

Fact 3. Any sequence of rewriting $s\overset{*}{\to}s'$ can be simulated by rules of the third group of rules, therefore for any t_1,t_2 such that $s\overset{*}{\to}t_1$ and $s\overset{*}{\to}t_2$, we can reach $(t_1,t_2,0,1)$ from $(s,s,0,1)$. The proof is by induction on the number of application of rules. Conversely firing a transition amounts to do some rewriting on the term represented by the component $1,\ldots,n$ or $n+1,\ldots,2n$. □

Now we state the main proposition which yields the decidability of confluence.

Proposition 27 *The rewrite system is confluent iff* $E=\{(z_1,\ldots,z_n,z_1,\ldots,z_n,0,1)\mid z_1,\ldots,z_n\in\mathbb{N}\}$ *is a home space for P.*

Proof: ($\Rightarrow$ direction) We assume the confluence of the rewrite system. Let M be some marking reachable from M_0. We show that E can be reached from M.

- Either the value of place $2n+1^{th}$ is 1 therefore $M=(m_1,m_1,1,0)$ (use fact 1). The transition τ_{q_0,q_1} is applicable yielding $M'=(m_1,m_1,0,1)\in E$.

– Or the value of place $2n+1^{th}$ is 0. This implies that $M = (t_1, t_2, 0, 1)$. Such marking can be reached only in the following way: reach some $(s, s, 1, 0)$ then reach $(s, s, 0, 1)$ and then reach $(t_1, t_2, 0, 1)$ where $s \xrightarrow{*} t_1$ and $s \xrightarrow{*} t_2$ are rewrite sequences. Since the rewrite system is confluent there exists t such that $t_1 \xrightarrow{*} t$ and $t_2 \xrightarrow{*} t$. Using previous propositions, we have that $(t, t, 0, 1) \in E$ is reachable from M.

The set E can be reached from any marking reachable from M_0, therefore it is a home space.

($\Leftarrow$ direction) We assume that E is a home space. Let s, t_1, t_2 be any terms such that $s \xrightarrow{*} t_1$ and $s \xrightarrow{*} t_2$. We know that the marking $M = (t_1, t_2, 0, 1)$ is reachable from M_0. Since E is a home state, there is some $(z, z, 0, 1)$ reachable from M, i.e. there is a z such that $t_1 \xrightarrow{*} z$ and $t_2 \xrightarrow{*} z$. This proves the confluence of the rewrite system. □

Since E is a linear set, the home space property for E is decidable [8] and:

Theorem 28 *Confluence for commutative semi-Thue systems is decidable.*

It is easy to see that we need only the decidability of the home space property for $E' = \{(0, \ldots, 0, 1)\}$ (subtract the period of E). The proof of decidability of the home-state property relies heavily on the reachability property in Petri nets. The complexity of this problem is high, but no exact bound is known in the general case. Therefore, we don't have precise information on the complexity of our decision method, except in special cases.

7 Conclusion

There are many directions for future work arising from this study. For example, more algorithms and polynomial-time reductions for the problems studied here and elsewhere are of considerable importance.

Acknowledgements. The first author thanks C. Kirchner and M. Rusinowitch for their generous support during his sabbatical year at INRIA Lorraine. The authors thank the reviewers and the committee for constructive comments.

References

1. G. Bauer and F. Otto. Finite complete rewriting systems and the complexity of the word problem. *Acta Informatica*, 21:521–540, 1984.
2. R. Book and F. Otto. String Rewriting Systems *Springer-Verlag*, 1992.
3. M. Dauchet, T. Heuillard, P. Lescanne, and S. Tison. Decidability of the confluence of finite ground term rewrite systems. *Info. & Comp.*, 88:187–201, 1990.
4. M. Dauchet and S. Tison. The theory of ground rewrite systems is decidable. *In Proc. LICS*, 1990.
5. N. Dershowitz and J.P. Jouannaud. Rewrite systems. In *Handbook of Theoretical Computer Science*, volume 2, chapter 6. North-Holland, 1990.
6. A. Deruyver and R. Gilleron. The reachability problem for ground TRS and some extensions. In *Lecture Notes in Computer Science*, volume 351, pages 227–243, 1989.

7. P.J. Downey, R. Sethi, and R.E. Tarjan. Variations on the common subexpression problem. *Journal of the ACM*, 27(4):758–771, 1980.
8. D.F. Escrig and C. Johnen. Decidability of home space property. Tech. Rep. 503, LRI, Université de Paris Sud (Fr.), July 1989.
9. J.W. Klop. Rewrite systems. In *H'book of Logic in Computer Science*. Oxford, 1992.
10. D. Kozen. Complexity of finitely presented algebras. In *Proc. Ninth ACM Symposium on Theory of Computing*, pages 164–177, 1977.
11. P. Narendran and F. Otto. The word matching problem is undecidable for finite special confluent string rewriting systems. In *Proc. ICALP*, 1997.
12. E. W. Mayr and A. R. Meyer. The complexity of the word problem for commutative semigroups and polynomial ideals. *Adv. in Mathematics*, 46:305–329, 1982.
13. G. Nelson and D.C. Oppen. Fast decision algorithms based on congruence closure. *Journal of the ACM*, 27:356–364, 1980. Also in the 18th IEEE FOCS, 1977.
14. M. Oyamaguchi. The church rosser property for ground term rewriting systems is decidable. *Theoretical Computer Science*, 49:43–79, 1987.
15. M. Oyamaguchi. The reachability and joinability problems for right-ground term rewriting systems. *Journal of Information Processing*, 13(3):347–354, 1990.
16. M. Oyamaguchi. On the word problem for right-ground term-rewriting systems. *Trans. IEICE Japam E73*, 1990.
17. D.A. Plaisted. Polynomial time termination and constraint satisfaction tests. In *Proc. Conf. on Rewriting Techniques & Applications*, 1993.

Appendix

Proposition 29 *Let $R' = R \cup S$, where $S = \{s \to c, t \to d\}$ and c, d are new constants not in R, s, t and s, t are ground terms. If $c =_{R'} d$ then $s =_R t$.*

Proof: An equational proof $c =_{R'} d$ can be decomposed as $c =_I A_1 =_J B_1 =_I \ldots = A_n =_J B_n =_I d$ or $c =_I A_1 =_J B_1 =_I \ldots = A_n =_I B_n =_J d$, where $I = S$ and $J = R$ or vice-versa. Wlog, we assume $c =_S A_1 =_R B_1 =_S \ldots = A_n =_S B_n =_R d$. Since c, d do not occur in R, using lemma 5 we have $A_{i+1}[c \parallel s, d \parallel t] =_R B_{i+1}[c \parallel s, d \parallel t]$. Since proofsteps between B_i and A_{i+1} amounts to replace c or d by s or t respectively and since c, d do not occur in s, t, we have also $B_i[c \parallel s, d \parallel t] = A_{i+1}[c \parallel s, d \parallel t]$. Doing the parallel replacements for all A_i and B_i gives a proof of the required type. □

The Decidability of Simultaneous Rigid *E*-Unification with One Variable

Anatoli Degtyarev[1,⋆], Yuri Gurevich[2,⋆⋆], Paliath Narendran[3,⋆⋆⋆], Margus Veanes[4], and Andrei Voronkov[1,†]

[1] Uppsala University Computing Science Department, P.O. Box 311, S-751 05 Uppsala, Sweden
[2] EECS Department, University of Michigan, Ann Arbor, MI 48109-2122, USA
[3] Department of Computer Science, University at Albany – SUNY, Albany, New York 12222, USA
[4] Max-Planck-Institut für Informatik Im Stadtwald, 66123 Saarbrücken, Germany

Abstract. We show that simultaneous rigid E-unification, or SREU for short, is decidable and in fact EXPTIME-complete in the case of one variable. This result implies that the $\forall^*\exists\forall^*$ fragment of intuitionistic logic with equality is decidable. Together with a previous result regarding the undecidability of the $\exists\exists$-fragment, we obtain *a complete classification of decidability of the prenex fragment of intuitionistic logic with equality, in terms of the quantifier prefix.* It is also proved that SREU with one variable and a constant bound on the number of rigid equations is P-complete.

1 Introduction

In Gallier, Raatz and Snyder [20] and Degtyarev, Gurevich and Voronkov [10], it is explained why simultaneous rigid E-unification, or SREU for short, plays such a fundamental role in automatic proof methods in classical logic with equality that are based on the Herbrand theorem, like semantic tableaux [17], the connection method [2] or the mating method [1], model elimination [31], and others.

It was shown recently in Degtyarev and Voronkov [11] that SREU is undecidable. The strong connections between SREU and intuitionistic logic with equality have led to new important decidability results in the latter area [12,44]. It follows, for example, that the $\exists^*$-fragment of intuitionistic logic with equality is undecidable [13,14]. This result is improved in Veanes [42] to the following.

The $\exists\exists$-fragment of intuitionistic logic with equality is undecidable.

⋆ Supported by grants from the Swedish Royal Academy of Sciences, INTAS and NUTEK.

⋆⋆ Partially supported by grants from NSF, ONR and the Faculty of Science and Technology of Uppsala University.

⋆⋆⋆ Supported by the NSF grants CCR-9404930 and INT-9401087.

† Supported by a TFR grant.

The decidability of the $\exists$-fragment of intuitionistic logic with equality, or equivalently SREU with one variable, has been an open problem which is settled in this paper. We prove the following.

SREU with one variable is decidable, in fact EXPTIME-complete.

This result is obtained by a polynomial time reduction of SREU with one variable to the intersection nonemptiness problem of finite tree automata. The latter problem is EXPTIME-complete [41]. By using an analogue of a Skolemization result for intuitionistic logic [12] we can deduce the following result.

The $\forall^*\exists\forall^*$-fragment of intuitionistic logic with equality is decidable.

The above results imply the following main contribution of this paper.

A complete classification of decidability of the prenex fragment of intuitionistic logic with equality, in terms of the quantifier prefix.

We prove also that rigid E-unification with one variable is P-complete and that SREU with one variable and a constant bound on the number of rigid equations is P-complete. One conclusion we can draw from this is that the intractability of SREU with one variable is strongly related to the *number* of rigid equations and not their *size*. With *two* variables, SREU is undecidable already with *three* rigid equations [24].

2 Preliminaries

We will first establish some notation and terminology. We follow Chang and Keisler [4] regarding first order languages and structures. For the purposes of this paper it is enough to assume that the first order languages that we are dealing with are languages with equality and contain only function symbols and constants, so we will assume that from here on. We will in general use Σ, possibly with an index, to stand for a signature, i.e., Σ is a collection of function symbols with fixed arities. A function symbol of arity 0 is called a *constant*. We will always assume that Σ *contains at least one constant.*

2.1 Terms and Formulas

Terms and formulas are defined in the standard manner. We refer to terms and formulas collectively as *expressions*. In the following let X be an expression or a set of expressions or a sequence of such.

We write $\Sigma(X)$ for the *signature of* X, i.e., the set of all function symbols that occur in X, $\mathcal{V}(X)$ for the set of all free variables in X. We write $X(x_1, x_2, \ldots, x_n)$ to express that $\mathcal{V}(X) \subseteq \{x_1, x_2, \ldots, x_n\}$. Let $t_1, t_2, \ldots, t_n$ be terms, then $X(t_1, t_2, \ldots, t_n)$ denotes the result of replacing each (free) occurrence of x_i in X by t_i for $1 \leq i \leq n$. By a *substitution* we mean a function from variables to terms. We will use θ to denote substitutions. We write $X\theta$ for $X(\theta(x_1), \theta(x_2), \ldots, \theta(x_n))$.

We say that X is *closed* or *ground* if $\mathcal{V}(X) = \emptyset$. By $\mathcal{T}_\Sigma$ or simply $\mathcal{T}$ we denote the set of all ground terms over the signature Σ. A substitution is called *ground* if its range consists of ground terms. A closed formula is called a *sentence*. Since there are no relation symbols all the atomic formulas are *equations*, i.e., of the form $t \approx s$ where t and s are terms and '$\approx$' is the formal equality sign.

2.2 First Order Structures

First order structures will (in general) be denoted by upper case gothic letters like $\mathfrak{A}$ and $\mathfrak{B}$ and their domains by corresponding capital roman letters like A and B respectively. A first order structure in a signature Σ is called a *Σ-structure.* For $F \in \Sigma$ we write $F^{\mathfrak{A}}$ for the interpretation of F in $\mathfrak{A}$.

For X a sentence or a set of sentences, $\mathfrak{A} \models X$ means that the structure $\mathfrak{A}$ is a *model of* or *satisfies* X according to Tarski's truth definition. A set of sentences is called *satisfiable* if it has a model. If X and Y are (sets of) sentences then $X \models Y$ means that Y is a *logical consequence* of X, i.e., that every model of X is a model of Y. We write $X \equiv Y$ when $X \models Y$ and $Y \models X$. We write $\models X$ to say that X is *valid*, i.e., true in all models.

By the *free algebra over Σ* we mean the Σ-structure $\mathfrak{A}$, with domain $\mathcal{T}_\Sigma$, such that for each n-ary function symbol $f \in \Sigma$ and $t_1, \ldots, t_n \in \mathcal{T}_\Sigma$, $f^{\mathfrak{A}}(t_1, \ldots, t_n) = f(t_1, \ldots, t_n)$. We let $\mathcal{T}_\Sigma$ also stand for the free algebra over Σ.

Let E be a set of ground equations. Define the equivalence relation $=_E$ on $\mathcal{T}$ by $s =_E t$ iff $E \models s \approx t$. By $\mathcal{T}_{\Sigma/E}$ (or simply $\mathcal{T}_{/E}$) we denote the quotient of $\mathcal{T}_\Sigma$ over $=_E$. Thus, for all $s, t \in \mathcal{T}$,

$$\mathcal{T}_{/E} \models s \approx t \quad \Leftrightarrow \quad E \models s \approx t.$$

We call $\mathcal{T}_{/E}$ the *canonical model* of E. Structures that are isomorphic with the canonical model of a finite set of ground equations are sometimes called *finitely presented algebras.* Various problems that are related to finitely presented algebras, and their computational complexity, have been studied in Kozen [26,27]. Below, we will make use of some of those results.

2.3 Simultaneous Rigid E-Unification

A *rigid equation* is an expression of the form $E \models_\forall s \approx t$ where E is a finite set of equations, called the *left-hand side* of the rigid equation, and s and t are arbitrary terms. A *system* of rigid equations is a finite set of rigid equations. A substitution θ is a *solution of* or *solves* a rigid equation $E \models_\forall s \approx t$ if

$$\models \quad (\bigwedge_{e \in E} e\theta) \Rightarrow s\theta \approx t\theta,$$

and θ is a *solution of* or *solves* a system of rigid equations if it solves each member of that system. The problem of solvability of systems of rigid equations is called *simultaneous rigid E-unification* or SREU for short. Solvability of a single rigid equation is called *rigid E-unification.* Rigid E-unification is known to be decidable, in fact NP-complete [19].

2.4 Term Rewriting

In some cases it is convenient to consider a system of ground equations as a rewrite system. We will assume that the reader is familiar with basic notions regarding ground term rewrite systems [15]. We will only use very elementary properties. In particular, we will use the following property of canonical (or convergent) rewrite systems. Let R be a ground and canonical rewrite system and consider it also as a set of equations. For any ground term t, let $t{\downarrow_R}$ denote the normal form of t with respect to R. Then, for all ground terms t and s, (cf [15, Section 2.4])

$$R \models t \approx s \quad \Leftrightarrow \quad t{\downarrow_R} = s{\downarrow_R}.$$

A *reduced* set of rules R is such that for each rule $l \to r$ in R, l is irreducible with respect to $R \setminus \{l \to r\}$ and r is irreducible with respect to R. In the case of ground rules, a reduced set of rules is also canonical [38]. It is always possible to find a reduced set of ground rewrite rules that is equivalent to a given finite set of ground equations [29]. Moreover, this can be done in $O(n \log n)$ time [38].

2.5 Finite Tree Automata

Finite tree automata, or simply tree automata from here on, is a generalization of classical automata. Tree automata were introduced, independently, in Doner [16] and Thatcher and Wright [40]. The main motivation was to obtain decidability results for the weak monadic second-order logic of the binary tree. Here we adopt the following definition of tree automata, that is based on rewrite rules [5,7].

▶ A *tree automaton* or *TA* A is a quadruple (Q, Σ, R, F) where
- Q is a finite set of constants called *states*,
- Σ is a *signature* that is disjoint from Q,
- R is a set of *rules* of the form $f(q_1, \dots, q_n) \to q$, where $f \in \Sigma$ has arity $n \geq 0$ and $q, q_1, \dots, q_n \in Q$,
- $F \subseteq Q$ is the set of *final states.*

A is called a *deterministic* TA or DTA if there are no two different rules in R with the same left-hand side.

Note that if A is deterministic then R is a reduced set of ground rewrite rules and thus canonical [38]. Tree automata as defined above are usually also called *bottom-up* tree automata. Acceptance for tree automata or recognizability is defined as follows.

▶ The set of terms *recognized* by a TA $A = (Q, \Sigma, R, F)$ is the set

$$T(A) = \{\tau \in \mathcal{T}_\Sigma \mid (\exists q \in F)\, \tau \xrightarrow{*}_R q\}.$$

A set of terms is called *recognizable* if it is recognized by some TA.

Two tree automata are *equivalent* if they recognize the same set of terms. It is well known that the nondeterministic and the deterministic versions of TAs have the same expressive power [16,21,40], i.e., for any TA there is an equivalent DTA. For an overview of the notion of recognizability in general algebraic structures see Courcelle [6] and the fundamental paper by Mezei and Wright [32].

3 Decidability of SREU with One Variable

In this section we will formally establish the decidability of SREU with one variable. The proof has two parts.

1. First we prove that rigid E-unification with one variable can be reduced to the problem of testing membership in a finite union of congruence classes.
2. By using the property that any finite union of congruence classes is recognizable, we then reduce SREU with one variable to the intersection nonemptiness problem of finite tree automata.

The decidability of SREU with one variable follows then from the fact that recognizable sets are closed under boolean operations and that the nonemptiness problem of finite tree automata is decidable. In Section 4 we will address the computational complexity of this reduction.

3.1 Reduction to Membership in a Union of Congruence Classes

We start by proving two lemmas. Roughly, these lemmas allow us to reduce an arbitrary rigid equation $S(x)$ with one variable to a finite collection of rigid equations $\{ S_i(x) \mid i < n \}$ such that, for all substitutions θ, θ solves S iff θ solves some S_i. Furthermore, each of the S_i's has the form $E \models_\forall x = t_i$ where E is ground and t_i is some ground term. The set E is common to all the S_i's.

Let E be a set of ground equations and t a ground term. We denote by $[t]_E$ the interpretation of t in $\mathcal{T}_{/E}$, in other words $[t]_E$ is the congruence class induced by $=_E$ on $\mathcal{T}$ that includes t. For a set T of ground terms we will write $[T]_E$ for $\{ [t]_E \mid t \in T \}$. We write *Terms*(E) for the set of all terms that occur in E, in particular *Terms*(E) is closed under the subterm relation. We will use the following lemma. Lemma 1 follows also from a more general statement in de Kogel [8, Theorem 5.11].

Lemma 1. *Let t be a ground term, c a constant, E a finite set of ground equations and e a ground equation. Let $T =$ Terms$(E \cup \{e\})$. If $[t]_E \notin [T]_E$ and $E \cup \{t \approx c\} \models e$ then $E \models e$.*

Proof. Assume that $[t]_E \notin [T]_E$ and that $E \cup \{t \approx c\} \models e$. Let E' be a reduced set of rules equivalent to E, such that $c{\downarrow_{E'}} = c$. Let $t' = t{\downarrow_{E'}}$. If $t' = c$ then

$$E \cup \{t \approx c\} \equiv E' \cup \{t \approx c\} \equiv E' \cup \{t' \approx c\} \equiv E$$

and the statement follows immediately. So assume that $t' \neq c$. Let $R = E' \cup \{t' \to c\}$. Let $l \to r$ be a rule in E'. Neither l nor r can be reduced with the rule $t' \to c$ because $[t']_E = [t]_E \notin [T]_E$. Hence R is reduced, and thus canonical [38]. Also, $R \equiv E \cup \{t \approx c\}$. (Note that $t' \in [t]_E$ and $[T]_E = [T]_{E'}$.)

Let $e = t_0 \approx s_0$ and let $u = t_0{\downarrow_R} = s_0{\downarrow_R}$. We have that

$$t_0 \xrightarrow{*}_R u, \quad s_0 \xrightarrow{*}_R u.$$

Consider the reduction $t_0 \xrightarrow{*}_R u$ and let $t_i \longrightarrow t_{i+1}$ be any rewrite step in that reduction. Obviously, if each subterm of t_i is in some congruence class in $[T]_E$ then the rule $t' \to c$ is not applicable since $[t']_E \notin [T]_E$ and it follows also that each subterm of t_{i+1} is in some congruence class in $[T]_E$. It follows by induction on i that the rule $t' \to c$ is not used in the reduction. The same argument holds for $s_0 \xrightarrow{*}_R u$. Hence

$$t_0 \xrightarrow{*}_{E'} u, \quad s_0 \xrightarrow{*}_{E'} u,$$

and thus $E' \models t_0 \approx s_0$. Hence $E \models e$. □

Consider a system S of rigid equations. There is an extreme case of rigid equations that are easy to handle from the point of view of solvability of S, namely the redundant ones:

- ▶ A rigid equation is *redundant* if all substitutions solve it.

To decide if a rigid equation $E(x) \models_\forall s(x) \approx t(x)$ is redundant, it is enough to decide if $E(c) \models s(c) \approx t(c)$ where c is a new constant.

- ▶ The *uniform word problem for ground equations* is the following decision problem. Given a set of ground equations E and a ground equation e, is e a logical consequence of E?

We will use the following complexity result [26,27].

Theorem 2 (Kozen). *The uniform word problem for ground equations is P-complete.*

So redundancy of rigid equations is decidable in polynomial time.

Lemma 3. *Let $E(x) \models_\forall e(x)$ be a rigid equation, c be a new constant and t be a ground term not containing c. Then*

$$E(c) \cup \{t \approx c\} \models e(c) \quad \Leftrightarrow \quad E(t) \models e(t).$$

Proof. The only non-obvious direction is '⇒'. Since t does not include c, $E(c) \cup \{t \approx c\} \models e(c)$ holds with c replaced by t, but then the equation $t \approx t$ is simply superfluous. □

Clearly, S is solvable iff the set of rigid equations in S that are not redundant, is solvable. We will use the following lemma.

Lemma 4. *Let $E(x) \models_\forall s_0(x) \approx t_0(x)$ be a non-redundant rigid equation of one variable x and let c be a new constant. There exists a finite set of ground terms T such that, for any ground term t not containing c the following holds:*

$$E(t) \models s_0(t) \approx t_0(t) \quad \Leftrightarrow \quad E(c) \models t \approx s \text{ for some } s \in T.$$

Furthermore, T can be obtained in polynomial time.

Proof. Let T' be the set $Terms(E(c) \cup \{s_0(c) \approx t_0(c)\})$. Let

$$T = \{ s \in T' \mid E(c) \cup \{s \approx c\} \models s_0(c) \approx t_0(c) \}.$$

Note that T may be empty. Let t be any ground term that does not contain c. By using Lemma 3, it is enough to prove that the following statements are equivalent:

1. $E(c) \cup \{t \approx c\} \models s_0(c) \approx t_0(c)$,
2. $E(c) \models t \approx s$ for some $s \in T$.

$(2 \Rightarrow 1)$ Assume that statement 2 holds. Then there is a term s in T such that $[t]_{E(c)} = [s]_{E(c)}$. Since $s \in T$, we know that $E(c) \cup \{s \approx c\} \models s_0(c) \approx t_0(c)$. Hence $E(c) \cup \{t \approx c\} \models s_0(c) \approx t_0(c)$.

$(1 \Rightarrow 2)$ Assume that statement 1 holds. First we prove that $[t]_{E(c)} \in [T']_{E(c)}$. Suppose (by contradiction) that this is not so. But then it follows from Lemma 1 that $E(c) \models s_0(c) \approx t_0(c)$, contradicting that the rigid equation is not redundant. So there is a term s in T' such that $[t]_{E(c)} = [s]_{E(c)}$, and thus (by statement 1) $E(c) \cup \{s \approx c\} \models s_0(c) \approx t_0(c)$. Hence $s \in T$ and statement 2 follows.

Finally, to prove that T can be obtained in polynomial time, observe that the size of T' is proportional to the size of the rigid equation, and to decide if some term in T' belongs to T takes polynomial time by Theorem 2. □

Decidability of SREU with one variable can now be proved by combining Lemma 4 with a result by Brainerd [3] (that states that, given a set R of a ground rewrite rules and a set T of ground terms, then the set $\{ t \mid (\exists s \in T)\ t \xrightarrow{*}_R s \}$ is recognizable) and by using elementary finite tree automata theory. However, this proof would not give us the computational complexity result that is established below.

4 Computational Complexity of SREU with One Variable

In this section we show formally that SREU with one variable is decidable, and in fact EXPTIME-complete. We first introduce the following definition.

- ▶ The *intersection nonemptiness problem of DTAs* or *DTAI* is the following decision problem. Given a collection $\{ A_i \mid 1 \leq i \leq n \}$ of DTAs, is $\bigcap_{i=1}^{n} T(A_i)$ nonempty?

The following result has been observed by other authors [18,22,36] and strictly proved in Veanes [41].

Theorem 5 (Veanes). *DTAI is EXPTIME-complete.*

We will first show that SREU with one variable reduces to DTAI in polynomial time. This establishes the inclusion of SREU with one variable in EXPTIME. We then show that DTAI reduces to SREU with one variable, which shows the hardness part. The construction that we will use is in fact based on a construction in de Kogel [8, Theorems 4.1 and 4.2] that is based on Shostak's congruence closure algorithm [37].[1] A similar construction is used also in Gurevich and Voronkov [25].

4.1 SREU with one variable is in EXPTIME

In the following we will assume that none of the rigid equations are redundant. Lemma 4 tells us that the set of solutions of a rigid equation $E(x) \models_\forall e(x)$ with one variable is given by the union of a finite number of congruence classes

$$\bigcup_{s \in T} \{ t \mid E(c) \models s \approx t \},$$

where $T \subseteq \mathit{Terms}(E(c) \cup \{e(c)\})$ and c is a new constant. We will now give a polynomial time construction of a DTA that recognizes the above set of terms. Our considerations lead naturally to the following definition. Let E be a set of ground equations and T a subset of $\mathit{Terms}(E)$.

► A DTA $A = (Q, \Sigma, R, F)$ is *presented by* (E, T) if A has the following form (modulo renaming of states). First, let q_C be a new state for each $C \in [\mathit{Terms}(E)]_E$.

$$\begin{aligned} Q &= \{ q_C \mid C \in [\mathit{Terms}(E)]_E \}, \\ \Sigma &= \Sigma(E), \\ F &= \{ q_C \mid C \in [T]_E \}, \\ R &= \{ f(q_{[t_1]_E}, \ldots, q_{[t_n]_E}) \to q_{[t]_E} \mid t = f(t_1, \ldots, t_n) \in \mathit{Terms}(E) \}. \end{aligned}$$

It is clear that the above definition is well defined. It follows from elementary properties of congruence relations that A is deterministic and thus R is reduced. Note that for each constant c in $\Sigma(E)$, there is a rule $c \to q_{[c]_E}$ in R. Note also that for any equation $s \approx t$ in E, both s and t reduce to the same normal form $q_{[s]_E} = q_{[t]_E}$ with respect to R, since they belong to $\mathit{Terms}(E)$. We will use the following lemma.

Lemma 6. *Let E be a set of ground equations and $T \subseteq \mathit{Terms}(E)$. Let A be a DTA presented by (E, T). Then*

1. $T(A) = \{ t \in \mathcal{T}_{\Sigma(E)} \mid (\exists s \in T)\, E \models t \approx s \}$,
2. *A can be constructed in polynomial time from E and T.*

Proof. By using some results in de Kogel [8] and Theorem 2. (See [9].) □

We prove now that SREU with one variable is in EXPTIME.

Lemma 7. *SREU with one variable is in EXPTIME.*

[1] De Kogel does not use tree automata but the main idea is the same.

Proof. Let $S(x) = \{ S_i(x) \mid 1 \leq i \leq n \}$ be a system of rigid equations. Assume, without loss of generality, that none of the rigid equations is redundant. Let $S_i(x) = E_i(x) \models_\forall e_i(x)$. Let Σ be the signature of S. Use Lemma 4 to obtain, for each i, $1 \leq i \leq n$, a set of ground terms T_i in polynomial time such that, for all t in $\mathcal{T}_\Sigma$,

$$E_i(t) \models e_i(t) \quad \Leftrightarrow \quad E_i(c) \models t \approx s \text{ for some } s \in T_i.$$

Use now Lemma 6 to obtain (in polynomial time) a DTA A_i that presents $(E_i(c), T_i)$, for $1 \leq i \leq n$. It follows by Lemma 4 and the first part of Lemma 6 that

$$T(A_i) = \{ t \in \mathcal{T}_\Sigma \mid E_i(t) \models e_i(t) \} \quad (\text{for } 1 \leq i \leq n).$$

Thus, θ is a solution to $S(x)$ iff $x\theta$ is recognizable by all $T(A_i)$. Consequently, $S(x)$ is solvable iff $\bigcap_{i=1}^n T(A_i)$ is nonempty. The lemma follows, since DTAI is in EXPTIME. □

4.2 SREU with one variable is EXPTIME-complete

We will reduce DTAI to SREU with one variable to establish the hardness part. First, let us state some simple but useful facts.

Lemma 8. *Let $A = (Q, \Sigma, R, F)$ be a DTA, f be a unary function symbol not in Σ, and c be a constant not in Q or Σ. Let*

$$S(x) = (R \cup \{ f(q) \to c \mid q \in F \} \models_\forall x \approx c).$$

Then, for all θ such that $x\theta \in \mathcal{T}_{\Sigma \cup \{f\}}$,

$$\theta \text{ solves } S(x) \quad \Leftrightarrow \quad x\theta = f(t) \text{ for some } t \in T(A).$$

Proof. Let $E = R \cup \{ f(q) \to c \mid q \in F \}$. From the fact that R is reduced and that $f(q)$ is irreducible in R and c is irreducible in E, follows that E is reduced and thus canonical. So, for any $x\theta \in \mathcal{T}_{\Sigma \cup \{f\}}$, θ solves $S(x)$ iff (since E is ground) $E \models x\theta \approx c$ iff $x\theta \xrightarrow{*}_E c$. But

$$\begin{aligned} x\theta \xrightarrow{*}_E c &\Leftrightarrow x\theta \xrightarrow{*}_E f(q) \longrightarrow c \text{ for some } q \in F \\ &\Leftrightarrow x\theta = f(t) \text{ for some } t \in \mathcal{T}_\Sigma \text{ and } t \xrightarrow{*}_R q \\ &\Leftrightarrow x\theta = f(t) \text{ for some } t \in T(A). \end{aligned}$$

□

For a given signature Σ, and some constant c in it, let us denote by $S_\Sigma(x)$ the following rigid equation:

$$S_\Sigma(x) = (\{ \sigma(c, \ldots, c) \approx c \mid \sigma \in \Sigma \} \models_\forall x \approx c).$$

The following lemma is elementary [14].

Lemma 9. *For all θ, θ solves $S_\Sigma(x)$ iff $x\theta \in \mathcal{T}_\Sigma$.*

We have now reached the point where we can state and easily prove the following result.

Theorem 10. *SREU with one variable is EXPTIME-complete.*

Proof. Inclusion in EXPTIME follows by Lemma 7. Let $\{ A_i \mid 1 \le i \le n \}$ be a collection of DTAs with a signature Σ. Let f be a new unary function symbol and $\Sigma' = \Sigma \cup \{f\}$. For each A_i, let $S_i(x)$ be the rigid equation given by Lemma 8. So, for all θ such that $x\theta \in \mathcal{T}_{\Sigma'}$,

$$\theta \text{ solves } S_i(x) \quad \Leftrightarrow \quad x\theta = f(t) \text{ for some } t \in T(A_i).$$

Let

$$S(x) = \{ S_i(x) \mid 1 \le i \le n \} \cup \{S_{\Sigma'}(x)\}.$$

It follows by Lemma 9 that for any θ that solves $S(x)$, $x\theta$ is in $\mathcal{T}_{\Sigma'}$. Hence, by Lemma 8, $S(x)$ is solvable iff $\bigcap_{i=1}^n T(A_i)$ is nonempty. Obviously, $S(x)$ has been constructed in polynomial time. The statement follows, since DTAI is EXPTIME-hard.

□

So in the general case, SREU is already intractable with one variable. It should be noted however that the exponential behavior is strongly related to the unboundedness of the number of rigid equations. (See Section 4.3.)

4.3 Bounded SREU with One Variable

The exponential worst case behavior of SREU with one variable is strongly related to the unboundedness of the number of rigid equations, and not to the size or other parameters of the rigid equations. This behavior is explained by the fact that the intersection nonemptiness problem of a family of DTAs is in fact the nonemptiness problem of the corresponding direct product of the family. The size of a direct product of a family of DTAs is proportional to the product of the sizes of the members of the family, and the time complexity of the nonemptiness problem of a DTA is polynomial.

▶ *Bounded SREU* is SREU with a number of rigid equations that is bounded by some fixed positive integer.

We will use the following definition.

▶ The *nonemptiness* problem of TAs is the following decision problem. Given a TA A, is $T(A)$ nonempty?

The nonemptiness problem of DTAs is basically the problem of generability of finitely presented algebras. The latter problem is P-complete [27] and thus, by a very simple reduction, also the former problem [41].[2] For bounded SREU with one variable we get the following result.

[2] The book of Greenlaw, Hoover and Ruzzo [23] includes an excellent up-to-date survey of around 150 P-complete problems, including generability.

Theorem 11. *Bounded SREU with one variable is P-complete.*

Proof. Let the number of rigid equations be bounded by some fixed positive integer n. P-hardness follows immediately from Theorem 2. Without loss of generality consider a system

$$S(x) = \{ S_i(x) \mid 1 \leq i \leq n \}$$

of exactly n rigid equations. For each S_i construct a DTA A_i in polynomial time, like in Lemma 7. Let A be the DTA that recognizes $\bigcap_{i=1}^{n} T(A_i)$. For example, A can be the direct product of $\{ A_i \mid 1 \leq i \leq n \}$ (Gécseg and Steinby [21]). It is straightforward to construct A in time that is proportional to the product of the sizes of the A_i's. Hence A is obtained in polynomial time (because n is fixed) and $T(A)$ is nonempty iff $S(x)$ is solvable. □

4.4 Monadic SREU with One Variable

When we restrict the signature to consist of function symbols of arity ≤ 1, i.e., when we consider the so-called *monadic* SREU then the complexity bounds are different. We can note that DTAs restricted to signatures with just unary function symbols correspond to classical deterministic finite automata or DFAs. It was proved by Kozen that the computational complexity of the intersection nonemptiness problem of DFAs is PSPACE-complete [28]. So, by using this fact we can see that Theorem 10 proves that monadic SREU with one variable is PSPACE-complete.

Monadic SREU is studied in detail elsewhere [25]. We can note that, in general, the decidability of monadic SREU is still an open problem. There is also a very close connection between monadic SREU and the prenex fragment of intuitionistic logic with equality restricted to function symbols of arity ≤ 1 [12].

5 Implications to the Prenex Fragment of Intuitionistic Logic

The *prenex fragment* of intuitionistic logic is the collection of all intuitionistically provable prenex formulas. Many new decidability results about the prenex fragment have been obtained quite recently by Degtyarev and Voronkov [12–14] and Voronkov [43]. Some of these results are:

1. Decidability, and in particular PSPACE-completeness, of the prenex fragment of intuitionistic logic *without* equality [43].
2. Prenex fragment of intuitionistic logic *with* equality but *without* function symbols is PSPACE-complete [12]. Decidability of this fragment was proved in Orevkov [35].
3. Prenex fragment of intuitionistic logic with equality in the language with one unary function symbol is decidable [12].
4. $\exists^*$-fragment of intuitionistic logic with equality is undecidable [13,14].

In some of the above results, the corresponding result has first been obtained for a fragment of SREU with similar restrictions. For example, the proof of the last statement is based on the undecidability of SREU. The undecidability of the $\exists^*$-fragment is improved in Veanes [42] where it is proved that, already the

5. $\exists\exists$-fragment of intuitionistic logic with equality is undecidable.

With the following result we obtain a complete characterization of decidability of the prenex fragment of intuitionistic logic with equality with respect to quantifier prefix.

Theorem 12. *The $\forall^*\exists\forall^*$-fragment of intuitionistic logic with equality is decidable and EXPTIME-hard.*

Proof. Intuitionistic provability of any formula in the $\forall^*\exists\forall^*$-fragment can be reduced to solvability of SREU with one variable [12]. Conversely, solvability of a system of rigid equations with one variable reduces trivially to provability of a corresponding formula in the $\exists$-fragment [12]. The statement follows by Theorem 10. □

Remark The undecidability of the $\exists\exists$-fragment holds if there is one binary function symbol in the signature. The reduction in Theorem 12 from a $\forall^*\exists\forall^*$-formula to SREU with one variable may take exponential time, so the precise computational complexity for this fragment is unknown at this moment.

Other fragments Decidability problems for other fragments of intuitionistic logic have been studied by Orevkov [34,35], Mints [33], Statman [39] and Lifschitz [30]. Orevkov proves that the $\neg\neg\forall\exists$-fragment of intuitionistic logic with function symbols is undecidable [34]. Lifschitz proves that intuitionistic logic with equality and without function symbols is undecidable, i.e., that the pure constructive theory of equality is undecidable [30]. Orevkov shows decidability of some fragments (that are close to the prenex fragment) of intuitionistic logic with equality [35]. Statman proves that the intuitionistic propositional logic is PSPACE-complete [39].

6 Current Status of SREU and Open Problems

Some other fragments of SREU, besides the one variable case, have been shown to be decidable and there are some subfragments of the two variable case that have been shown to be undecidable. A list of such results, including some other results related to SREU, is given in Degtyarev, Gurevich, Narendran, Veanes and Voronkov [9]. The *unsolved cases* are:

1. Decidability of monadic SREU [25].
2. Decidability of SREU with *two* rigid equations.

Both problems are highly non-trivial.

References

1. P.B. Andrews. Theorem proving via general matings. *Journal of the Association for Computing Machinery*, 28(2):193–214, 1981.
2. W. Bibel. *Deduction. Automated Logic.* Academic Press, 1993.
3. W.S. Brainerd. Tree generating regular systems. *Information and Control*, 14:217–231, 1969.
4. C.C. Chang and H.J. Keisler. *Model Theory.* North-Holland, Amsterdam, third edition, 1990.
5. J.L. Coquidé, M. Dauchet, R. Gilleron, and S. Vágvölgyi. Bottom-up tree pushdown automata: classification and connection with rewrite systems. *Theoretical Computer Science*, 127:69–98, 1994.
6. B. Courcelle. On recognizable sets and tree automata. In M. Nivat and H. Ait-Kaci, editors, *Resolution of Equations in Algebraic Structures.* Academic Press, 1989.
7. M. Dauchet. Rewriting and tree automata. In H. Comon and J.P. Jouannaud, editors, *Term Rewriting (French Spring School of Theoretical Computer Science)*, volume 909 of *Lecture Notes in Computer Science*, pages 95–113. Springer Verlag, Font Romeux, France, 1993.
8. E. De Kogel. Rigid E-unification simplified. In P. Baumgartner, R. Hähnle, and J. Posegga, editors, *Theorem Proving with Analytic Tableaux and Related Methods*, number 918 in Lecture Notes in Artificial Intelligence, pages 17–30, Schloß Rheinfels, St. Goar, Germany, May 1995.
9. A. Degtyarev, Yu. Gurevich, P. Narendran, M. Veanes, and A. Voronkov. The decidability of simultaneous rigid E-unification with one variable. UPMAIL Technical Report 139, Uppsala University, Computing Science Department, March 1997.
10. A. Degtyarev, Yu. Gurevich, and A. Voronkov. Herbrand's theorem and equational reasoning: Problems and solutions. In *Bulletin of the European Association for Theoretical Computer Science*, volume 60. October 1996. The "Logic in Computer Science" column.
11. A. Degtyarev and A. Voronkov. Simultaneous rigid E-unification is undecidable. UPMAIL Technical Report 105, Uppsala University, Computing Science Department, May 1995.
12. A. Degtyarev and A. Voronkov. Decidability problems for the prenex fragment of intuitionistic logic. In *Eleventh Annual IEEE Symposium on Logic in Computer Science (LICS'96)*, pages 503–512, New Brunswick, NJ, July 1996. IEEE Computer Society Press.
13. A. Degtyarev and A. Voronkov. Simultaneous rigid E-unification is undecidable. In H. Kleine Büning, editor, *Computer Science Logic. 9th International Workshop, CSL'95*, volume 1092 of *Lecture Notes in Computer Science*, pages 178–190, Paderborn, Germany, September 1995, 1996.
14. A. Degtyarev and A. Voronkov. The undecidability of simultaneous rigid E-unification. *Theoretical Computer Science*, 166(1–2):291–300, 1996.
15. N. Dershowitz and J.-P. Jouannaud. Rewrite systems. In J. Van Leeuwen, editor, *Handbook of Theoretical Computer Science*, volume B: Formal Methods and Semantics, chapter 6, pages 243–309. North Holland, Amsterdam, 1990.
16. J. Doner. Tree acceptors and some of their applications. *Journal of Computer and System Sciences*, 4:406–451, 1970.
17. M. Fitting. First-order modal tableaux. *Journal of Automated Reasoning*, 4:191–213, 1988.

18. T. Frühwirth, E. Shapiro, M. Vardi, and E. Yardeni. Logic programs as types of logic programs. In *Proc. 6th Symposium on Logics in Computer Science (LICS)*, pages 300–309, 1991.
19. J.H. Gallier, P. Narendran, D. Plaisted, and W. Snyder. Rigid E-unification is NP-complete. In *Proc. IEEE Conference on Logic in Computer Science (LICS)*, pages 338–346. IEEE Computer Society Press, July 1988.
20. J.H. Gallier, S. Raatz, and W. Snyder. Theorem proving using rigid E-unification: Equational matings. In *Proc. IEEE Conference on Logic in Computer Science (LICS)*, pages 338–346. IEEE Computer Society Press, 1987.
21. F. Gécseg and M. Steinby. *Tree Automata.* Akadémiai Kiodó, Budapest, 1984.
22. J. Goubault. Rigid $\boldsymbol{E}$-unifiability is DEXPTIME-complete. In *Proc. IEEE Conference on Logic in Computer Science (LICS).* IEEE Computer Society Press, 1994.
23. R. Greenlaw, H.J. Hoover, and W.L. Ruzzo. *Limits to Parallel Computation: P-Completeness Theory.* Oxford University Press, 1995.
24. Y. Gurevich and M. Veanes. Some undecidable problems related to the Herbrand theorem. UPMAIL Technical Report 138, Uppsala University, Computing Science Department, March 1997.
25. Y. Gurevich and A. Voronkov. The monadic case of simultaneous rigid E-unification. UPMAIL Technical Report 137, Uppsala University, Computing Science Department, 1997. To appear in Proc. of *ICALP'97.*
26. D. Kozen. Complexity of finitely presented algebras. Technical Report TR 76-294, Cornell University, Ithaca, N.Y., 1976.
27. D. Kozen. Complexity of finitely presented algebras. In *Proc. of the 9th Annual Symposium on Theory of Computing*, pages 164–177, New York, 1977. ACM.
28. D. Kozen. Lower bounds for natural proof systems. In *Proc. 18th IEEE Symposium on Foundations of Computer Science (FOCS)*, pages 254–266, 1977.
29. D.S. Lankford. Canonical inference. Technical report, Department of Mathematics, South-Western University, Georgetown, Texas, 1975.
30. V. Lifschitz. Problem of decidability for some constructive theories of equalities (in Russian). *Zapiski Nauchnyh Seminarov LOMI*, 4:78–85, 1967. English Translation in: Seminars in Mathematics: Steklov Math. Inst. 4, Consultants Bureau, NY-London, 1969, p.29–31.
31. D.W. Loveland. Mechanical theorem proving by model elimination. *Journal of the Association for Computing Machinery*, 15:236–251, 1968.
32. J. Mezei and J.B. Wright. Algebraic automata and context-free sets. *Information and Control*, 11:3–29, 1967.
33. G.E. Mints. Choice of terms in quantifier rules of constructive predicate calculus (in Russian). *Zapiski Nauchnyh Seminarov LOMI*, 4:78–85, 1967. English Translation in: Seminars in Mathematics: Steklov Math. Inst. 4, Consultants Bureau, NY-London, 1969, p.43–46.
34. V.P. Orevkov. Unsolvability in the constructive predicate calculus of the class of the formulas of the type $\neg\neg\forall\exists$ (in Russian). *Soviet Mathematical Doklady*, 163(3):581–583, 1965.
35. V.P. Orevkov. Solvable classes of pseudo-prenex formulas (in Russian). *Zapiski Nauchnyh Seminarov LOMI*, 60:109–170, 1976. English translation in: Journal of Soviet Mathematics.
36. H. Seidl. Haskell overloading is DEXPTIME-complete. *Information Processing Letters*, 52(2):57–60, 1994.
37. R. Shostak. An algorithm for reasoning about equality. *Communications of the ACM*, 21:583–585, July 1978.

38. W. Snyder. Efficient ground completion: An $O(n\log n)$ algorithm for generating reduced sets of ground rewrite rules equivalent to a set of ground equations E. In G. Goos and J. Hartmanis, editors, *Rewriting Techniques and Applications*, volume 355 of *Lecture Notes in Computer Science*, pages 419–433. Springer-Verlag, 1989.
39. R. Statman. Lower bounds on Herbrand's theorem. *Proc. American Mathematical Society*, 75(1):104–107, 1979.
40. J.W. Thatcher and J.B. Wright. Generalized finite automata theory with an application to a decision problem of second-order logic. *Mathematical Systems Theory*, 2(1):57–81, 1968.
41. M. Veanes. On computational complexity of basic decision problems of finite tree automata. UPMAIL Technical Report 133, Uppsala University, Computing Science Department, January 1997.
42. M. Veanes. The undecidability of simultaneous rigid E-unification with two variables. To appear in *Proc. Kurt Gödel Colloquium KGC'97*, 1997.
43. A. Voronkov. Proof search in intuitionistic logic based on constraint satisfaction. In P. Miglioli, U. Moscato, D. Mundici, and M. Ornaghi, editors, *Theorem Proving with Analytic Tableaux and Related Methods. 5th International Workshop, TABLEAUX '96*, volume 1071 of *Lecture Notes in Artificial Intelligence*, pages 312–329, Terrasini, Palermo Italy, May 1996.
44. A. Voronkov. Proof search in intuitionistic logic with equality, or back to simultaneous rigid E-unification. In M.A. McRobbie and J.K. Slaney, editors, *Automated Deduction — CADE-13*, volume 1104 of *Lecture Notes in Computer Science*, pages 32–46, New Brunswick, NJ, USA, 1996.

Ordering Constraints over Feature Trees Expressed in Second-Order Monadic Logic

Martin Müller Joachim Niehren

Programming Systems Lab, Universität des Saarlandes
66041 Saarbrücken, Germany, {mmueller,niehren}@ps.uni-sb.de

Abstract. The system $FT_{\leq}$ of ordering constraints over feature trees has been introduced as an extension of the system FT of equality constraints over feature trees. We investigate decidability and complexity questions for fragments of the first-order theory of $FT_{\leq}$. It is well-known that the first-order theory of FT is decidable and that several of its fragments can be decided in quasi-linear time, including the satisfiability problem of FT and its entailment problem with existential quantification $\varphi \models \exists x_1 \ldots \exists x_n \varphi'$. Much less is known on the first-order theory of $FT_{\leq}$. The satisfiability problem of $FT_{\leq}$ can be decided in cubic time, as well as its entailment problem without existential quantification. Our main result is that the entailment problem of $FT_{\leq}$ with existential quantifiers is decidable but PSPACE-hard. Our decidability proof is based on a new technique where feature constraints are expressed in second-order monadic logic with countably many successors SωS. We thereby reduce the entailment problem of $FT_{\leq}$ with existential quantification to Rabin's famous theorem on tree automata.
Keywords Feature logic, tree orderings, entailment, decidability, complexity, second-order monadic logic.

1 Introduction

Feature constraints have been used for describing records in constraint programming [1, 24, 23] and record like structures in computational linguistics [12, 11, 22, 18, 20]. Following [2, 4, 3], we consider feature constraints as predicate logic formulae interpreted in the structure of feature trees. We consider the system $FT_{\leq}$ of ordering constraints over feature trees [17, 13] which is an extension of the system FT of equality constraints over feature trees. Ordering constraints in $FT_{\leq}$ are interpreted with respect to the weak subsumption ordering [7] on feature trees. Here, we investigate decidability and complexity questions for fragments of the first-order theory of $FT_{\leq}$.

Ordering Constraints over Feature Trees. A feature tree is a tree with unordered edges labeled by features, and with possibly labeled nodes. Features are functional: All features at edges departing from the same node are pairwise distinct. A feature tree τ_1 is smaller than a feature tree τ_2 in the weak subsumption ordering if τ_1 has fewer edges and node labels than τ_2. In this case we write $\tau_1 \leq \tau_2$. An example is given in the picture.

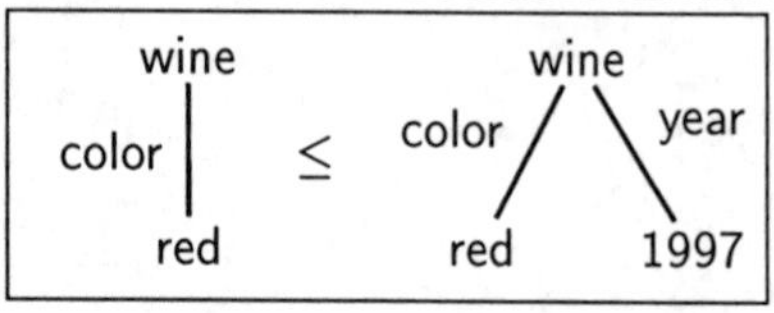

We focus on possibly infinite trees but we also consider the case of finite trees. The particular choice will be made explicit whenever necessary.

The constraints φ of $FT_{\leq}$ are defined as follows (where x and x' are variables).

$$\varphi ::= x{\leq}x' \mid x[f]x' \mid a(x) \mid \varphi \wedge \varphi'$$

The semantics of $FT_{\leq}$ is given by the interpretation over feature trees where the symbol $\leq$ is interpreted as weak subsumption ordering. The constraints of FT have the same syntax as those of $FT_{\leq}$ but with equalities $x{=}y$ instead of ordering constraints $x{\leq}y$. Equalities are expressible in $FT_{\leq}$ since $x{=}y \leftrightarrow x{\leq}y \wedge y{\leq}x$ holds. The semantics of selection $x[f]y$ and labeling constraints $a(x)$ is the same in FT and in $FT_{\leq}$. For instance, both trees in the picture above are possible denotations for x in solutions of the constraint $\mathsf{wine}(x) \wedge x[\mathsf{color}]x' \wedge \mathsf{red}(x')$.

Decidability and Complexity. The first-order theory of FT is well-known to be decidable [4] but has non-elementary complexity [26]. Several of its fragments can be decided in quasi-linear time [24], including the satisfiability problem of FT and its entailment problem with existential quantification $\varphi \models \exists x_1 \ldots \exists x_n \varphi'$. Much less is known on the first-order theory of $FT_{\leq}$. The entailment problem $\varphi \models \varphi'$ of $FT_{\leq}$ has been shown to have cubic time complexity in [17]. It is however not known whether more expressive fragments of the first-order theory of $FT_{\leq}$ are decidable.

We consider the entailment problem $\varphi \models \exists x_1 \ldots \exists x_n \varphi'$ of $FT_{\leq}$ with existential quantifiers. In the case of infinite trees, we show that this problem is at least PSPACE-hard. We prove this result by reducing the inclusion problem between regular word languages to it. Our proof makes essential use of infinite trees for encoding the Kleene star. When interpreted over the structure of finite trees, the entailment problem of $FT_{\leq}$ with existential quantifiers is at least coNP-hard. This follows by restricting our reduction to the inclusion between star-free regular languages. In the full paper [15] we give an alternative proof based on reducing the complement of the SAT-Problem of Boolean formulas. There, we adapt a proof idea introduced by Henglein and Rehof [9].

Relation to Second-Order Monadic Logic. We prove that the entailment problem of $FT_{\leq}$ with existential quantifiers $\varphi \models \exists x_1 \ldots \exists x_n \varphi'$ is decidable, both for finite and for infinite trees. We show this result for a countably infinite set of features f and a finite set of node labels a. In the case of finite trees, we give a reduction to the weak second-order monadic logic WSωS with countably many successors [25] and in the case of infinite trees to the full second-order monadic logic SωS [19]. The idea to encode trees as sets of words is well-known, for instance from [5]. Feature constraints, however, have not yet been encoded in SωS. The reason is that it is impossible in SωS to express prefix closedness of tree domains and direct subtree relation $\tau[f]\tau'$ simultaneously. In this paper, we avoid the need to express prefix closedness by a change of semantics, which we prove correct independently of our encoding. In more detail: First, we move from the model of (possibly labeled) feature trees to the model of so-called *sufficiently labeled* feature trees, and encode the first-order theory of ordering constraints over sufficiently labeled feature in SωS. An inverse encoding also exists as we prove in a follow-up paper [16]. Second, we encode the entailment problem of $FT_{\leq}$ with existential quantification into the corresponding problem for ordering constraints over sufficiently labeled feature trees This encoding cannot be lifted to the first-order theory of $FT_{\leq}$, as we show in [16].

Plan of the Paper. Section 2 introduces the syntax and semantics of the constraint system $FT_{\leq}$. Section 3 illustrates the expressiveness of entailment with existential quantification and gives the lower bound complexity results. Section 4 defines second-order monadic logic and gives our reduction of entailment in $FT_{\leq}$ to validity in SωS resp. WSωS. Section 5 contains the correctness proof of our reduction. Section 6 summarizes. The full paper [15] extends the conference version with two appendices that contain all omitted proofs.

2 Syntax and Semantics of $FT_{\leq}$

The constraint system $FT_{\leq}$ is defined by a set of constraints together with an interpretation over feature trees. We assume an infinite set of *variables* ranged over by x, y, z, a countably infinite sets $\mathcal{F}$ of *features* ranged over by f, g, and a set $\mathcal{L}$ with a least two elements of *labels* ranged over by a, b.

Feature Trees. A *path* π is a word over features. The *empty path* is denoted by ε and the free-monoid concatenation of paths π and π' as $\pi\pi'$; we have $\varepsilon\pi = \pi\varepsilon = \pi$. Given paths π and π', π' is called a *prefix of* π if $\pi = \pi'\pi''$ for some path π''. A *tree domain* is a non-empty prefix closed set of paths. A *feature tree* τ is a pair (D, L) consisting of a tree domain D and a partial function $L : D \rightharpoonup \mathcal{L}$ that we call *labeling function* of τ. Given a feature tree τ, we write D_τ for its tree domain and L_τ for its labeling function. For instance, $\tau_0 = (\{\varepsilon, f\}, \{(f, a)\})$ is a feature tree with domain $D_{\tau_0} = \{\varepsilon, f\}$ and $L_{\tau_0} = \{(f, a)\}$. The set of features occuring in some feature tree τ is denoted by $\mathcal{F}(\tau)$, i.e. $\mathcal{F}(\tau) = \{f \mid \pi f \pi' \in D_\tau\}$. We write as $\tau.\pi$ the subtree of τ at path π, if $\pi \in D_\tau$; formally $D_{\tau.\pi} = \{\pi' \mid \pi\pi' \in D_\tau\}$ and $L_{\tau.\pi} = \{(\pi', a) \mid (\pi\pi', a) \in L_\tau\}$. A feature tree is *finite* if its tree domain is finite, and *infinite* otherwise. A *node of* τ is an element of D_τ. A *leaf of* τ is a maximal node of τ. A node π of τ is *labeled with a* if $(\pi, a) \in L_\tau$. A node of τ is unlabeled if it is not labeled by any a. The *root* of τ is the node ε. For example, τ_0 as defined above is a finite feature tree with a single leaf f that is labeled with a. The root of τ_0 is unlabeled.

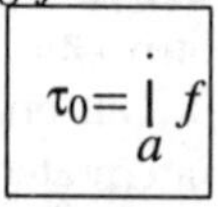

Ordering Constraints. An $FT_{\leq}$ *constraint* φ is defined by the abstract syntax

$$\varphi ::= x \leq y \mid a(x) \mid x[f]y \mid \varphi_1 \wedge \varphi_2$$

An $FT_{\leq}$ constraint is a conjunction of *basic constraints* which are either *inclusion constraints* $x \leq y$, *labeling constraints* $a(x)$, or *selection constraints* $x[f]y$.

We interpret $FT_{\leq}$ constraints in the structure $FT_{\leq}$ over feature trees: The signature of $FT_{\leq}$ contains the binary relation symbol $\leq$, for every label a a unary relation symbol $a()$, and for every feature f a binary relation symbol $[f]$. In $FT_{\leq}$ these relation symbols are interpreted such:

$$\begin{aligned} \tau_1 \leq \tau_2 \quad &\text{iff} \quad D_{\tau_1} \subseteq D_{\tau_2} \text{ and } L_{\tau_1} \subseteq L_{\tau_2} \\ \tau_1[f]\tau_2 \quad &\text{iff} \quad D_{\tau_2} = \{\pi \mid f\pi \in D_{\tau_1}\} \text{ and } L_{\tau_2} = \{(\pi, a) \mid (f\pi, a) \in L_{\tau_1}\} \\ a(\tau) \quad &\text{iff} \quad (\varepsilon, a) \in L_\tau \end{aligned}$$

First-Order Formulas. Let Φ and Φ' be first-order formulas built from $FT_{\leq}$ constraints with the usual first-order connectives. We call Φ *satisfiable* (valid) if Φ is satisfiable (valid) in the structure $FT_{\leq}$. We say that Φ *entails* Φ', written $\Phi \models \Phi'$, if $\Phi \rightarrow \Phi'$ is valid, and that Φ is *equivalent* to Φ' if $\Phi_1 \leftrightarrow \Phi_2$ is valid. We denote with $\mathcal{V}(\Phi)$ the set of variables occurring free in Φ and with $\mathcal{F}(\Phi)$ the set of features occurring in Φ.

We use the notation $x{\sim}a$ as an abbreviation for the formula $\exists y(x{\leq}y \wedge a(y))$. The formula $x{=}a$ means that x denotes the tree $(\{\varepsilon\}, \{(\varepsilon, a)\})$ and is defined as a short hand for the first-order formula $a(x) \wedge \forall y(a(y) \rightarrow x{\leq}y)$. We write $\bar{x}$ for a possibly empty word of variables $x_1 \ldots x_n$ and $\exists\bar{x}\varphi$ instead of $\exists x_1 \ldots \exists x_n \varphi$.

As additional notation, we define extended constraints for non-immediate subtree relations. Generalizing $[f]$, we introduce a binary relation symbol $[\pi]$ for every path π. We also define *extended constraints* $x[\pi]y$ for every π. We interpret extended constraints over feature trees such that the following equations hold:

$$x[\varepsilon]y \leftrightarrow x{\leq}y \wedge y{\leq}x \qquad \text{and} \qquad x[\pi_1\pi_2]y \leftrightarrow \exists z(x[\pi_1]z \wedge z[\pi_2]y)$$

The relation $\tau[\pi]\tau'$ holds whenever for all x, y every variable assignment α with $\alpha(x) = \tau$ and $\alpha(y) = \tau'$ is a solution of $x[\pi]y$. We will make use of the notations $x[\pi]y$, $x[\pi]_{\geq}y$, $x[\pi]_{\leq}y$, and $x[\pi]_{\geq}a$, which we consider as abbreviations for the following first-order formulas over extended constraints:

$$x[\pi]_{\geq}y \leftrightarrow \exists z(x[\pi]z \wedge y{\leq}z) \qquad \text{and} \qquad x[\pi]_{\leq}y \leftrightarrow \exists z(x{\leq}z \wedge z[\pi]y)$$
$$x[\pi]_{\geq}a \leftrightarrow \exists y(x[\pi]_{\geq}y \wedge a(y))$$

The formula $x[\pi]_{\geq}y$ (*resp.*, $x[\pi]_{\geq}a$) requires that x denote a tree τ_x with path π such that $\tau_x.\pi$ is greater than y (than a) with respect to $\leq$. The formula $x[\pi]_{\leq}y$ requires that the denotation of x either does not have the path π or that its subtree at π be smaller than y. Notice that $x[\pi]_{\leq}y$ is *not* equivalent to $\exists z x[\pi]z \wedge z{\leq}y$.

Alternative Definitions of Feature Trees. In the literature, there are two alternative definitions of feature trees [2, 3] distinct from ours. According to [2], every node must be labeled, and [3] requires exactly the leaves to be labeled. However, we follow previous work of ours [17] and *allow* labels at all nodes but do *not require* any.

For equality constraints as in FT, the particular definition of feature trees does not matter. The reason is that the first-order theory of FT is completely axiomatizable [4]. Each definition of feature trees yields a model of the axiomatization of FT. All these models are distinct but their first-order theories coincide due to complete axiomatization.

With respect to ordering constraints as in $FT_{\leq}$ the particular definition of feature trees does matter. For example, lets consider the formulas Φ_1 and Φ_2 where $a_1 \neq a_2$:

$$\Phi_1 = x{\sim}a_1 \wedge x{\sim}a_2 \qquad \text{and} \qquad \Phi_2 = \exists x \forall y\, x{\leq}y$$

The formula Φ_1 says that the label at the root of the denotation of x is compatible both with a_1 and a_2. Since $a_1 \neq a_2$, this is equivalent to saying that the root node of the denotation of x is unlabeled. Thus, Φ_1 is satisfiable in $FT_{\leq}$, but not in a structure of feature trees where every node has to be labeled. The formula Φ_2 says that there exists a smallest feature tree with respect to the weak subsumption ordering. Such a tree exists in $FT_{\leq}$, namely $(\{\varepsilon\}, \emptyset)$. However, there is no smallest tree in structures over feature trees that requires all nodes or all leaves to be labeled. Thus, Φ_2 distinguishes the structure $FT_{\leq}$ from those proposed for FT in [2, 3, 4].

3 Expressiveness of Entailment in $FT_{\leq}$

We consider the expressiveness of the entailment problem of $FT_{\leq}$ with existential quantification $\varphi \models \exists\bar{x}\varphi'$.[1] Without existential quantifiers, the expressiveness is quite low.

Theorem 1. *The entailment problem $\varphi \models \varphi'$ of $FT_{\leq}$ can be tested in cubic time (both over finite and over infinite trees).*

This result is proved in [17]. There, we also show that ordering constraints of $FT_{\leq}$ (without existential quantification) have the independence property: If $\varphi \models \varphi_1 \vee \ldots \vee \varphi_n$ then there exists $1 \leq i \leq n$ with $\varphi \models \varphi_i$. Independence fails in the presence of existential quantifiers since certain disjunctions can now be expressed by an entailment problem. *E.g.*, if $a_1 \neq a_2$ then $x{\sim}a_1 \models x{\sim}a_2 \vee a_1(x)$ but neither $x{\sim}a_1 \models x{\sim}a_2$ nor $x{\sim}a_1 \models a_1(x)$.

Infinite Trees. We show how to linearly reduce the inclusion problem between regular languages over finite words to the entailment problem $\varphi \models \exists\bar{x}\varphi'$ over $FT_{\leq}$. Since this problem is well-known to be PSPACE-complete [10, 21], we obtain PSPACE-hardness of entailment.

Theorem 2. *Over infinite trees, the entailment problem of $FT_{\leq}$ with existential quantification $\varphi \models \exists\bar{x}\varphi'$ is PSPACE-hard.*

Proof. Follows from Proposition 3 below. □

Lets fix a finite set of features F of $\mathcal{F}$. The idea of the proof of Theorem 2 is to encode regular sets of words over F as feature trees. For instance, if $f \in F$ then the set $\{f, fff\}$ can be described by the feature tree τ with

$$D_\tau = \{f, ff, fff\} \quad \text{and} \quad L_\tau = \{(f,a), (fff,a)\}$$

in that $\{f, fff\}$ is equal to the set of nodes of τ that are labeled with a. We consider regular expressions over a F as usual.

$$S ::= \varepsilon \mid f \mid S^* \mid S_1 \cup S_2 \mid S_1 S_2 \qquad \text{where } f \in F$$

Each regular expression S defines a non-empty set $\mathcal{L}(S)$ of finite words in F^*.

Proposition 3. *Let x and y be arbitrary variables. For every pair of regular expressions S_1 and S_2 there exist existential formulas $\Theta(x, S_1, y)$ and $\Theta(x, S_2, y)$ with size linear in the size of S_1 and S_2, respectively, such that*

$$\Theta(x, S_1, y) \models \Theta(x, S_2, y) \quad \textit{if and only if} \quad \mathcal{L}(S_2) \subseteq \mathcal{L}(S_1).$$

[1] Quantifiers on the left-hand side of entailment judgements can be used since $\exists\bar{y}\varphi \models \exists\bar{x}\varphi'$ is equivalent to $\varphi \models \exists\bar{x}\varphi'$ if $\{\bar{y}\} \cap \mathcal{V}(\exists\bar{x}\varphi') = \emptyset$. This can be ensured by α-conversion.

The proof is given below after the necessary definitions and two auxiliary Lemmas. We define the formula $\Theta(x,S,y)$ inductively over the form of S. Informally, the formula $\Theta(x,S,y)$ constrains x to denote some tree τ_x with all paths in $\mathcal{L}(S)$ and such that the subtrees of τ_x at these paths are greater than the denotation of y (see Lemma 5).

$$\begin{array}{lcl}
\Theta(x,\varepsilon,y) & = & y \leq x \\
\Theta(x,f,y) & = & \exists z(x[f]z \wedge y \leq z) \\
\Theta(x,S_1 \cup S_2,y) & = & \Theta(x,S_1,y) \wedge \Theta(x,S_2,y) \\
\Theta(x,S^*,y) & = & \exists z(y \leq z \wedge \Theta(z,S,z) \wedge z \leq x) \\
\Theta'(x,S_1S_2,y) & = & \exists z(\Theta(x,S_1,z) \wedge \Theta(z,S_2,y))
\end{array}$$

Lemma 4. *The formula $\Theta(x,S,y)$ has size linear in the size of S.*

Lemma 5. *Let α be a variable assignment, and S be a regular expression. Then α is a solution of $\Theta(x,S,y)$ if and only if $\forall \pi \in \mathcal{L}(S) : \alpha(x)[\pi] \geq \alpha(y)$.*

Proof. Structural induction over S. For the proof see Appendix A of the full paper. □

Proof of Proposition 3. ($\Rightarrow$) Assume that $\mathcal{L}(S_2) \not\subseteq \mathcal{L}(S_1)$. Then there exists $\pi_0 \in \mathcal{L}(S_2)$ such that $\pi_0 \notin \mathcal{L}(S_1)$. Assume $a \neq b$ and define a valuation α with

$$\begin{array}{ll}
D_{\alpha(y)} = \{\varepsilon\} & L_{\alpha(y)} = \{(\varepsilon, a)\} \\
D_{\alpha(x)} = pr(\mathcal{L}(S_1 \cup S_2)) & L_{\alpha(x)} = \{(\pi, a) \mid \pi \in \mathcal{L}(S_1)\} \cup \{(\pi, b) \mid \pi \in \mathcal{L}(S_2) \setminus \mathcal{L}(S_1)\}
\end{array}$$

where $pr(\mathcal{L}(S_1 \cup S_2))$ is the prefix-closure of $\mathcal{L}(S_1 \cup S_2)$. Clearly, both $\alpha(x)$ and $\alpha(y)$ define a feature tree. One easily checks that $\forall \pi \in \mathcal{L}(S_1) : \alpha(x)[\pi] \geq \alpha(y)$ holds, and that $\alpha(x)[\pi_0] \geq \alpha(y)$ does not hold. Hence we obtain from Lemma 5 that $\alpha \models \Theta(x,S_1,y)$ but also that $\alpha \not\models \Theta(x,S_2,y)$. Hence $\Theta(x,S_1,y) \not\models \Theta(x,S_2,y)$.

($\Leftarrow$) Assume $\mathcal{L}(S_2) \subseteq \mathcal{L}(S_1)$. Then apparently for all α the following property holds: If $\forall \pi \in \mathcal{L}(S_1) : \alpha(x)[\pi] \geq \alpha(y)$ then $\forall \pi \in \mathcal{L}(S_2) : \alpha(x)[\pi] \geq \alpha(y)$. By Lemma 5, this is equivalent to saying that every a solution of $\Theta(x,S_1,y)$ is a solution of $\Theta(x,S_2,y)$, *i.e.*, $\Theta(x,S_1,y) \models \Theta(x,S_2,y)$. □

Finite Trees. First notice that the encoding of regular expressions given above fails over finite trees. For instance $\Theta(x,f^*,y)$ is unsatisfiable since it constrains x to have all paths $\mathcal{L}(f^*)$ which is infinite. However, if S does not contain the Kleene star, then $\mathcal{L}(S)$ is finite. Therefore, we obtain the following corollary.

Corollary 6. *Over finite trees, the entailment problem of* $\mathrm{FT}_{\leq}$ *with existential quantification is co*NP*-hard.*

Proof. It suffices to restrict consideration to star-free languages, whose inclusion problem is known to be coNP-complete (*e.g.*, see problem set AL9 in [8]).

We owe this corollary to Franz Baader. In the full paper [15], we present a direct proof of this result, using a reduction of the complement of the propositional satisfiability problem SAT [6]. This proof is based on an idea by Henglein and Rehof [9] but modifies their technique in a non-obvious manner; it may therefore be of independent interest.

4 Deciding Entailment with Existential Quantifiers

In this Section we prove the decidability of the entailment problem of $FT_{\leq}$ by reduction to Rabin's decidability result for second-order monadic logic. Our proof is based on a model change from (possibly labeled) feature trees to sufficiently labeled feature trees. From now on, we will assume the set $\mathcal{L}$ of labels to be finite.

Second-Order Monadic Logic (SωS and WSωS). We recall the definitions of *second-order monadic logic with countably many successors* SωS [19] and of *weak second-order monadic logic with countably many successors* WSωS [25]. Syntactically, SωS and WSωS coincide. We assume an additional infinite set of *path variables* denoted by p that is disjoint from the variables denoted by x. Formulas ψ of SωS and WSωS are built from variables x and p and features f.

$$w ::= p \mid \varepsilon \mid fw$$
$$\psi ::= w{\in}x \mid w{=}w' \mid \psi \wedge \psi' \mid \neg\psi \mid \forall p \psi \mid \forall x \psi$$

The semantics of SωS is defined as follows. A path variable p is interpreted as a path (a word over features) and a variable x as a set of words over features. The denotation of ε is the empty path and the denotation of fw is the path obtained by concatenation f in front of the denotation of w. The membership constraint $w{\in}x$ holds if the denotation of w is a member of the denotation of x. The equality constraint $w{=}w'$ holds if the denotations of w and w' are equal. The semantics of WSωS coincides with the semantics of SωS except that in WSωS a variable x denotes a *finite* set of paths.

As derived forms we will use the following formulas with their usual semantics:

$$\exists p \psi,\ \exists x \psi,\ \psi \rightarrow \psi'\ \psi \leftrightarrow \psi'$$

Theorem 7 (Rabin,Thatcher,Wright [19, 25]). *The satisfiability problems of* WSωS *and* SωS *are decidable.*

Theorem 8. *The entailment problem of* $FT_{\leq}$ *with existential quantification* $\varphi \models \exists \bar{x} \varphi'$ *is decidable, both when interpreted over finite feature trees or over infinite feature trees.*

The proof is developed in this section and given at its end. The details of the proof cover the rest of the paper. The underlying idea is quite simple. If we ignore labels then a feature tree coincides with its domain which is a set of paths. Therefore, ordering constraints over trees can be translated into monadic second-order logic. This idea is well known for constructor trees [5]. The pitfall here is that only prefix closed sets of paths correspond to feature trees. Prefix-closedness can be expressed in SωS but not simultaneously with the direct subtree relation $\tau[f]\tau'$. The reason is that SωS allows for concatenation to the left πf but not for concatenation to the right $f\pi$ (or vice versa).

We avoid the need to express prefix closedness by first changing semantics (Sect. 4.1). We define the structure $FT^{-}_{\leq}$ of sufficiently labeled feature trees and reduce the entailment problem of $FT_{\leq}$ to the entailment problem of $FT^{-}_{\leq}$. In a second step (Sect. 4.2), we encode entailment relative to $FT^{-}_{\leq}$ into formulas of second-order monadic logic with countably many successors (WSωS for finite trees and SωS for infinite trees).

4.1 Changing Semantics

Definition 9. We call a feature tree τ *sufficiently labeled* if for every $\pi \in D_\tau$ there exists a path π' and a label a such $(\pi\pi', a) \in L_\tau$.

Note that a finite feature tree is sufficiently labeled if and only if all its leaves are labeled. Every sufficiently labeled feature tree can be identified with a unique n-tuple of pairwise disjoint non-empty sets of paths with non-empty union, and vice versa, if n is the number of labels in $\mathcal{L}$. For every label a we define a function γ_a from feature trees to non-empty sets of paths:

$$\gamma_a(\tau) = \{\pi \mid L_\tau(\pi) = a\}$$

For $\mathcal{L}=\{a_1,\ldots,a_n\}$ we define $\gamma(\tau)$ as the following n-tuple of sets of paths:

$$\gamma(\tau) = (\gamma_{a_1}(\tau),\ldots,\gamma_{a_n}(\tau))$$

Proposition 10. *The mapping γ from sufficiently labeled feature trees to n-tuples of pairwise disjoint sets of words with non-empty union is one-to-one and onto. Furthermore, τ is a finite tree if and only if every component of $\gamma(\tau)$ is finite.*

The proof is given in Appendix B.1 of the full paper. Note that we need not require prefix closedness for the sets in the domain of γ, since the domain of a sufficiently labeled feature tree τ is uniquely determined by its labeling function L_τ. This observation is crucial for our reduction to second-order monadic logic. Note also that the notion of sufficient labeling does not make sense for the alternative notions of feature trees mentioned above [2, 3].

Definition 11. The structure $\mathrm{FT}_{\leq}^{-}$ is the restriction of the structure $\mathrm{FT}_{\leq}$ to the domain of sufficiently labeled feature trees.

We may interpret $\mathrm{FT}_{\leq}^{-}$ either over finite trees or over possibly infinite trees. Whenever this choice matters, we will make it explicit.

The first-order theories of $\mathrm{FT}_{\leq}$ and $\mathrm{FT}_{\leq}^{-}$ differ. For instance, consider the following existential formula Φ_3 (or alternatively the formula Φ_2 from above):

$$\Phi_3 = \exists x(x{\leq}x_1 \wedge x{\leq}x_2)$$

Φ_3 requires for all τ_1 and τ_2 that there exists τ such that $\tau{\leq}\tau_1$ and $\tau{\leq}\tau_2$. Φ_3 is valid over $\mathrm{FT}_{\leq}$ but not valid over $\mathrm{FT}_{\leq}^{-}$. In $\mathrm{FT}_{\leq}$ one may always chose $\tau = (\varepsilon, \emptyset)$. This is impossible in $\mathrm{FT}_{\leq}^{-}$ since $(\varepsilon, \emptyset)$ is not sufficiently labeled. Even worse, if $\tau_1 = (\{\varepsilon\}, \{(\varepsilon, a_1)\})$, $\tau_2 = (\{\varepsilon\}, \{(\varepsilon, a_2)\})$, and $a_1 \neq a_2$ then we cannot find any appropriate tree τ in $\mathrm{FT}_{\leq}^{-}$.

We need distinct notations for entailment with respect to $\mathrm{FT}_{\leq}$ and $\mathrm{FT}_{\leq}^{-}$. For this purpose, we write $\Phi \models_{\mathrm{FT}_{\leq}} \Phi'$ and $\Phi \models_{\mathrm{FT}_{\leq}^{-}} \Phi'$. The next proposition claims that, in some sense, this distinction is not necessary for the formulas of interest.

We fix a label $b \in \mathcal{L}$ for the rest of the paper. Given some feature $g \in \mathcal{F}$ we define a function η_g that maps a constraint φ to a first-order formula over constraints as follows:

$$\eta_{g,b}(\varphi) = \varphi \wedge \bigwedge_{y\in\mathcal{V}(\varphi)} \exists y'(y[g]y' \wedge y'{=}b)$$

Lemma 12. *Let* φ *and* φ' *be constraints such that* $\mathcal{V}(\varphi') \subseteq \mathcal{V}(\varphi)$, $\bar{x}$ *a sequence of variables, and* g *a feature. If* $g \notin \mathcal{F}(\varphi \wedge \varphi')$ *then* $\varphi \models_{\mathrm{FT}_\leq} \exists\bar{x}\varphi'$ *is equivalent to* $\eta_{g,b}(\varphi) \models_{\mathrm{FT}_{\overline{\leq}}} \exists\bar{x}\varphi'$, *both over finite trees and over infinite trees.*

Proof. This proof of this proposition is technically involved. It is given in Section 5, the two implications being subject of Propositions 20 and 24. □

Note that Lemma 12 fails when $\eta_{g,b}(\varphi)$ is replaced by φ. This can also illustrated by formula Φ_3. As argued above, Φ_3 is valid over $\mathrm{FT}_\leq$ but not over $\mathrm{FT}_{\overline{\leq}}$. In order to relate this fact to Lemma 12, let φ_3 be the tautological constraint $x_1 \leq x_1 \wedge x_2 \leq x_2$ such that $\mathcal{V}(\Phi_3) \subseteq \mathcal{V}(\varphi_3)$. Now, $\varphi_3 \models_{\mathrm{FT}_\leq} \Phi_3$ but $\varphi_3 \not\models_{\mathrm{FT}_{\overline{\leq}}} \Phi_3$. However, $\eta_{g,b}(\varphi_3) \models_{\mathrm{FT}_{\overline{\leq}}} \Phi_3$ where $\eta_{g,b}(\varphi_3) = \exists x_1'(x_1[g]x_1' \wedge x_1'{=}b) \wedge \exists x_2'(x_2[g]x_2' \wedge x_2'{=}b)$.

4.2 Reduction to SωS or WSωS

We next define a mapping from first-order formulas over ordering constraints (interpreted over $\mathrm{FT}_{\overline{\leq}}$) to formulas of second-order monadic logic with countably many successors. We will make use of the following abbreviations:

$$x \cap y = \emptyset = \neg\exists p(p \in x \wedge p \in y) \quad \text{and} \quad x \subseteq y = \forall p(p \in x \rightarrow p \in y)$$

For every variable x and label a let x_a be a fresh variable. Suppose that $\mathcal{L} = \{a_1, \ldots, a_n\}$. Here comes the definition of the mapping $[\![\ \text{-}\ ]\!]$:

$$\begin{array}{lcl}
[\![a(x)]\!] & = & \varepsilon \in x_a \\
[\![x[f]y]\!] & = & \bigwedge_{i=1}^n \forall p(fp \in x_{a_i} \leftrightarrow p \in y_{a_i}) \\
[\![x \leq y]\!] & = & \bigwedge_{i=1}^n x_{a_i} \subseteq y_{a_i} \\
[\![\varphi \wedge \varphi']\!] & = & [\![\varphi]\!] \wedge [\![\varphi']\!] \\
[\![\neg\varphi]\!] & = & \neg[\![\varphi]\!] \\
[\![\exists x\varphi]\!] & = & \exists x_{a_1} \ldots \exists x_{a_n} ((\bigwedge_{i,j=1, i \neq j}^n x_{a_i} \cap x_{a_j} = \emptyset) \wedge \exists p(\bigvee_{i=1}^n p \in x_{a_i}) \wedge [\![\varphi]\!])
\end{array}$$

Proposition 13. *A first-order formula* Φ *is valid over* $\mathrm{FT}_{\overline{\leq}}$ *interpreted over finite (resp. infinite) trees if and only if its translation* $[\![\varphi]\!]$ *is valid over* WSωS *(resp.* SωS*).*

Proof. If α is a solution of Φ then α' with $\alpha'(x_{a_i}) = \gamma_{a_i}(\alpha(x))$ for all i, $1 \leq i \leq n$, is a solution of $[\![\Phi]\!]$. If β is a solution of $[\![\Phi]\!]$ then the mapping β' with $\beta'(x) = \gamma^{-1}(\beta(x_{a_1}), \ldots, \beta(x_{a_n}))$ is a solution of Φ. The existence of the inverse mapping γ^{-1} of γ is proved by Proposition 10.

An immediate corollary of this proposition is the following one:

Corollary 14. *The first-order theory of* $\mathrm{FT}_{\overline{\leq}}$ *is decidable.*

In [16] we show that there also exists an embedding from (W)SωS to the first-order of $\mathrm{FT}_\leq$. In the same paper we show that the first-order theory of $\mathrm{FT}_\leq$ is undecidable. Hence, the inverse of Corollary 14 does not hold.

Proof of Theorem 8. We wish to decide an entailment problem of the form $\varphi \models_{FT_{\leq}} \exists\bar{x}\varphi'$. We choose a feature g not occurring in φ or φ' (this exists since the set of all features $\mathcal{F}$ is infinite). By Lemma 12 it is sufficient to decide the entailment propositions $\eta_{g,b}(\varphi) \models_{FT_{\leq}^{-}} \exists\bar{x}\varphi'$ over $FT_{\leq}^{-}$. By Proposition 13, $\eta_{g,b}(\varphi) \models_{FT_{\leq}^{-}} \exists\bar{x}\varphi'$ holds if and only if the translation $[\![\eta_{g,b}(\varphi) \rightarrow \exists\bar{x}\varphi']\!]$ is a valid formula of WSωS in the case of finite trees and of SωS in the case of infinite trees. The validity of these formulas is decidable by Rabin's Theorem 7. □

5 Changing Semantics is Correct

We prove that the semantics change from $FT_{\leq}$ to $FT_{\leq}^{-}$ is correct in the sense of Lemma 12. All omitted proofs can be found in Appendix B of the full paper.

5.1 Entailment in $FT_{\leq}^{-}$ Implies Entailment in $FT_{\leq}$

Throughout this Section we are interested in entailment propositions $\varphi \models \exists\bar{x}\varphi'$ where $g \notin \mathcal{F}(\varphi \wedge \varphi')$ for a fixed feature g.

Adding Labels. We define a mapping δ_g from feature trees to feature trees. Intuitively, $\delta_g(\tau)$ is obtained by adding a leaf $(\pi g, b)$ to every node π of τ. Formally, we assume a feature tree τ such that $g \notin \mathcal{F}(\tau)$.

$$D_{\delta_g(\tau)} = D_\tau \cup \{\pi g \mid \pi \in D_\tau\}$$
$$L_{\delta_g(\tau)} = L_\tau \cup \{(\pi g, b) \mid \pi \in D_\tau\}$$

Lemma 15. *If τ is a feature tree such that $g \notin \mathcal{F}(\tau)$ for all π then $\delta_g(\tau)$ is a feature tree that is sufficiently labeled.*

Proof. The assumption $g \notin \mathcal{F}(\tau)$ implies that $L_{\delta_g(\tau)}$ is a partial function such that $\delta_g(\tau)$ is indeed a feature tree. □

Lemma 16. *Assume $g \notin \mathcal{F}(\alpha(x))$ for x and $g \notin \mathcal{F}(\varphi)$. If α is a solution of φ in $FT_{\leq}$ then $\delta_g \circ \alpha$ is a solution of φ in $FT_{\leq}^{-}$.*

Deleting Labels. There exists a left-inverse δ_g^{-1} of the function δ_g. For arbitrary τ, we define a feature tree $\delta_g^{-1}(\tau)$ as follows:

$$D_{\delta_g^{-1}(\tau)} = D_\tau \setminus \{\pi g \pi' \mid \pi, \pi' \in \mathcal{F}^*\}$$
$$L_{\delta_g^{-1}(\tau)} = L_\tau \setminus \{(\pi, a) \mid \pi = \pi' g \pi'',\ a \in \mathcal{L}\}$$

Lemma 17. *If $g \notin \mathcal{F}(\tau)$ then $\delta_g^{-1}(\delta_g(\tau)) = \tau$.*

Lemma 18. *Let $g \notin \mathcal{F}(\varphi)$. If α is a solution of φ in $FT_{\leq}^{-}$ then $\delta_g^{-1} \circ \alpha$ is a solution of φ in $FT_{\leq}$.*

Lemma 19. *Let φ be a constraint with $g \notin \mathcal{F}(\varphi)$, $\bar{x}$ a sequence of variables, and α a variable assignment such that $g \notin \mathcal{F}(\alpha(y))$ for all $y \in \mathcal{V}(\exists\bar{x}\varphi)$. Then α is a solution of $\exists\bar{x}\varphi$ over* $\mathrm{FT}_{\leq}$ *if and only if $\delta_g \circ \alpha$ is a solution of $\exists\bar{x}\varphi$ over* $\mathrm{FT}^-_{\leq}$.

Proof. Let α be a solution of $\exists\bar{x}\varphi$ over $\mathrm{FT}_{\leq}$. There exists a sequence of trees $\bar{\tau}$ such that $\alpha[\bar{\tau}/\bar{x}]$ is a solution of φ over $\mathrm{FT}_{\leq}$. (Here, $\alpha[\bar{\tau}/\bar{x}]$ denotes the valuation that maps $\bar{x}$ pointwise to $\bar{\tau}$ and coincides with α everywhere else.) Since $g \notin \mathcal{F}(\varphi)$, the mapping $\delta_g^{-1} \circ (\alpha[\bar{\tau}/\bar{x}])$ is also a solution of φ by Lemma 18. The latter variable assignment coincides with $\alpha[\delta_g^{-1}(\bar{\tau})/\bar{x}]$ since we have assumed $g \notin \mathcal{F}(\alpha(y))$ for all y. Thus $\delta_g \circ (\alpha[\delta_g^{-1}(\bar{\tau})/\bar{x}])$ is a solution of φ over $\mathrm{FT}^-_{\leq}$ (Lemma 16), which implies that $\delta_g \circ \alpha$ is a solution of $\exists\bar{x}\varphi$ over $\mathrm{FT}^-_{\leq}$.

For the converse, assume that $\delta_g \circ \alpha$ is a solution of $\exists\bar{x}\varphi$ over $\mathrm{FT}^-_{\leq}$. There exists a sequence of trees $\bar{\tau}$ such that $(\delta_g \circ \alpha)[\bar{\tau}/\bar{x}]$ is a solution of φ over $\mathrm{FT}^-_{\leq}$. Hence, $\delta_g^{-1} \circ ((\delta_g \circ \alpha)[\bar{\tau}/\bar{x}])$ is a solution of φ over $\mathrm{FT}_{\leq}$(Lemma 18). Also, $\delta_g^{-1} \circ \delta_g \circ \alpha = \alpha$ due to Lemma 17 and $g \notin \alpha(y)$ for all y. Thus:

$$\delta_g^{-1} \circ ((\delta_g \circ \alpha)[\bar{\tau}/\bar{x}]) = (\delta_g^{-1} \circ \delta_g \circ \alpha)[\delta_g^{-1}(\bar{\tau})/\bar{x}] = \alpha[\delta_g^{-1}(\bar{\tau})/\bar{x}]$$

This proves that α is a solution of $\exists\bar{x}\varphi$ over $\mathrm{FT}_{\leq}$. □

Proposition 20 Correctness. *Let $g \notin \mathcal{F}(\varphi \wedge \varphi')$ and $\mathcal{V}(\exists\bar{x}\varphi') \subseteq \mathcal{V}(\varphi)$. If $\eta_{g,b}(\varphi) \models_{\mathrm{FT}^-_{\leq}} \exists\bar{x}\varphi'$ then $\varphi \models_{\mathrm{FT}_{\leq}} \exists\bar{x}\varphi'$.*

Proof. Let $g \notin \mathcal{F}(\varphi \wedge \varphi')$, $\mathcal{V}(\exists\bar{x}\varphi') \subseteq \mathcal{V}(\varphi)$, and $\eta_{g,b}(\varphi) \models_{\mathrm{FT}^-_{\leq}} \exists\bar{x}\varphi'$. We have to show that every solution α of φ in $\mathrm{FT}_{\leq}$ is also a solution of $\exists\bar{x}\varphi'$. Since the number of features in $\mathcal{F}$ is infinite and $g \notin \mathcal{F}(\varphi \wedge \varphi')$, we only need to consider solutions α of φ such that $g \notin \mathcal{F}(\alpha(x))$ for all $x \in \mathcal{V}(\varphi)$. For the proof of this fact see the full paper [15].

Since α is a solution of φ, and $g \notin \mathcal{F}(\alpha(x))$ for all $x \in \mathcal{V}(\varphi \wedge \exists\bar{x}\varphi')$, $\delta_g \circ \alpha$ is a solution of φ (Lemma 19). From the definition of δ_g it follows that $\delta_g \circ \alpha$ is also a solution of $\bigwedge_{y \in \mathcal{V}(\varphi)} \exists y' \, (y[g]y' \wedge y'{=}b)$, *i.e.*, $\delta_g \circ \alpha$ is a solution of $\eta_{g,b}(\varphi)$ over $\mathrm{FT}^-_{\leq}$. Entailment as assumed implies that $\delta_g \circ \alpha$ is a solution of $\exists\bar{x}\varphi'$ over $\mathrm{FT}^-_{\leq}$. Thus α is also a solution of $\exists\bar{x}\varphi'$ over $\mathrm{FT}_{\leq}$ (Lemma 19). □

5.2 Least Solutions

We will exploit the completeness of the satisfiability test for ordering constraints over feature trees given in [17]. This test also computes the least solution of a satisfiable constraint. The form of these least solutions can be interpreted as a criterion for entailment. We call a constraint φ F1F2-closed if it satisfies the following properties:

F1.1	$x{\leq}x$ *in* φ	if $x \in \mathcal{V}(\varphi)$
F1.2	$x{\leq}z$ *in* φ	if $x{\leq}y$ *in* φ and $y{\leq}z$ *in* φ
F2	$x'{\leq}y'$ *in* φ	if $x[f]x'$ *in* φ, $x{\leq}y$ *in* φ and $y[f]y'$ *in* φ

Lemma 21. *There exists a cubic time algorithm that given a constraint* φ *either detects the unsatisfiability of* φ, *or proves its satisfiability and returns an* F1F2*-closed constraint equivalent to* φ.

Proof. We first consider an extended set of rules that defines a satisfiability test. Let *false* and $x{\sim}y$ be auxiliary constraints where *false* denotes inconsistency and the semantics of $x{\sim}y$ is given by the equivalence $x_1{\sim}x_2 \leftrightarrow \exists x(x_1{\leq}x \wedge x_2{\leq}x)$. We call a constraint φ F-*closed* if it satisfies all properties F1–F5 where:

F3.1	$x{\sim}y$ *in* φ	if $x{\leq}y$ *in* φ
F3.2	$x{\sim}z$ *in* φ	if $x{\leq}y$ *in* φ and $y{\sim}z$ *in* φ
F3.3	$x{\sim}y$ *in* φ	if $y{\sim}x$
F4	$x'{\sim}y'$ *in* φ	if $x[f]x'$ *in* φ, $x{\sim}y$ *in* φ and $y[f]y'$ *in* φ
F5	*false in* φ	if $a(x) \wedge x{\sim}y \wedge a'(y)$ *in* φ and $a \neq a'$

For every constraint φ an equivalent F-closed constraint φ' can be computed in cubic time applying the rules in F exhaustively. Furthermore, φ is satisfiable if and only if φ' does not contain *false*. Deleting the auxiliary constraints $x_1{\sim}x_2$ from φ' transforms φ' into an equivalent F1F2-closed constraint in linear time. □

We will use a syntactic description of the least solution of a satisfiable constraint in order to derive properties of entailment in $FT_{\leq}$ (see [17]). We define *syntactic entailment* judgements of the form $\varphi \vdash x[\pi]{\geq}y$ and $\varphi \vdash x[\pi]{\geq}a$ as follows.

$$\begin{array}{rl} \varphi \vdash x[\varepsilon]{\geq}y & \text{if } y{\leq}x \textit{ in } \varphi \\ \varphi \vdash x[f]{\geq}y & \text{if } x[f]y \textit{ in } \varphi \\ \varphi \vdash x[\pi_1\pi_2]{\geq}y & \text{if exists } z \text{ such that } \varphi \vdash x[\pi_1]{\geq}z \text{ and } \varphi \vdash z[\pi_2]{\geq}y \\ \varphi \vdash x[\pi]{\geq}a & \text{if exists } z \text{ such that } \varphi \vdash x[\pi]{\geq}z \text{ and } a(z) \textit{ in } \varphi \end{array}$$

Proposition 22 Least Solutions. *Let* φ *be satisfiable and* F1F2*-closed. For every variable* $x \in \mathcal{V}(\varphi)$, *and all* a, π, z *the following two equivalences hold:*

$$\begin{array}{rl} \varphi \models_{FT_{\leq}} \exists z\, x[\pi]{\geq}z & \textit{iff} \quad \textit{exists } z' \textit{ such that } \varphi \vdash x[\pi]{\geq}z' \\ \varphi \models_{FT_{\leq}} x[\pi]{\geq}a & \textit{iff} \quad \varphi \vdash x[\pi]{\geq}a \end{array}$$

Proof. The implications from the right to the left hold because syntactic entailment is correct with respect to semantic entailment in $FT_{\leq}$. For the converse implication, we define the least solution least_φ of φ such that for all $x \in \mathcal{V}(\varphi)$.

$$\begin{array}{rl} D_{\mathsf{least}_\varphi(x)} & = \{\pi \mid \text{exists } z \text{ such that } \varphi \vdash x[\pi]{\geq}z\} \\ L_{\mathsf{least}_\varphi(x)} & = \{(\pi, a) \mid \varphi \vdash x[\pi]{\geq}a\} \end{array}$$

Without loss of generality, we can assume that φ is F-complete. First note that completion with F1–F5 does never derive *false* since φ is satisfiable. Second, since φ is F1F2-closed, completion may only add auxiliary constraints $x{\sim}y$, which does not affect the validity of judgements $\varphi \vdash x[\pi]{\geq}z$ and $\varphi \vdash x[\pi]{\geq}a$. In [17] it is proved that least_φ is a solution of φ whenever φ is F1-F5-closed.

Since syntactic entailment is correct with respect to semantic entailment in $\mathrm{FT}_{\leq}$, it is clear that least_φ is smaller than every solution of φ, *i.e.*, least_φ is the least solution of φ: If $\varphi \not\vdash x[\pi]{\geq}z'$ for all z' then $\pi \notin D_{\mathsf{least}_\varphi(x)}$, *i.e.*, $\varphi \not\models_{\mathrm{FT}_{\leq}} \exists z\, x[\pi]{\geq}z$. In analogy, if $\varphi \not\vdash x[\pi]{\geq}a$ then $(\pi, a) \notin L_{\mathsf{least}_\varphi(x)}$, *i.e.*, $\varphi \not\models_{\mathrm{FT}_{\leq}} x[\pi]{\geq}a$. □

5.3 Entailment in $\mathrm{FT}_{\leq}$ Implies Entailment in $\mathrm{FT}^{-}_{\leq}$

For every constraint φ let $-\varphi$ be the constraint that is obtained from φ by inverting all its ordering constraints, *i.e.*, by replacing $x{\leq}y$ with $y{\leq}x$. We define:

$$\varphi \vdash x[\pi]{\leq}y \quad \text{iff} \quad -\varphi \vdash x[\pi]{\geq}y$$

Lemma 23 Mountain Chains. *Let α be a solution of φ and assume variables y, z, z', paths π_0, π_1, π_2 and a label a. If $\varphi \models \exists z'\,(z[\pi_0]{\leq}z' \wedge y[\pi_1]{\geq}z')$ and $(\pi_0\pi_2, a) \in L_{\alpha(z)}$ then $(\pi_1\pi_2, a) \in L_{\alpha(y)}$.*

Proof. By induction on the length of π_0 one shows that $(\pi_2, a) \in L_{\alpha(z')}$, hence $(\pi_1\pi_2, a) \in L_{\alpha(y)}$. (The picture on the right illustrates the situation. The vertical dimension (top to bottom) corresponds to feature selection, the horizontal dimension (left to right) to the ordering $\leq$.) □

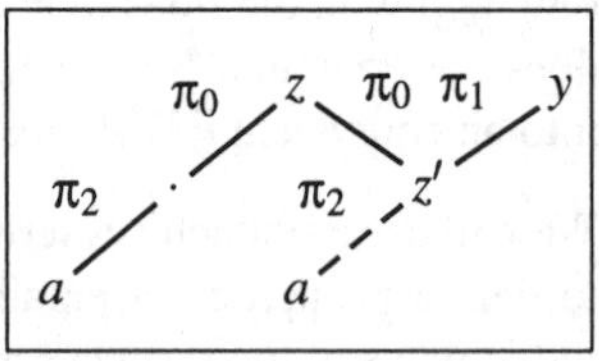

Proposition 24 Completeness. *Suppose $\mathcal{V}(\exists\bar{x}\varphi') \subseteq \mathcal{V}(\varphi)$ and $g \notin \mathcal{F}(\varphi \wedge \varphi')$. If $\varphi \models_{\mathrm{FT}_{\leq}} \exists\bar{x}\varphi'$ then $\eta_{g,b}(\varphi) \models_{\mathrm{FT}^{-}_{\leq}} \exists\bar{x}\varphi'$.*

Proof. First notice that the assumption $\mathcal{V}(\exists\bar{x}\varphi') \subseteq \mathcal{V}(\varphi)$ does not restrict generality, since we can always assume that φ contains trivial inequalities $y{\leq}y$ for all $y \in \mathcal{V}(\exists\bar{x}\varphi')$. Now suppose $\mathcal{V}(\varphi) \cap \{\bar{x}\} = \emptyset$ and let α be a solution of $\eta_{g,b}(\varphi)$ over $\mathrm{FT}^{-}_{\leq}$. Note that $\mathcal{V}(\varphi') \subseteq \{\bar{x}\} \cup \mathcal{V}(\varphi)$. We have to construct a solution α' of φ' over $\mathrm{FT}^{-}_{\leq}$ which coincides with α on the variables in $\mathcal{V}(\varphi)$.

We define $\alpha'(y)$ for all $y \in \{\bar{x}\}$. Since $\alpha'(y)$ must be sufficiently labeled, it suffices to define its labeling function. Let φ and φ' be F1F2 closed. For $y \in \{\bar{x}\}$ we define:

L1	$(\pi, a) \in L_{\alpha'(y)}$	if $\varphi' \vdash y[\pi]{\geq}a$
L2	$(\pi g, b) \in L_{\alpha'(y)}$	if exists z such that $\varphi' \vdash y[\pi]{\geq}z$
L3	$(\pi_1\pi_2, a) \in L_{\alpha'(y)}$	if $\begin{cases} \text{exists } z \in \mathcal{V}(\varphi) \text{ and } z' \text{ such that} \\ \varphi' \vdash y[\pi_1]{\geq}z',\ \varphi' \vdash z[\pi_0]{\leq}z' \text{ and } (\pi_0\pi_2, a) \in L_{\alpha(z)} \end{cases}$

Compare condition L3 with the mountain situation depicted above. For all $y \in \{\bar{x}\}$, we define $D_{\alpha'(y)} = \{\pi \mid \pi \text{ is a prefix of } \pi' \text{ and } (\pi', a) \in L_{\alpha'(y)} \text{ for some } a\}$. For all $y \notin \{\bar{x}\}$ we set $\alpha'(y) = \alpha(y)$. It is clear that $\alpha'(y)$ is sufficiently labeled because of L2. It remains to show that α' is a solution of φ', *i.e.*, that α' satisfies all basic constraints in φ'. Here, we only consider a single case. The complete case distinction covers three pages and is given in the proof of Lemma 31 in the full paper [15]. Consider the case $x[f]y$ *in* φ', $x \in \mathcal{V}(\varphi)$, $y \in \{\bar{x}\}$, and $(\pi, a) \in L_{\alpha'(y)}$ because of L2 or L3. We show $(f\pi, a) \in L_{\alpha'(x)}$.

L2 exists π' and z such that $\varphi' \vdash y[\pi']_{\geq} z$, $\pi = \pi' g$ and $a = b$. Since $x[f]y \in \varphi'$, $\varphi' \vdash x[f\pi']_{\geq} z$ such that $\varphi \models_{FT_{\leq}} \exists z(x[f\pi']_{\geq} z)$ since $x \in \mathcal{V}(\varphi)$. Proposition 22 implies the existence of $z' \in \mathcal{V}(\varphi)$ such that $\varphi \vdash x[f\pi]_{\geq} z'$ and thus $\eta_{g,b}(\varphi) \vdash x[f\pi g]_{\geq} b$. Since α is a solution of $\eta_{g,b}(\varphi)$, we have $(f\pi' g, b) \in L_{\alpha(x)}$, *i.e.*, $(f\pi, a) \in L_{\alpha'(x)}$.

L3 Let $(\pi, a) \in L_{\alpha'(y)}$ since there exists $z \in \mathcal{V}(\varphi)$, z', π_0, π_1, and π_2 with $\varphi' \vdash y[\pi_1]_{\geq} z'$, $\varphi' \vdash z[\pi_0]_{\leq} z'$, $(\pi_0\pi_2, a) \in L_{\alpha(z)}$, and $\pi = \pi_1\pi_2$. Since $x[f]y$ we also have $\varphi' \vdash x[f\pi_1]_{\geq} z'$. Since $x, z \in \mathcal{V}(\varphi)$ this implies $\varphi \models_{FT_{\leq}} \exists z' (z[\pi_0]_{\leq} z' \wedge x[f\pi_1]_{\geq} z')$. Now, Lemma 23 and $(\pi_0\pi_2, a) \in L_{\alpha(z)}$ imply $(f\pi_1\pi_2, a) \in L_{\alpha(x)}$, *i.e.*, $(f\pi, a) \in L_{\alpha'(x)}$. □

6 Conclusion and Future Work

We have investigated decidability and complexity questions for fragments of the first-order theory of ordering constraints over feature trees ($FT_{\leq}$). We have proved that the entailment problem of $FT_{\leq}$ with existential quantifiers is coNP-hard over finite trees, PSPACE-hard over infinite trees, and decidable in both cases. Our main contribution in this paper is that we have related $FT_{\leq}$ to the monadic second-order logic with multiple successors, that enabled us to solve the first non-trivial decision problem over feature trees by reduction to Rabin's famous decidability result for (W)SωS.

In more recent work [16] we have proved PSpace-completeness for the entailment problem of $FT_{\leq}$ with existential quantification, both over finite trees and over infinite trees. There, a new decision algorithm is presented based on finite automata instead of Rabin automata. It is also shown that the first-order theory of $FT_{\leq}$ is undecidable in contrast the first-order theory of ordering constraints over sufficiently labeled feature trees.

Acknowledgments. We thank Jean-Marc Talbot, Sophie Tison, and Marc Tommasi for inspiring discussions and Franz Baader for contributing Corollary 6. We thank Ralf Treinen for discussion of Henglein's and Rehof's result, and Ralf Treinen and the anonymous referees for careful reading and invaluable remarks on this paper. The research reported here has been supported by the the Esprit Working Group CCL II (EP 22457) and the DFG through SFB 378 at the Universität des Saarlandes.

References

1. H. Aït-Kaci and A. Podelski. Towards a Meaning of Life. *The Journal of Logic Programming*, 16(3 and 4):195–234, July, Aug. 1993.
2. H. Aït-Kaci, A. Podelski, and G. Smolka. A Feature-based Constraint System for Logic Programming with Entailment. *Theoretical Computer Science*, 122(1–2):263–283, 1994.
3. R. Backofen. A Complete Axiomatization of a Theory with Feature and Arity Constraints. *The Journal of Logic Programming*, 1995. Special Issue on Computational Linguistics and Logic Programming.
4. R. Backofen and G. Smolka. A Complete and Recursive Feature Theory. *Theoretical Computer Science*, 146(1–2):243–268, July 1995.
5. H. Comon. Sequentiality, second-order monadic logic and tree automata. In K. Dexter, editor, *Proceedings of the Logic in Computer Science Conference*, pages 508–517, 1995.

6. S. A. Cook. The Complexity of Theorem-Proving Procedures. In *Annual ACM Symposium on Theory of Computing*, pages 151–158, New York, 1971. ACM.
7. J. Dörre. Feature logics with weak subsumption constraints. In *Annual Meeting of the ACL (Association of Computational Logics)*, pages 256–263, 1991.
8. M. R. Garey and D. S. Johnson. *Computers and Intractability: A Guide to the Theory of NP-Completeness*. W.H. Freeman and Company, New York, 1979.
9. F. Henglein and J. Rehof. The Complexity of Subtype Entailment for Simple Types. In *12th IEEE Symposium on Logic in Computer Science*, Warsaw, Poland, 1997.
10. J. D. Hopcroft and J. D. Ullman. *Introduction to Automata Theory, Languages and Computation*. Addison-Wesley, Reading, MA, 1979.
11. R. M. Kaplan and J. Bresnan. Lexical-Functional Grammar: A Formal System for Grammatical Representation. pages 173–381. The MIT Press, Cambridge, MA, 1982.
12. M. Kay. Functional Grammar. In C. Chiarello et al., editor, *Proc. of the 5th Annual Meeting of the Berkeley Linguistics Society*, pages 142–158, 1979.
13. M. Müller. Ordering Constraints over Feature Trees with Ordered Sorts. In P. Lopez, S. Manandhar, and W. Nutt, editors, *Computational Logic and Natural Language Understanding*, Lecture Notes in Artificial Intelligence, to appear, 1997.
14. M. Müller and J. Niehren. Entailment for Set Constraints is not Feasible. Technical report, Programming Systems Lab, Universität des Saarlandes, 1997. `http://www.ps.uni-sb.de/Papers/abstracts/inesInfeas.html`
15. M. Müller and J. Niehren. Ordering Constraints over Feature Trees Expressed in Second-order Monadic Logic. Full version available at `http://www.ps.uni-sb.de/Papers/abstracts/SWS97.html` 1997.
16. M. Müller, J. Niehren, and R. Treinen. The First-order Theory of Ordering Constraints over Feature Trees. Submitted. `http://www.ps.uni-sb.de/Papers/abstracts/FTSubTheory-98.html`. 1997.
17. M. Müller, J. Niehren, and A. Podelski. Ordering Constraints over Feature Trees. In *3rd Int. Conf. on Principles and Practice of Constraint Programming*, vol 1330 of *LNCS*, 1997.
18. C. Pollard and I. Sag. *Head-Driven Phrase Structure Grammar*. Studies in Contemporary Linguistics. Cambridge University Press, Cambridge, England, 1994.
19. M. O. Rabin. Decidability of second-order theories and automata on infinite trees. *Transactions of the American Mathematical Society*, 141:1–35, 1969.
20. W. C. Rounds. Feature Logics. In J. v. Benthem and A. ter Meulen, editors, *Handbook of Logic and Language*. Elsevier Science Publishers B.V. (North Holland), 1997.
21. H. Seidl. Deciding Equivalence of Finite Tree Automata. *SIAM Journal of Computing*, 19(3):424–437, June 1990.
22. S. Shieber. *An Introduction to Unification-based Approaches to Grammar*. CSLI Lecture Notes No. 4. Center for the Study of Language and Information, 1986.
23. G. Smolka. The Oz Programming Model. In J. van Leeuwen, editor, *Computer Science Today*, LNCS, vol. 1000, pages 324–343. Springer-Verlag, Berlin, Germany, 1995.
24. G. Smolka and R. Treinen. Records for Logic Programming. *The Journal of Logic Programming*, 18(3):229–258, Apr. 1994.
25. J. W. Thatcher and J. B. Wright. Generalized finite automata theory with an application to a decision problem of second-order logic. *Mathematical Systems Theory*, 2(1):57–81, 1967.
26. S. Vorobyov. An improved lower bound for the elementary theories of trees. In *Internat. Conf. on Automated Deduction*, vol. 1104 of *LNCS*, 275–287, 1996.

Co-definite Set Constraints

Witold Charatonik* Andreas Podelski

Max-Planck-Institut für Informatik
Im Stadtwald, D-66123 Saarbrücken, Germany
{witold;podelski}@mpi-sb.mpg.de

Abstract. In this paper, we introduce the class of co-definite set constraints. This is a natural subclass of set constraints which, when satisfiable, have a *greatest* solution. It is practically motivated by the set-based analysis of logic programs with the greatest-model semantics. We present an algorithm solving co-definite set constraints and show that their satisfiability problem is DEXPTIME-complete.

1 Introduction

Set constraints and set-based analysis form an established research topic. It combines theoretical investigations ranging from expressiveness and decidability to program semantics and domain theory, with direct practical applications to type inference, optimization and verification of imperative, functional, logic and reactive programs (see [1,14,20] for overviews).

In set-based analysis, the problem of reasoning about runtime properties of programs is transferred to the problem of solving set constraints. The design of a system for a particular program analysis problem (for a particular class of programs) involves two steps: (1) single out a subclass of set constraints and devise an algorithm for solving set constraints in this subclass, and (2) define a mapping $P \mapsto \varphi_P$ from programs into this subclass and show the soundness of the abstraction of P by a distinguished solution of φ_P. The advantage with respect to other static-analysis methods is the common to all constraint-based approaches: the logical formulation of the problems allows for their classification and for the reuse of optimized implementations. It is thus important to classify the arising constraint-solving problems and devise algorithms for them.

In this paper, we define the subclass of *co-definite* set constraints. This is a natural subclass of set constraints which, when satisfiable, have a *greatest* solution. We present an algorithm solving co-definite set constraints in DEXPTIME. The algorithm involves some novel adaptations of standard techniques for solving set constraints to the new situation where the solutions range over sets of *infinite* trees and must be constructed by co-induction (and not by induction as with least solutions). We show how one can encode the problem of emptiness of intersection of tree automata in a direct way. Thus, the satisfiability problem is DEXPTIME-complete.

---

* On leave from Wrocław University. Partially supported by Polish KBN grant 8T11C02913.

The new class of co-definite set constraints is practically motivated by the set-based analysis of *reactive logic programs* (called perpetual processes in [16]). Their semantics is defined by the greatest fixpoint of the immediate consequence operator T_P, which at the same time is the greatest model. The semantics is defined not over finite but over infinite trees.[1] Our algorithm accounts for either case. In [21], we show that the greatest solution of the co-definite set constraint φ_P that we assign to the program P is larger than the greatest model of P. The error diagnosis for concurrent constraint programs (the static prediction of the inevitability of failure or deadlock), which is presented in [21], is based on that fact and employs the algorithm presented here.

Related work. Heintze and Jaffar [11,12] formulated the general problem of solving set constraints and gave the first decidability result for a subclass of set constraints which they called *definite*, for the reason that all satisfiable constraints in the class have a least solution. They have singled out this subclass for the analysis of logic programs with the (standard) least model semantics. The present authors [7] have recently characterized the complexity for this subclass (DEXPTIME). The general problem is NEXPTIME-complete [4,5].

Definite and co-definite set constraints are not dual with respect to their syntax. We must exclude constraints of the form $f(x,y) \subseteq f(a,a) \cup f(b,b)$ which do not have a greatest solution. They are also not dual with respect to the constraint solving problem (although the two complexity characterizations might suggest this). Although one can directly dualize the Boolean set operators and also the tree constructors, this is not the case for the projection operator. The complement of the application of the projection is generally not the application of the projection of the complement. The algorithm given in Section 4.2 in [11] does therefore *not* compute the greatest solution. (The greatest solution of $x = f_{(1)}^{-1}(f(a,a))$ is $\{a\}$, but starting from this constraint that algorithm yields $x = \bot$ whose greatest solution is $\emptyset$. This is because $dual(f_{(1)}^{-1}(f(a,a))) = f_{(1)}^{-1}(dual(f(a,a))) = f_{(1)}^{-1}(\Omega_f \cup f(\Omega_a, \top) \cup f(\top, \Omega_a)) = \top$ and thus $x = dual(\top) = \bot$).

In definite set constraints, union is expressed via conjunction (e.g., $a \cup b \subseteq y$ by $a \subseteq y \wedge b \subseteq y$ and need not be dealt with explicitly. Co-definite set constraints employ union as an operator over terms, and conjunction introduces intersection additionally. The next example shows that our algorithm must combine ("multiply out") intersections of unions of terms. (How can this be done in single

---

[1] The reactive logic program $P \equiv p(f(x)) \leftrightarrow p(x)$ illustrates the difference between infinite and finite trees. When interpreted over *finite* trees, the greatest model is the empty set; otherwise, it is the singleton containing the infinite tree $f(f(f(\ldots)))$. In either case, the execution of the call of $p(x)$ does not fail. More generally, one can characterize finite failure by the greatest model in the case of infinite trees, but not in the case of finite trees. In [21] we use co-definite set constraints to approximate the greatest model; we have to interpret them over sets of *infinite trees* in order to apply this approximation to the prediction of finite failure of logic programs and of errors in concurrent constraint programs.

exponential time? - There are exponentially many union terms.) The co-definite set constraint

$$y \subseteq f(a,c) \cup f(c,b) \wedge \qquad (1)$$
$$y \subseteq f(a,c) \cup f(d,b)$$

is satisfiable in conjunction with $a \subseteq f^{-1}_{(1)}(y)$ but unsatisfiable in conjunction with $b \subseteq f^{-1}_{(2)}(y)$. In the last case, however, the naive method would derive only $b \subseteq c \cup b$, which does not give unsatisfiability.

Analyzing logic programs with the *least* model semantics, Mishra [18] has used a class of set constraints with a non-standard interpretation over non-empty *path-closed* sets of finite trees, which also have a greatest solution. In that interpretation, $f(x,y) \subseteq f(a,a) \cup f(b,b)$ has a greatest solution (which assigns both variables x and y the set $\{a,b\}$). Heintze and Jaffar [13] have shown that Mishra's analysis is less accurate than theirs in two ways, due to the choice of the greatest solution and due to the choice of the non-standard interpretation, respectively. The choice of the non-standard interpretation over path-closed sets of trees is not traded with by a lower complexity. Our hardness proof for co-definite set constraints carries over to Mishra's set constraints. This is because the tree automata used in the reduction can be chosen deterministic [22].

2 Definitions

A (general) set expression e is built up by: variables, tree constructors, the Boolean set operators and the *projection* operator [11]. If e does not contain the complement operator, then e is called a *positive* set expression. A (general) set constraint is a conjunction of inclusions of the form $e \subseteq e'$.

Definition 1. A *co-definite* set constraint is a conjunction of inclusions $e_l \subseteq e_r$ between positive set expressions, where the set expressions e_l on the left-hand side of $\subseteq$ are furthermore restricted to contain only variables, constants, unary function symbols and the union operator (that is, no projection, intersection or function symbol of arity greater than one).

We assume given a ranked alphabet Σ fixing the arity $n \geq 0$ of its function symbols $f, g, \ldots$ and constant symbols $a, b, \ldots$, and an infinite set Var of variables $x, y, z, u, v, w, \ldots$. The formulations and results in this paper apply to either case: finite trees, or infinite trees. We then say simply *trees* and use the notation T_Σ. By infinite trees we mean trees, whose branches are infinite or finite.

We interpret set constraints over $\mathcal{P}(T_\Sigma)$, the domain of sets of trees over the signature Σ. That is, the values of variables are sets of trees, or: a valuation is a mapping $\alpha : \mathsf{Var} \to \mathcal{P}(T_\Sigma)$. Tree constructors are interpreted as functions over sets of trees: the constant a is interpreted as $\{a\}$, the function f applied to the sets $S_1, \ldots, S_n$ yields the set $\{f(t_1, \ldots, t_n) \mid t_1 \in S_1, \ldots, t_n \in S_n\}$. The application of the projection operator for a function symbol f and the k-th argument position on a set S of trees is defined by $f^{-1}_{(k)}(S) = \{t \mid \exists t_1, \ldots t_n : t_k = t,\ f(t_1, \ldots, t_k, \ldots, t_n) \in S\}$.

The next remark (which is proven by checking all cases of possible inclusions) implies an important property: if a co-definite set constraint is satisfiable, then it has a greatest solution.

Remark 2. The solutions of co-definite set constraints are closed under arbitrary union. □

For the formal treatment, we will use co-definite set constraints in a restricted form, which we will simply call constraints.

Definition 3 (restricted syntax: constraints φ). A constraint φ is a co-definite set constraint in the syntax given below.

$$\begin{aligned} \tau &::= \; x \mid f(\bar{u}) \mid \tau_1 \cup \tau_2 \mid \bot \\ \varphi &::= \; a \subseteq x \mid x \subseteq \tau \mid x \subseteq f_{(k)}^{-1}(u) \mid \varphi_1 \wedge \varphi_2 \end{aligned}$$

For better manipulation of constraints, we have added the symbol $\bot$, which is interpreted as the empty set and is the neutral element wrt. $\cup$. By convention, the empty union is $\bot$ (i.e., $\bigcup \emptyset = \bot$); similarly, $\bigcap \emptyset = \top$.

We write $\bar{u}$ for the tuple $(u_1, \ldots, u_n)$ of variables and $\bar{t}$ for the tuple $(t_1, \ldots, t_n)$ of trees, where $n \geq 0$ is given implicitly (*e.g.*, in $x \subseteq f(\bar{u})$ by the arity of the function symbol f). We write $\bar{u} \subseteq \bar{v}$ for $\{u_1 \subseteq v_1, \ldots, u_n \subseteq v_n\}$. As is usual, we identify a conjunction of constraints with the set of all conjuncts.

We use $\mathsf{Var}(E)$ for the set of variables contained in the expression E, and $\mathsf{Terms}(\varphi)$ for the sets of all *flat* terms τ (*i.e.*, without union) occurring in φ. We use $\Sigma(\varphi)$ for the set of all function symbols occurring in φ; this set is finite.[2]

Given a co-definite set constraint, we can transform it into an equivalent one of restricted syntax easily. We eliminate function and union symbols on the left-hand side by using the equivalences $f(e) \subseteq e'$ iff $e \subseteq f_{(1)}^{-1}(e')$ and $e_1 \cup e_2 \subseteq e$ iff $e_1 \subseteq e \wedge e_2 \subseteq e$. We flatten the terms on the right-hand side by replacing intersection with conjunction and by introducing a fresh variable for each subexpression occurring on the right-hand side of inclusions. Since we are interested in the greatest solution of the initial constraint, it is enough to write only one inclusion (instead of equality) between the new variable and the expression. For example, we replace the inclusion $x \subseteq f_{(1)}^{-1}(y_1 \cap y_2)$ by $x \subseteq f_{(1)}^{-1}(y) \wedge y \subseteq y_1 \wedge y \subseteq y_2$. The transformation does not change the complexity measure. The number of new variables is linear in the size of the initial constraint.

3 Algorithm

The algorithm for solving a constraint φ_0 computes the fixpoint under the operator that, applied to a constraint φ, adds the direct consequences of φ under the axioms given in Table 1 to φ. The control of the algorithm is presented in Table 2. Compared to the usual fixpoint saturation, the only difference is that in the

[2] We do not want to assume that the signature Σ is finite. This is important for the use of set constraints in (modular) program analysis: the constructor alphabet is never fully known, or is assumed to be extensible.

1. $x \subseteq y \wedge y \subseteq z \rightarrow x \subseteq z$	
2. $x \subseteq \tau_1 \cup y,\ y \subseteq \tau_2 \rightarrow x \subseteq \tau_1 \cup \tau_2$	
3. $\gamma \wedge u \subseteq f_{(k)}^{-1}(x) \rightarrow \bigwedge_i u \subseteq \bigcup_j u_{ij}$	where $\bigcap_i \bigcup_j u_{ij} = f_{(k)}^{-1}(x,\gamma)$
4. $\gamma \wedge u \subseteq f_{(k)}^{-1}(x) \rightarrow u \subseteq \bot$	if $f_{(k)}^{-1}(x,\gamma) = \bot$
5. $a \subseteq x \wedge x \subseteq \bigcup_i f_i(\bar{u}_i) \rightarrow \mathit{false}$	if $a \neq f_i$ for all i
6. $a \subseteq x \wedge x \subseteq \bot \rightarrow \mathit{false}$	

Table 1. Satisfiability-complete axiom scheme for constraints φ

case of Axioms 3 and 4, the operator adds only the direct consequences that are obtained by applications where the constraint γ is instantiated to φ (as opposed to: a subpart of φ). Applying the axioms to subparts of a constraint with, say, m conjuncts would amount to applying the axioms 2^m times. All applications to proper subparts are redundant. For example, $u \subseteq v$ could be inferred from $\varphi \equiv u \subseteq f_{(1)}^{-1}(x), x \subseteq f(v), x \subseteq a$ under Axiom 3; the algorithm infers only a stronger consequence $u \subseteq \bot$ by instantiating γ with φ. Note that conjunction $\wedge$ is idempotent; the conjunct $u \subseteq f_{(k)}^{-1}(x)$ in the axioms is, of course, instantiated to a conjunct of φ. Computing the expressions $f_{(k)}^{-1}(x,\gamma)$ in Axioms 3 and 4 is involved; we will discuss this in Section 3.3.

A constraint obtained as the fixpoint under the operator of the algorithm is in *closed form*, and φ^C is the closed form of φ. Note that, as we mentioned above, φ^C is not closed under all (possibly redundant) consequences under the axioms in Table 1.

We will next introduce automaton constraints ψ (Section 3.1). These form a subclass of co-definite set constraints which directly exhibit their greatest solution (Remark 7). We can construct, with each constraint φ, an automaton constraint $\Psi(\varphi)$ (Section 3.2). We use $\Psi(\varphi)$ for computing the expressions $f_{(k)}^{-1}(x,\gamma)$ in Axioms 3 and 4 (Section 3.3). (To give some intuition: As indicated by the example (1) in the introduction, we cannot apply the projection operator on terms τ directly but we have to first combine them and transform them into expressions with intersections below the function symbol. This leads us out of the restricted syntax of constraints φ.) Furthermore, if the constraint φ is in closed form then it has the same greatest solution as $\Psi(\varphi)$.

Before going into more detail, we summarize the main results of the paper.

Theorem 4. The algorithm in Table 2 computes the closed form φ^C of the input constraint φ in deterministic, single exponential time. The constraint φ is unsatisfiable if and only if φ^C contains *false*; otherwise, the greatest solution of φ is presented by $\Psi(\varphi^C)$.

Proof. See Propositions 17, 18 and 19 in Section 4. □

$\varphi := \varphi_0$
Repeat
  apply Axioms 1 and 2 to φ
  apply Axioms 3 and 4 to φ where γ is instantiated to φ
  apply Axioms 5 and 6 to φ
  add all direct direct consequences to φ
Until φ does not change or φ contains *false*
If φ contains *false*
  then "φ_0 is unsatisfiable"
  else "φ_0 is satisfiable" and $\varphi_0^C := \varphi$ ("φ is closed form of φ_0")

Table 2. Algorithm solving a constraint φ_0

Theorem 5. The satisfiability problem for co-definite set constraints is DEXPTIME-complete.

Proof. See Propositions 19 and 22 in Section 5. □

3.1 Automaton constraints ψ

We assume given a set q-Var of variables $q, q', \ldots$ which we want to distinguish from variables $x, y, \ldots$ in Var. Later we will take variables q that stand for intersections $x_1 \cap \ldots \cap x_k$ of variables $x_i \in$ Var.

Definition 6 (automaton constraint ψ). An automaton constraint ψ is a conjunction of the form $\psi \equiv \bigwedge_i q_i \subseteq E_i$ such that

- the variables q_i are pairwise different, and
- each expression E_i is either $\perp$ or of the form $\bigcup_j f_j(\bar{q}_j)$.

A variable q is *unbounded in* ψ if q is different from all q_i's on the left-hand side in ψ.

The interpretation of automaton constraints is as usual. A valuation is now a mapping from q-Var to sets. The next remark justifies the name automaton constraint.

Remark 7. The value of a variable q in the greatest solution of an automaton constraint ψ (in both cases of finite and infinite trees) is the language $\mathcal{L}(\mathcal{A}^\psi(q))$ of the top-down tree automaton $\mathcal{A}^\psi(q)$ constructed directly from ψ; in particular, the emptiness of the value of q can be tested in polynomial time.

We give the construction of the automata and the proof of the remark in the Appendix since we did not find it in the literature; it must, however, be folklore (cf. also [2]).

3.2 Constructing $\Psi(\varphi)$

Given a constraint φ, we can extract an automaton constraint $\Psi(\varphi)$ from φ which is equivalent to its subpart consisting of the conjuncts of the form $x \subseteq \bigcup_j f_j(\bar{u}_j)$. The variables q in $\Psi(\varphi)$ stand for intersections $x_1 \cap \ldots \cap x_n$ of variables $x_i \in \mathsf{Var}$. We note $\cap$-Var the set that these *intersection variables* q form. We use also $\bigcap S$ as another notation for q that stands for the intersection of the variables in $S \subseteq \mathsf{Var}$. The *proper upper bounds* τ of a variable x in φ are the terms of the form $\tau = \bigcup_j f_j(\bar{u}_j)$ such that $x \subseteq \tau$ lies in φ. Note that τ may be $\bot$.

We next define their combination, for variables x as well as for intersections q.

Definition 8 ($lub(x,\varphi)$, $lub(q,\varphi)$). The least upper bound of the variable x in the constraint φ is an intersection of terms τ,

$$lub(x,\varphi) = \bigcap\{\bigcup_j f_j(\bar{u}_j) \mid x \subseteq \bigcup_j f_j(\bar{u}_j) \text{ lies in } \varphi\}.$$

The least upper bound of an intersection $q = x_1 \cap \ldots \cap x_n$ is $lub(q,\varphi) = \bigcap_{i=1}^n lub(x_i,\varphi)$.

If x does not have proper upper bounds in φ then $lub(x,\varphi) = \top$. Also, note that $\top \cap \ldots \cap \top = \top$ and $\tau \cap \top = \tau$.

The expression $E = lub(q,\varphi)$ is an intersection of unions of proper terms $f(\bar{u})$. We transform such an expression E into a union of terms $f(\bar{q})$ over intersections of variables q, hereby using a variant of the disjunctive normal form (the computation of the standard one would here require doubly-exponential time).

Definition 9 (FDNF). The *full disjunctive normal form* of $E = \bigcap_{i\in I}\bigcup_{j\in J_i} f_{ij}(\bar{u}_{ij})$ is a union of terms $f(\bar{q})$ over intersection variables q,

$$\begin{aligned} FDNF(E) = \bigcup\{f(\bigcap S_1,\ldots,\bigcap S_n) \mid\ & f \in \Sigma,\ n = \text{arity}(f), \\ & S_1 \subseteq \mathsf{Var}(E),\ldots,S_n \subseteq \mathsf{Var}(E), \\ & \forall i \in I\ \exists j \in J_i : f = f_{ij} \wedge u_{ij,1} \in S_1 \wedge \ldots \wedge u_{ij,n} \in S_n\}. \end{aligned}$$

We set $FDNF(\top) = \top$ and $FDNF(\bot) = \bot$.

Example 10. If $E = (f(u,v_1) \cup f(u,v_2)) \cap f(u,u)$ then $FDNF(E)$ is the expression $f(u, u\cap v_1) \cup f(u, u \cap v_2) \cup \ldots \cup f(u\cap v_1 \cap v_2, u \cap v_1 \cap v_2)$ which contains redundant disjuncts. Using the convention that $\bigcup\emptyset = \bot$, we take $E = a \cap b$ and have $FDNF(E) = \bot$.

Given a constraint φ, we note $\cap$-$\mathsf{Var}(\varphi)$ the set of all q standing for intersections $x_1 \cap \ldots \cap x_n$ of variables $x_i \in \mathsf{Var}(\varphi)$ occurring in φ. We now can give the construction of the automaton constraint $\Psi(\varphi)$ from the constraint φ.

Definition 11 ($\Psi(\varphi)$). The automaton constraint corresponding to the constraint φ is

$$\Psi(\varphi) \equiv \bigwedge_{q\in\cap\text{-}\mathsf{Var}(\varphi)} q \subseteq FDNF(lub(q,\varphi)).$$

We discard from $\Psi(\varphi)$ all inclusions of the form $q \subseteq \top$.

3.3 Projection $f_{(k)}^{-1}(x, \varphi)$

Given a conjunct $u \subseteq f_{(k)}^{-1}(x)$ in the constraint φ, and the (unique) expression E_x such that $x \subseteq E_x$ lies in $\Psi(\varphi)$, we want to express $f_{(k)}^{-1}(E_x)$ (the projection $f_{(k)}^{-1}$ applied to E_x) as an expression E_u such that we can add $u \subseteq E_u$ to φ.

Assume that E_x is of the form $E_x = f(q_1, \ldots, q_n)$. Then one can infer $u \subseteq \perp$ if the value of at least one of $q_1, \ldots, q_n$ is the empty set in the greatest solution of $\Psi(\varphi)$. This is the case if one of the automata $\mathcal{A}^{\Psi(\varphi)}(q_i)$ constructed from $\Psi(\varphi)$ recognizes the empty set. This again can be expressed as

$$\mathcal{L}(\mathcal{A}^{\Psi(\varphi)}(f(q_1, \ldots, q_n))) = \emptyset \tag{2}$$

where we set $\mathcal{L}(\mathcal{A}^{\psi}(f(q_1, \ldots, q_n))) = f(\mathcal{L}(\mathcal{A}^{\psi}(q_1)), \ldots, \mathcal{L}(\mathcal{A}^{\psi}(q_n)))$. Otherwise (i.e., if the values of $q_1, \ldots, q_n$ are all nonempty, and condition (2) does not hold), one can infer $u \subseteq q_k$.

In general, E_x is of the form $E_x = \bigcup_i f_i(q_{i1}, \ldots, q_{in_i})$. Now, assume that $f(q_1, \ldots, q_n)$ is a member of this union. If condition (2) is satisfied, then this member can be discarded from the union. Otherwise, q_k becomes a member of the union on the right-hand side of the infered constraint $u \subseteq q_k \cup \ldots$.

Definition 12 (pre-$f_{(k)}^{-1}(E, \psi)$). The k-th *pre-projection* of f applied to an expression $E = \bigcup_i f_i(q_{i1}, \ldots, q_{in_i})$ with respect to the automaton constraint ψ, is the union of intersections

$$\text{pre-}f_{(k)}^{-1}(E, \psi) = \bigcup\{q_{ik} \mid f = f_i, \mathcal{L}(\mathcal{A}^{\psi}(f_i(q_{i1}, \ldots, q_{in_i}))) \neq \emptyset\}.$$

By applying the pre-projection we obtain expressions E such that the inclusions $u \subseteq E$ are not yet directly expressible in the restricted syntax of constraints φ. We can, however, transform a union of intersection variables into an intersection of unions using a variant of the conjunctive normal form (the computation of the standard one would here require doubly-exponential time). We then obtain an expression of the form $E' = \bigcap_i \bigcup_j u_{ij}$. We can express $u \subseteq E'$ as the conjunction $\bigwedge_i u \subseteq \bigcup_j u_{ij}$, which we then can add to φ, remaining within the restricted syntax of constraints.

Definition 13 (FCNF). The *full conjunctive normal form* of a union of intersection variables $E = \bigcup_{i \in I} \bigcap_{j \in J_i} u_{ij}$ is an intersection of unions of variables $x \in \mathsf{Var}$,

$$FCNF(E) = \bigcap\{\bigcup S \mid S \subseteq \mathsf{Var}(E),\ \forall i \in I\ \exists j \in J_i :\ \ u_{ij} \in S\}.$$

We set $FCNF(\perp) = \perp$.

Now we compose the operations above and obtain the full projection operation.

Definition 14 ($f_{(k)}^{-1}(x, \varphi)$). The k-th projection of $f \in \Sigma$ applied to the variable $x \in \mathsf{Var}$ wrt. to the constraint φ is an intersection of unions of variables $u_{ij} \in \mathsf{Var}$,

$$f_{(k)}^{-1}(x, \varphi) = FCNF(\text{pre-}f_{(k)}^{-1}(FDNF(lub(x, \varphi)), \Psi(\varphi))).$$

for all $q \in \cap\text{-Var}(\varphi)$

$E_q := lub(q, \varphi)$ (Definition 8)

$E'_q := FDNF(E_q)$ (Definition 9)

$\Psi(\varphi) := \bigwedge_{q \in \cap\text{-Var}(\varphi)} q \subseteq E'_q$ (Definition 11)

construct transition table of automata $\mathcal{A}^{\Psi(\varphi)}(q)$ *(same for all q; Definition 23 in Appendix)*

for all $q \in \cap\text{-Var}(\varphi)$

test emptiness of $\mathcal{L}(\mathcal{A}^{\Psi(\varphi)}(q))$ (Remark 7)

for all inclusions $u \subseteq f^{-1}_{(k)}(x)$ in φ

$E^x_{(k)} := \text{pre-}f^{-1}_{(k)}(E'_x, \Psi(\varphi))$ (Definition 12)

$f^{-1}_{(k)}(x, \varphi) := FCNF(E^x_{(k)})$ (Definition 13)

Table 3. Subprocedure computing $f^{-1}_{(k)}(x, \varphi)$ for all inclusions $u \subseteq f^{-1}_{(k)}(x)$ in constraint φ

Given a constraint φ, we compute the projections $f^{-1}_{(k)}(x, \varphi)$ simultaneously for all variables x such that an inclusion $u \subseteq f^{-1}_{(k)}(x)$ exists in φ. The corresponding subprocedure is presented in Table 3.

4 Correctness of the algorithm

The next two lemmas simply express that both full normal forms preserve the meaning of an expression.

Lemma 15 (FDNF). For any expression E of the form $\bigcap_{i \in I} \bigcup_{j \in J_i} f_{ij}(\bar{u}_{ij})$, the equality $\alpha(E) = \alpha(FDNF(E))$ holds for every valuation α.

Proof. To see that $\alpha(E) \subseteq \alpha(FDNF(E))$, transform E into a disjunctive normal form. Now, using the equality $\alpha(f(\bar{u}) \cap g(\bar{v})) = \emptyset$ for $f \neq g$ and the equality $\alpha(f(u_1, \ldots, u_n) \cap f(v_1, \ldots, v_n)) = \alpha(f(u_1 \cap v_1, \ldots, u_n \cap v_n))$, we can transform the result to an expression such that it is in disjunctive normal form and each disjunct satisfies the condition from the definition of $FDNF(E)$.

To see that $\alpha(FDNF(E)) \subseteq \alpha(E)$, take the partial ordering on tuples of intersections defined by $(\bigcap S_1, \ldots, \bigcap S_n) \prec (\bigcap S'_1, \ldots, \bigcap S'_n)$ (which we abbreviate by $\overline{\bigcap S} \prec \overline{\bigcap S'}$) if $S_i \subseteq S'_i$ holds for all $i = 1, \ldots, n$. We observe that, if $\overline{\bigcap S} \prec \overline{\bigcap S'}$, then $\alpha(f(\overline{\bigcap S}) \cup f(\overline{\bigcap S'})) \subseteq \alpha(f(\overline{\bigcap S}))$. Discard from $FDNF(E)$ all disjuncts that are not minimal in this ordering, and call the result F. By the observation above, $\alpha(FDNF(E)) = \alpha(F)$. We have to show that $\alpha(F) \subseteq \alpha(E)$. Take any disjunct $f(\overline{\bigcap S})$ from F. We will show that the value of this disjunct under α is equal to the value of some disjunct from the disjunctive normal form of E. We know that for all $i \in I$ there exists a $j_i \in J_i$ such that $f_{ij_i} = f$ and $\bar{u}_{ij_i} \in \overline{S}$. Hence, for all $k = 1, \ldots, \text{arity}(f)$, it holds that $\bigcup_{i \in I}\{u_{ij_i,k}\} \subseteq S_k$, and by the minimality of $\overline{\bigcap S}$ these two sets are equal. The expression $\bigcap_{i \in I} f(\bar{u}_{ij_i}))$ occurs in the disjunctive normal form of E and $\alpha(f(\overline{\bigcap S})) = \alpha(\bigcap_{i \in I} f(\bar{u}_{ij_i}))$. □

Lemma 16 (FCNF). For any expression E of the form $\bigcup_{i \in I} \bigcap_{j \in J_i} u_{ij}$, the equality $\alpha(E) = \alpha(FCNF(E))$ holds for every valuation α.

Proof. The proof is similar to the proof of Lemma 15; we can take the expression dual to E (replace unions with intersections and vice versa), compute the full disjunctive normal form (this time over variables, not terms $f(\bar{u})$) and then take once more the dual, which is in conjunctive normal form. □

Proposition 17 (Soundness). The axioms in Table 1 are valid. In particular, if a constraint φ is satisfiable then its closed form φ^C does not contain *false*.

Proof. The proof is done by inspection of each axiom. The validity of Axioms 3 and 4 follows from consecutive applications of Lemma 15, Remark 7, and Lemma 16. □

Proposition 18 (Completeness). If the closed form φ^C of a constraint φ does not contain *false* then φ is satisfiable. Moreover, the greatest solution of φ is the greatest solution of the automaton constraint $\Psi(\varphi^C)$.

Proof. Let α be the valuation defined by $\alpha(x) = \mathcal{L}(\mathcal{A}(x))$, where $\mathcal{A}(x)$ is the automaton corresponding to $\Psi(\varphi^C)$ and the variable x. By Remark 7, the unique extension of α to $\cap\text{-Var}(\varphi)$ is the greatest solution of $\Psi(\varphi^C)$. Below we show that α satisfies each conjunct in φ. Since φ implies $\Psi(\varphi^C)$, this will show that α is the greatest solution of φ.

Since $FDNF(lub(x, \varphi^C))$ is equivalent to an intersection of the expressions $\bigcup_i f_i(\bar{u}_i)$, the conjuncts of the form $x \subseteq \bigcup_i f_i(\bar{u}_i)$ are trivially satisfied.

We will show the satisfaction of the constraints $x \subseteq \bigcup_i f_i(\bar{u}_i) \cup \bigcup_j y_j$ (this includes the case $x \subseteq y$) indirectly. Suppose $t \notin \alpha(\bigcup_i f_i(\bar{u}_i) \cup \bigcup_j y_j)$; we will show $t \notin \alpha(x)$. Since $t \notin \alpha(y_j)$ for all j, the variables y_j cannot be unbounded in $\Psi(\varphi^C)$. Hence, every variable y_j occurs in a constraint of the form $y_j \subseteq \bigcup_k f_{jk}(\bar{u}_{jk})$ in φ^C (which includes the case of empty union $y_j \subseteq \bot$). Since $t \notin \alpha(FDNF(lub(y_j, \varphi^C)))$ for all j, there is a constraint of the above form in φ^C such that $t \notin \alpha(\bigcup_k f_{jk}(\bar{u}_{jk}))$. By Axiom 2, φ^C contains a constraint $x \subseteq \bigcup_i f_i(\bar{u}_i) \cup \bigcup_j \bigcup_k f_{jk}(\bar{u}_{jk})$ such that t does not belong to the value of the right-hand side of the inclusion under α. Hence, $t \notin \alpha(x)$.

The proof for the constraints $u \subseteq f_{(k)}^{-1}(x)$ is similar. If $t \notin \alpha(f_{(k)}^{-1}(x))$, then, by the definition of projection, for all trees $t_1, \ldots, t_n$ such that $t_k = t$ and n is the arity of f, $f(t_1, \ldots, t_n) \notin \alpha(x)$. Let $FDNF(lub(x, \varphi^C)) = \bigcup_i f_i(\bar{q}_i)$. Then, for all $t_1, \ldots, t_n$ as above, $f(t_1, \ldots, t_n) \notin \alpha(\bigcup_{\{i \mid f = f_i, \mathcal{L}(\mathcal{A}(f_i(\bar{q}_i))) \neq \emptyset\}} f(\bar{q}_i))$. Hence,

$t \notin \alpha(\bigcup_{\{i \mid f=f_i, \mathcal{L}(\mathcal{A}(f_i(\bar{q}_i)))\neq\emptyset\}} q_{i,k})$. By Axioms 3,4 and Lemma 16, φ^C contains a sequence of constraints equivalent to $u \subseteq \bigcup_{\{i \mid f=f_i, \mathcal{L}(\mathcal{A}(f_i(\bar{q}_i)))\neq\emptyset\}} q_{i,k}$. Hence, $t \notin \alpha(u)$.

The last case are the constraints of the form $a \subseteq x$. Again, if $a \notin \alpha(x)$ then x is bounded in $\Psi(\varphi^C)$ and a does not occur in $FDNF(lub(x, \varphi^C))$. But then, by Axiom 5 or 6, *false*$\in \varphi^C$, which is a contradiction. □

5 Complexity

Proposition 19 (upper bound). The algorithm in Table 2 computes the closed form φ^C of the input constraint φ in deterministic, single exponential time.

Proof. For an input φ of size n, the number of flat terms and variables that occur in φ is bounded by n. Each derived inclusion involves a variable in $\mathcal{V}(\varphi)$ on the left-hand side and a union of variables and flat terms on the right-hand side. All these flat terms occur in φ. Thus, the number of derived inclusions is bounded by $n \cdot 2^n$. At each iteration of the algorithm, the consequences of all (pairwise combinations of) inclusions under Axioms 1–2 are computed. This amounts to a cost of $O((n2^n)^2)$. Adding consequences of Axioms 3 and 4 is done in exponential time (say, $O(2^{n^c})$) by the lemmas below and by the polynomial time complexity of the emptiness test for tree automata (also in the case of Büchi tree automata [24]). There may be at most $n2^n$ iterations. Adding consequences of Axioms 5 and 6 costs at most $n2^n$, since the number of inclusions $a \subseteq x$ is bounded by n and number of inclusions with x on the left-hand side is bounded by 2^n. Hence, the whole algorithm runs in time $O(((n2^n)^2 + 2^{n^c}) \cdot n2^n + n2^n)$.

Lemma 20. For any intersection q, $FDNF(lub(q, \varphi))$ can be computed in time exponential in the size of φ.

Proof. Let $E = lub(q, \varphi)$ and n be the size of φ. To compute $FDNF(E)$, we check, for all $f \in \Sigma(\varphi)$ and all sequences $(S_1, \ldots, S_{a(f)})$ such that $S_i \subseteq \mathcal{V}(E)$ for $i = 1, \ldots, a(f)$, if the condition from the definition of *FDNF* is satisfied. The size of $\mathcal{V}(E)$ is at most n, so the number of terms $f(\bigcap S_1, \ldots, \bigcap S_{a(f)})$ is bounded by $|\Sigma(\varphi)|(2^n)^k$, where k is the maximal arity of a symbol in $\Sigma(\varphi)$. Hence, the number of these terms is bounded by 2^{n^c} for some constant c (note that $k < n$). To check the condition, we have to run through the constraints $x \subseteq \bigcup_j f_j(\bar{u}_j)$ such that x occurs in the intersection q. The number of such constraints is bounded by $n2^n$. For each such constraint, checking if there exists a j such that $f = f_j$ and $u_{j,1} \in S_1, \ldots, u_{j,a(f)} \in S_{a(f)}$ can be done in time polynomial in the size of the constraint and the sequence $(S_1, \ldots, S_{a(f)})$ (which is polynomial in n). Therefore, the whole procedure runs in time $O(2^{n^c} \cdot n2^n \cdot poly(n))$, which is single exponential. □

Lemma 21. For any expression $E = \bigcup_{i \in I} \bigcap_{j \in J_i} u_{ij}$, the expression $FCNF(E)$ can be computed in time exponential in the number of variables in $\mathcal{V}(E)$.

Proof. The proof is analogous to the proof of the lemma above. □

Proposition 22 (lower bound). The problem of the satisfiability of the codefinite set constraints is DEXPTIME-hard.

Proof. The proof follows by the reduction of the problem of the emptiness of the intersection of tree automata [9].[3] For given n tree automata, let $\varphi_1, \ldots, \varphi_n$ be the constraints bounding the variables $X_1, \ldots, X_n$ to the languages of the automata. Then, the constraint

$$a \subseteq f_{(1)}^{-1}(f(a, X_1 \cap \ldots \cap X_n))$$

is satisfiable if and only if the intersection of the languages is nonempty. □

Since intersection corresponds to conjunction, one can expect the DEXPTIME lower bound for every formalism of set constraints that can express regular sets of trees.

6 Conclusion

We have defined a class of set constraints which arises in program analysis and error diagnosis, and we have given the complexity-theoretic characterization of its constraint-solving problem. We have applied our techniques also to the already existing class of path-closed set constraints and characterized its complexity too.

We now need to refine the abstract fixpoint strategy of our algorithm in order to improve its practical efficiency. In succession to the technical report [6] on which this paper is based, Devienne, Talbot and Tison [8] have already given a strategy for our algorithm which can achieve an exponential speedup. Unfortunately, their setup relies on bottom-up tree automata (in bit-vector representation) and thus, as the authors point out, applies to the case of finite trees only. Our algorithm uses top-down tree automata and accounts for both cases (where, again, the case of infinite trees is the only relevant one for analyzing the operational semantics).

Kozen has given an equational axiomatization of the algebra of sets of trees in [15]. It would be useful to modify this axiomatization in order to account for the projection operator and thus fix the algebraic laws underlying our algorithm.

To our knowledge, this is the first time that automata over infinite trees have been used to represent solutions of set constraints. The represented sets of infinite trees appear in the ν-level in the hierarchy of the fixpoint calculus of Niwiński [19]. The essential difference between the fixpoint expressions on the ν-level and our set constraints formalisms seems to be the projection operator; for the addition of intersection to the fixpoint expressions see [3]. The question arises whether the formalism of set constraints can be extended to have solutions in all levels, *i.e.*, to be able to express all Rabin-recognizable sets. This is related to the addition of fixpoint operators as in [17] (there, however, not over infinite trees but arbitrary first-order domains).

[3] The lower bound requires that the signature contains at least two function symbols, one of them having arity ≥ 2.

Appendix: Automaton constraints and automata

A (finite non-deterministic top-down tree) automaton is a tuple $\mathcal{A} = \langle \Sigma', Q, \delta, q_{sfinit}, Q_\top \rangle$ consisting of its finite alphabet $\Sigma' \subseteq \Sigma$, finite set of states Q, (non-deterministic) transition function $\delta : Q \times \Sigma' \to \mathcal{P}(\overline{Q})$ (where $\overline{Q}$ stands for the set of all tuples over Q), initial state q_{sfinit} and the set $Q_\top$ of *all-accept* states. The tree automaton $\mathcal{A}$ *accepts* a tree t (or: t lies in the language $\mathcal{L}(\mathcal{A})$ *recognized* by $\mathcal{A}$) iff there exists a run of $\mathcal{A}$ on t; this acceptance condition works for finite as well as for infinite trees. In the case of infinite trees, the automaton corresponds to a Büchi tree automaton where all states are final states. In both cases of finite and infinite trees, the emptiness of the automaton can be tested in polynomial time [23]. A run of $\mathcal{A}$ on the tree t assigns to the root the initial state and to each node of t a state q such that: if t is labeled with the function symbol $f \in \Sigma'$ of arity k, then the states assigned to the k successor nodes form a tuple that lies in the set $\delta_\psi(q, f)$. If the label f is a constant symbol, then the set $\delta_\psi(q, f)$ must contain the empty tuple. If the state assigned to the node is an all-accept state, $q \in Q_\top$, then the successor nodes are assigned any states (whether the node label f lies in the alphabet Σ' or not).

Given an automaton constraint ψ, we first define the family of automata $\mathcal{A}^\psi(q)$ (one for each variable $q \in \mathsf{q\text{-}Var}$, all with the same transition table δ_ψ) and then show that it recognizes exactly the greatest solution of ψ.

Definition 23 ($\mathcal{A}^\psi(q)$). The *automaton* $\mathcal{A}^\psi(q_0)$ *corresponding to* the automaton constraint ψ and the variable $q_0 \in \mathsf{q\text{-}Var}(\psi)$ is the tuple $\langle \Sigma(\psi), \mathsf{Var}(\psi), \delta_\psi, q_0 \rangle$ where

- the alphabet is the set $\Sigma(\psi)$ of function symbols occurring in ψ;
- the states are the variables q occurring in ψ;
- the set $\delta_\psi(q, f)$, i.e., the transition function δ_ψ applied on a state q and a function symbol f, is
  - the set $\{\bar{q}_j \mid f_j = f\}$ if $q \subseteq \bigcup_j f_j(\bar{q}_j)$ is a conjunct in ψ (which is then unique),
  - the empty set $\emptyset$ if $q \subseteq \bot$ is a conjunct in ψ;
- the initial state is q_0;
- the all-accept states are the unbounded variables in ψ.

If the variable $q_0 \in \mathsf{q\text{-}Var}$ does not occur at all in ψ, then $\mathcal{A}^\psi(q_0)$ is the tuple $\langle \emptyset, \{q_0\}, \emptyset, q_0, \{q_0\} \rangle$ (an automaton accepting T_Σ).

It is clear that $\mathcal{L}(\mathcal{A}^\psi(q))$ is the empty set if $q \subseteq \bot$ is in ψ and the set T_Σ of all trees if q is unbounded in ψ. More generally, for both finite and infinite trees, the statement below holds.

Observation 1. The valuation $\alpha : q \mapsto \mathcal{L}(\mathcal{A}^\psi(q))$ is the greatest solution of the automaton constraint ψ.

Proof. We will first show that any solution β of ψ is smaller than the valuation α. We extend β to a mapping over all states of the automata by setting $\beta(\top) = T_\Sigma$. We will show that $\beta(q) \subseteq \alpha(q)$ for all states q. If $\beta(q)$ is empty, then the inclusion is trivially satisfied; otherwise, take any tree $t \in \beta(q)$. By induction of the depth of the positions p in t, we will construct a run of $\mathcal{A}^\psi(q)$ on t that satisfies the following invariant: If $\mathcal{A}^\psi(q)$ is in state q' at position p, then the subtree $t|_p$ of t rooted at p belongs to $\beta(q')$.

For the root position, the initial state is q and $t \in \beta(q)$. Let $\mathcal{A}^\psi(q)$ be in state q' at position p such that $t|_p \in \beta(q')$. If $t|_p$ is a tree of the form $f(t_1, \ldots, t_n)$, then we will continue the construction of the run at the positions $p.1, \ldots, p.n$. If q' is $\top$ or an unbounded variable in ψ, then the automaton goes to the state $\top$ in all positions $p.1, \ldots, p.n$. Since $\mathcal{A}^\psi(\top)$ recognizes the set T_Σ of all trees, our invariant is satisfied. Now suppose q' is not unbounded. Since β is a solution, on the right-hand side of the inclusion constraining q in ψ must occur an expression of the form $f(q_1, \ldots, q_n)$, with $t|_p \in \beta(f(q_1, \ldots, q_n))$, that is, $t_i \in \beta(q_i)$ for $i = 1, \ldots, n$. But then, by the definition of $\mathcal{A}^\psi(q)$, $(q_1, \ldots, q_n) \in \delta_\psi(q', f)$. By taking this transition we satisfy the invariant and are thus able to extend the definition of the run to all positions in t. Hence, $t \in \alpha(q)$.

For the other direction of the proof, we will show that α satisfies every inclusion $q \subseteq E$ in ψ. Again, if $\mathcal{L}(\mathcal{A}^\psi(q))$ is empty, nothing is to show; otherwise, we take an element $t = f(t_1, \ldots, t_n)$ from $\mathcal{L}(\mathcal{A}^\psi(q))$. By the definition of $\mathcal{L}(\mathcal{A}^\psi(q))$, there exists a run of $\mathcal{A}^\psi(q)$ on t, starting from q. Let $(q, f, q_1, \ldots, q_n)$ be the first transition used in this run. By the definition of a run, there are runs on t_i starting from q_i, and, hence, $t_i \in \mathcal{L}(\mathcal{A}^\psi(q_i))$. That is, $t \in f(\mathcal{L}(\mathcal{A}^\psi(q_1)), \ldots, \mathcal{L}(\mathcal{A}^\psi(q_n))) = f(\alpha(q_1), \ldots, \alpha(q_n))$. By the definition of $\mathcal{A}^\psi(q)$, the expression E is of the form $E = f(q_1, \ldots, q_n) \cup E'$. Therefore, $t \in \alpha(E)$. □

References

1. A. Aiken. Set constraints: Results, applications and future directions. In *Proceedings of the Workshop on Principles and Practice of Constraint Programming*, LNCS 874, pages 326–335. Springer-Verlag, 1994.
2. A. Arnold and M. Nivat. Formal computations of non deterministic recursive program schemes. *Mathematical Systems Theory*, 13:219–236, 1980.
3. A. Arnold and D. Niwiński. Fixed point characterization of weak monadic logic definable sets of trees. In M. Nivat and A. Podelski, editors, *Tree Automata and Languages*, pages 159–188. North Holland, 1992.
4. L. Bachmair, H. Ganzinger, and U. Waldmann. Set constraints are the monadic class. In *Eighth Annual IEEE Symposium on Logic in Computer Science*, pages 75–83, 1993.
5. W. Charatonik and L. Pacholski. Set constraints with projections are in NEXPTIME. In *Proceedings of the* 35th *Symposium on Foundations of Computer Science*, pages 642–653, 1994.
6. W. Charatonik and A. Podelski. Set constraints for greatest models. Technical Report MPI-I-97-2-004, Max-Planck-Institut für Informatik, April 1997. www.mpi-sb.mpg.de/~podelski/papers/greatest.html.

7. W. Charatonik and A. Podelski. Set constraints with intersection. In G. Winskel, editor, *Twelfth Annual IEEE Symposium on Logic in Computer Science (LICS)*, pages 362–372. IEEE, June 1997.
8. P. Devienne, J.-M. Talbot, and S. Tison. Solving classes of set constraints with tree automata. Technical Report IT-303, Laboratoire d'Informatique Fondamentale de Lille, May 1997.
9. T. Frühwirth, E. Shapiro, M. Vardi, and E. Yardeni. Logic programs as types for logic programs. In *Sixth Annual IEEE Symposium on Logic in Computer Science*, pages 300–309, July 1991.
10. F. Gécseg and M. Steinby. *Tree Automata.* Akademiai Kiado, 1984.
11. N. Heintze and J. Jaffar. A decision procedure for a class of set constraints (extended abstract). In *Fifth Annual IEEE Symposium on Logic in Computer Science*, pages 42–51, 1990.
12. N. Heintze and J. Jaffar. A finite presentation theorem for approximating logic programs. In *Seventeenth Annual ACM Symposium on Principles of Programming Languages*, pages 197–209, January 1990.
13. N. Heintze and J. Jaffar. Semantic types for logic programs. In F. Pfenning, editor, *Types in Logic Programming*, pages 141–156. MIT Press, 1992.
14. N. Heintze and J. Jaffar. Set constraints and set-based analysis. In *Proceedings of the Workshop on Principles and Practice of Constraint Programming*, LNCS 874, pages 281–298. Springer-Verlag, 1994.
15. D. Kozen. Logical aspects of set constraints. In *1993 Conference on Computer Science Logic*, LNCS 832, pages 175–188. Springer-Verlag, Sept. 1993.
16. J. W. Lloyd. *Foundations of Logic Programming.* Symbolic Computation. Springer-Verlag, Berlin, Germany, second, extended edition, 1987.
17. D. A. McAllester, R. Givan, C. Witty, and D. Kozen. Tarskian set constraints. In *Proceedings, 11th Annual IEEE Symposium on Logic in Computer Science*, pages 138–147, New Brunswick, New Jersey, July 1996. IEEE Computer Society Press.
18. P. Mishra. Towards a theory of types in Prolog. In *IEEE International Symposium on Logic Programming*, pages 289–298, 1984.
19. D. Niwiński. On fixed-point clones. In L. Kott, editor, *Proceedings of the 13th International Conference on Automata, Languages and Programming*, volume 226 of *Lecture Notes in Computer Science*, pages 464–473. Springer-Verlag, 1986.
20. L. Pacholski and A. Podelski. Set constraints - a pearl in research on constraints. In G. Smolka, editor, *Proceedings of the Third International Conference on Principles and Practice of Constraint Programming - CP97*, volume 1330 of *Springer LNCS*, Berlin, Germany, October 1997. Springer-Verlag.
21. A. Podelski, W. Charatonik, and M. Müller. Set-based error diagnosis of concurrent constraint programs. submitted for publication, 1997.
22. H. Seidl. Haskell overloading is DEXPTIME-complete. *Information Processing Letters*, 52:57–60, 1994.
23. W. Thomas. *Handbook of Theoretical Computer Science*, volume B, chapter Automata on Infinite Objects, pages 134–191. Elsevier, 1990.
24. M. Vardi and P. Wolper. Automata-theoretic techniques for modal logics of programs. *Journal of Computer and System Sciences*, 32, 1986.

Modularity of Termination Using Dependency Pairs*

Thomas Arts[1] and Jürgen Giesl[2]

[1] Computer Science Laboratory, Ericsson Telecom AB, 126 25 Stockholm, Sweden, E-mail: thomas@cslab.ericsson.se

[2] FB Informatik, Darmstadt University of Technology, Alexanderstr. 10, 64283 Darmstadt, Germany, E-mail: giesl@informatik.tu-darmstadt.de

Abstract. The framework of dependency pairs allows *automated* termination and innermost termination proofs for many TRSs where such proofs were not possible before. In this paper we present a refinement of this framework in order to prove termination in a *modular* way. Our modularity results significantly increase the class of term rewriting systems where termination resp. innermost termination can be proved automatically. Moreover, the modular approach to dependency pairs yields new modularity criteria which extend previous results in this area. In particular, existing results for modularity of innermost termination can easily be obtained as direct consequences of our new criteria.

1 Introduction

Termination is one of the most important properties of a term rewriting system (TRS). While in general this problem is undecidable [HL78], several methods for proving termination have been developed (for surveys see e.g. [Der87, Ste95, DH95]). However, most methods that are amenable to automation are restricted to the generation of *simplification orderings* and there exist numerous important TRSs whose termination cannot be proved by orderings of this restricted class.

For that reason we developed the framework of *dependency pairs* [Art96, AG96, AG97a, AG97b, Art97] which allows to apply standard methods for termination proofs to such TRSs where they failed up to now. In this way, termination of many (also non-simply terminating) systems could be proved automatically.

When proving termination, one benefits from *modularity* results that ensure termination of the whole TRS as soon as it is proved for parts of the TRS. The aim of this paper is to refine the dependency pair approach in order to allow *modular* termination proofs using dependency pairs.

Although in general, termination is not modular for the *direct sum* [Toy87, Dro89, TKB95], i.e. the partition of a TRS into subsystems with disjoint signatures, this modularity property holds for TRSs of a special form [Rus87, Mid89, Gra94, TKB95, SMP95]. For a survey see e.g. [Mid90, Ohl94, Gra96a].

However, a TRS often cannot be split into subsystems with disjoint signatures. Therefore, partitions into subsystems which may at least have constructors

* This work was partially supported by the Deutsche Forschungsgemeinschaft under grants no. Wa 652/7-1,2 as part of the focus program 'Deduktion'.

in common have also been considered [KO92, MT93, Gra95, MZ97]. Nevertheless, in practice these results often cannot be applied for automated termination proofs, either. For example, many systems are hierarchical combinations of TRSs that do not only share constructors, but where one subsystem contains defined symbols of the other subsystem. Termination is only proved modular for hierarchical combinations of several restricted forms [Der94, FJ95].

The modularity results for *innermost* termination are less restrictive than those for termination. Innermost termination is modular for direct sums and for TRSs with shared constructors [Gra95], for composable constructor systems [MT93], for composable TRSs [Ohl95], and for proper extensions [KR95], which are special hierarchical combinations. As innermost termination implies termination for several classes of TRSs [Gra95, Gra96b], these results can also be used for termination proofs of such systems. For example, this holds for locally confluent overlay systems (and in particular for non-overlapping TRSs).

In this paper we show that the modular approach using dependency pairs extends previous modularity results and we demonstrate that in our framework the existing modularity results for innermost termination of composable TRSs and proper extensions are obtained as easy consequences.

In Sect. 2 we present the dependency pair approach and introduce a new termination criterion to use this framework in a modular way. Similarly, in Sect. 3 we present a modular approach for innermost termination proofs using dependency pairs. As shown in Sect. 4, these results imply new modularity criteria (which can also be used independently from the dependency pair technique). See [AG97c] for a collection of examples to demonstrate the power of these results. In Sect. 5 we give a comparison with related work and we conclude in Sect. 6.

2 Modular Termination with Dependency Pairs

In [AG97a] we introduced the dependency pair technique to prove termination automatically. In this section we briefly recapitulate its basic concepts and present a new modular approach for automated termination proofs.

In the following, the *root* of a term $f(\ldots)$ is the leading function symbol f. For a TRS $\mathcal{R}$ with the rules R over a signature $\mathcal{F}$, $D = \{\mathrm{root}(l) | l \to r \in R\}$ is the set of the *defined symbols* and $C = \mathcal{F} \setminus D$ is the set of *constructors* of $\mathcal{R}$. To stress the splitting of the signature we denote a TRS by $\mathcal{R}(D, C, R)$. For example consider the following TRS with the constructors s and c and the defined symbol f.

$$\mathsf{f}(x, \mathsf{c}(y)) \to \mathsf{f}(x, \mathsf{s}(\mathsf{f}(y, y)))$$
$$\mathsf{f}(\mathsf{s}(x), y) \to \mathsf{f}(x, \mathsf{s}(\mathsf{c}(y)))$$

Most methods for automated termination proofs are restricted to *simplification orderings* [Der87, Ste95]. Hence, these methods cannot prove termination of TRSs like the one above, as $\mathsf{f}(x, \mathsf{c}(\mathsf{s}(x)))$ can be reduced to the term $\mathsf{f}(x, \mathsf{s}(\mathsf{f}(x, \mathsf{s}(\mathsf{c}(\mathsf{s}(x))))))$ where it is embedded in.

In contrast to previous approaches we do not compare left- and right-hand sides of rules, but we only compare left-hand sides with those *subterms* that may possibly start a new reduction. Hence, we focus on those subterms of right-hand sides which have a defined root symbol.

More precisely, if $f(s_1, \ldots, s_n)$ rewrites to $C[g(t_1, \ldots, t_m)]$ (where g is a defined symbol and C is some context), then we only compare the argument tuples $s_1, \ldots, s_n$ and $t_1, \ldots, t_m$. To avoid the handling of *tuples*, a new *tuple symbol* $F \notin \mathcal{F}$ is introduced for every defined symbol f in D. Instead of comparing *tuples*, now the *terms* $F(s_1, \ldots, s_n)$ and $G(t_1, \ldots, t_m)$ are compared. To ease readability we assume that the signature $\mathcal{F}$ consists of lower case function symbols only and denote the tuple symbols by the corresponding upper case symbols.

Definition 1 (Dependency Pair). Let $\mathcal{R}(D, C, R)$ be a term rewriting system. If $f(s_1, \ldots, s_n) \to C[g(t_1, \ldots, t_m)]$ is a rewrite rule of R with $g \in D$, then $\langle F(s_1, \ldots, s_n), G(t_1, \ldots, t_m)\rangle$ is a *dependency pair* of $\mathcal{R}$.

In the above example we obtain the following dependency pairs:

$$\langle \mathsf{F}(x, \mathsf{c}(y)), \mathsf{F}(x, \mathsf{s}(\mathsf{f}(y, y)))\rangle \tag{1}$$
$$\langle \mathsf{F}(x, \mathsf{c}(y)), \mathsf{F}(y, y)\rangle \tag{2}$$
$$\langle \mathsf{F}(\mathsf{s}(x), y), \mathsf{F}(x, \mathsf{s}(\mathsf{c}(y)))\rangle. \tag{3}$$

To trace newly introduced redexes in a reduction, we consider special sequences of dependency pairs. Here, the right-hand side of every dependency pair corresponds to the redex being traced.

Definition 2 (Chain). Let $\mathcal{R}$ be a TRS. A sequence of dependency pairs $\langle s_1, t_1\rangle$ $\langle s_2, t_2\rangle \ldots$ is an *$\mathcal{R}$-chain* if there exists a substitution σ, such that $t_j\sigma \to_{\mathcal{R}}^* s_{j+1}\sigma$ holds for every two consecutive pairs $\langle s_j, t_j\rangle$ and $\langle s_{j+1}, t_{j+1}\rangle$ in the sequence.

We always assume that different (occurrences of) dependency pairs have disjoint sets of variables and we always regard substitutions whose domains may be infinite. Hence, in our example we have the $\mathcal{R}$-chain (resp. 'chain' for short)

$$\langle \mathsf{F}(x_1, \mathsf{c}(y_1)), \mathsf{F}(y_1, y_1)\rangle\, \langle \mathsf{F}(x_2, \mathsf{c}(y_2)), \mathsf{F}(y_2, y_2)\rangle\, \langle \mathsf{F}(x_3, \mathsf{c}(y_3)), \mathsf{F}(y_3, y_3)\rangle,$$

as $\mathsf{F}(y_1, y_1)\sigma \to_{\mathcal{R}}^* \mathsf{F}(x_2, \mathsf{c}(y_2))\sigma$ and $\mathsf{F}(y_2, y_2)\sigma \to_{\mathcal{R}}^* \mathsf{F}(x_3, \mathsf{c}(y_3))\sigma$ hold for the substitution σ replacing y_1 and x_2 by $\mathsf{c}(\mathsf{c}(y_3))$ and both y_2 and x_3 by $\mathsf{c}(y_3)$. In fact any finite sequence of the dependency pair (2) is a chain. As proved in [AG97a], absence of infinite chains is a sufficient and necessary criterion for termination.

Theorem 3 (Termination Criterion). *A TRS $\mathcal{R}$ is terminating if and only if there exists no infinite $\mathcal{R}$-chain.*

Some dependency pairs can never occur twice in any chain and hence, they need not be considered when proving that no infinite chain exists. Recall that a dependency pair $\langle v, w\rangle$ may only follow $\langle s, t\rangle$ in a chain if $t\sigma$ reduces to $v\sigma$ for some substitution σ. For a term t with a constructor root symbol c, $t\sigma$ can only be reduced to terms which have the same root symbol c. If the root symbol of t is defined, then this does not give us any direct information about those terms $t\sigma$ can be reduced to. Let CAP(t) result from replacing all subterms of t

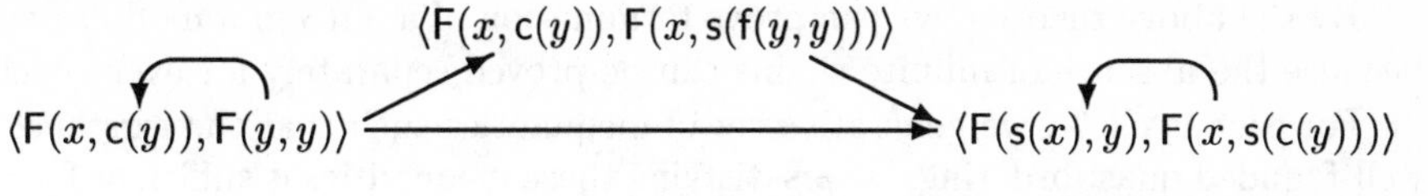

Fig. 1. The estimated dependency graph in our example.

that have a defined root symbol by different fresh variables and let $\text{REN}(t)$ result from replacing all variables in t by different fresh variables. Then, to determine whether $\langle v, w\rangle$ can follow $\langle s, t\rangle$ in a chain, we check whether $\text{REN}(\text{CAP}(t))$ unifies with v. Here, the function REN is needed to rename multiple occurrences of the same variable x in t, because when instantiated with σ, two occurrences of $x\sigma$ could reduce to different terms.

So for instance we have $\text{REN}(\text{CAP}(\mathsf{F}(y,y))) = \text{REN}(\mathsf{F}(y,y)) = \mathsf{F}(y_1,y_2)$ and $\text{REN}(\text{CAP}(\mathsf{F}(x,\mathsf{s}(\mathsf{f}(y,y))))) = \text{REN}(\mathsf{F}(x,\mathsf{s}(z))) = \mathsf{F}(x_1,\mathsf{s}(z_1))$. Hence, (1) can never follow itself in a chain, because $\mathsf{F}(x_1,\mathsf{s}(z_1))$ does not unify with $\mathsf{F}(x,\mathsf{c}(y))$. To estimate which dependency pairs may occur consecutive, the *estimated dependency graph* has been introduced, cf. [AG97a].

Definition 4 (Estimated Dependency Graph). The *estimated dependency graph* of a TRS $\mathcal{R}$ is the directed graph whose nodes are the dependency pairs and there is an arc from $\langle s, t\rangle$ to $\langle v, w\rangle$ if $\text{REN}(\text{CAP}(t))$ and v are unifiable.

In our example, we obtain the estimated dependency graph in Fig. 1. As usual, a subset $\mathcal{P}$ of dependency pairs is called a *cycle* if for any two dependency pairs $\langle s, t\rangle, \langle v, w\rangle$ in $\mathcal{P}$ there is a path from $\langle s, t\rangle$ to $\langle v, w\rangle$ and from $\langle v, w\rangle$ to $\langle s, t\rangle$ in the estimated dependency graph. (In particular, there must also be a path from $\langle s, t\rangle$ to itself for every $\langle s, t\rangle$ in $\mathcal{P}$.) In our example we have two non-empty cycles, viz. $\{(2)\}$ and $\{(3)\}$.

Using the estimated dependency graph, we develop a new *modular* refinement of Thm. 3. In the following we always restrict ourselves to finite TRSs. Then any infinite chain corresponds to a cycle. Dependency pairs that do not occur on cycles (such as (1)) can be ignored. Hence, it suffices to prove that there is no infinite chain *from any cycle.*

Theorem 5 (Modular Termination Criterion). *A TRS $\mathcal{R}$ is terminating if and only if for each cycle $\mathcal{P}$ in the estimated dependency graph there exists no infinite $\mathcal{R}$-chain of dependency pairs from $\mathcal{P}$.*

Proof. The 'only if' direction is a direct consequence of Thm. 3. For the other direction, suppose that $\mathcal{R}$ is not terminating. Then by Thm. 3 there exists an infinite $\mathcal{R}$-chain. As $\mathcal{R}$ is finite, there are only finitely many dependency pairs and hence, one dependency pair occurs infinitely many times in the chain (up to renaming of the variables). Thus the infinite chain has the form $\ldots\langle s\rho_1, t\rho_1\rangle\ldots \langle s\rho_2, t\rho_2\rangle\ldots\langle s\rho_3, t\rho_3\rangle\ldots$, where $\rho_1, \rho_2, \rho_3, \ldots$ are renamings. Hence, the tail $\langle s\rho_1, t\rho_1\rangle\ldots\langle s\rho_2, t\rho_2\rangle\ldots$ is an infinite $\mathcal{R}$-chain which consists of dependency pairs from one cycle in the estimated dependency graph only. □

By the above theorem we can prove termination of a TRS in a modular way, because the absence of infinite chains can be proved separately for every cycle.

For each cycle $\mathcal{P}$, we generate a set of inequalities such that the existence of well-founded quasi-orderings[1] $\geq_{\mathcal{P}}$ satisfying these inequalities is sufficient for the absence of infinite chains. For that purpose we have to ensure that the dependency pairs from $\mathcal{P}$ are decreasing w.r.t. $\geq_{\mathcal{P}}$. More precisely, for any sequence of dependency pairs $\langle s_1, t_1\rangle\langle s_2, t_2\rangle\langle s_3, t_3\rangle \ldots$ from $\mathcal{P}$ and for any substitution σ with $t_j\sigma \to_{\mathcal{R}}^* s_{j+1}\sigma$ (for all j) we demand

$$s_1\sigma \geq_{\mathcal{P}} t_1\sigma \geq_{\mathcal{P}} s_2\sigma \geq_{\mathcal{P}} t_2\sigma \geq_{\mathcal{P}} s_3\sigma \geq_{\mathcal{P}} t_3\sigma \geq_{\mathcal{P}} \ldots,$$

and for at least one $\langle s, t\rangle$ in $\mathcal{P}$ we demand the *strict* inequality $s\sigma >_{\mathcal{P}} t\sigma$. Then there exists no chain of dependency pairs from $\mathcal{P}$ which traverses all dependency pairs in $\mathcal{P}$ infinitely many times.

In the following we restrict ourselves to *weakly monotonic* quasi-orderings $\geq_{\mathcal{P}}$ where both $\geq_{\mathcal{P}}$ and its strict part $>_{\mathcal{P}}$ are *closed under substitution.* (A quasi-ordering $\geq_{\mathcal{P}}$ is *weakly monotonic* if $s \geq_{\mathcal{P}} t$ implies $f(\ldots s \ldots) \geq_{\mathcal{P}} f(\ldots t \ldots)$.) Then, to guarantee $t_j\sigma \geq_{\mathcal{P}} s_{j+1}\sigma$ whenever $t_j\sigma \to_{\mathcal{R}}^* s_{j+1}\sigma$ holds, it is sufficient to demand $l \geq_{\mathcal{P}} r$ for all rules $l \to r$ of the TRS. Moreover, $s_j \geq_{\mathcal{P}} t_j$ and $s_j >_{\mathcal{P}} t_j$ ensure $s_j\sigma \geq_{\mathcal{P}} t_j\sigma$ and $s_j\sigma >_{\mathcal{P}} t_j\sigma$, respectively, for all substitutions σ.

Theorem 6 (Modular Termination Proofs). *A TRS $\mathcal{R}(D, C, R)$ is terminating if for each cycle $\mathcal{P}$ in the estimated dependency graph there exists a well-founded weakly monotonic quasi-ordering $\geq_{\mathcal{P}}$ where both $\geq_{\mathcal{P}}$ and $>_{\mathcal{P}}$ are closed under substitution, such that*

- *$l \geq_{\mathcal{P}} r$ for all rules $l \to r$ in R,*
- *$s \geq_{\mathcal{P}} t$ for all dependency pairs from $\mathcal{P}$, and*
- *$s >_{\mathcal{P}} t$ for at least one dependency pair from $\mathcal{P}$.*

Proof. Suppose there exists an infinite $\mathcal{R}$-chain of dependency pairs from a cycle $\mathcal{P}$. Without loss of generality let $\mathcal{P}$ be *minimal*, i.e. if $\mathcal{P}$ contains a cycle $\mathcal{P}'$ as proper subset, then there is no infinite chain of dependency pairs from $\mathcal{P}'$.

For one dependency pair $\langle s, t\rangle$ in $\mathcal{P}$ we have the strict inequality $s >_{\mathcal{P}} t$. Due to the minimality of $\mathcal{P}$, $\langle s, t\rangle$ occurs infinitely many times in the chain (up to variable renaming), i.e. the chain has the form

$$\langle v_{1,1} w_{1,1}\rangle \ldots \langle v_{1,n_1} w_{1,n_1}\rangle\ \langle s\rho_1, t\rho_1\rangle\ \langle v_{2,1} w_{2,1}\rangle \ldots \langle v_{2,n_2} w_{2,n_2}\rangle\ \langle s\rho_2, t\rho_2\rangle \ldots,$$

where $\rho_1, \rho_2, \ldots$ are renamings. Hence, there exists a substitution σ such that $w_{i,j}\sigma \to_{\mathcal{R}}^* v_{i,j+1}\sigma$, $w_{i,n_i}\sigma \to_{\mathcal{R}}^* s\rho_i\sigma$, and $t\rho_i\sigma \to_{\mathcal{R}}^* v_{i+1,1}\sigma$. As $l \geq_{\mathcal{P}} r$ holds for all rules of $\mathcal{R}$ and as $\geq_{\mathcal{P}}$ is weakly monotonic, we have $\to_{\mathcal{R}}^* \subseteq \geq_{\mathcal{P}}$. Moreover, all dependency pairs from $\mathcal{P}$ are weakly decreasing. Thus, we obtain

[1] A quasi-ordering $\geq_{\mathcal{P}}$ is a reflexive and transitive relation and $\geq_{\mathcal{P}}$ is called *well-founded* if its strict part $>_{\mathcal{P}}$ is well founded.

$$v_{1,1}\sigma \gtrsim_{\mathcal{P}} w_{1,1}\sigma \gtrsim_{\mathcal{P}} \dots v_{1,n_1}\sigma \gtrsim_{\mathcal{P}} w_{1,n_1}\sigma \gtrsim_{\mathcal{P}} s\rho_1\sigma >_{\mathcal{P}} t\rho_1\sigma \gtrsim_{\mathcal{P}}$$
$$v_{2,1}\sigma \gtrsim_{\mathcal{P}} w_{2,1}\sigma \gtrsim_{\mathcal{P}} \dots v_{2,n_2}\sigma \gtrsim_{\mathcal{P}} w_{2,n_2}\sigma \gtrsim_{\mathcal{P}} s\rho_2\sigma >_{\mathcal{P}} t\rho_2\sigma \gtrsim_{\mathcal{P}} \dots$$

But this is a contradiction to the well-foundedness of $>_{\mathcal{P}}$. Hence, no infinite chain of dependency pairs from $\mathcal{P}$ exists and by Thm. 5, $\mathcal{R}$ is terminating. □

With this theorem, termination of our example can easily be proved automatically. After computing the estimated dependency graph in Fig. 1, two quasi-orderings $\gtrsim_1, \gtrsim_2$ have to be generated which satisfy

$$\mathsf{f}(x,\mathsf{c}(y)) \gtrsim_1 \mathsf{f}(x,\mathsf{s}(\mathsf{f}(y,y))) \quad (4) \qquad \mathsf{f}(x,\mathsf{c}(y)) \gtrsim_2 \mathsf{f}(x,\mathsf{s}(\mathsf{f}(y,y))) \quad (7)$$
$$\mathsf{f}(\mathsf{s}(x),y) \gtrsim_1 \mathsf{f}(x,\mathsf{s}(\mathsf{c}(y))) \quad (5) \qquad \mathsf{f}(\mathsf{s}(x),y) \gtrsim_2 \mathsf{f}(x,\mathsf{s}(\mathsf{c}(y))) \quad (8)$$
$$\mathsf{F}(x,\mathsf{c}(y)) >_1 \mathsf{F}(y,y) \quad (6) \qquad \mathsf{F}(\mathsf{s}(x),y) >_2 \mathsf{F}(x,\mathsf{s}(\mathsf{c}(y))). \quad (9)$$

Note that in contrast to *direct* termination proofs, here we only need *weakly* monotonic quasi-orderings $\gtrsim_1, \gtrsim_2$. Hence, before synthesizing a suitable ordering some of the arguments of function symbols may be eliminated, cf. [AG97a]. For instance, in the inequalities (4) - (6) one may eliminate the second argument of the function symbol f. Then every term $\mathsf{f}(s,t)$ in the inequalities is replaced by $\mathsf{f}'(s)$ (where f' is a new unary function symbol). So instead of (4) we obtain the inequality $\mathsf{f}'(x) \gtrsim_1 \mathsf{f}'(x)$. By comparing the terms resulting from this replacement (instead of the original terms) we can take advantage of the fact that f does not have to be strongly monotonic in its second argument. Now the inequalities resulting from (4) - (6) are satisfied by the lexicographic path ordering (lpo) where subterms are compared right-to-left [KL80]. For the inequalities (7) - (9) we again delete the second argument of f. Then these inequalities are also satisfied by the lpo (with the precedence $\mathsf{F} \rhd \mathsf{s}, \mathsf{F} \rhd \mathsf{c}$), but this time subterms are compared left-to-right. Note that there exist only finitely many (and only few) possibilities to eliminate arguments of function symbols. Therefore all these possibilities can be checked automatically. As path orderings like the lpo can also be generated automatically, this enables a fully automatic termination proof of our TRS, whereas a direct termination proof with simplification orderings was not possible.

So Thm. 6 allows us to use *different* quasi-orderings to prove the absence of chains for different cycles. In our example this is essential, because there exists no quasi-simplification ordering satisfying *all* inequalities (4) - (9) (not even after elimination of arguments). Hence, without our modularity result, an automated termination proof with the dependency pair approach fails.

3 Modular Innermost Termination with Dependency Pairs

In [AG97b] we showed that the dependency pair approach can also be modified in order to verify *innermost* termination. Unlike previous methods, this technique can also prove innermost termination of non-terminating systems automatically.

Similar to the preceding section, our technique for innermost termination proofs can also be used in a modular way. As an example consider the following TRS:

$$\begin{aligned} \mathsf{f}(x,\mathsf{c}(x),\mathsf{c}(y)) &\to \mathsf{f}(y,y,\mathsf{f}(y,x,y)) & \mathsf{g}(x,y) &\to x \\ \mathsf{f}(\mathsf{s}(x),y,z) &\to \mathsf{f}(x,\mathsf{s}(\mathsf{c}(y)),\mathsf{c}(z)) & \mathsf{g}(x,y) &\to y \\ \mathsf{f}(\mathsf{c}(x),x,y) &\to \mathsf{c}(y) \end{aligned}$$

By applying the first f-rule to $\mathsf{f}(x,\mathsf{c}(x),\mathsf{c}(\mathsf{g}(x,\mathsf{c}(x))))$, we obtain an infinite (cycling) reduction. However, it is not an innermost reduction, because this term contains a redex $\mathsf{g}(\ldots)$ as a proper subterm. It turns out that the TRS is not terminating, but it is innermost terminating.

To develop a criterion for innermost termination similar to the termination criterion of Sect. 2, we have to restrict the notion of chains. Since we now consider innermost reductions, arguments of a redex must be in normal form before the redex is contracted. Therefore we demand that all instantiated left-hand sides $s_j\sigma$ of dependency pairs have to be normal. Moreover, the reductions of the arguments to normal forms must be innermost reductions (denoted by '$\stackrel{\mathsf{i}}{\to}$').

Definition 7 (Innermost Chain). Let $\mathcal{R}$ be a TRS. A sequence of dependency pairs $\langle s_1,t_1\rangle\,\langle s_2,t_2\rangle\ldots$ is an *innermost $\mathcal{R}$-chain* if there exists a substitution σ, such that all $s_j\sigma$ are in normal form and $t_j\sigma \stackrel{\mathsf{i}}{\to}^*_{\mathcal{R}} s_{j+1}\sigma$ holds for every two consecutive pairs $\langle s_j,t_j\rangle$ and $\langle s_{j+1},t_{j+1}\rangle$ in the sequence.

Of course, every innermost chain is also a chain, but not vice versa. In our example, we have the following dependency pairs.

$$\langle \mathsf{F}(x,\mathsf{c}(x),\mathsf{c}(y)),\mathsf{F}(y,y,\mathsf{f}(y,x,y))\rangle \tag{10}$$
$$\langle \mathsf{F}(x,\mathsf{c}(x),\mathsf{c}(y)),\mathsf{F}(y,x,y)\rangle \tag{11}$$
$$\langle \mathsf{F}(\mathsf{s}(x),y,z),\mathsf{F}(x,\mathsf{s}(\mathsf{c}(y)),\mathsf{c}(z))\rangle \tag{12}$$

The infinite sequence consisting of the dependency pair (10) is an infinite chain, but no *innermost* chain, because $\mathsf{F}(y_1,y_1,\mathsf{f}(y_1,x_1,y_1))\sigma$ can only reduce to $\mathsf{F}(x_2,\mathsf{c}(x_2),\mathsf{c}(y_2))\sigma$ for substitutions σ where $y_1\sigma$ is not a normal form. In [AG97b] we proved that absence of infinite innermost chains is a sufficient and necessary criterion for innermost termination.

Theorem 8 (Innermost Termination Criterion). *A TRS $\mathcal{R}$ is innermost terminating if and only if there exists no infinite innermost $\mathcal{R}$-chain.*

Analogous to Sect. 2, we introduce the estimated *innermost* dependency graph to approximate whether a dependency pair $\langle v,w\rangle$ can follow $\langle s,t\rangle$ in an innermost chain. Again we replace subterms in t with defined root symbols by new variables and check whether this modification of t unifies with v, but in contrast to Sect. 2 we do not have to rename multiple occurrences of the same variable. The reason is that we restrict ourselves to normal substitutions σ, i.e. all variables x are instantiated with normal forms and therefore, occurrences of $x\sigma$ cannot be reduced. Hence, there is no arc from (10) to itself, because

$\text{CAP}(\mathsf{F}(y_1, y_1, \mathsf{f}(y_1, x_1, y_1))) = \mathsf{F}(y_1, y_1, z)$ does not unify with $\mathsf{F}(x_2, \mathsf{c}(x_2), \mathsf{c}(y_2))$. Furthermore, we also demand that the most general unifier of $\text{CAP}(t)$ and v instantiates the left-hand sides s and v to normal forms.

Definition 9 (Estimated Innermost Dependency Graph). The *estimated innermost dependency graph* of a TRS $\mathcal{R}$ is the directed graph whose nodes are the dependency pairs and there is an arc from $\langle s, t\rangle$ to $\langle v, w\rangle$ if $\text{CAP}(t)$ and v are unifiable by a most general unifier μ such that $s\mu$ and $v\mu$ are normal forms.

In the estimated innermost dependency graph of our example, there are arcs from (11) to each dependency pair and there are arcs from (10) to (12) and from (12) to itself. Hence, the only non-empty cycles are $\{(11)\}$ and $\{(12)\}$. Analogous to Thm. 5 one can show that it suffices to prove the absence of infinite innermost chains separately for every cycle.

Theorem 10 (Modular Innermost Termination Criterion). *A TRS $\mathcal{R}$ is innermost terminating iff for each cycle $\mathcal{P}$ in the estimated innermost dependency graph there exists no infinite innermost $\mathcal{R}$-chain of dependency pairs from $\mathcal{P}$.*

To prove innermost termination in a modular way, we again generate a set of inequalities for every cycle $\mathcal{P}$ and search for a well-founded quasi-ordering $\geq_\mathcal{P}$ satisfying them. However, to ensure $t\sigma \geq_\mathcal{P} v\sigma$ whenever $t\sigma$ reduces to $v\sigma$, we do not have to demand $l \geq_\mathcal{P} r$ for *all* rules of the TRS any more. As we restrict ourselves to *normal* substitutions σ, not all rules are *usable* in a reduction of $t\sigma$. For example, *no* rule can be used to reduce a normal instantiation of $\mathsf{F}(y, x, y)$, because F is no defined symbol. In general, if t contains a defined symbol f, then all f-rules are *usable* and moreover, all rules that are *usable* for right-hand sides of f-rules are also *usable* for t.

Definition 11 (Usable Rules). Let $\mathcal{R}(D, C, R)$ be a TRS. For any symbol f let $\text{Rls}_R(f) = \{l \to r \in R \mid \text{root}(l) = f\}$. For any term we define the *usable rules*:

- $\mathcal{U}_R(x) = \emptyset$,
- $\mathcal{U}_R(f(t_1, \ldots, t_n)) = \text{Rls}_R(f) \;\cup\; \bigcup_{l \to r \in \text{Rls}_R(f)} \mathcal{U}_{R'}(r) \;\cup\; \bigcup_{j=1}^{n} \mathcal{U}_{R'}(t_j)$,

where $R' = R \setminus \text{Rls}_R(f)$. Moreover, for any set $\mathcal{P}$ of dependency pairs we define $\mathcal{U}_R(\mathcal{P}) = \bigcup_{\langle s,t\rangle \in \mathcal{P}} \mathcal{U}_R(t)$.

So we have $\mathcal{U}_R(\mathsf{F}(y, y, \mathsf{f}(y, x, y))) = \text{Rls}_R(\mathsf{f})$ and $\mathcal{U}_R(\{(11)\}) = \mathcal{U}_R(\{(12)\}) = \emptyset$, i.e. there are no usable rules for the cycles. Note that $\text{Rls}_R(f) = \emptyset$ for any constructor f. Now our theorem for automatic[2] modular verification of innermost termination can be proved analogously to Thm. 6.

Theorem 12 (Modular Innermost Termination Proofs). *A TRS $\mathcal{R}(D, C, R)$ is innermost terminating if for each cycle $\mathcal{P}$ in the estimated innermost dependency graph there exists a well-founded weakly monotonic quasi-ordering $\geq_\mathcal{P}$ where both $\geq_\mathcal{P}$ and $>_\mathcal{P}$ are closed under substitution, such that*

[2] Additional refinements for the automated checking of our innermost termination criterion can be found in [AG97b].

- *$l \geq_{\mathcal{P}} r$ for all rules $l \to r$ in $\mathcal{U}_R(\mathcal{P})$,*
- *$s \gtrsim_{\mathcal{P}} t$ for all dependency pairs from $\mathcal{P}$, and*
- *$s >_{\mathcal{P}} t$ for at least one dependency pair from $\mathcal{P}$.*

In this way, we obtain the following constraints for our example:

$$\mathsf{F}(x, \mathsf{c}(x), \mathsf{c}(y)) >_1 \mathsf{F}(y, x, y) \qquad \mathsf{F}(\mathsf{s}(x), y, z) >_2 \mathsf{F}(x, \mathsf{s}(\mathsf{c}(y)), \mathsf{c}(z)).$$

For $>_1$ we may use the lpo comparing subterms right-to-left and for $>_2$ we may use the lpo comparing subterms left-to-right. Hence, innermost termination of this example can easily be proved automatically. Without our modularity result, this proof would not be possible, because there exists no simplification ordering satisfying *both* inequalities (not even after elimination of arguments).

4 Modularity Criteria

In this section we present two corollaries of our results from the preceding sections which are particularly useful in practice. Moreover, these corollaries also allow a comparison with existing modularity results, as will be shown in Sect. 5.

4.1 Hierarchical Combinations

A straightforward corollary of Thm. 10 and 12 can be obtained for *hierarchical combinations.* Two term rewriting systems $\mathcal{R}_0(D_0, C_0, R_0)$ and $\mathcal{R}_1(D_1, C_1, R_1)$ form a *hierarchical combination* if $D_0 \cap D_1 = C_0 \cap D_1 = \emptyset$, i.e. defined symbols of $\mathcal{R}_0$ may occur as constructors in $\mathcal{R}_1$, but not vice versa. As an example consider the following TRS. Here, nil denotes the empty list and $n{\bullet}x$ represents the insertion of a number n into a list x, where '$n{\bullet}m{\bullet}x$' abbreviates '$n{\bullet}(m{\bullet}x)$'. The function $\mathsf{sum}(x, y)$ adds all elements of x to the first element of y, i.e. $\mathsf{sum}(n_0{\bullet}n_1{\bullet}\dots{\bullet}n_k{\bullet}\mathsf{nil}, m{\bullet}y) = (m + \sum_{i=0}^{k} n_i){\bullet}y$. The function weight computes the weighted sum, i.e. $\mathsf{weight}(n_0{\bullet}n_1{\bullet}\dots{\bullet}n_k{\bullet}\mathsf{nil}) = n_0 + \sum_{i=1}^{k} i\, n_i$.

$$\begin{aligned}
\mathsf{sum}(\mathsf{s}(n){\bullet}x, m{\bullet}y) &\to \mathsf{sum}(n{\bullet}x, \mathsf{s}(m){\bullet}y)\\
\mathsf{sum}(0{\bullet}x, y) &\to \mathsf{sum}(x, y)\\
\mathsf{sum}(\mathsf{nil}, y) &\to y\\
\mathsf{weight}(n{\bullet}m{\bullet}x) &\to \mathsf{weight}(\mathsf{sum}(n{\bullet}m{\bullet}x, 0{\bullet}x))\\
\mathsf{weight}(n{\bullet}\mathsf{nil}) &\to n
\end{aligned}$$

Let $\mathcal{R}_0$ consist of the three sum-rules and let $\mathcal{R}_1$ be the system consisting of the two weight-rules. Then these two systems form a hierarchical combination, where sum is a defined symbol of $\mathcal{R}_0$ and a constructor of $\mathcal{R}_1$.

Note that tuple symbols from dependency pairs of $\mathcal{R}_0$ do not occur in left-hand sides of $\mathcal{R}_1$-dependency pairs. Hence, a cycle in the estimated innermost dependency graph either consists of $\mathcal{R}_0$-dependency pairs or of $\mathcal{R}_1$-dependency pairs only. So in our example, every cycle either contains just SUM- or just WEIGHT-dependency pairs. Thus, we obtain the following corollary.

Corollary 13 (Innermost Termination for Hierarchical Combinations). *Let $\mathcal{R}$ be the hierarchical combination of $\mathcal{R}_0(D_0, C_0, R_0)$ and $\mathcal{R}_1(D_1, C_1, R_1)$.*

(a) $\mathcal{R}$ is innermost terminating iff $\mathcal{R}_0$ is innermost terminating and there exists no infinite innermost $\mathcal{R}$-chain of $\mathcal{R}_1$-dependency pairs.
(b) $\mathcal{R}$ is innermost terminating if $\mathcal{R}_0$ is innermost terminating and if there exists a well-founded weakly monotonic quasi-ordering $\geq$ where both $\geq$ and $>$ are closed under substitution, such that for all dependency pairs $\langle s, t\rangle$ of $\mathcal{R}_1$
- *$l \geq r$ for all rules $l \rightarrow r$ in $\mathcal{U}_{R_0 \cup R_1}(t)$ and*
- *$s > t$.*

Proof. The corollary is a direct consequence of Thm. 10 and 12, since for any dependency pair $\langle s, t\rangle$ of $\mathcal{R}_0$ the only rules that can be used to reduce a normal instantiation of t are the rules from $\mathcal{R}_0$ (i.e. $\mathcal{U}_{R_0 \cup R_1}(t) \subseteq R_0$). □

(Innermost) termination of the sum-system ($\mathcal{R}_0$) is easily proved (e.g. by the lpo with the precedence $\mathsf{sum} \rhd \bullet$ and $\mathsf{sum} \rhd \mathsf{s}$). For the weight-subsystem ($\mathcal{R}_1$) we obtain the following constraints. (Note that $\langle \mathsf{WEIGHT}(\ldots), \mathsf{SUM}(\ldots)\rangle$ is no dependency pair of $\mathcal{R}_1$, since $\mathsf{sum} \notin D_1$.)

$$\begin{aligned} \mathsf{sum}(\mathsf{s}(n)\bullet x, m\bullet y) &\geq \mathsf{sum}(n\bullet x, \mathsf{s}(m)\bullet y) \\ \mathsf{sum}(\mathsf{0}\bullet x, y) &\geq \mathsf{sum}(x, y) \\ \mathsf{sum}(\mathsf{nil}, y) &\geq y \\ \mathsf{WEIGHT}(n\bullet m\bullet x) &> \mathsf{WEIGHT}(\mathsf{sum}(n\bullet m\bullet x, \mathsf{0}\bullet x)) \end{aligned}$$

After eliminating the first arguments of sum and '$\bullet$' (i.e. after replacing each term $\mathsf{sum}(s,t)$ and $s\bullet t$ by $\mathsf{sum}'(t)$ and $\bullet'(t)$, respectively), the inequalities are also satisfied by the lpo, but now we have to use the precedence $\bullet' \rhd \mathsf{sum}'$.

In this way, innermost termination of the example can be proved automatically. Moreover, as the system is non-overlapping, this also proves its termination. Note that this system is not simply terminating and without modularity, no quasi-simplification ordering would have satisfied the constraints resulting from the dependency pair approach (even when using elimination of arguments).

A corollary like Cor. 13 can also be formulated for termination instead of innermost termination, because in the termination case there cannot be a cycle consisting of dependency pairs from both $\mathcal{R}_0$ and $\mathcal{R}_1$ either. But in contrast to the innermost termination case, rules of $\mathcal{R}_1$ can be used to reduce instantiated right-hand sides of $\mathcal{R}_0$-dependency pairs (as we cannot restrict ourselves to normal substitutions then). Hence, to prove the absence of infinite $\mathcal{R}_0$-chains we have to use a quasi-ordering where the rules of $\mathcal{R}_1$ are also weakly decreasing. Therefore, the constraints for the termination proof of the sum and weight-example (according to Sect. 2) are not satisfied by any quasi-simplification ordering amenable to automation, whereas the constraints for *innermost* termination are fulfilled by such an ordering. Hence, for non-overlapping systems, it is always advantageous to verify termination by proving *innermost* termination only.

4.2 Splitting into Subsystems

The modularity results presented so far were all used in the context of dependency pairs. However, the classical approach to modularity is to split a TRS into subsystems and to prove their (innermost) termination separately. The following corollary of Thm. 10 shows that the consideration of cycles in the estimated innermost dependency graph can also be used to decompose a TRS into modular subsystems. In the following, let $\mathcal{O}(\mathcal{P})$ denote the *origin* of the dependency pairs in $\mathcal{P}$, i.e. $\mathcal{O}(\mathcal{P})$ is a set of those rules where the dependency pairs of $\mathcal{P}$ stem from[3]. So for the example of Sect. 3 we have $\mathcal{O}(\{(11)\}) = \{\mathsf{f}(x, \mathsf{c}(x), \mathsf{c}(y)) \rightarrow \mathsf{f}(y, y, \mathsf{f}(y, x, y))\}$ and $\mathcal{O}(\{(12)\}) = \{\mathsf{f}(\mathsf{s}(x), y, z) \rightarrow \mathsf{f}(x, \mathsf{s}(\mathsf{c}(y)), \mathsf{c}(z))\}$.

Corollary 14 (Modularity for Subsystems). *Let $\mathcal{R}(D, C, R)$ be a TRS, let $\mathcal{P}_1, \ldots, \mathcal{P}_n$ be the cycles in its estimated innermost dependency graph, and let $\mathcal{R}_j(D_j, C_j, R_j)$ be subsystems of $\mathcal{R}$ such that $\mathcal{U}_R(\mathcal{P}_j) \cup \mathcal{O}(\mathcal{P}_j) \subseteq R_j$ (for all $j \in \{1, \ldots, n\}$). If $\mathcal{R}_1, \ldots, \mathcal{R}_n$ are innermost terminating, then $\mathcal{R}$ is also innermost terminating.*

Proof. As $\mathcal{P}_j$ is a cycle, every dependency pair from $\mathcal{P}_j$ is an $\mathcal{R}_j$-dependency pair. (The reason is that for every[4] $\langle F(\mathbf{s}), G(\mathbf{t})\rangle$ in $\mathcal{P}_j$ there is also a dependency pair $\langle G(\mathbf{v}), H(\mathbf{w})\rangle$ in $\mathcal{P}_j$. Hence, g must be a defined symbol of $\mathcal{R}_j$.) Thus, every innermost $\mathcal{R}$-chain of dependency pairs from $\mathcal{P}_j$ is also an innermost $\mathcal{R}_j$-chain. Now the corollary is a direct consequence of Thm. 10. □

For instance, in the example of Sect. 3 we only have two non-empty cycles, viz. $\{(11)\}$ and $\{(12)\}$. As these dependency pairs have no defined symbols on their right-hand sides, their sets of usable rules are empty. Hence, to prove innermost termination of the whole system, by Cor. 14 it suffices to prove innermost termination of the two one-rule subsystems $\mathsf{f}(x, \mathsf{c}(x), \mathsf{c}(y)) \rightarrow \mathsf{f}(y, y, \mathsf{f}(y, x, y))$ and $\mathsf{f}(\mathsf{s}(x), y, z) \rightarrow \mathsf{f}(x, \mathsf{s}(\mathsf{c}(y)), \mathsf{c}(z))$.

In fact, both subsystems are even terminating as can easily be proved automatically. For the first system one can use a polynomial interpretation mapping $\mathsf{f}(x, y, z)$ to $x+y+z$ and $\mathsf{c}(x)$ to $5x+1$ [Lan79]. Methods for the automated generation of polynomial orderings have for instance been developed in [Ste94, Gie95]. For the second system one can use the lpo with the precedence $\mathsf{f} \rhd \mathsf{s}$ and $\mathsf{f} \rhd \mathsf{c}$.

Hence, the modularity criterion of Cor. 14 allows the use of well-known simplification orderings for innermost termination proofs of non-terminating systems, because it guarantees that innermost termination of the two simply terminating subsystems is sufficient for innermost termination of the original TRS.

A similar splitting is also possible for the example in Sect. 2. Even better, if we modify the TRS into a non-overlapping one

$$\begin{aligned} \mathsf{f}(x, \mathsf{c}(y)) &\rightarrow \mathsf{f}(x, \mathsf{s}(\mathsf{f}(y, y))) \\ \mathsf{f}(\mathsf{s}(x), \mathsf{s}(y)) &\rightarrow \mathsf{f}(x, \mathsf{s}(\mathsf{c}(\mathsf{s}(y)))), \end{aligned}$$

---

[3] If a dependency pair of $\mathcal{P}$ may stem from *several* rules, then it is sufficient if $\mathcal{O}(\mathcal{P})$ just contains one of them.

[4] Here, $\mathbf{s}$ and $\mathbf{t}$ denote *tuples* of terms $s_1, \ldots, s_n$ and $t_1, \ldots, t_m$, respectively.

then Cor. 14 allows to conclude termination of the whole system from termination of the two one-rule subsystems. Innermost termination of the original example resp. termination of the above modified example can be proved by the lpo, but for the first rule one needs the precedence $\mathsf{c} \triangleright \mathsf{s}$ and $\mathsf{c} \triangleright \mathsf{f}$, whereas for the second rule the precedence $\mathsf{f} \triangleright \mathsf{s}$ and $\mathsf{f} \triangleright \mathsf{c}$ is required.

5 Comparison with Related Work

Now we show that in the case of finite TRSs, existing modularity results for innermost termination are obtained as easy consequences of our criteria and that our criteria extend previously developed results. Sect. 5.1 focuses on composable TRSs and Sect. 5.2 gives a comparison with results on hierarchical combinations.

5.1 Shared Constructors and Composable TRSs

By the framework of the previous sections we can easily prove that innermost termination is modular for composable TRSs [Ohl95] and hence also for TRSs with disjoint sets of defined symbols and shared constructors [Gra95]. Two TRSs $\mathcal{R}_0(D_0, C_0, R_0)$ and $\mathcal{R}_1(D_1, C_1, R_1)$ are *composable* if $C_0 \cap D_1 = C_1 \cap D_0 = \emptyset$ and if both systems contain all rewrite rules that define a defined symbol whenever that symbol is shared, i.e. $\{l \to r \mid \mathrm{root}(l) \in D_0 \cap D_1\} \subseteq R_0 \cap R_1$. Now Cor. 14 immediately implies[5] the following result of Ohlebusch [Ohl95].

Theorem 15 (Modularity for Composable TRSs). *Let $\mathcal{R}_0(D_0, C_0, R_0)$ and $\mathcal{R}_1(D_1, C_1, R_1)$ be composable TRSs. If $\mathcal{R}_0$ and $\mathcal{R}_1$ are innermost terminating, then $\mathcal{R}_0 \cup \mathcal{R}_1$ is also innermost terminating.*

Proof. Let $\langle F(\mathbf{s}), G(\mathbf{t})\rangle$ be a dependency pair of $\mathcal{R}_0 \cup \mathcal{R}_1$. If $f \in D_0$, then there exists a rule $f(\mathbf{t}) \to C[g(\mathbf{t})]$ in R_0. (This rule cannot be from $R_1 \setminus R_0$, because $\mathcal{R}_0$ and $\mathcal{R}_1$ are composable.) Hence, $g \in D_0$, because constructors of $\mathcal{R}_0$ are not defined symbols of $\mathcal{R}_1$. Similarly, $f \in D_1$ implies $g \in D_1$. So any dependency pair of $\mathcal{R}_0 \cup \mathcal{R}_1$ is an $\mathcal{R}_0$-dependency pair or an $\mathcal{R}_1$-dependency pair.

Moreover, there can only be an arc from $\langle F(\mathbf{s}), G(\mathbf{t})\rangle$ to a dependency pair of the form $\langle G(\mathbf{v}), H(\mathbf{w})\rangle$. Hence, if $\langle F(\mathbf{s}), G(\mathbf{t})\rangle$ is an $\mathcal{R}_j$-dependency pair, then $g \in D_j$ and therefore, $\langle G(\mathbf{v}), H(\mathbf{w})\rangle$ is also an $\mathcal{R}_j$-dependency pair (for $j \in \{0,1\}$). So every cycle $\mathcal{P}$ in the estimated innermost dependency graph of $\mathcal{R}_0 \cup \mathcal{R}_1$ either consists of $\mathcal{R}_0$-dependency pairs or of $\mathcal{R}_1$-dependency pairs only.

If a cycle $\mathcal{P}$ only contains $\mathcal{R}_0$-dependency pairs, then R_0 is a superset of $\mathcal{U}_{R_0 \cup R_1}(\mathcal{P}) \cup \mathcal{O}(\mathcal{P})$, as the defined symbols of $\mathcal{R}_1 \setminus \mathcal{R}_0$ do not occur as constructors in $\mathcal{R}_0$. Similarly, for a cycle $\mathcal{P}$ of $\mathcal{R}_1$-dependency pairs, we have $\mathcal{U}_{R_0 \cup R_1}(\mathcal{P}) \cup \mathcal{O}(\mathcal{P}) \subseteq R_1$. Hence by Cor. 14, $\mathcal{R}_0 \cup \mathcal{R}_1$ is innermost terminating if $\mathcal{R}_0$ and $\mathcal{R}_1$ are innermost terminating. □

[5] A direct proof of Thm. 15 is not too difficult either, but our alternative proof serves to illustrate the connections between our criteria and existing modularity results.

Note that our results extend modularity to a much larger class of TRSs, e.g. they also allow a splitting into non-composable subsystems which share defined symbols as demonstrated in Sect. 4.2.

5.2 Proper Extensions

Krishna Rao [KR95] proved that innermost termination is modular for a certain form of hierarchical combinations, viz. so-called *proper* extensions. In this section we show that for finite TRSs this is also a direct consequence of our results.

For a TRS $\mathcal{R}(D, C, R)$, the dependency relation $\succeq_d$ is the smallest quasi-ordering satisfying the condition $f \succeq_d g$ whenever there is a rewrite rule $f(\ldots) \to C[g(\ldots)] \in R$. So $f \succeq_d g$ holds if the function f depends on the definition of g.

Let $\mathcal{R}_0(D_0, C_0, R_0)$ and $\mathcal{R}_1(D_1, C_1, R_1)$ form a hierarchical combination. Now the defined symbols D_1 of $\mathcal{R}_1$ are split in two sets D_1^0 and D_1^1, where D_1^0 contains all defined symbols which depend on a defined symbol of $\mathcal{R}_0$, i.e. $D_1^0 = \{f | f \in D_1, f \succeq_d g \text{ for some } g \in D_0\}$ and $D_1^1 = D_1 \setminus D_1^0$. Then $\mathcal{R}_1$ is a *proper extension* of $\mathcal{R}_0$ if each rewrite rule $l \to r \in R_1$ satisfies the following condition: For every subterm t of r, if $root(t) \in D_1^0$ and $root(t) \succeq_d root(l)$, then t contains no symbols from $D_0 \cup D_1^0$ except at the root position, cf. [KR95].

For instance, in the sum and weight-example from Sect. 4.1 we have $D_0 = \{\mathsf{sum}\}$, $D_1^0 = \{\mathsf{weight}\}$ (because weight depends on the definition of sum), and $D_1^1 = \emptyset$. This example is not a proper extension, because there is a weight-rule where the D_0-symbol sum occurs below the D_1^0-symbol weight. Thus, in a proper extension functions depending on $\mathcal{R}_0$ are never called within a recursive call of $\mathcal{R}_1$-functions. Cor. 13 and 14 imply the following result of [KR95]

Theorem 16 (Modularity for Proper Extensions). *Let $\mathcal{R}_1(D_1, C_1, R_1)$ be a proper extension of $\mathcal{R}_0(D_0, C_0, R_0)$. The TRS $\mathcal{R}_0 \cup \mathcal{R}_1$ is innermost terminating if $\mathcal{R}_0$ and $\mathcal{R}_1$ are innermost terminating.*

Proof. As in the proof of Cor. 13, since $\mathcal{R}_0$ and $\mathcal{R}_1$ form a hierarchical combination, every cycle in the innermost dependency graph of $\mathcal{R}_0 \cup \mathcal{R}_1$ consists solely of $\mathcal{R}_0$-dependency pairs or of $\mathcal{R}_1$-dependency pairs. If a cycle $\mathcal{P}$ consists of dependency pairs of $\mathcal{R}_0$, we have $\mathcal{U}_{R_0 \cup R_1}(\mathcal{P}) \cup \mathcal{O}(\mathcal{P}) \subseteq R_0$, because dependency pairs of $\mathcal{R}_0$ do not contain any defined symbols of $\mathcal{R}_1$.

Otherwise, the cycle $\mathcal{P}$ consists of $\mathcal{R}_1$-dependency pairs. If $\langle F(\mathbf{s}), G(\mathbf{t}) \rangle$ is an $\mathcal{R}_1$-dependency pair in $\mathcal{P}$, then there exists a rule $f(\mathbf{s}) \to C[g(\mathbf{t})]$ in R_1 and $f, g \in D_1$. In addition, we have $f \succeq_d g$ and $g \succeq_d f$ (as $\mathcal{P}$ is a cycle).

If $g \in D_1^1$, then f also belongs to D_1^1, hence no defined symbol of $D_0 \cup D_1^0$ occurs in $\mathbf{t}$. Otherwise, if $g \in D_1^0$, then by definition of a proper extension again all defined symbols in $\mathbf{t}$ are from D_1^1. Thus, in both cases, all defined symbols of $\mathcal{U}_{R_0 \cup R_1}(G(\mathbf{t}))$ belong to D_1^1. Hence, $\mathcal{U}_{R_0 \cup R_1}(G(\mathbf{t}))$ is a subsystem of $\mathcal{R}_1$.

So for any cycle $\mathcal{P}$ of $\mathcal{R}_1$-dependency pairs, we have $\mathcal{U}_{R_0 \cup R_1}(\mathcal{P}) \cup \mathcal{O}(\mathcal{P}) \subseteq R_1$. Hence, by Cor. 14 innermost termination of $\mathcal{R}_0$ and $\mathcal{R}_1$ implies innermost termination of $\mathcal{R}_0 \cup \mathcal{R}_1$. □

However, apart from proper extensions, we can also handle certain hierarchical combinations where $\mathcal{R}_1$ contains defined symbols of $\mathcal{R}_0$ in the arguments of its recursive calls, cf. the sum and weight-example. Such systems occur frequently in practice. Hence, our results significantly extend the class of TRSs where innermost termination can be proved in a modular way.

Another modularity criterion for hierarchical combinations is due to Dershowitz [Der94]. Here, occurrences of D_0-symbols in recursive calls of D_1-symbols are allowed, but only if $\mathcal{R}_1$ is *oblivious* of the $\mathcal{R}_0$-rules, i.e. termination of $\mathcal{R}_1$ must not depend on the $\mathcal{R}_0$-rules. However, this criterion is not applicable for the sum and weight-example, because termination of the weight-rules in fact depends on the result of $\mathsf{sum}(n{\scriptstyle\bullet}m{\scriptstyle\bullet}x, \mathsf{0}{\scriptstyle\bullet}x)$.

An alternative modularity result for hierarchical combinations was presented by Fernandez and Jouannaud [FJ95]. However, their result is restricted to systems where the arguments of recursive calls in $\mathcal{R}_1$ decrease w.r.t. the subterm relation (compared as multisets or lexicographically). Hence, their result is not applicable to the sum and weight-example either.

6 Conclusion

In this paper we introduced a refinement of the dependency pair approach in order to perform termination and innermost termination proofs in a modular way. This refinement allows automated termination and innermost termination proofs for many TRSs where such proofs were not possible before, cf. [AG97c]. We showed that our new modularity results extend previous results for modularity of innermost termination. Due to the framework of dependency pairs, we also obtain easy proofs for existing modularity theorems.

References

[AG96] T. Arts & J. Giesl, Termination of constructor systems. In *Proc. RTA-96*, LNCS 1103, pp. 63–77, New Brunswick, NJ, 1996.

[AG97a] T. Arts & J. Giesl, Automatically proving termination where simplification orderings fail. *TAPSOFT '97*, LNCS 1214, pp. 261–273, Lille, France, 1997.

[AG97b] T. Arts & J. Giesl, Proving innermost normalisation automatically. In *Proc. RTA-97*, LNCS 1232, pp. 157–172, Sitges, Spain, 1997.

[AG97c] T. Arts & J. Giesl, Modularity of termination using dependency pairs. Tech. Rep. IBN 97/45, TU Darmstadt, 1997. `http://www.inferenzsysteme.informatik.tu-darmstadt.de/~reports/notes/ibn-97-45.ps`

[Art96] T. Arts, Termination by absence of infinite chains of dependency pairs. In *Proc. CAAP '96*, LNCS 1059, pp. 196–210, Linköping, Sweden, 1996.

[Art97] T. Arts, *Automatically proving termination and innermost normalisation of term rewriting systems.* PhD Thesis, Utrecht Univ., The Netherlands, 1997.

[Der87] N. Dershowitz, Termination of rewriting. *JSC*, 3:69–116, 1987.

[Der94] N. Dershowitz, *Hierarchical Termination.* In *Proc. CTRS-94*, LNCS 968, pp. 89–105, Jerusalem, Israel, 1994.

[DH95] N. Dershowitz & C. Hoot, Natural termination. *TCS*, 142(2):179–207, 1995.

[Dro89] K. Drosten, *Termersetzungssysteme*. Springer, Berlin, 1989.

[FJ95] M. Fernández & J.-P. Jouannaud, Modular termination of term rewriting systems revisited. In *Proc. 10th Workshop on Specification of Abstract Data Types*, LNCS 906, pp. 255–273, S. Margherita, Italy, 1995.

[Gie95] J. Giesl, Generating polynomial orderings for termination proofs. In *Proc. RTA-95*, LNCS 914, pp. 426–431, Kaiserslautern, Germany, 1995.

[Gra94] B. Gramlich, Generalized sufficient conditions for modular termination of rewriting. *Appl. Algebra in Engineering, Comm. & Comp.*, 5:131–158, 1994.

[Gra95] B. Gramlich, Abstract relations between restricted termination and confluence properties of rewrite systems. *Fundamenta Informaticae*, 24:3–23, 1995.

[Gra96a] B. Gramlich, *Termination and confluence properties of structured rewrite systems*. PhD Thesis, Universität Kaiserslautern, Germany, 1996.

[Gra96b] B. Gramlich, On proving termination by innermost termination. In *Proc. RTA-96*, LNCS 1103, pp. 93–107, New Brunswick, NJ, 1996.

[HL78] G. Huet & D. Lankford, On the uniform halting problem for term rewriting systems. Technical Report 283, INRIA, Le Chesnay, France, 1978.

[KL80] S. Kamin & J.-J. Levy, Two generalizations of the recursive path ordering. Department of Computer Science, University of Illinois, IL, 1980.

[KR95] M. R. K. Krishna Rao, Modular proofs for completeness of hierarchical term rewriting systems. *TCS*, 151:487–512, 1995.

[KO92] M. Kurihara & A. Ohuchi, Modularity of simple termination of term rewriting systems with shared constructors. *TCS*, 103:273–282, 1992.

[Lan79] D. S. Lankford, On proving term rewriting systems are noetherian. Technical Report Memo MTP-3, Louisiana Tech. University, Ruston, LA, 1979.

[Mid89] A. Middeldorp, A sufficient condition for the termination of the direct sum of term rewriting systems. *LICS '89*, pp. 396–401, Pacific Grove, CA, 1989.

[Mid90] A. Middeldorp, *Modular properties of term rewriting systems*. PhD Thesis, Free University Amsterdam, The Netherlands, 1990.

[MT93] A. Middeldorp & Y. Toyama, Completeness of combinations of constructor systems. *JSC*, 15:331–348, 1993.

[MZ97] A. Middeldorp & H. Zantema, Simple termination of rewrite systems. *TCS*, 175:127–158, 1997.

[Ohl94] E. Ohlebusch, On the modularity of termination of term rewriting systems. *TCS*, 136:333–360, 1994.

[Ohl95] E. Ohlebusch, Modular properties of composable term rewriting systems. *JSC*, 1:1–42, 1995.

[Rus87] M. Rusinowitch, On termination of the direct sum of term-rewriting systems. *Information Processing Letters*, 26:65–70, 1987.

[SMP95] M. Schmidt-Schauß, M. Marchiori, & S. E. Panitz, Modular termination of r-consistent and left-linear term rewriting systems. *TCS*, 149:361–374, 1995.

[Ste94] J. Steinbach, Generating polynomial orderings. *Information Processing Letters*, 49:85–93, 1994.

[Ste95] J. Steinbach, Simplification orderings: history of results. *Fundamenta Informaticae*, 24:47–87, 1995.

[Toy87] Y. Toyama, Counterexamples to the termination for the direct sum of term rewriting systems. *Information Processing Letters*, 25:141–143, 1987.

[TKB95] Y. Toyama, J. W. Klop, & H. P. Barendregt, Termination for direct sums of left-linear complete term rewriting systems. *J. ACM*, 42:1275–1304, 1995.

Termination of Associative-Commutative Rewriting by Dependency Pairs*

Claude Marché and Xavier Urbain

LRI, CNRS URA 410
Bât. 490, Université Paris-Sud, Centre d'Orsay
91405 Orsay Cedex, France
Phone:+33 1 69 15 64 85, Fax: +33 1 69 15 65 86
Email: {marche,urbain}@lri.fr

Abstract. A new criterion for termination of rewriting has been described by Arts and Giesl in 1997. We show how this criterion can be generalized to rewriting modulo associativity and commutativity. We also show how one can build *weak* AC-compatible reduction orderings which may be used in this criterion.

1 Introduction

For almost all applications of rewrite systems, one is interested in finite termination of reductions, a property called *termination* or *strong normalization* of a system. When rewriting is used as a model of computation, for example in relation to functional programming, termination insures the existence of the result of computation. When rewriting is used for automated reasoning in equational logic, existence of normal forms is also important, and a terminating rewrite system can also be the basis of a Noetherian induction.

Thus, checking termination of a given rewrite system is a crucial issue. Although this property is known to be computationally unsolvable in general, it is important to have automatic tools, correct but incomplete, that can prove termination of rewrite systems one meets in practice.

Until 1997, among the methods known for proving termination of a TRS, the only implementable one was the technique consisting in building a *reduction ordering* on terms with which we can check automatically that the rewrite rules we consider all strictly decrease with respect to this ordering. Thus, implementing a program for checking termination consisted mainly as implementing ways of building reduction orderings.

In 1997, Arts and Giesl [1,2] proposed a new method for proving termination which does not consist anymore in checking that rules strictly decrease but in a more careful check on structure of the rules, the important point being that this remains a purely syntactic finite test (the so-called *dependency pairs*) thus it is still implementable.

* This research was supported in part by the EWG CCL II, the HCM Network CONSOLE, and the "GDR de programmation du CNRS".

In this paper, we propose a generalization of the dependency pairs method to rewriting modulo AC (associativity and commutativity) which is an important issue in practice.

In fact, there are quite different notions of AC rewriting, and in Section 2 we will first describe clearly which kind of AC rewriting we are going to consider: *flat* AC rewriting, and Theorem 4 shows why using flat AC rewriting is important for us. In Section 3, we will then show how the dependency pairs criterion for standard rewriting is lifted up to flat AC rewriting (Theorem 8 and Corollary 9).

Arts and Giesl also have shown that the dependency pairs criterion can be made a bit stronger by use of a *dependency graph*, again a deeper analysis of the rewrite rules structure, and we show in Section 4 how this improvement is lifted up to AC rewriting.

The dependency pairs criterion still amounts the use of a term ordering for proving termination, but a slightly larger class of orderings is allowed: *weak* reduction orderings, that is orderings that do not need to fulfill the *strict* monotonicity property ($t_1 > t_2$ implies $f(\dots, t_1, \dots) > f(\dots, t_2, \dots)$) but only the *weak* monotonicity one ($t_1 \geq t_2$ implies $f(\dots, t_1, \dots) \geq f(\dots, t_2, \dots)$). In the case of AC rewriting, one needs *AC-compatible* weak reduction orderings and in Section 5, we will show how such orderings can be built.

Finally, in Section 6 we give a small but important remark on the use of this new criterion in completion procedures.

In the course of the paper, we will give some examples, for which the termination has been proved automatically by a new termination toolbox that we implemented inside the C*i*ME system [10] developed in our research team.

2 Basic notions

We recall first the usual notions about rewriting and give our notations. We then focus on the definition of rewriting modulo AC.

A signature $\mathcal{F}$ is a finite set of symbols with arities, the set of terms on $\mathcal{F}$ and variables X is denoted by $T(\mathcal{F}, X)$. The root position is denoted by Λ, the root symbol of a term t is denoted $\text{Head}(t)$, $t|_p$ denotes the subterm of t at position p, and $C[t]_p$ denotes the term obtained by putting the term t in the context C at position p.

We denote $s \xrightarrow[l \to r]{p} t$ if s rewrites to t at position p by a rule $l \to r$, and $s \xrightarrow[R]{>p\ *} t$ if s rewrites to t by a set of rules R in 0, 1 or several steps *below* p. We say that a rewrite system R *terminates* if no infinite sequence of reduction by $\to_R$ exists. We say that a term t is *strongly normalizable* (SN for short) if there is no infinite sequence of reduction by $\to_R$ starting from t.

We say that f is a *defined* symbol of R if it occurs at the top of a left-hand side, and is a *constructor* otherwise.

2.1 Terms with AC symbols, flat terms

In practice, we often have symbols f that fulfill the commutativity axiom $f(x,y) = f(y,x)$. This cannot be turned into a rule without losing termination, and the standard technique to deal with it is to consider classes of terms modulo commutativity. Moreover, if f also fulfills associativity $f(x, f(y,z)) = f(f(x,y),z)$, this equation itself cannot be turned into a rule without losing termination modulo commutativity $(f(x, f(y,z)) \rightarrow f(f(x,y),z) =_C f(z, f(x,y)) \rightarrow \cdots)$, hence in such a case we also consider class of terms modulo associativity. In the following, we call such an f an AC symbol, and we denote by $\mathcal{F}_{AC}$ the set of AC symbols. We do not consider symbols which are only commutative in the following, but everything can be specialized to that case.

Formally speaking, we consider the set of *varyadic* terms: AC symbols may have any arity greater than 2. The *flat form* of a term t, denoted $\bar{t}$, is the normal form of t for the rules $f(x_1, \ldots, x_{i-1}, f(y_1, \ldots, y_k), x_{i+1}, \ldots, x_n) \rightarrow f(x_1, \ldots, x_{i-1}, y_1, \ldots, y_k, x_{i+1}, \ldots, x_n)$ for each AC symbol f. It is clear that $s =_{AC} t$ iff $\bar{s} \equiv \bar{t}$ where $\equiv$ is the equivalence up to permutations of AC symbols arguments.

2.2 Rewriting modulo AC

Definition 1. We say that s rewrites to t by a rule $l \rightarrow r$ modulo AC, denoted by $s \xrightarrow[l \rightarrow r/AC]{} t$, iff there are s' and t' s.t. $s =_{AC} s'$, $t =_{AC} t'$ and $s' \xrightarrow[l \rightarrow r]{} t'$.

Deciding whether a term s rewrites to another modulo AC amounts to searching for an instance of a rule in any term AC-equivalent to s, thus it is not efficient at all. Peterson and Stickel [24] proposed the use of a restricted definition where when one wants to match a rule on a given term at a given position p, AC-equivalence is allowed only at or below position p.

Definition 2. The *AC-extended* rewrite relation is defined by $s \xrightarrow[AC\backslash l \rightarrow r]{p} t$ iff there is a substitution σ s.t. $s|_p =_{AC} l\sigma$ and $t = s[r\sigma]_p$.

Using AC-extended rewriting may lead to pathological examples where for a rewrite system R, $\rightarrow_{AC\backslash R}$ terminates whereas $\rightarrow_{R/AC}$ does not. This is the case for example with $R = \{a + b \rightarrow a + (b + c)\}$, where $+$ is an AC symbol: the term $a + (b + c)$ is irreducible by AC-extended rewriting, but it is reducible infinitely by class rewriting because $a + (b + c) =_{AC} (a + b) + c$ which contains a redex.

This phenomenon arises because a system like R above is not *coherent* with AC in the sense defined by Jouannaud and Kirchner [17]: $s =_{AC} t$ and $t \xrightarrow[AC\backslash R]{} u$ should imply that there exist v and w s.t. $s \xrightarrow[AC\backslash R]{}^{\star} v =_{AC} w \xleftarrow[AC\backslash R]{}^{\star} u$.

In fact, one may make any rewrite system R AC-coherent by adding to R the so-called *extended rules*, that is for each rule $f(s,t) \rightarrow r$ where f is AC, add the rule $f(f(s,t),x) \rightarrow f(r,x)$ where x is a new variable.

There is a variant of AC-extended rewriting that, in some sense, builds in the rewriting relation the extended rules, by considering flat terms.

Definition 3. The *flat-AC* rewrite relation [21] is defined on flat terms by $s \xrightarrow[l \to r/\overline{\mathrm{AC}}]{p} t$ iff there is a substitution σ s.t. either

- $s|_p \equiv \overline{l\sigma}$ and $t = \overline{s[r\sigma]_p}$ or
- $l = f(l_1, \ldots, l_n)$ where f is AC and for a new variable x, $s|_p \equiv \overline{f(l_1, \ldots, l_n, x)\sigma}$ and $t = \overline{s[f(r,x)\sigma]_p}$.

The second case is called *extended matching.* Note that as for extension rules *à la* Peterson-Stickel, this case is not necessary if l has already a linear variable as an immediate subterm, like $x + 0 \to x$.

This definition seems not to be very different from using AC-extended rewriting with always coherent systems, but in fact it is important because the notion of position in a term has changed a bit: with flat AC rewriting, it does not mean anything to rewrite a term like $a + (b + c)$ at position 2, it is a rewrite of $a+b+c$ at position Λ with extended matching. This change in the definition was shown quite powerful for example for AC-completion: the completeness proof is far much simpler [21].

The termination criterion described in the next section is based on detecting on which position a term can be rewritten indefinitely, so the notion of position discussed above is really important.

Flat-AC rewriting is always coherent. More generally, it is Church-Rosser if it is locally confluent and terminating. As shown by Jouannaud and Kirchner [17], this Church-Rosser property is valid only if class rewriting is proved terminating. Fortunately, the following proposition allows to check only termination of flat AC-rewriting.

Theorem 4. *Termination of flat AC rewriting is equivalent to termination of class rewriting.*

Proof. Assume that a term s is not SN w.r.t. $\to_{R/\mathrm{AC}}$. An infinite reduction of s starts with $s \xrightarrow[l \to r/\mathrm{AC}]{p} t$, that is $s =_{\mathrm{AC}} s'$ where $s'|_p = l\sigma$. But then it is easy to see that $\overline{s}$ also contains an (extended) instance of l, hence rewrites with $\to_{R/\overline{\mathrm{AC}}}$ to a flat term AC-equivalent to t. By repeating this construction we build an infinite reduction for $R/\overline{\mathrm{AC}}$. The reciprocal is trivial.

2.3 Term orderings and termination

For general notions about term orderings and termination, we refer to [14]. We briefly recall what we need.

Definition 5. A *quasi-ordering* on terms is any reflexive and transitive relation $\succeq$ on $T(\mathcal{F}, X)$. The *strict part* of $\succeq$ is defined by $s \succ t$ iff $s \succeq t$ and $t \not\succeq s$, and the *equivalence part* is defined by $s \simeq t$ iff $s \succeq t$ and $t \succeq s$. $\succeq$ is said to be *well-founded* if $\succ$ is Noetherian; that is no infinite decreasing sequence $t_1 \succ t_2 \succ t_3 \succ \cdots$ exists. $\succeq$ is said *weakly monotonic* if $s \succeq t$ implies $C[s] \succeq C[t]$

for any context C, and (strictly) *monotonic* if $s \succ t$ implies $C[s] \succ C[t]$ also. It is said *stable* if $s \succeq t$ implies $s\sigma \succeq t\sigma$ and $s \succ t$ implies $s\sigma \succ t\sigma$ for any substitution σ.

A (strict) *reduction ordering* is a well-founded, monotonic and stable quasi-ordering, it is a *weak* reduction ordering if it is only weakly monotonic.

A quasi-ordering is *AC-compatible* if $s \succ t$, $s =_{\mathrm{AC}} s'$ and $t =_{\mathrm{AC}} t'$ imply $s' \succ t'$.

The following well-known proposition gives a termination criterion for rewriting by a system R modulo AC by showing that all rules of R decrease w.r.t. an AC-compatible reduction ordering.

Proposition 6. *$\rightarrow_{R/\overline{AC}}$ terminates iff there is an AC-compatible reduction ordering $\succeq$ such that $l \succ r$ for all rules $l \rightarrow r \in R$.*

3 The dependency pairs termination criterion

3.1 The basic criterion

The key idea is to show that there are infinite reductions of a particular form, involving the so-called dependency pairs.

Definition 7. For a given rewrite rule $f(t_1, \ldots, t_n) \rightarrow r$, its set of *dependency pairs* is the set of pairs of terms of the form $\langle f(t_1, \ldots, t_n), g(u_1, \ldots, u_m)\rangle$ where $g(u_1, \ldots, u_m)$ is any subterm of r with a defined root symbol g. If f is AC, the set of *AC-extended* dependency pairs is the union of the set of standard dependency pairs and the set of dependency pairs of $f(t_1, \ldots, t_n, x) \rightarrow \overline{f(r, x)}$ where x is a new variable that does not occur in the rule. Note that flattening is necessary if $\mathrm{Head}(r) = f$ and that extended pair is not necessary if one of t_i is already a linear variable of the initial rule, that is if the rule does not need extended matching as remark in Definition 3.

An R-chain of dependency pairs is a sequence $\langle s_1, t_1\rangle, \langle s_2, t_2\rangle, \ldots$ of pairs of R such that for a substitution σ and for each i, $t_i\sigma \xrightarrow[R/\overline{AC}]{>\Lambda}{}^{\star} s_{i+1}\sigma$. (We consider only one substitution for all i because we can assume by renaming that sets of variables of two different pairs are disjoint.)

Example: Peàno arithmetic. Addition and multiplication on natural numbers presented by 0 and successor s can be defined by

$$x + 0 \rightarrow x \qquad\qquad x \times 0 \rightarrow 0$$
$$x + s(y) \rightarrow s(x + y) \qquad\qquad x \times s(y) \rightarrow x \times y + x$$

where $+$ and $\times$ are AC. The constructors are 0 and s so the dependency pairs of this TRS are $\langle x + s(y), x + y\rangle$, $\langle x + s(y) + z, x + y\rangle$ and $\langle x + s(y) + z, s(x + y) + z\rangle$ from the second rule; $\langle x \times 0 \times z, 0 \times z\rangle$ from the third; and $\langle x \times s(y), x \times y\rangle$, $\langle x \times s(y), (x \times y) + x\rangle$, $\langle x \times s(y) \times z, x \times y\rangle$, $\langle x \times s(y) \times z, ((x \times y) + x)\rangle$ and $\langle x \times s(y) \times z, ((x \times y) + x) \times z\rangle$ from the fourth.

Theorem 8. *For a given rewrite system R, the relation $\xrightarrow[R/\overline{AC}]{}$ terminates if and only if there is no infinite R-chain.*

We do not give the proof here due to lack of space, see [26]. It does not differ essentially from the one of Arts and Giesl, one simply has to take care where flattening may occur.

Corollary 9. *If there is an AC-compatible weak reduction ordering $\succeq$ such that $l \succeq r$ for each rule of R and $s \succ t$ for each dependency pair $\langle s,t\rangle$ of R, then $R/\overline{AC}$ is terminating.*

Note that weak monotonicity is enough because each dependency pair corresponds to a rewrite step without context in R-chains.

3.2 Dependency pairs criterion with marked symbols

Arts and Giesl have shown an enhancement of the criterion by marking, in R-chains, the top symbols of the non-SN subterms. Because of flattening, the definition of R-chains have to be modified a bit, and the reason for this modification will be shown by the counter-example below.

Definition 10. Let $\check{\mathcal{F}} = \mathcal{F} \cup \{\check{f}, f \text{ defined symbol}\}$ an extended signature where each defined symbol is copied with a mark. For a rule $f(t_1, \ldots, t_n) \to r$, its set of *dependency pairs with marks* is the set of pairs of terms in $T(\check{\mathcal{F}}, X)$ of the form $\langle \check{f}(t_1, \ldots, t_n), \check{g}(u_1, \ldots, u_m)\rangle$ where $g(u_1, \ldots, u_m)$ is any subterm of r with a defined root symbol g. The marked symbols only occur at root position. The set of marked *AC-extended* dependency pairs is defined accordingly.

Let $\check{R}$ be R augmented with rules $l \to \check{f}(\mathbf{t})$ for each R-rule of the form $l \to f(\mathbf{t})$ where f is a defined AC symbol and is not at root position in l. A *marked R-chain* of dependency pairs is a sequence $\langle s_1, t_1\rangle, \langle s_2, t_2\rangle, \ldots$ of pairs of R s.t. for a substitution σ and for each i, $t_i\sigma \xrightarrow[\check{R}/\overline{AC}]{>\Lambda}{}^{*} s_{i+1}\sigma$.

Proposition 11. *For a given rewrite system R, there are infinite marked R-chains iff there are infinite non-marked R-chains.*

Proof. Given a chain with marks, you can remove the marks and reciprocally, given a chain without marks, you can put marks, possibly introducing rewrite steps with rules in $\check{R}$.

Marked symbols will allow a larger class of orderings to be used in the following corollary.

Corollary 12. *If there is an AC-compatible weak reduction ordering $\succeq$ such that $l \succeq r$ for each rule of $\check{R}$ and $s \succ t$ for each marked dependency pair $\langle s,t\rangle$ of R, then $R/\overline{AC}$ is terminating.*

Counter-example (if we do not consider rules of $\check{R}$). The rewrite system $R = \{a \to b + c, d + b \to d + a\}$ is not terminating, because of the infinite sequence $d{+}a \to d{+}b{+}c \to d{+}a{+}c$ etc. which appears to be an infinite non-marked chain. However the correponding marked chain is $d\check{+}a \to d\check{+}b\check{+}c \to d\check{+}a\check{+}c$ involving a rewrite step $a \to b\check{+}c$.

One can see that in some sense, this modification is made because we need to "flatten" $d\check{+}(b + c)$ into $d\check{+}b\check{+}c$.

Example: Binary arithmetic. For computing with natural numbers, a binary notation is much more efficient than Peàno notation. Such a binary notation can be defined by a constant # and two unary postfixed functions (_)0 and (_)1. For example, 4 is written (((#)1)0)0 that is a # followed by the usual binary notation. (Semantically, # denotes 0, $(x)0$ denotes $2x$ and $(x)1$ denotes $2x + 1$.) Addition and multiplication can be defined by the following TRS:

$$
\begin{array}{lll}
(\#)0 \to \# & (x)0 + (y)0 \to (x + y)0 & x \times (y)0 \to (x \times y)0 \\
x + \# \to x & (x)0 + (y)1 \to (x + y)1 & x \times (y)1 \to x + (x \times y)0 \\
x \times \# \to \# & (x)1 + (y)1 \to (x + y + (\#)1)0 &
\end{array}
$$

where $+$ and $\times$ are AC. This TRS has 26 marked AC-extended dependency pairs, we only give here the dependency pairs of the third rule in the middle.

$$
\begin{array}{cc}
\langle (x)1\check{+}(y)1, x\check{+}y\check{+}(\#)1\rangle & \langle z\check{+}(x)1\check{+}(y)1, x\check{+}y\check{+}(\#)1\rangle \\
\langle (x)1\check{+}(y)1, (x + y + (\#)1)\check{0}\rangle & \langle z\check{+}(x)1\check{+}(y)1, (x + y + (\#)1)\check{0}\rangle \\
 & \langle z\check{+}(x)1\check{+}(y)1, z\check{+}(x + y + (\#)1)0\rangle
\end{array}
$$

When computing $2+3$, that is $((\#)1)0+((\#)1)1$, with this system we obtain an R-chain: $\langle (x)0\check{+}(y)1, x\check{+}y\rangle$ $\langle (x')1\check{+}(y')1, x'\check{+}y'\check{+}(\#)1\rangle$ with substitution $\{x \mapsto (\#)1, y \mapsto (\#)1, x' \mapsto \#, y' \mapsto \#\}$.

4 Dependency graphs

Dependency graphs allow one to not check that *all* dependency pairs strictly decrease. We now lift this refinement up to AC-TRS by defining AC-dependency graphs.

Definition 13. The *AC-dependency graph* of a rewrite system R is a graph having AC-extended dependency pairs as vertices and s.t. there is an edge from $\langle s_1, t_1\rangle$ to $\langle s_2, t_2\rangle$ iff there is a substitution σ that allows $t_1\sigma \xrightarrow[\check{R}/\overline{AC}]{>\Lambda}{}^{\star} s_2\sigma$.

We know that R terminates modulo AC iff there is no R-chain, but an R-chain corresponds to an infinite path in the graph. Since the graph is always finite, infinite paths depend on cycles. Hence, we just have to check that the pairs decrease in cycles.

Corollary 14. *Let R be an AC-TRS and $\mathcal{G}$ be its AC-dependency graph. If there is an AC-compatible weak reduction quasi-ordering $\succeq$ such that:*

– *for each rule* $l \to r \in \check{R}$, $l \succeq r$,
– *for each cycle* $\mathcal{C} \in \mathcal{G}$ *and each* $\langle s, t\rangle \in \mathcal{C}$, $s \succeq t$,
– *each cycle* $\mathcal{C} \in \mathcal{G}$ *contains a strictly decreasing pair,*

then $R/\overline{AC}$ *is terminating.*

Unfortunately, existence of a σ s.t. $t_1\sigma \xrightarrow[\check{R}/\overline{AC}]{>\Lambda}{}^{\star} s_2\sigma$ for some pairs $\langle s_1, t_1\rangle$ and $\langle s_2, t_2\rangle$ is undecidable in general. An approximation of the dependency graph must be computed in such a way that the criterion soundness is preserved. Again we adapt the approximation proposed by Arts and Giesl for standard rewriting.

Definition 15. For any term t, let $\textsc{cap}(t)$ be t where all subterms headed by defined symbols (or variables) have been replaced by new distinct variables. A term t is said *connectable* to a term t' if $\textsc{cap}(t)$ and t' unify modulo AC. The *approximated AC-dependency graph* of an AC-TRS R is a graph whose vertices are the AC-extended dependency pairs and s.t. there is an edge from $\langle s_1, t_1\rangle$ to $\langle s_2, t_2\rangle$ iff t_1 and s_2 are connectable.

Our aim now is to show that this approximated graph indeed contains the AC-dependency graph.

Lemma 16. *If* $s \xrightarrow[\check{R}/\overline{AC}]{}{}^{\star} t$ *then* $\textsc{cap}(s)$ *and* t *are AC-unifyable.*

Proof. By induction on the structure of s. If s is a variable or Head(s) is a defined symbol then $\textsc{cap}(s)$ is a variable, hence it unifies with t.

If Head(s) is a constructor, say c, then $s = c(s_1, \ldots, s_n)$, hence $\textsc{cap}(s) = \underline{c(\textsc{cap}(s_1), \ldots, \textsc{cap}(s_n))}$. Since c is a constructor no rewriting at root can occur, thus $t = c(t_1, \ldots, t_n)$ where for each i $s_i \xrightarrow[\check{R}/\overline{AC}]{}{}^{\star} t_i$, the flattening being necessary if some t_i has c as root symbol. By induction hypothesis $\textsc{cap}(s_i)$ AC-unifies with t_i for each i, but since all $\textsc{cap}(s_i)$ have pairwise distinct variables all these unifiers for each i can be combined into a single AC-unifier of s and t.

Corollary 17. *If there is a substitution* σ *s.t.* $s\sigma \xrightarrow[\check{R}/\overline{AC}]{>\Lambda}{}^{\star} t\sigma$ *for two terms* s *and* t, *then* s *is connectable to* t.

Example. Let $R = \{f(x, x) \to f(0, 1)\}$. If f is free then R terminates because the only dependency pair is $\langle \check{f}(x, x), \check{f}(0, 1)\rangle$ and $\textsc{cap}(\check{f}(0, 1)) = \check{f}(0, 1)$ does not unify with $\check{f}(x, x)$, hence there are no cycle at all in the dependency graph.

If f is AC, the situation is different since there is the extended pair $\langle \check{f}(x, x, y), \check{f}(0, 1, y)\rangle$ and $\textsc{cap}(\check{f}(0, 1, y)) = \check{f}(0, 1, y')$ AC-unifies with both left-hand sides of the two pairs. The dependency graph is then

$$\langle \check{f}(x, x), \check{f}(0, 1)\rangle \longrightarrow \langle \check{f}(x, x, y), \check{f}(0, 1, y)\rangle \circlearrowleft$$

In fact, one unifier of $\check{f}(0, 1, y')$ and $\check{f}(x, x, y)$ is $\{x \mapsto 0, y \mapsto 1, y' \mapsto 0\}$ and this substitution gives us a counter-example for termination: $f(0, 0, 1) \to f(0, 1, 1) \to f(0, 0, 1) \to \cdots$. We see on this example how the graph criterion may also give some diagnostic when it cannot conclude to termination.

Example: Peàno arithmetic continued. The shape of the AC-dependency graph is the following

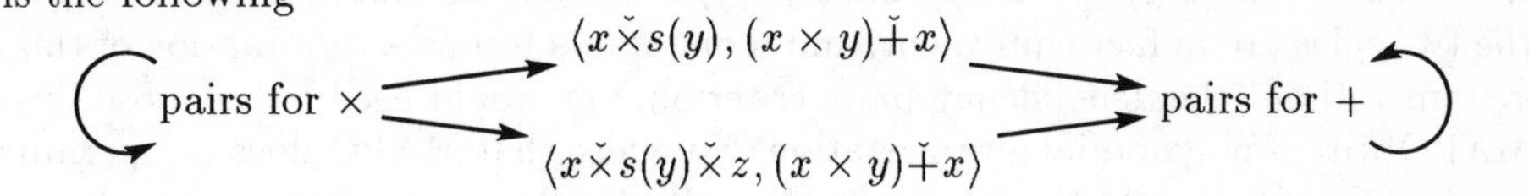

Thus, there is no need for an orientation for $\langle x \check{\times} s(y), (x \times y) \check{+} x \rangle$ and $\langle x \check{\times} s(y) \check{\times} z, (x \times y) \check{+} x \rangle$.

5 Building AC-compatible weak reduction orderings

As shown in the previous sections, proving termination via the dependency pair criterion amounts to building weak reduction orderings. Several constructions exist and here we are going to describe first some AC-compatible path orderings (which actually are always strictly monotonic); then we focus on how one can build weak orderings from polynomial interpretations; and finally we will show orderings built by *recursive program schemes* (which allow in particular building weak reduction orderings based on RPO).

5.1 Path orderings

We assume the reader familiar with the *Recursive Path Orderings with status* [13, 20]. We use this ordering on flat terms by requiring that AC symbols always have multiset status. But such an ordering is not always monotonic: we need high restrictions on the precedence, as shown by the following proposition.

Proposition 18. *If all AC-symbols are minimal in the precedence, then* $\succeq_{rpo}$ *is an AC-compatible simplification ordering.*

Associative path orderings are extensions of RPO that allow a more general precedence, they are built on RPO by comparing normal forms of terms w.r.t. some well-chosen rewrite systems [3,4,7,12]. We will use in this paper the *modified associative path ordering* defined by Delor and Puel [12]. More elaborated AC-compatible path orderings are known [12,18,23] but we don't need these powerful orderings in this paper.

Assume given a precedence s.t. for each AC symbol f, the precedence of symbols less than f has the form $f > u_1 > \cdots > u_k > g > a$ where the u_i are unary, g is AC and a is a constant (g and/or a may of course be absent). Let D be the following set of rules:

$$
\begin{array}{ll}
f(x, g(y, z)) \rightarrow g(f(x, y), f(x, z)) & u_i(g(x, y)) \rightarrow g(u_i(x), u_i(y)) \text{ for all } i \\
f(x, u_i(y)) \rightarrow u_i(f(x, y)) \text{ for all } i & u_i(u_j(x)) \rightarrow u_j(u_i(x)) \text{ for all } i > j
\end{array}
$$

Proposition 19 (Delor and Puel [12]). *D is AC-convergent. Let $s \succeq_{mapo} t$ iff $s\downarrow_D \succeq_{rpo} t\downarrow_D$. $\succeq_{mapo}$ is an AC-compatible simplification ordering, called* modified associative path ordering.

Example: Addition in Peàno arithmetic. The precedence $+ > s$ suffices to prove that $x+0 \succeq x$, $x+s(y) \succeq s(x+y)$ and $x\check{+}s(y) \succ x\check{+}y$. Note that in the second rule the two sides are in fact equivalent, and if one wants to prove termination of this system without the dependency pairs criterion, one would need to compose this MAPO with a polynomial interpretation. Note also that MAPO does not permit proving termination of the system with multiplication, because the precedence $\times > + > s$ is not allowed.

Remark that the orderings defined here are always strictly monotonic. In Section 5.3, we will see a construction aimed at making weak reduction orderings from them.

5.2 Weak polynomial interpretations

Definition 20. A polynomial interpretation is given by, for each symbol f of arity n, a polynomial $[\![f]\!]$ with integral coefficients and n variables. Such an interpretation defines an interpretation of any term with variables by

- $[\![x]\!] = x$ for any variable x;
- $[\![f(t_1, \ldots, t_n)]\!] = [\![f]\!]([\![t_1]\!], \ldots, [\![t_n]\!])$;

thus for any term t, $[\![t]\!]$ is a polynomial function of the variables of t.

A polynomial interpretation defines a strict ordering on terms by $s \succ t$ iff $[\![s]\!] >_\mu [\![t]\!]$ where μ is any integer and $>_\mu$ is the ordering on polynomial functions given by $P >_\mu Q$ iff for every $\mathbf{x} \geq \mu$, $P(\mathbf{x}) > Q(\mathbf{x})$ ($\mathbf{x}$ abbreviates $x_1, \ldots, x_n$ and $\mathbf{x} \geq \mu$ means $x_i \geq \mu$ for all i). We say $s \succeq t$ iff $[\![s]\!] \geq_\mu [\![t]\!]$ where $P \geq_\mu Q$ iff for every $\mathbf{x} \geq \mu$, $P(\mathbf{x}) \geq Q(\mathbf{x})$. (Be careful that $\succ$ is *not* the strict part of $\succeq$.)

Proposition 21. *If for each $f \in \mathcal{F}$ of arity n, $\mathbf{x} \geq \mu$ implies $[\![f]\!](\mathbf{x}) \geq \mu$ and $x_i \geq y_i$ for all i implies $[\![f]\!](\mathbf{x}) \geq [\![f]\!](\mathbf{y})$ then $\succeq$ is a weak reduction ordering.*

If for any AC symbol f, $[\![f]\!](x, y)$ has the form $axy + b(x + y) + c$ with $b^2 = b + ac$, then it is AC-compatible.

The first part of this proposition is almost straightforward and the second part is already known [8]. Note that several methods exist for checking whether a polynomial is greater than another w.r.t. $>_\mu$ [8, 15, 25].

Example: Peàno arithmetic continued. Since there is no need for an orientation for $\langle x\check{\times}s(y), (x \times y)\check{+}x\rangle$, the following simple interpretation does the trick ($\mu = 1$).

$$[\![\times]\!](x, y) = xy + x + y \qquad [\![\check{\times}]\!](x, y) = x + y$$
$$[\![+]\!](x, y) = x + y \qquad [\![\check{+}]\!](x, y) = x + y$$
$$[\![s]\!](x) = x + 1 \qquad [\![0]\!] = 1$$

Example: Binary arithmetic continued. There is no infinite chain because we can apply Corollary 9 with the polynomial interpretation ($\mu = 2$):

$$
\begin{array}{ll}
[\![\times]\!](x,y) = xy - x - y + 2 & [\![\check{\times}]\!](x,y) = xy + 2(x+y) + 2 \\
[\![+]\!](x,y) = x + y - 2 & [\![\check{+}]\!](x,y) = x + y - 1 \\
[\![(\_)0]\!](x) = x + 1 & [\![(\_)\check{0}]\!](x) = x + 1 \\
[\![(\_)1]\!](x) = x + 2 & [\![\#]\!] = 2
\end{array}
$$

Example: Ternary integral arithmetic. Contejean, Marché and Rabehasaina proposed the following system for addition and multiplication of integers in *balanced ternary notation* [11, 19] (semantically, $(x)0$, $(x)1$ and $(x)j$ denote respectively $3x$, $3x+1$ and $3x-1$):

$$
\begin{array}{rcl}
(\#)0 & \to & \# \\
x + \# & \to & x \\
(x)0 + (y)0 & \to & (x+y)0 \\
(x)0 + (y)1 & \to & (x+y)1 \\
(x)0 + (y)j & \to & (x+y)j \\
(x)1 + (y)j & \to & (x+y)0 \\
(x)1 + (y)1 & \to & (x + y + (\#)1)j \\
(x)j + (y)j & \to & (x + y + (\#)j)1
\end{array}
\qquad
\begin{array}{rcl}
x - y & \to & x + \mathrm{opp}(y) \\
\mathrm{opp}(\#) & \to & \# \\
\mathrm{opp}((x)0) & \to & (\mathrm{opp}(x))0 \\
\mathrm{opp}((x)1) & \to & (\mathrm{opp}(x))j \\
\mathrm{opp}((x)j) & \to & (\mathrm{opp}(x))1 \\
x \times \# & \to & \# \\
x \times (y)0 & \to & (x \times y)0 \\
x \times (y)1 & \to & x + (x \times y)0 \\
x \times (y)j & \to & (x \times y)0 + \mathrm{opp}(x)
\end{array}
$$

To prove this system terminates modulo AC they need a quite complicated ad-hoc demonstration, and no automatic proof of termination was known. With the dependency pairs criterion, termination can be checked automatically by the interpretation:

$$
\begin{array}{lll}
[\![\#]\!] = 2 & [\![\check{\times}]\!](x,y) = xy + 2(x+y) + 2 & [\![\check{-}]\!](x,y) = x + y + 2 \\
[\![(\_)0]\!](x) = x + 1 & [\![\check{+}]\!](x,y) = x + y - 1 & [\![\check{\mathrm{opp}}]\!](x) = x + 2 \\
[\![(\_)1]\!](x) = x + 2 & [\![\times]\!](x,y) = xy - (x+y) + 2 & [\![-]\!](x,y) = x + y \\
[\![(\_)j]\!](x) = x + 2 & [\![+]\!](x,y) = x + y - 2 & [\![\mathrm{opp}]\!](x) = x \\
[\![(\_)\check{0}]\!](x) = x + 1 & &
\end{array}
$$

5.3 Recursive program schemes

Definition 22. A rewrite system R is called a *recursive program scheme* (RPS for short) if:

- each rule is of the form $f(x_1, \ldots, x_n) \to r$ where x_i are pairwise distinct variables and r is any term;
- each defined symbol occurs at most once on left-hand sides;

It is an *AC-RPS* if for each defined AC symbol f, the rule for f has the form $f(x,y) \to r$ where $r =_{AC} r\{x \mapsto y, y \mapsto x\}$.

Note that it is easy to check whether a rewrite system of this form terminates: there should be neither recursive nor mutually recursive rules. Further note that we do not allow rules for AC symbols having more than 2 variables on the left-hand side, see the counter-example below.

Proposition 23. *Any terminating AC-RPS is AC-confluent.*

Proof. It suffices to check confluence of the AC-critical pairs. There are no superpositions with rules for free symbols. Let us consider an AC symbol f, a rule $f(x,y) \to r$ and a renaming $f(x',y') \to r\tau$ where $\tau = \{x \mapsto x', y \mapsto y'\}$. AC-unification of $f(x,y)$ and $f(x',y')$ yields 7 solutions, only two of them have to be considered since the others are reducible (unblocked criterion [5]): $\sigma_1 = \{x \mapsto x', y \mapsto y'\}$ and $\sigma_2 = \{x \mapsto y', y \mapsto x'\}$. Then $r\sigma_1 =_{AC} r\tau\sigma_1$ trivially and $r\sigma_2 =_{AC} r\tau\sigma_2$ by the hypothesis.

The following proposition is a reformulation of Theorem 5.3.3 of [1] in terms of weak reduction orderings, and generalized to the AC case.

Proposition 24. *Let P be a terminating AC-RPS and $\succeq$ be any AC-compatible weak reduction ordering. The relation $\succeq_P$ given by $s \succeq_P t$ iff $s{\downarrow_P} \succeq t{\downarrow_P}$ is an AC-compatible weak reduction ordering.*

Proof. It is obviously a well-founded quasi-ordering. In order to prove stability and monotonicity we need a lemma:

Lemma 25. *Let P be an AC-RPS. For each term t and each substitution σ: $(t\sigma){\downarrow_P} = (t{\downarrow_P})(\sigma{\downarrow_P})$.*

Proof. We have $(t\sigma){\downarrow_P} = (t{\downarrow_P})(\sigma{\downarrow_P}){\downarrow_P}$ thus we just have to prove $(t{\downarrow_P})(\sigma{\downarrow_P})$ is a normal form. Since the AC-symbol arity on left-hand sides is always 2, $(t{\downarrow_P})$ cannot contain any defined symbol, similarly with $x(\sigma{\downarrow_P})$ for all x. Hence $(t{\downarrow_P})(\sigma{\downarrow_P})$ contains no defined symbol so cannot be reduced, thus it is a normal form.

For stability now, if $s \succeq_P t$ then $s{\downarrow_P} \succeq t{\downarrow_P}$ hence $(s{\downarrow_P})(\sigma{\downarrow_P}) \succeq (t{\downarrow_P})(\sigma{\downarrow_P})$ since $\succeq$ is a reduction ordering. That is, by Lemma 25, $(s\sigma){\downarrow_P} \succeq (t\sigma){\downarrow_P}$ thus $s\sigma \succeq_P t\sigma$. Same process for strict part.

For monotonicity, if $s \succeq_P t$ then $s{\downarrow_P} \succeq t{\downarrow_P}$ hence $f(\ldots,x,\ldots){\downarrow_P}\{x \mapsto s{\downarrow_P}\} \succeq f(\ldots,x,\ldots){\downarrow_P}\{x \mapsto t{\downarrow_P}\}$ since $\succeq$ is a reduction ordering. By Lemma 25 we get $(f(\ldots,x,\ldots)\{x \mapsto s\}){\downarrow_P} \succeq (f(\ldots,x,\ldots)\{x \mapsto t\}){\downarrow_P}$ i.e. $f(\ldots,s,\ldots) \succeq_P f(\ldots,t,\ldots)$.

We now prove AC-compatibility. Let us suppose that $s \succeq_P t$ with $s =_{AC} s'$ and $t =_{AC} t'$. Since P is AC-confluent, we get $s{\downarrow_P} =_{AC} s'{\downarrow_P}$ and $t{\downarrow_P} =_{AC} t'{\downarrow_P}$, but $\succeq$ is AC-compatible then $s'{\downarrow_P} \succeq t'{\downarrow_P}$ that is $s' \succeq_P t'$.

Counter-example. If we allow RPS rules for AC symbols having more than 2 variables in the lhs, Proposition 24 is wrong: with the RPS $\{x + y + z \to a\}$ we would have $a + a \succeq_P a$ but when putting another a in the context we would get $a + a + a \to_P a \preceq a + a$, a contradiction to monotonicity.

Example. Let us consider the following system for computing intersection of multisets of numbers. Multisets are presented by the constant $\emptyset$, the unary symbol $\{\_\}$ that builds a singleton, and the union.

$$\begin{array}{lll} \emptyset \cup x \rightarrow x & eq(0,0) \rightarrow true & x \cap \emptyset \rightarrow \emptyset \\ if(true,x,y) \rightarrow x & eq(0,s(x)) \rightarrow false & x \cap (y \cup z) \rightarrow (x \cap y) \cup (x \cap z) \\ if(false,x,y) \rightarrow y & eq(s(x),s(y)) \rightarrow eq(x,y) & \{x\} \cap \{y\} \rightarrow if(eq(x,y),\{x\},\emptyset) \end{array}$$

where $\cup$ and $\cap$ are AC and eq is commutative. To prove termination of this system, the idea is to use MAPO with precedence $\cap > \cup$, $eq > false$ and $eq > true$ which will do the trick for all rules except the last one: we would like $\cap > if$ but it is not allowed in MAPO. The idea is then to say that in some sense, the first argument of *if* is not important, but only that $\{x\}$ and $\emptyset$ are smaller than the left-hand side. This is made evident when computing the dependency graph of this system, in which the only problem is to make the pair $\langle z \check{\cap} \{x\} \check{\cap} \{y\}, z \check{\cap}\, if(eq(x,y),\{x\},\emptyset)\rangle$ decrease. This can be done by the RPS $\{if(x,y,z) \rightarrow y \cap z, x \check{\cap} y \rightarrow x \cap y, x \check{\cup} y \rightarrow x \cup y\}$ which satisfies the AC-RPS conditions. Note that without the graph improvement, the basic criterion can also be used if we add in the RPS $\check{if}(x,y,z) \rightarrow y \cap z$, $\check{eq}(x,y) \rightarrow x \cap y$.

6 Dependency pairs criterion and theorem proving

An important question is how the dependency pairs criterion can be used in theorem proving. We explain in this section how it can be used in Knuth-Bendix completion or AC-completion, in which in the inference rule

$$\textbf{(Orient)}\ E \cup \{u = v\}, R \ \vdash\ E, R \cup \{u \rightarrow v\} \text{ if } u > v$$

we would like to replace $u > v$ by $u \geq v$ and $s > t$ for each $\langle s,t\rangle$ dependency pair of rule $u \rightarrow v$. For this to be correct, we have to make sure that the set of constructors or defined symbols does not change during completion, so we will assume that the set of constructors is fixed at once by the user, and any attempt to make a new rule with a constructor at root of left-hand side will raise a failure.

The main question is whether the *completeness* of the procedure is preserved that is: is the following theorem [6, 16] still true?

Theorem 26. *Let E_0 be the initial set of equations and $\succeq$ a weak reduction ordering. Let E_i (resp. R_i) be the set of equations (resp. rules) at completion step i, $E_\infty = \bigcap_n \bigcup_{i=n}^{\infty} E_i$, $R^\star = \bigcup_i R_i$, $R_\infty = \bigcap_n \bigcup_{i=n}^{\infty} R_i$.*

If completion succeeds ($E_\infty = \emptyset$) and is fair ($CP(R_\infty) \subseteq \bigcup_i E_i$), then $s =_{E_0} t$ iff $s\downarrow_{R_\infty} =_{AC} t\downarrow_{R_\infty}$.

The proof of this theorem uses in a fundamental way the strict orientation $l > r$ for each rule $l \rightarrow r \in R^\star$. Nevertheless there is a nice way of proving that completeness is preserved when using dependency pairs by reducing to the usual case: actually, if completion is successful we know that $\rightarrow_{R^\star}$ terminates by the dependency pairs criterion, hence is a reduction ordering by Proposition 6. But then we may suppose that completion was run with this ordering, and the same inferences would be computed, thus we obtain the same R_∞.

7 Conclusion and future work

The dependency pair criterion has been shown powerful for proving termination of rewrite systems, but several problems still have to be studied.

Firstly, if this criterion can be used in completion procedures, it is not clear at all how it can be used for other first-order theorem proving methods, where usually a *simplification* ordering is necessary.

Secondly, we have seen that the dependency pair criterion allows the use of a wider class of term orderings, but it means that it is getting more difficult to build systems that will try to *find* a suitable ordering, via RPS in particular. This question has been already studied by Arts [1] who proposed to try RPS of a certain form.

Thirdly, there is the question of termination of *modular* or *hierarchical* rewrite systems, which have been studied by Arts in the case of innermost strategy, and seems to be very promising since the hierarchy of rules corresponds to some hierarchy between parts of the dependency graphs. It could be interesting also to study *Normalized* rewriting [22] which is a special case of hierarchical rewriting.

Acknowledgments

We would like to thank Keiichirou Kusakari for his useful comments about a preliminary version of this paper.

References

1. T. Arts. *Automatically proving termination and innermost normalisation of term rewriting systems.* PhD thesis, Universiteit Utrecht, 1997.
2. T. Arts and J. Giesl. Automatically proving termination where simplification orderings fail. In M. Bidoit and M. Dauchet, editors, *Theory and Practice of Software Development*, volume 1214 of *Lecture Notes in Computer Science*, Lille, France, Apr. 1997. Springer-Verlag.
3. L. Bachmair. Associative-commutative reduction orderings. *Info Proc. Letters*, 1992.
4. L. Bachmair and N. Dershowitz. Commutation, transformation, and termination. In J. H. Siekmann, editor, *Proc. 8th Int. Conf. on Automated Deduction, Oxford, England, LNCS 230*, pages 5–20, July 1986.
5. L. Bachmair and N. Dershowitz. Critical pair criteria for completion. *Journal of Symbolic Computation*, 6(1):1–18, 1988.
6. L. Bachmair, N. Dershowitz, and J. Hsiang. Orderings for equational proofs. In *Proc. 1st IEEE Symp. Logic in Computer Science, Cambridge, Mass.*, pages 346–357, June 1986.
7. L. Bachmair and D. A. Plaisted. Termination orderings for associative-commutative rewriting systems. *Journal of Symbolic Computation*, 1(4):329–349, Dec. 1985.

8. A. Ben Cherifa and P. Lescanne. An actual implementation of a procedure that mechanically proves termination of rewriting systems based on inequalities between polynomial interpretations. In *Proc. 8th Int. Conf. on Automated Deduction, Oxford, England, LNCS 230*, pages 42–51. Springer-Verlag, July 1986.
9. H. Comon, editor. *8th International Conference on Rewriting Techniques and Applications*, volume 1232 of *Lecture Notes in Computer Science*, Barcelona, Spain, June 1997. Springer-Verlag.
10. E. Contejean and C. Marché. CiME: Completion Modulo *E*. In H. Ganzinger, editor, *7th International Conference on Rewriting Techniques and Applications*, volume 1103 of *Lecture Notes in Computer Science*, pages 416–419, New Brunswick, NJ, USA, July 1996. Springer-Verlag. System Description available at `http://www.lri.fr/~demons/cime.html`.
11. E. Contejean, C. Marché, and L. Rabehasaina. Rewrite systems for natural, integral, and rational arithmetic. In Comon [9].
12. C. Delor and L. Puel. Extension of the associative path ordering to a chain of associative-commutative symbols. In *Proc. 5th Rewriting Techniques and Applications, Montréal, LNCS 690*, pages 389–404, 1993.
13. N. Dershowitz. Orderings for term rewriting systems. *Theoretical Computer Science*, 17(3):279–301, Mar. 1982.
14. N. Dershowitz. Termination of rewriting. *Journal of Symbolic Computation*, 3(1):69–115, Feb. 1987.
15. J. Giesl. Generating polynomial orderings for termination proofs. In J. Hsiang, editor, *6th International Conference on Rewriting Techniques and Applications*, volume 914 of *Lecture Notes in Computer Science*, Kaiserslautern, Germany, Apr. 1995. Springer-Verlag.
16. G. Huet. A complete proof of correctness of the Knuth-Bendix completion algorithm. *J. Comput. Syst. Sci.*, 23:11–21, 1981.
17. J.-P. Jouannaud and H. Kirchner. Completion of a set of rules modulo a set of equations. *SIAM Journal on Computing*, 15(4):1155–1194, 1986.
18. D. Kapur and G. Sivakumar. A total, ground path ordering for proving termination of AC-rewrite systems. In Comon [9].
19. D. E. Knuth. *The art of computer programming*, volume 2. Addison-Wesley, 2nd edition, 1981.
20. P. Lescanne. Uniform termination of term rewriting systems with status. In *Proc. 9th CAAP*, Cambridge, 1984. Cambridge University Press.
21. C. Marché. *Réécriture modulo une théorie présentée par un système convergent et décidabilité des problèmes du mot dans certaines classes de théories équationnelles.* Thèse de doctorat, Université Paris-Sud, Orsay, France, Oct. 1993.
22. C. Marché. Normalized rewriting: an alternative to rewriting modulo a set of equations. *Journal of Symbolic Computation*, 21(3):253–288, 1996.
23. R. Nieuwenhuis and A. Rubio. A precedence-based total AC-compatible ordering. In C. Kirchner, editor, *Proc. 5th Rewriting Techniques and Applications, Montréal, LNCS 690*. Springer-Verlag, June 1993.
24. G. E. Peterson and M. E. Stickel. Complete sets of reductions for some equational theories. *J. ACM*, 28(2):233–264, Apr. 1981.
25. J. Steinbach. Proving polynomials positive. In R. Shyamasundar, editor, *Foundations of Software Technology and Theoretical Computer Science*, volume 652 of *Lecture Notes in Computer Science*, pages 191–202, New Delhi, India, Dec. 1992. Springer-Verlag.
26. X. Urbain. Preuves de terminaison à l'aide de paires de dépendances. Rapport de DEA, Université Paris-Sud, Orsay, France, 1997.

Termination Transformation by Tree Lifting Ordering

Takahito Aoto and Yoshihito Toyama

School of Information Science, JAIST
Tatsunokuchi, Ishikawa 923-12, Japan
{aoto, toyama}@jaist.ac.jp

Abstract. An extension of a modular termination result for term rewriting systems (TRSs, for short) by A. Middeldorp (1989) is presented. We intended to obtain this by adapting the dummy elimination transformation by M. C. F. Ferreira and H. Zantema (1995) under the presence of a non-collapsing non-duplicating terminating TRS whose function symbols are all to be eliminated. We propose a tree lifting ordering induced from a reduction order and a set $\mathcal{G}$ of function symbols, and use this ordering to transform a TRS $\mathcal{R}$ into $\mathcal{R}'$; termination of $\mathcal{R}'$ implies that of $\mathcal{R} \cup \mathcal{S}$ for any non-collapsing non-duplicating terminating TRS $\mathcal{S}$ whose function symbols are contained in $\mathcal{G}$, provided that for any $l \to r$ in $\mathcal{R}$ (1) the root symbol of r is in $\mathcal{G}$ whenever that of l is in $\mathcal{G}$; and (2) no variable appears directly below a symbol from $\mathcal{G}$ in l when $\mathcal{G}$ contains a constant. Because of conditions (1) and (2), our technique covers only a part of the dummy elimination technique; however, even when $\mathcal{S}$ is empty, there are cases that our technique has an advantage over the dummy elimination technique.

1 Introduction

Modular aspects of properties of term rewriting systems (TRSs, for short), in particular those of termination of TRSs, have been investigated in many papers. In this paper, we present a technique that is useful to infer termination of the union of two TRSs provided that one of them does not contain collapsing rules nor duplicating rules. It is worth mentioning that our result also admits the hierarchical combination (i.e. a combination such that defined symbols in one TRS may occur in the other TRS), for which a few results have been obtained; see [2][3][8][10].

We propose a *tree lifting ordering induced from a reduction order and a set $\mathcal{G}$ of function symbols*, and use this ordering to transform a TRS $\mathcal{R}$ into $\hat{\mathcal{R}}$; termination of $\hat{\mathcal{R}}$ implies that of $\mathcal{R} \cup \mathcal{S}$ for any non-collapsing non-duplicating terminating TRS $\mathcal{S}$ whose function symbols are contained in $\mathcal{G}$, provided that for any $l \to r \in \mathcal{R}$ (1) $\mathrm{root}(l) \in \mathcal{G}$ implies $\mathrm{root}(r) \in \mathcal{G}$ and (2) no variable appears directly below a symbol from $\mathcal{G}$ in l when $\mathcal{G}$ contains a constant. Here $\mathrm{root}(t)$ denotes the root symbol of a term t.

Our approach to infer termination of the union of two TRSs $\mathcal{R}$ and $\mathcal{S}$ is different from others in a point that we infer the termination of $\mathcal{R} \cup \mathcal{S}$ not directly from that of $\mathcal{R}$ and $\mathcal{S}$, but instead we first transform $\mathcal{R}$ to $\hat{\mathcal{R}}$ (depending on the set of function symbols appearing in $\mathcal{S}$) and infer the termination of $\mathcal{R} \cup \mathcal{S}$ from that of $\hat{\mathcal{R}}$ and $\mathcal{S}$. In particular, when the sets of function symbols appearing in $\mathcal{R}$ and $\mathcal{S}$ are disjoint, we can take $\mathcal{R}$ as $\hat{\mathcal{R}}$ and hence our result covers a modular termination result [9] by A. Middeldorp.

We intended to obtain our result by adapting the dummy elimination transformation [6] by M. C. F. Ferreira and H. Zantema under the presence of a non-collapsing non-duplicating terminating TRS whose function symbols are all to be eliminated. Indeed, under conditions (1) and (2), the dummy elimination transformation is a particular case of the transformation from $\mathcal{R}$ to $\hat{\mathcal{R}}$ above. Although it is naturally expected that the dummy elimination transformation is sound without these conditions, but this is not true under the presence of a non-empty $\mathcal{S}$; see Example 2 in Section 3. Thus, to admit non-empty $\mathcal{S}$ we require the conditions also on $\mathcal{R}$, and consequently our technique covers only a part of the dummy elimination technique. However, even when $\mathcal{S}$ is empty, there are cases that our technique has an advantage over the dummy elimination technique; see Example 5 in Section 4.

The rest of this paper is organized as follows. In Section 2, we fix notations on term rewriting and orderings used in this paper. In Section 3, we introduce a tree lifting ordering and show its properties. Also, our main theorem (Theorem 13) is established at the end of Section 3. In Section 4, we discuss our result in the light of related works.

2 Preliminaries

Let $\mathcal{V}$ be a set of countably infinite variables (denoted by $x, y, z, \ldots$) and $\mathcal{F}$ a set of arity-fixed function symbols (denoted by $f, g, h, \ldots$). Then the set of *terms* (denoted by $s, t, u, \ldots$) built from $\mathcal{F}$ and $\mathcal{V}$ is written as $\mathcal{T}(\mathcal{F}, \mathcal{V})$, or as just $\mathcal{T}$ when $\mathcal{F}$ is obvious from the context. $\mathcal{V}(t)$ and $\mathrm{Var}(t)$ are the set and the multiset of variables occurring in t, respectively. We denote by $\equiv$ the syntactical equality.

We assume familiarity with the context formalism. A term s is a *subterm* of a term t (denoted by $s \trianglelefteq t$) if $t \equiv C[s]$ for some context C with precisely one hole; it is a *proper* subterm (denoted by $s \triangleleft t$) if $s \not\equiv t$.

A *substitution* σ on $\mathcal{T}$ is a mapping from $\mathcal{V}$ to $\mathcal{T}$ with finite $\mathrm{dom}(\sigma) = \{x \in \mathcal{V} \mid \sigma(x) \not\equiv x\}$. A substitution σ is extended to a homomorphism from $\mathcal{T}$ to $\mathcal{T}$. For a substitution σ and a term t, we conventionally write $t\sigma$ instead of $\sigma(t)$. A substitution σ with $\mathrm{dom}(\sigma) = \{x_1 \ldots, x_n\}$ and $\sigma(x_i) = u_i$ for $i = 1, \ldots, n$ is also denoted by $\{x_1 \mapsto u_1, \ldots, x_n \mapsto u_n\}$.

Given a binary relation R on a set W, the reflexive transitive closure and transitive closure of R are denoted by R^* and R^+, respectively. A binary relation R on $\mathcal{T}$ is said to be *closed under contexts* if sRt implies $C[s]RC[t]$ for every terms s, t and every context C; it is *closed under substitutions* if sRt implies $(s\sigma)R(t\sigma)$ for every terms s, t and every substitution σ. A *rewrite relation* is a binary relation on $\mathcal{T}$ closed under contexts and substitutions. A *strict partial order* $>$ on a set W is a transitive irreflexive relation on W; it is *well-founded* if there is no infinite sequence $a_1 > a_2 > \cdots$ of elements from W. A *rewrite order* is a rewrite relation that is a strict partial order. A rewrite order is a *reduction order* if it is well-founded. Note that, for any reduction order $>$, $s > t$ implies $\mathcal{V}(s) \supseteq \mathcal{V}(t)$.

A pair $\langle l, r \rangle$ of terms is called a *rewrite rule*. We conventionally write $l \to r$ instead of $\langle l, r \rangle$. A rewrite rule $l \to r$ is *collapsing* if $r \in \mathcal{V}$; it is *duplicating* if $\mathrm{Var}(l) \not\supseteq \mathrm{Var}(r)$. A *term rewriting system* is a set of rewrite rules[1]. A TRS is said to be *non-collapsing* (*non-duplicating*) if it contains no collapsing (duplicating, respectively) rules. Given a TRS $\mathcal{R}$, we denote by $\to_{\mathcal{R}}$ the smallest rewrite relation on $\mathcal{T}$ that includes $\mathcal{R}$. A TRS $\mathcal{R}$ is *compatible* with a rewrite order $>$ if $l > r$ for any $l \to r \in \mathcal{R}$. A TRS $\mathcal{R}$ is said to be *terminating* if it is compatible with a reduction order.

We denote by $\mathrm{Mult}(W)$ the set of finite multisets of elements taken from a set W. We use $\in$ for the membership relation and $[\,]$ for the empty multiset. The *multiset sum* and the *multiset minus* are denoted by $+$ and $\setminus$, respectively; e.g. $[a,b,b,c] + [b,c,c,d] = [a,b,b,b,c,c,c,d]$, and $[a,b,b,c] \setminus [b,c,c,d] = [a,b]$. For a natural number n and a multiset M, $M * n$ stands for $\underbrace{M + \cdots + M}_{n\text{-times}}$. For $M \in \mathrm{Mult}(W)$ and $a \in W$, $\#(a, M)$ is the number of a in M, and $|M|$ the number of elements in M; e.g. $\#(a, [a,b,a,b,a,c]) = 3$, and $|[a,b,a,b,a,c]| = 6$.

A *quasi-order* $\gtrsim$ on a set W is a transitive reflexive relation on W. Given a quasi-order $\gtrsim$, the equivalence relation $\sim = (\gtrsim \cap \lesssim)$ is called the *equivalence relation of* $\gtrsim$, and the strict partial order $> = (\gtrsim \cap \not\lesssim)$ is called the *strict part of* $\gtrsim$. The quasi-order is well-founded if so is its strict part.

Given a quasi-order $\gtrsim$ on a set W, we denote by $\gtrsim^{mult}$ the *multiset quasi-ordering on* $\mathrm{Mult}(W)$ *induced from* $\gtrsim$; we denote by $\gtrsim\!\!\!\!/^{mult}$ its strict part; see e.g. [11].

We denote by $\mathrm{Tr}(W)$ the set of finite (unordered) trees whose nodes are labeled with elements taken from a set W. Formally, $a\langle\Gamma\rangle \in \mathrm{Tr}(W)$ if $a \in W$ and $\Gamma \in \mathrm{Mult}(\mathrm{Tr}(W))$. We abbreviate $a\langle[T_1, \ldots, T_n]\rangle$ to $a\langle T_1, \ldots, T_n\rangle$ and $a\langle[\,]\rangle$ to a.

Given a strict partial order $>$ on a set W, we denote by $>_{rpo}$ the *recursive path ordering on* $\mathrm{Tr}(W)$ *induced from* $>$; see e.g. [11].

3 Tree lifting ordering

In this section, let us fix sets $\mathcal{F}$, $\mathcal{G}$ of function symbols such that $\mathcal{G} \subseteq \mathcal{F}$, a constant symbol $\Diamond$ such that $\Diamond \notin \mathcal{F}$, and a reduction order $>$ on $\mathcal{T}(\mathcal{F} \backslash \mathcal{G} \cup \{\Diamond\}, \mathcal{V})$.

Definition 1. 1. The *root* of a term $t \in \mathcal{T}(\mathcal{F}, \mathcal{V})$ is defined by

$$\mathrm{root}(t) = \begin{cases} t & \text{if } t \in \mathcal{V}, \\ f & \text{if } t \equiv f(t_1, \ldots, t_n). \end{cases}$$

[1] We omit the usual restrictions on a TRS $\mathcal{R}$. For, they are implied by termination of $\mathcal{R}$.

2. Let $t \equiv C[t_1, \ldots, t_n]$ $(n \geq 0)$ be a term in $\mathcal{T}(\mathcal{F}, \mathcal{V})$ with $\mathrm{root}(t) \in \mathcal{G}$. We write $t \equiv C[\![t_1, \ldots, t_n]\!]$ if $C \in \mathcal{C}(\mathcal{G}, \emptyset)$ and $\mathrm{root}(t_1), \ldots, \mathrm{root}(t_n) \notin \mathcal{G}$.
3. Let $t \equiv C[t_1, \ldots, t_n]$ $(n \geq 0)$ be a term in $\mathcal{T}(\mathcal{F}, \mathcal{V})$ such that $C \in \mathcal{C}(\mathcal{F} \backslash \mathcal{G}, \mathcal{V})$ and $\mathrm{root}(t_1), \ldots, \mathrm{root}(t_n) \in \mathcal{G}$. Suppose that $t_i \equiv C_i[\![s_{i,1}, \ldots, s_{i,m_i}]\!]$ for $i = 1, \ldots, n$. We denote by $\mathrm{suc}(t)$ the multiset $[s_{1,1}, \ldots, s_{n,m_n}]$, and by $\mathrm{cap}(t)$ the term $C[\diamondsuit, \ldots, \diamondsuit]$.
4. The *rank* of a term $t \in \mathcal{T}(\mathcal{F}, \mathcal{V})$ is defined by

$$\mathrm{rank}(t) = \begin{cases} 1 & \text{if } \mathrm{suc}(t) = [\,], \\ 1 + \max\{\mathrm{rank}(t_i) \mid t_i \in \mathrm{suc}(t)\} & \text{otherwise.} \end{cases}$$

5. For any term $t \in \mathcal{T}(\mathcal{F}, \mathcal{V})$, put

$$\mathrm{tree}(t) = \begin{cases} \mathrm{cap}(t) & \text{if } \mathrm{suc}(t) = [\,], \\ \mathrm{cap}(t)\langle \mathrm{tree}(t_1), \ldots, \mathrm{tree}(t_n) \rangle & \\ \qquad \text{if } \mathrm{suc}(t) = [t_1, \ldots, t_n] \text{ with } n \geq 1, \end{cases}$$

$$\mathrm{nodes}(t) = \begin{cases} [\mathrm{cap}(t)] & \text{if } \mathrm{suc}(t) = [\,], \\ [\mathrm{cap}(t)] + \sum_{1 \leq i \leq n} \mathrm{nodes}(t_i) & \\ \qquad \text{if } \mathrm{suc}(t) = [t_1, \ldots, t_n] \text{ with } n \geq 1. \end{cases}$$

Definition 2. Let $\succ = (> \cup \rhd)^+$. We define binary relations $\sim_{tlo(\mathcal{G})}$ and $>_{tlo'(\mathcal{G})}$ on $\mathcal{T}(\mathcal{F}, \mathcal{V})$ by

$$s \sim_{tlo(\mathcal{G})} t \text{ if and only if } \mathrm{tree}(s) = \mathrm{tree}(t);$$
$$s >_{tlo'(\mathcal{G})} t \text{ if and only if } \mathrm{tree}(s) \succ_{rpo} \mathrm{tree}(t).$$

Clearly, $\sim_{tlo(\mathcal{G})}$ is an equivalence relation. Also, since $\succ$ is a well-founded strict partial order, so is $>_{tlo'(\mathcal{G})}$.

We put $\gtrsim_{tlo'(\mathcal{G})} = (>_{tlo'(\mathcal{G})} \cup \sim_{tlo(\mathcal{G})})$. It follows from $s \sim_{tlo(\mathcal{G})} s' >_{tlo'(\mathcal{G})} t' \sim_{tlo(\mathcal{G})} t$ implies $s >_{tlo'(\mathcal{G})} t$ that $\gtrsim_{tlo'(\mathcal{G})}$ is a quasi-order with $\sim_{tlo(\mathcal{G})}$ as its equivalence relation and $>_{tlo'(\mathcal{G})}$ as its strict part. Thus, alternatively, we can define $>_{tlo'(\mathcal{G})}$ as follows: $s >_{tlo'(\mathcal{G})} t$ if and only if either

TLO':1 $\mathrm{cap}(s)(> \cup \rhd)^+ \mathrm{cap}(t)$, and $s >_{tlo'(\mathcal{G})} v$ for any $v \in \mathrm{suc}(t)$,
TLO':2 $u \gtrsim_{tlo'(\mathcal{G})} t$ for some $u \in \mathrm{suc}(s)$, or
TLO':3 $\mathrm{cap}(s) \equiv \mathrm{cap}(t)$, and $\mathrm{suc}(s) \gtrsim\!\!\!\!/^{mult}_{tlo'(\mathcal{G})} \mathrm{suc}(t)$.

Here, $\gtrsim\!\!\!\!/^{mult}_{tlo'(\mathcal{G})}$ is the strict part of multiset quasi-ordering induced from $\gtrsim_{tlo'(\mathcal{G})}$. We will mainly use this definition in the proofs hereafter.

The following lemma is readily checked.

Lemma 3. *1.* $s >_{tlo'(\mathcal{G})} u$ *for any* $u \in \mathrm{suc}(s)$*;*
2. $s >_{tlo'(\mathcal{G})} a$ *for any* $a \in \mathrm{nodes}(s)$ *provided that* $|\mathrm{nodes}(s)| > 1$.

At first sight, it may seem strange to include the (proper) subterm relation "$\rhd$" in Definition 2 (and in **TLO':1**). But without "$\rhd$", we can not generally guarantee that the tree lifting ordering is closed under contexts; see Figure 1.

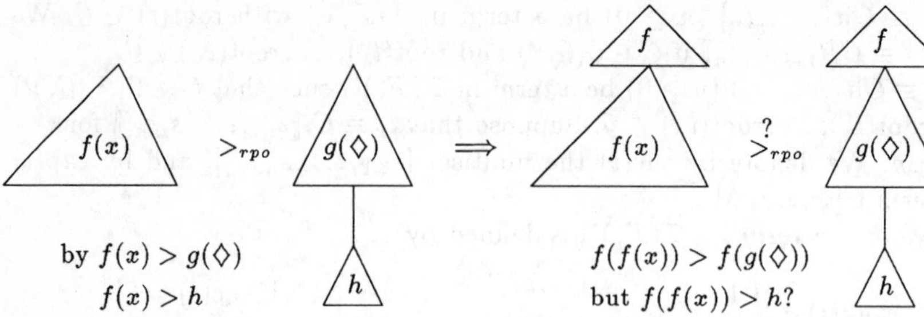

Fig. 1. Why is "$\triangleright$" required?

Definition 4. The *tree lifting ordering* $>_{tlo(\mathcal{G})}$ *on* $\mathcal{T}(\mathcal{F},\mathcal{V})$ *induced from* $>$ *and* $\mathcal{G}$ is defined as follows: $s >_{tlo(\mathcal{G})} t$ if and only if either

TLO:1 $\text{cap}(s) > \text{cap}(t)$, and $s >_{tlo'(\mathcal{G})} v$ for any $v \in \text{suc}(t)$, or
TLO:2 $\text{cap}(s) \equiv \text{cap}(t)$, and $\text{suc}(s) \gtrsim^{mult}_{tlo'(\mathcal{G})} \text{suc}(t)$.

It is clear that $>_{tlo(\mathcal{G})} \subseteq >_{tlo'(\mathcal{G})}$.

We are going to give here an explanation why $>_{tlo'(\mathcal{G})}$ itself is not suitable for our purpose and have introduced an alternative relation $>_{tlo(\mathcal{G})}$. The first difference is that **TLO':2** is omitted in $>_{tlo(\mathcal{G})}$ at the first stage of comparison. For, otherwise the tree lifting ordering can not be closed under contexts; see Figure 2. We also can not use " $\triangleright$ " at the first stage of comparison to make sure that the tree lifting ordering is closed under contexts; see Figure 3. Accordingly, we omit " $\triangleright$ " in **TLO:1**.

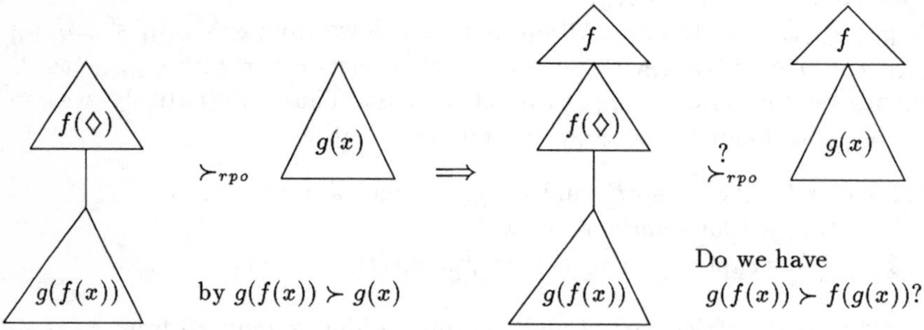

Fig. 2. Why is "$>_{tlo'(\mathcal{G})}$" not suitable? (1)

Lemma 5. *The relation* $>_{tlo(\mathcal{G})}$ *is transitive.*

Proof. The claim can be proved using the fact $>_{tlo(\mathcal{G})} \subseteq >_{tlo'(\mathcal{G})}$. □

Lemma 6. *The relation* $>_{tlo(\mathcal{G})}$ *is a well-founded strict partial order.*

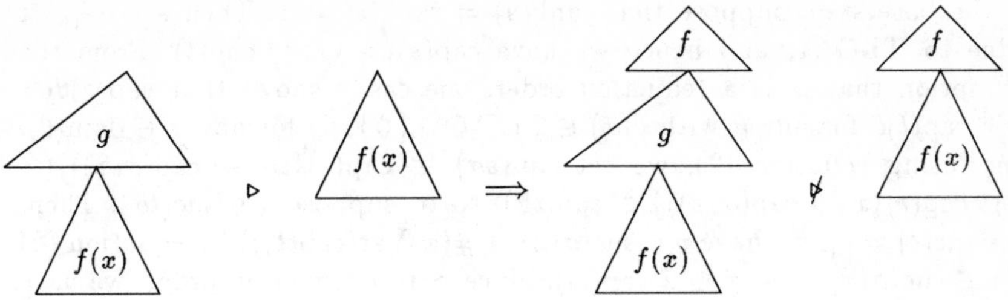

Fig. 3. Why is "$>_{tlo'(\mathcal{G})}$" not suitable? (2)

Proof. By Lemma 5, it suffices to show $>_{tlo(\mathcal{G})}$ is well-founded (because irreflexivity follows from well-foundedness). But this follows from $>_{tlo(\mathcal{G})} \subseteq >_{tlo'(\mathcal{G})}$ since $>_{tlo'(\mathcal{G})}$ is well-founded. □

Henceforth, $\sim_{tlo(\mathcal{G})} \cup >_{tlo(\mathcal{G})}$ is denoted by $\gtrsim_{tlo(\mathcal{G})}$. It is readily checked that $\gtrsim_{tlo(\mathcal{G})}$ is a quasi-order with $\sim_{tlo(\mathcal{G})}$ as its equivalence relation and $>_{tlo(\mathcal{G})}$ as its strict part.

Lemma 7. *For any substitution σ, if $s \sim_{tlo(\mathcal{G})} t$ then $s\sigma \sim_{tlo(\mathcal{G})} t\sigma$.*

Proof. The claim can be proved from the definition of $\sim_{tlo(\mathcal{G})}$ and the fact that $\mathrm{tree}(u) = x$ iff $u \equiv x$. □

Lemma 8. *Let θ be a substitution such that $\theta(x) \in \mathcal{T}(\mathcal{F}\backslash\mathcal{G}, \mathcal{V})$ for any $x \in \mathrm{dom}(\theta)$. Then $s >_{tlo'(\mathcal{G})} t$ implies $s\theta >_{tlo'(\mathcal{G})} t\theta$.*

Proof. One easily shows that $u(> \cup \triangleright)^+ v$ implies $u\theta(> \cup \triangleright)^+ v\theta$ for any $u, v \in \mathcal{T}(\mathcal{F}\backslash\mathcal{G} \cup \{\Diamond\}, \mathcal{V})$. Also, note that $\mathrm{cap}(w\theta) \equiv \mathrm{cap}(w)\theta$ and $\mathrm{suc}(w\theta) = [u\theta \mid u \in \mathrm{suc}(w)]$ for any term $w \in \mathcal{T}(\mathcal{F}, \mathcal{V})$. Now the statement can be proved by induction on $\mathrm{rank}(s) + \mathrm{rank}(t)$. □

Lemma 9. *Suppose $s >_{tlo'(\mathcal{G})} t$, and let σ be a substitution with $\mathrm{dom}(\sigma) = \{x\}$. Suppose $x \notin \mathrm{nodes}(s)$ when $\mathcal{G}$ contains a constant. Then $s\sigma \gtrsim_{tlo'(\mathcal{G})} t\sigma$; furthermore, $s\sigma >_{tlo'(\mathcal{G})} t\sigma$ if either $\mathrm{root}(\sigma(x)) \notin \mathcal{G}$, $\mathrm{root}(s) \notin \mathcal{G}$ or $t \not\equiv x$.*

Proof. We only consider here the case $\sigma(x) \notin \mathcal{V}$. Also, we can assume $\mathrm{root}(s) \notin \mathcal{V}$ (because otherwise there is no such t). We distinguish two cases:

1. $\mathrm{root}(\sigma(x)) \in \mathcal{F}\backslash\mathcal{G}$. Then we have

$$\mathrm{cap}(w\sigma) \equiv \mathrm{cap}(w)\{x \mapsto \mathrm{cap}(\sigma(x))\} \tag{1}$$
$$\mathrm{suc}(w\sigma) = [u\sigma \mid u \in \mathrm{suc}(w)] + \mathrm{suc}(\sigma(x)) * \#(x, \mathrm{Var}(\mathrm{cap}(w))) \tag{2}$$

for any term $w \in \mathcal{T}(\mathcal{F}, \mathcal{V})$.
We show that $s >_{tlo'(\mathcal{G})} t$ implies $s\sigma >_{tlo'(\mathcal{G})} t\sigma$, by induction on $\mathrm{rank}(s) + \mathrm{rank}(t)$. Suppose that $s >_{tlo'(\mathcal{G})} t$.

For the base step, suppose that $\text{rank}(s) = \text{rank}(t) = 1$. Then $s >_{tlo'(\mathcal{G})} t$ is due to **TLO':1**, and hence we have $\text{cap}(s)(> \cup \rhd)^+ \text{cap}(t)$. From the assumption that $>$ is a reduction order, one easily shows that $\text{cap}(s)\theta(> \cup \rhd)^+ \text{cap}(t)\theta$ for any θ with $\theta(x) \in \mathcal{T}(\mathcal{F}\backslash\mathcal{G} \cup \{\Diamond\}, \mathcal{V})$ for any $x \in \text{dom}(\theta)$. Thus, using equation (2), we get $\text{cap}(s\sigma) \equiv \text{cap}(s)\{x \mapsto \text{cap}(\sigma(x))\}(> \cup \rhd)^+ \text{cap}(t)\{x \mapsto \text{cap}(\sigma(x))\} \equiv \text{cap}(t\sigma)$. Next, suppose $v \in \text{suc}(t\sigma)$. Then, since $\text{suc}(t) = [\,]$, we have $v \in \text{suc}(\sigma(x)) * \#(x, \text{Var}(\text{cap}(t)))$ by equation (2) i.e. $v \in \text{suc}(\sigma(x))$ and $x \in \mathcal{V}(\text{cap}(t))$. Since $>$ is a reduction order, we have $\mathcal{V}(\text{cap}(t)) \subseteq \mathcal{V}(\text{cap}(s))$, and so $x \in \mathcal{V}(\text{cap}(s))$. Then, again using equation (2), we obtain $v \in \text{suc}(\sigma(x)) * \#(x, \text{Var}(\text{cap}(s))) = \text{suc}(s\sigma)$, and so $s\sigma >_{tlo'(\mathcal{G})} v$ by Lemma 3. Thus we have shown $s\sigma >_{tlo'(\mathcal{G})} v$ for any $v \in \text{suc}(t\sigma)$. Hence by **TLO':1** we conclude that $s\sigma >_{tlo'(\mathcal{G})} t\sigma$.

For the induction step, we distinguish three cases:

(a) $s >_{tlo'(\mathcal{G})} t$ by **TLO':1**. Then $\text{cap}(s)(> \cup \rhd)^+ \text{cap}(t)$ and $s >_{tlo'(\mathcal{G})} u$ for any $u \in \text{suc}(t)$. We have $\text{cap}(s\sigma)(> \cup \rhd)^+ \text{cap}(t\sigma)$ using equation (2). Let $v \in \text{suc}(t\sigma)$. To show $s\sigma >_{tlo'(\mathcal{G})} v$, we distinguish two cases according to equation (2):
   i. $v \equiv u\sigma$ for some $u \in \text{suc}(t)$. Since $\text{root}(\sigma(x)) \notin \mathcal{G}$, we can apply the induction hypothesis to $s >_{tlo'(\mathcal{G})} u$; thus, we obtain $s\sigma >_{tlo'(\mathcal{G})} u\sigma \equiv v$.
   ii. $v \in \text{suc}(\sigma(x)) * \#(x, \text{Var}(\text{cap}(t)))$. Then $x \in \mathcal{V}(\text{cap}(s))$, and so $v \in \text{suc}(s\sigma)$. Hence $s\sigma >_{tlo'(\mathcal{G})} v$ by Lemma 3.

   Thus, $s\sigma >_{tlo'(\mathcal{G})} t\sigma$ by **TLO':1**.

(b) $s >_{tlo'(\mathcal{G})} t$ by **TLO':2**. Then $u \gtrsim_{tlo'(\mathcal{G})} t$ for some $u \in \text{suc}(s)$. Then, by induction hypothesis and Lemma 7, $u\sigma \gtrsim_{tlo'(\mathcal{G})} t\sigma$. Since $\text{root}(\sigma(x)) \in \mathcal{F}\backslash\mathcal{G}$, we have $u\sigma \in \text{suc}(s\sigma)$, and so $s\sigma >_{tlo'(\mathcal{G})} t\sigma$ by **TLO':2**.

(c) $s >_{tlo'(\mathcal{G})} t$ by **TLO':3**. Then $\text{cap}(s) \equiv \text{cap}(t)$ and $\text{suc}(s) \mathrel{{>}\!\!\!\!{\not\sim}}^{mult}_{tlo'(\mathcal{G})} \text{suc}(t)$. We have $\text{cap}(s\sigma) \equiv \text{cap}(s)\{x \mapsto \text{cap}(\sigma(x))\} \equiv \text{cap}(t)\{x \mapsto \text{cap}(\sigma(x))\} \equiv \text{cap}(t\sigma)$. Also, by induction hypothesis and Lemma 7, we get $[u\sigma \mid u \in \text{suc}(s)] \mathrel{{>}\!\!\!\!{\not\sim}}^{mult}_{tlo'(\mathcal{G})} [v\sigma \mid v \in \text{suc}(t)]$. Thus,

$$\begin{aligned}\text{suc}(s\sigma) &= [u\sigma \mid u \in \text{suc}(s)] + \text{suc}(\sigma(x)) * \#(x, \text{Var}(\text{cap}(s)))\\ &= [u\sigma \mid u \in \text{suc}(s)] + \text{suc}(\sigma(x)) * \#(x, \text{Var}(\text{cap}(t)))\\ &\mathrel{{>}\!\!\!\!{\not\sim}}^{mult}_{tlo'(\mathcal{G})} [v\sigma \mid v \in \text{suc}(t)] + \text{suc}(\sigma(x)) * \#(x, \text{Var}(\text{cap}(t)))\\ &= \text{suc}(t\sigma).\end{aligned}$$

   Hence, $s\sigma >_{tlo'(\mathcal{G})} t\sigma$ by **TLO':3**.

2. $\text{root}(\sigma(x)) \in \mathcal{G}$. Then we have

$$\text{cap}(w\sigma) \equiv \text{cap}(w)\{x \mapsto \Diamond\} \tag{3}$$

$$\begin{aligned}\text{suc}(w\sigma) = {} & [u\sigma \mid u \in \text{suc}(w), u \not\equiv x]\\ & + \text{suc}(\sigma(x)) * \#(x, \text{Var}(\text{cap}(w)) + \text{suc}(w)).\end{aligned} \tag{4}$$

for any term $w \in \mathcal{T}(\mathcal{F}, \mathcal{V})$.

For the base step, suppose that $\text{rank}(s) = \text{rank}(t) = 1$. Then $s >_{tlo'(\mathcal{G})} t$ is due to **TLO':1**, and hence we have $\text{cap}(s)(> \cup \rhd)^+ \text{cap}(t)$. From this follows

$\text{cap}(s\sigma)(> \cup \triangleright)^+ \text{cap}(t\sigma)$ using equation (4). Furthermore, since $\text{suc}(t) = [\,]$ and $\mathcal{V}(\text{cap}(t)) \subseteq \mathcal{V}(\text{cap}(s))$, $v \in \text{suc}(t\sigma)$ implies $v \in \text{suc}(s\sigma)$ by equation (4). Hence, $s\sigma >_{tlo'(\mathcal{G})} v$ for any $v \in \text{suc}(t\sigma)$ by Lemma 3. Thus, $s\sigma >_{tlo'(\mathcal{G})} t\sigma$ by **TLO':1**.

For the induction step, we distinguish three cases:

(a) $s >_{tlo'(\mathcal{G})} t$ by **TLO':1**. Then $\text{cap}(s)(> \cup \triangleright)^+ \text{cap}(t)$ and $s >_{tlo'(\mathcal{G})} u$ for any $u \in \text{suc}(t)$. We have $\text{cap}(s\sigma)(> \cup \triangleright)^+ \text{cap}(t\sigma)$ using equation (4). Let $v \in \text{suc}(t\sigma)$. To show $s\sigma >_{tlo'(\mathcal{G})} v$, we distinguish three cases according to equation (4):
   i. $v \equiv u\sigma$ for some $u \in \text{suc}(t)$ with $u \not\equiv x$. Since $u \not\equiv x$, we can apply the induction hypothesis to $s >_{tlo'(\mathcal{G})} u$; thus, we obtain $s\sigma >_{tlo'(\mathcal{G})} u\sigma \equiv v$.
   ii. $v \in \text{suc}(\sigma(x)) * \#(x, \text{Var}(\text{cap}(t)))$. Then $x \in \mathcal{V}(\text{cap}(t))$, and so $x \in \mathcal{V}(\text{cap}(s))$ by $\text{cap}(s)(> \cup \triangleright)^+ \text{cap}(t)$, since $(> \cup \triangleright)^+$ is well-founded and closed under substitutions. Thus, $v \in \text{suc}(s\sigma)$ by equation (4). Hence $s\sigma >_{tlo'(\mathcal{G})} v$ by Lemma 3.
   iii. $v \in \text{suc}(\sigma(x)) * \#(x, \text{suc}(t))$. Since $x \in \text{nodes}(t)$, there exists $u \in \text{nodes}(s)$ such that $x \in \mathcal{V}(u)$. Thus $v \in \text{nodes}(s\sigma)$ and $|\text{nodes}(s\sigma)| > 1$, and hence $s\sigma >_{tlo'(\mathcal{G})} v$ by Lemma 3.

   Thus, $s\sigma >_{tlo'(\mathcal{G})} t\sigma$ by **TLO':1**.

(b) $s >_{tlo'(\mathcal{G})} t$ by **TLO':2**. Then $v \gtrsim_{tlo'(\mathcal{G})} t$ for some $v \in \text{suc}(s)$. By Lemma 7 and induction hypothesis, we get $v\sigma \gtrsim_{tlo'(\mathcal{G})} t\sigma$. If $v \not\equiv x$, then $v\sigma \in \text{suc}(s\sigma)$, and so $s\sigma >_{tlo'(\mathcal{G})} t\sigma$ by **TLO':2**. Suppose otherwise i.e. $v \equiv x$. Then, since $v \gtrsim_{tlo'(\mathcal{G})} t$, we have $t \equiv x$. Also, since $\text{suc}(s) \neq [\,]$, $\text{cap}(s) \trianglerighteq \Diamond$.
   i. $\text{root}(s) \in \mathcal{F}\backslash\mathcal{G}$. Then $\text{cap}(s\sigma) \not\equiv \Diamond$, and hence we have $\text{cap}(s\sigma) \triangleright \Diamond$. By $\text{suc}(\sigma(x)) = \text{suc}(t\sigma) \subseteq \text{suc}(s\sigma)$, we have $s\sigma >_{tlo'(\mathcal{G})} t\sigma$ by **TLO':1** and Lemma 3.
   ii. $\text{root}(s) \in \mathcal{G}$. Then $\text{cap}(s\sigma) \equiv \Diamond \equiv \text{cap}(t\sigma)$. By $\text{suc}(\sigma(x)) = \text{suc}(t\sigma) \subseteq \text{suc}(s\sigma)$, we have $s\sigma \gtrsim_{tlo'(\mathcal{G})} t\sigma$. (Note that in this case it is not necessary to have $s\sigma >_{tlo'(\mathcal{G})} t\sigma$ because the condition of the lemma is not satisfied.)

(c) $s >_{tlo'(\mathcal{G})} t$ by **TLO':3**. Then $\text{cap}(s) \equiv \text{cap}(t)$ and $\text{suc}(s) \gtrsim\!\!\!\!/^{mult}_{tlo'(\mathcal{G})} \text{suc}(t)$. From the former, we obtain $\text{cap}(s\sigma) \equiv \text{cap}(t\sigma)$ using equation (4). We are going to show $\text{suc}(s\sigma) \gtrsim\!\!\!\!/^{mult}_{tlo'(\mathcal{G})} \text{suc}(t\sigma)$. Since $\text{cap}(s) \equiv \text{cap}(t)$, we have

$$\text{suc}(\sigma(x)) * \#(x, \text{Var}(\text{cap}(s))) = \text{suc}(\sigma(x)) * \#(x, \text{Var}(\text{cap}(t))).$$

Thus, it remains to show

$$\begin{aligned} &[v\sigma \mid v \in \text{suc}(s), v \not\equiv x] + \text{suc}(\sigma(x)) * \#(x, \text{suc}(s)) \\ &\gtrsim\!\!\!\!/^{mult}_{tlo'(\mathcal{G})} [v\sigma \mid v \in \text{suc}(t), v \not\equiv x] + \text{suc}(\sigma(x)) * \#(x, \text{suc}(t)). \end{aligned}$$

Let $\|\text{suc}(w)\|$ be $[\text{tree}(u) \mid u \in \text{suc}(w)]$ for any $w \in \mathcal{T}(\mathcal{F}, \mathcal{V})$. When $\|\text{suc}(s)\| \not\supseteq \|\text{suc}(t)\|$, it suffices to show that for any $\text{tree}(w) \in \|\text{suc}(s)\| \backslash$

$\|\mathrm{suc}(t)\|$ and $\mathrm{tree}(v) \in \|\mathrm{suc}(t)\| \setminus \|\mathrm{suc}(s)\|$ such that $w >_{tlo'(\mathcal{G})} v$, (i) $w\sigma >_{tlo'(\mathcal{G})} v\sigma$ when $x \not\equiv v$, and (ii) $[w\sigma] \gtrsim^{mult}_{\not\sim tlo'(\mathcal{G})} \mathrm{suc}(v\sigma)$ when $x \equiv v$.

i. $x \not\equiv v$. Since $x \not\equiv v$, it follows from the induction hypothesis.

ii. $x \equiv v$. Since $\mathrm{root}(w) \notin \mathcal{G}$, $w\sigma >_{tlo'(\mathcal{G})} v\sigma$ by induction hypothesis; hence $[w\sigma] \gtrsim^{mult}_{\not\sim tlo'(\mathcal{G})} [v\sigma] \gtrsim^{mult}_{\not\sim tlo'(\mathcal{G})} \mathrm{suc}(v\sigma)$ using Lemma 3.

Otherwise, since $\mathrm{suc}(s) \gtrsim^{mult}_{\not\sim tlo'(\mathcal{G})} \mathrm{suc}(t)$, we have $\|\mathrm{suc}(s)\| \supsetneq \|\mathrm{suc}(t)\|$. Then, since $\mathrm{cap}(s\sigma) \equiv \mathrm{cap}(t\sigma)$, we clearly have $\|\mathrm{suc}(s\sigma)\| \supseteq \|\mathrm{suc}(t\sigma)\|$; hence, it remains to show $\|\mathrm{suc}(s\sigma)\| \setminus \|\mathrm{suc}(t\sigma)\| \neq [\ ]$. If there exists $\mathrm{tree}(v) \in \|\mathrm{suc}(s)\| \setminus \|\mathrm{suc}(t)\|$ with $v \not\equiv x$, then $\mathrm{tree}(v\sigma) \in \|\mathrm{suc}(s\sigma)\| \setminus \|\mathrm{suc}(t\sigma)\|$; otherwise, i.e. when $v \equiv x$ for any $\mathrm{tree}(v) \in \|\mathrm{suc}(s)\| \setminus \|\mathrm{suc}(t)\|$, we need to show $\mathrm{suc}(\sigma(x)) \neq [\ ]$. But if $x \notin \mathrm{nodes}(s)$ then there is no such case; and if $\mathcal{G}$ contains no constant then $\sigma(x)$ contains at least one non-$\mathcal{G}$-symbol, and so $\mathrm{suc}(\sigma(x)) \neq [\ ]$ holds. Thus, we conclude $\|\mathrm{suc}(s\sigma)\| \supsetneq \|\mathrm{suc}(t\sigma)\|$. Hence, $s\sigma >_{tlo'(\mathcal{G})} t\sigma$ by **TLO':3**.

□

Remark. The conditions in Lemma 9 can not be dropped: Suppose $\mathcal{G} = \{a, b\}$. Then $a(x) >_{tlo'(\mathcal{G})} x$ and $a(a(y)) \sim_{tlo(\mathcal{G})} a(y)$. If $\mathcal{G}$ additionally contains a constant c then $f(a(x,y)) >_{tlo'(\mathcal{G})} f(b(y))$ and $f(a(c,y)) \sim_{tlo(\mathcal{G})} f(b(y))$; also note that $x \in \mathrm{nodes}(f(a(x,y)))$.

Lemma 10. *Let σ be a substitution, and suppose that* $\mathrm{nodes}(s)$ *contains no variable when $\mathcal{G}$ contains a constant. Then, $s >_{tlo(\mathcal{G})} t$ implies $s\sigma >_{tlo(\mathcal{G})} t\sigma$.*

Proof. Similar to the proof of Lemma 9. (Replace "induction hypothesis" by "Lemma 9" and omit the cases 1(b) and 2(b).) □

Lemma 11. *For any $C \in \mathcal{C}(\mathcal{F}\backslash\mathcal{G}, \mathcal{V})$, if $s >_{tlo'(\mathcal{G})} t$ then $C[s] >_{tlo'(\mathcal{G})} t$.*

Proof. Let $C \in \mathcal{C}(\mathcal{F}\backslash\mathcal{G}, \mathcal{V})$ and suppose $s >_{tlo'(\mathcal{G})} t$. The case where $C \equiv \Box$ is trivial; so we assume $C \not\equiv \Box$. Then, $\mathrm{cap}(C[s]) \equiv C[\mathrm{cap}(s)](> \cup \rhd)^+ \mathrm{cap}(s)$. Also, since $\mathrm{suc}(C[s]) = \mathrm{suc}(s)$, $C[s] >_{tlo'(\mathcal{G})} v$ for any $v \in \mathrm{suc}(s)$ by Lemma 3. Thus, by **TLO':1** and our assumption, we get $C[s] >_{tlo'(\mathcal{G})} s >_{tlo'(\mathcal{G})} t$. □

Lemma 12. *If* $\mathrm{root}(s) \in \mathcal{G}$ *implies* $\mathrm{root}(t) \in \mathcal{G}$, *then, for any $C \in \mathcal{C}(\mathcal{F}, \mathcal{V})$, $s >_{tlo(\mathcal{G})} t$ implies $C[s] >_{tlo(\mathcal{G})} C[t]$.*

Proof. Let $w_1, \ldots, w_n \in \mathcal{T}(\mathcal{F}, \mathcal{V})$. We prove this by showing that $s >_{tlo(\mathcal{G})} t$ implies $e(w_1, \ldots, s, \ldots, w_n) >_{tlo(\mathcal{G})} e(w_1, \ldots, t, \ldots, w_n)$ for any $e \in \mathcal{F}$. Then by induction on the length of C, the statement of the lemma follows. Suppose $s >_{tlo(\mathcal{G})} t$, and let $\tilde{s} \equiv e(w_1, \ldots, s, \ldots, w_n)$ and $\tilde{t} \equiv e(w_1, \ldots, t, \ldots, w_n)$.

1. $e = f \in \mathcal{F}\backslash\mathcal{G}$. Then we have

$$\mathrm{cap}(f(w_1, \ldots, u, \ldots, w_n)) \equiv f(\mathrm{cap}(w_1), \ldots, \mathrm{cap}(u), \ldots, \mathrm{cap}(w_n)) \quad (5)$$

$$\mathrm{suc}(f(w_1, \ldots, u, \ldots, w_n)) = \mathrm{suc}(u) + \sum_{1 \leq i \leq n} \mathrm{suc}(w_i) \quad (6)$$

for any $u \in \mathcal{T}(\mathcal{F}, \mathcal{V})$. We distinguish two cases:

(a) $s >_{tlo(\mathcal{G})} t$ by **TLO:1**. Then $\mathrm{cap}(s) > \mathrm{cap}(t)$ and $s >_{tlo'(\mathcal{G})} v$ for any $v \in \mathrm{suc}(t)$. Since $>$ is a reduction order, we have $\mathrm{cap}(\tilde{s}) > \mathrm{cap}(\tilde{t})$ using equation (6). Suppose $v \in \mathrm{suc}(\tilde{t})$.
   i. $v \in \mathrm{suc}(t)$. Then we have $s >_{tlo'(\mathcal{G})} v$, and so, by Lemma 11, $f(z_1, \ldots, s, \ldots, z_n) >_{tlo'(\mathcal{G})} v$ with new variables $z_1, \ldots, z_n$. Let $\sigma_i = \{z_i \mapsto w_i\}$ $(i = 1, \ldots, n)$. Since $f \notin \mathcal{G}$ and $z_i \not\equiv v$ for all i, we can apply Lemma 9 to obtain $(\cdots(f(z_1, \ldots, s, \ldots, z_n)\sigma_1)\cdots)\sigma_n >_{tlo'(\mathcal{G})} (\cdots(v\sigma_1)\cdots)\sigma_n$. Thus, $\tilde{s} \equiv f(w_1, \ldots, s, \ldots, w_n) >_{tlo'(\mathcal{G})} v$.
   ii. $v \in \mathrm{suc}(w_i)$ for some i. Then $v \in \mathrm{suc}(\tilde{s})$ by equation (6), and so $\tilde{s} >_{tlo'(\mathcal{G})} v$ by Lemma 3.

(b) $s >_{tlo(\mathcal{G})} t$ by **TLO:2**. Then $\mathrm{cap}(s) \equiv \mathrm{cap}(t)$ and $\mathrm{suc}(s) \gtrsim^{mult}_{tlo'(\mathcal{G})} \mathrm{suc}(t)$. From the former $\mathrm{cap}(\tilde{s}) \equiv \mathrm{cap}(\tilde{t})$ follows using equation (6), and from the latter $\mathrm{suc}(\tilde{s}) \gtrsim^{mult}_{tlo'(\mathcal{G})} \mathrm{suc}(\tilde{t})$ using equation (6). Thus $\tilde{s} >_{tlo(\mathcal{G})} \tilde{t}$ by **TLO:2**.

2. $e = a \in \mathcal{G}$. Then we have

$$\mathrm{cap}(a(w_1, \ldots, u, \ldots, w_n)) \equiv \Diamond \tag{7}$$

$$\mathrm{suc}(a(w_1, \ldots, u, \ldots, w_n)) = [\delta \in [w_1, \ldots, u, \ldots, w_n] \mid \mathrm{root}(\delta) \notin \mathcal{G}] + \sum [\mathrm{suc}(\delta) \mid \delta \in [w_1, \ldots, u, \ldots, w_n] \text{ and } \mathrm{root}(\delta) \in \mathcal{G}] \tag{8}$$

for any $u \in \mathcal{T}(\mathcal{F}, \mathcal{V})$. By equation (8), it remains to show $\mathrm{suc}(\tilde{s}) \gtrsim^{mult}_{tlo'(\mathcal{G})} \mathrm{suc}(\tilde{t})$. By the condition, it suffices to distinguish three cases:

(a) $\mathrm{root}(s), \mathrm{root}(t) \notin \mathcal{G}$. Since $s >_{tlo(\mathcal{G})} t$, we have $s >_{tlo'(\mathcal{G})} t$, and so $\mathrm{suc}(\tilde{s}) \gtrsim^{mult}_{tlo'(\mathcal{G})} \mathrm{suc}(\tilde{t})$ by equation (8).

(b) $\mathrm{root}(s), \mathrm{root}(t) \in \mathcal{G}$. Then $s >_{tlo(\mathcal{G})} t$ is due to **TLO:2**, and so we have $\mathrm{suc}(s) \gtrsim^{mult}_{tlo'(\mathcal{G})} \mathrm{suc}(t)$. Hence $\mathrm{suc}(\tilde{s}) \gtrsim^{mult}_{tlo'(\mathcal{G})} \mathrm{suc}(\tilde{t})$ by equation (8).

(c) $\mathrm{root}(s) \notin \mathcal{G}$ and $\mathrm{root}(t) \in \mathcal{G}$. Since $s >_{tlo(\mathcal{G})} t$, we have $s >_{tlo'(\mathcal{G})} t$, and so $[s] \gtrsim^{mult}_{tlo'(\mathcal{G})} [t]$. Also, by Lemma 3, $t >_{tlo'(\mathcal{G})} u$ for any $u \in \mathrm{suc}(t)$, and so $[t] \gtrsim^{mult}_{tlo'(\mathcal{G})} \mathrm{suc}(t)$. Thus, we obtain $[s] \gtrsim^{mult}_{tlo'(\mathcal{G})} [t] \gtrsim^{mult}_{tlo'(\mathcal{G})} \mathrm{suc}(t)$, and hence $\mathrm{suc}(\tilde{s}) \gtrsim^{mult}_{tlo'(\mathcal{G})} \mathrm{suc}(\tilde{t})$ by equation (8).

□

Remark. The condition in Lemma 12 can not be dropped: Let $\mathcal{G} = \{a\}$ and suppose $\Diamond > 1 > 0$. Then $a(0,0) >_{tlo(\mathcal{G})} 1$ and $a(1,0) >_{tlo(\mathcal{G})} a(a(0,0), 0)$.

Theorem 13. *Let $\mathcal{G}$ be a set of function symbols and $\mathcal{R}$ a TRS such that*

1. $\mathrm{root}(l) \in \mathcal{G}$ *implies* $\mathrm{root}(r) \in \mathcal{G}$ *and*
2. *no variable appears directly below a symbol from $\mathcal{G}$ in l when $\mathcal{G}$ contains a constant*

for any $l \to r \in \mathcal{R}$. If there exists a reduction order $>$ on $\mathcal{T}(\mathcal{F} \backslash \mathcal{G} \cup \{\Diamond\}, \mathcal{V})$ such that $l >_{tlo(\mathcal{G})} r$ for any $l \to r \in \mathcal{R}$, then $\mathcal{S} \cup \mathcal{R}$ is terminating for any non-collapsing non-duplicating terminating TRS $\mathcal{S}$ whose function symbols are contained in $\mathcal{G}$.

Proof. We show that $s \to_{\mathcal{S}} t$ implies $s \gtrsim_{tlo(\mathcal{G})} t$. Suppose $s \to_{\mathcal{S}} t$, then $s \equiv C[l\sigma]$ and $t \equiv C[r\sigma]$ for some $l \to r \in \mathcal{S}$. Since $\mathcal{S}$ is non-collapsing and non-duplicating, $l \to r$ must be of the form $C_l[x_1, \ldots, x_n] \to C_r[x_{i_1}, \ldots, x_{i_m}]$ with some subsequence $i_1, \ldots, i_m$ $(m \leq n)$ of a permutation of $1, \ldots, n$ and non-empty contexts C_l, C_r. Thus, if $m = n$ then $\mathrm{tree}(s) = \mathrm{tree}(t)$; and if $m < n$ then $s >_{tlo(\mathcal{G})} t$, since $\mathrm{tree}(s) \succ_{rpo} \mathrm{tree}(t)$ and $\mathrm{cap}(s) \equiv \mathrm{cap}(t)$.

Suppose there exists an infinite reduction sequence of $\mathcal{S} \cup \mathcal{R}$. Since $\mathcal{S}$ is terminating, the sequence has the form $t_0 \xrightarrow{*}_{\mathcal{S}} s_0 \to_{\mathcal{R}} t_1 \xrightarrow{*}_{\mathcal{S}} s_1 \to_{\mathcal{R}} t_2 \xrightarrow{*}_{\mathcal{S}} s_2 \to_{\mathcal{R}} \cdots$. Then we have $t_0 \gtrsim_{tlo(\mathcal{G})} s_0 >_{tlo(\mathcal{G})} t_1 \gtrsim_{tlo(\mathcal{G})} s_1 >_{tlo(\mathcal{G})} t_2 \gtrsim_{tlo(\mathcal{G})} s_2 >_{tlo(\mathcal{G})} \cdots$ by Lemma 10, Lemma 12 and our assumptions; thus $t_0 >_{tlo(\mathcal{G})} t_1 >_{tlo(\mathcal{G})} t_2 >_{tlo(\mathcal{G})} \cdots$. This contradicts Lemma 6 that states $>_{tlo(\mathcal{G})}$ is well-founded. □

The following example addresses how to apply our result.

Example 1. Let

$$\begin{cases} f(f(x)) \to f(a(b(f(x)))) & (1) \\ f(a(g(x))) \to b(x) & (2) \\ b(x) \to a(x) & (3) \end{cases}$$

be a TRS. We show this TRS is terminating via our technique. For this, let $\mathcal{G} = \{a, b\}$. Since (3) is a non-duplicating non-collapsing rule consisting of $\mathcal{G}$-symbols, we let $\mathcal{R} = \{(1), (2)\}$, $\mathcal{S} = \{(3)\}$. Now, what we want is a reduction order $>$ satisfying

$$\begin{array}{ll} f(f(x)) >_{tlo(\mathcal{G})} f(a(b(f(x)))) & (4) \\ f(a(g(x))) >_{tlo(\mathcal{G})} b(x) & (5). \end{array}$$

We first compare $\mathrm{cap}(f(f(x))) \equiv f(f(x))$ and $f(\Diamond) \equiv \mathrm{cap}(f(a(b(f(x)))))$. Since $f(f(x)) \not\equiv f(\Diamond)$, in order to satisfy (4), we have to impose $f(f(x)) > f(\Diamond)$ to our ordering $>$. Thus, (4) should be due to **TLO:1**; and so we require $f(f(x)) >_{tlo'(\mathcal{G})} f(x)$, which immediately follows from **TLO':1**.

To satisfy (5), we first need $\mathrm{cap}(f(a(g(x)))) \equiv f(\Diamond) > \Diamond \equiv \mathrm{cap}(b(x))$. Thus, (5) should be due to **TLO:1**, and so we require $f(a(g(x))) >_{tlo'(\mathcal{G})} x$. For this, by **TLO':2**, it suffices to show $g(x) >_{tlo'(\mathcal{G})} x$, which immediately follows from **TLO':1**.

To conclude, we need a reduction order satisfying $f(f(x)) > f(\Diamond)$ and $f(\Diamond) > \Diamond$. But this can be obtained by taking $\hat{\mathcal{R}} = \{f(f(x)) \to f(\Diamond), f(\Diamond) \to \Diamond\}$ and $> = \xrightarrow{+}_{\hat{\mathcal{R}}}$, since $\hat{\mathcal{R}}$ is terminating.

Thus, we can apply Theorem 13, and we conclude the termination of the TRS $\{(1), (2), (3)\}$.

Remark. As the example above, we can conclude termination of a given TRS $\mathcal{R}$ by choosing a reduction order $> = \xrightarrow{+}_{\hat{\mathcal{R}}}$ for some suitable terminating TRS $\hat{\mathcal{R}}$. Further, such $\hat{\mathcal{R}}$ can be computed effectively (when $\mathcal{R}$ is finite); below we describe this taking a single rule TRS for an example.

Suppose now that we are requested to detect termination of a TRS $\mathcal{R} = \{l \to r\}$. It is clear that the number of choices of $\mathcal{G}$ is finite. Fix an arbitrary $\mathcal{G}$;

make tree(l) and tree(r). Let P and Q be the sets of nodes of tree(l) and tree(r), respectively. Since P and Q are finite, and so is the powerset $\mathcal{P}(P \times Q)$ of the direct product of P and Q. This set $\mathcal{P}(P \times Q)$ is the set of candidates for $\hat{\mathcal{R}}$. Indeed, if $\hat{\mathcal{R}} \notin \mathcal{P}(P \times Q)$ then we can alternatively take as $\hat{\mathcal{R}}$ the terminating TRS $\{l' \to r' \mid l' \xrightarrow{+}_{\hat{\mathcal{R}}} r', l' \in P, r' \in Q\} \in \mathcal{P}(P \times Q)$, which has the same effect as $\hat{\mathcal{R}}$ for our technique.

Further, if we are given with a decision procedure $\mathcal{D}$ which returns 'yes' for inputs of some terminating TRSs (e.g. a procedure which returns 'yes' for TRSs whose rules are all directed w.r.t. rpo induced from a given precedence), then we have a procedure which decides whether our technique is effective: For each member $\mathcal{R}'$ of $\mathcal{P}(P \times Q)$, check that whether $\mathcal{D}(\mathcal{R}')$ is 'yes.' If positive then check whether $l >_{tlo(\mathcal{G})} r$ for $> = \xrightarrow{+}_{\mathcal{R}'}$; this check can be done in a finite time when $\mathcal{R}'$ is terminating, because the set $\{t \mid s(> \cup \rhd)^+ t\}$ is finite for each s by König's Lemma. Thus, through this procedure, we can find $\hat{\mathcal{R}}$ $(= \mathcal{R}')$ if it exists.

The conditions in Theorem 13 can not be dropped:

Example 2. Let $\mathcal{G} = \{a, b\}$,

$$\mathcal{R}\left\{ a(x) \to f, \qquad\qquad \mathcal{S}\left\{ b(x,x) \to b(x, a(x)). \right.\right.$$

Note that $\mathcal{S}$ is a terminating non-collapsing non-duplicating TRS and that $\mathcal{R}$ satisfies condition 2 of the theorem but not condition 1. Now, let $\hat{\mathcal{R}} = \{\Diamond \to f\}$ and take $> = \xrightarrow{+}_{\hat{\mathcal{R}}}$. Clearly, $\hat{\mathcal{R}}$ is terminating, and so $>$ is a reduction order. Also, we have $a(x) >_{tlo(\mathcal{G})} f$. But $\mathcal{R} \cup \mathcal{S}$ is not terminating since $b(f,f) \to_{\mathcal{R}\cup\mathcal{S}} b(f, a(f)) \to_{\mathcal{R}\cup\mathcal{S}} b(f,f)$.

Example 3. Let $\mathcal{G} = \{a, b, c\}$,

$$\mathcal{R}\left\{ a(x) \to c, \qquad\qquad \mathcal{S}\left\{ b(c,x) \to b(a(x), a(c)). \right.\right.$$

Note that $\mathcal{S}$ is a terminating non-collapsing non-duplicating TRS and that $\mathcal{R}$ satisfies condition 1 of the theorem but not condition 2. Now, take $>$ as the empty order, which is a reduction order. Then, we have $a(x) >_{tlo(\mathcal{G})} c$. But $\mathcal{R} \cup \mathcal{S}$ is not terminating, since $b(c, a(c)) \to_{\mathcal{R}\cup\mathcal{S}} b(a(c), a(c)) \to_{\mathcal{R}\cup\mathcal{S}} b(c, a(c))$.

4 Related works

A modular termination result in [9] follows from our result.

Corollary 14. *Let $\mathcal{R}$, $\mathcal{S}$ be terminating TRSs, and suppose that $\mathcal{S}$ is non-collapsing non-duplicating and that the sets of function symbols that appear in $\mathcal{R}$ and $\mathcal{S}$ are disjoint. Then $\mathcal{R} \cup \mathcal{S}$ is also terminating.*

Proof. Let $\mathcal{G}$ be the set of function symbols that appear in $\mathcal{S}$. By assumption, no symbol from $\mathcal{G}$ appears in $\mathcal{R}$. Thus, by putting $> = \xrightarrow{+}_{\mathcal{R}}$, we obviously have $l >_{tlo(\mathcal{G})} r$ for all $l \to r \in \mathcal{R}$. Since l does not contain function symbols from $\mathcal{G}$, the conditions of Theorem 13 are trivially satisfied, and hence the corollary follows. □

Now we compare our result with hierarchical modular termination results [2][3] [8][10].

Example 4. Let

$$\mathcal{R}\begin{cases} f(f(x)) \to f(f(x)+f(x)) \\ g(0,x,f(x)) \to f(x+s(x)), \end{cases} \qquad \mathcal{S}\begin{cases} 0+0 \to 0 \\ s(x)+0 \to s(x) \\ x+s(y) \to s(x+y). \end{cases}$$

Then let $\mathcal{G} = \{+, s, 0\}$ and we can conclude $\mathcal{R} \cup \mathcal{S}$ is terminating by Theorem 13, since $\hat{\mathcal{R}} = \{f(f(x)) \to f(\diamondsuit), g(\diamondsuit, x, f(x)) \to f(\diamondsuit)\}$ is terminating. Neither hierarchical modularity results in [2] (Theorems 15-17), which require $\mathcal{R}$ to be left-linear, nor those in [2] (Theorems 19-21), [3] and [10], which can not have nested occurrences of f in the left-hand sides of rules in $\mathcal{R}$, apply to the combination of $\mathcal{R}$ and $\mathcal{S}$. In addition, $\mathcal{R}$ is not a proper extension of $\mathcal{S}$, and so [8] also does not apply.

Our work was inspired from the dummy elimination [6], which was further developed in [4][5]. Unfortunately, our result does not cover even the result in [6]. In our notation, the dummy elimination transformation $\mathcal{E}$ for a given a set $\mathcal{G}$ of function symbols to be eliminated can be defined as follows:

$$\mathcal{E}(\mathcal{R}) = \bigcup_{l \to r \in \mathcal{R}} \{\mathrm{cap}(l) \to s \mid s \in \mathrm{nodes}(r)\}.$$

(Note that $\mathrm{cap}(l)$ and $\mathrm{nodes}(r)$ depend on $\mathcal{G}$.)

Corollary 15. *Let $\mathcal{R}$ be a TRS and $\mathcal{G}$ the set of function symbols to be eliminated. Suppose that for any $l \to r \in \mathcal{R}$*

1. $\mathrm{root}(l) \in \mathcal{G}$ *implies* $\mathrm{root}(r) \in \mathcal{G}$, *and*
2. *no variable appears directly below a symbol from $\mathcal{G}$ in l when $\mathcal{G}$ contains a constant.*

Then, if $\mathcal{E}(\mathcal{R})$ is terminating then $\mathcal{R}$ is terminating.

Proof. Let $> = \xrightarrow{+}_{\mathcal{E}(\mathcal{R})}$. It is easy to see $l >_{tlo(\mathcal{G})} r$ for all $l \to r \in \mathcal{R}$, by applying **TLO:1** and **TLO':1**. □

Because of additional restrictions 1 and 2, our result covers only a part of dummy elimination technique. But the next example shows that even when $\mathcal{S} = \emptyset$, there are cases that our technique has an advantage over the (general) dummy elimination.

Example 5. Let $\mathcal{G} = \{a\}$ and

$$\mathcal{R}_1\begin{cases} f(f(x)) \to f(a(f(x))) \\ f(a(g(x))) \to a(x), \end{cases} \qquad \mathcal{R}_1'\begin{cases} f(f(x)) \to f(\diamondsuit) \\ f(\diamondsuit) \to \diamondsuit, \end{cases}$$

$$\mathcal{R}_2\,\{\, a(g(g(x))) \to a(g(a(g(x)))), \qquad \mathcal{R}_2'\,\{\, g(g(x)) \to g(\diamondsuit),$$

$$\mathcal{R}_3\,\{\, f(f(a(x))) \to f(a(f(a(x)))), \qquad \mathcal{R}_3'\,\{\, f(f(\diamondsuit)) \to f(\diamondsuit).$$

Termination of $\mathcal{R}_i$ can be shown using termination of $\mathcal{R}'_i$ ($i \in \{1,2,3\}$), respectively, while the (general) dummy elimination is not effective.

Similar to the dummy elimination technique, when $\mathcal{S} = \emptyset$ eliminating more than one symbol simultaneously just weaken the power of the technique; successive elimination of symbols works better; see [4]. But in Example 1 it is witnessed that there is a case that successive elimination of a and b is not effective, while the simultaneous elimination of a and b is.

It should be remarked here that a much simpler proof for the correctness of the dummy elimination transformation appeared in [10]. But it seems that their proof technique is not directly applicable to our case.

Very recently, M. C. F. Ferreira et al. showed in [7] that the dummy elimination is sound under the AC-theory for dummy symbols, i.e. termination of $\mathcal{H}(\mathcal{R})$ implies termination of $\mathcal{R}$ modulo $\mathcal{E}$ where $\mathcal{H}(\mathcal{R}) = \bigcup_{l \to r \in \mathcal{R}} \{\mathrm{cap}(l) \to s \mid s \in \mathrm{nodes}(r)\}$ and $\mathcal{E} = \{a(x,y) \approx a(y,x) \mid a \in \mathcal{G}\} \cup \{a(x,a(y,z)) \approx a(a(x,y),z) \mid a \in \mathcal{G}\}$.

Our technique can be applied in a similar way:

Theorem 16. *Let $\mathcal{G}$ be a set of function symbols and $\mathcal{R}$ a TRS such that*

1. $\mathrm{root}(l) \in \mathcal{G}$ *implies* $\mathrm{root}(r) \in \mathcal{G}$ *and*
2. *no variable appears directly below a symbol from $\mathcal{G}$ in l when $\mathcal{G}$ contains a constant*

for any $l \to r \in \mathcal{R}$. If there exists a reduction order $>$ on $\mathcal{T}(\mathcal{F} \backslash \mathcal{G} \cup \{\Diamond\}, \mathcal{V})$ such that $l >_{tlo(\mathcal{G})} r$ for any $l \to r \in \mathcal{R}$, then $\mathcal{R}$ is terminating modulo $\mathcal{E}$ for any non-collapsing variable-preserving equational system $\mathcal{E}$ whose function symbols are contained in $\mathcal{G}$.

5 Concluding remarks

We have proposed a tree lifting ordering $>_{tlo(\mathcal{G})}$ induced from a reduction order $>$ and a set $\mathcal{G}$ of function symbols. We have shown that for a given TRS $\mathcal{R}$ such that for any $l \to r \in \mathcal{R}$ (1) $\mathrm{root}(l) \in \mathcal{G}$ implies $\mathrm{root}(r) \in \mathcal{G}$ and (2) no variable appears directly below a symbol from $\mathcal{G}$ in l when $\mathcal{G}$ contains a constant, if $l >_{tlo(\mathcal{G})} r$ for any $l \to r \in \mathcal{R}$ then $\mathcal{R} \cup \mathcal{S}$ is terminating for any non-collapsing non-duplicating terminating TRS $\mathcal{S}$ whose function symbols are contained in $\mathcal{G}$. Furthermore, the reduction order $>$ that works on our technique is effectively obtained from $\mathcal{R}$ when $\mathcal{R}$ is finite.

Also, we have shown that our result covers a modular termination result [9] and that there are cases that our technique has an advantage over the (general) dummy elimination [6]([4]).

Acknowledgments

The authors are grateful to Maria C. F. Ferreira and Aart Middeldorp for detecting errors in the previous version [1] of the paper. Thanks are due to the RTA'98 referees for corrections, improvements and constructive comments.

This work is partially supported by Grants from Ministry of Education, Science and Culture of Japan, #09245212 and #07680347.

References

1. T. Aoto and Y. Toyama. Tree lifting orderings for termination transformations of term rewriting systems. Research Report IS-RR-97-0033F, School of Information Science, JAIST, 1997.
2. N. Dershowitz. Hierarchical termination. In *Proceedings of the 4th International Workshop on Conditional (and Typed) Rewriting Systems (CTRS-94), LNCS 968*, pages 89–105. Springer-Verlag, 1995.
3. M. Fernández and J.-P. Jouannaud. Modular termination of term rewriting systems revisited. In *Proceedings of the 10th Workshop on Specification of Abstract Data Types, LNCS 906*, pages 255–272. Springer-Verlag, 1995.
4. M. C. F. Ferreira. *Termination of term rewriting.* PhD thesis, Utrecht University, 1995.
5. M. C. F. Ferreira. Dummy elimination in equational rewriting. In *Proceedings of the 7th International Conference on Rewriting Techniques and Applications (RTA-96), LNCS 1103*, pages 78–92. Springer-Verlag, 1996.
6. M. C. F. Ferreira and H. Zantema. Dummy elimination: making termination easier. In *Proceedings of the 10th International Conference on Foundamentals of Computation Theory (FCT'95), LNCS 965*, pages 243–252. Springer-Verlag, 1995.
7. M. C. F. Ferreira, D. Kesner and L. Puel. Reducing AC-termination to termination. Manuscript, 1997.
8. M. R. K. Krishna Rao. Modular proofs for completeness of hierarchical term rewriting systems. *Theoretical Computer Science*, 151:487–512, 1995.
9. A. Middeldorp. A sufficient condition for the termination of the direct sum of term rewriting systems. In *Proceedings of the 4th annual IEEE Symposium on Logic in Computer Science (LICS'89)*, pages 396–401. IEEE Computer Society Press, 1989.
10. A. Middeldorp, H. Ohsaki and H. Zantema. Transforming termination by self-labelling. In *Proceedings of the 13th International Conference on Automated Deduction (CADE-13), LNAI 1104*, pages 373–387. Springer-Verlag, 1996.
11. A. Middeldorp and H. Zantema. Simple termination of rewrite systems. *Theoretical Computer Science*, 175(1):127–158, 1997.

Towards Automated Termination Proofs through "Freezing"

Hongwei Xi

Department of Mathematical Sciences
Carnegie Mellon University
Pittsburgh, PA 15213, USA

e-mail: hwxi+@cs.cmu.edu

Abstract. We present a transformation technique called *freezing* to facilitate automatic termination proofs for left-linear term rewriting systems. The significant merits of this technique lie in its simplicity, its amenability to automation and its effectiveness, especially, when combined with other well-known methods such as recursive path orderings and polynomial interpretations. We prove that applying the freezing technique to a left-linear term rewriting system always terminates. We also show that many interesting TRSs in the literature can be handled with the help of *freezing* while they elude a lot of other approaches aiming for generating termination proofs automatically for term rewriting systems. We have mechanically verified all the left-linear examples presented in this paper.

1 Introduction

It is an ever present task to decide whether a given term rewriting system(TRS) is (strongly) terminating. While this problem is not decidable in general [14], many approaches, such as path orderings [18, 5, 7], Knuth-Bendix ordering [15], semantic interpretations[17, 21, 12], transformation orderings[4, 20] and semantic labelling [22], have been developed to give termination proofs. See [6, 19] for surveys.

In this paper, we are primarily concerned with automatic termination proofs for TRSs. We propose a technique called *freezing*, which transforms a given left-linear TRS into a family of left-linear TRSs such that the termination of any TRS in this family implies the termination of the original TRS. The significant merits of the freezing technique lie in its simplicity, its amenability to automation and its effectiveness. Also we prove that the transformation terminates for all left-linear TRSs.

In practice, we know that most automatic termination proving methods rely on *simplification* orderings. [1] On the other hand, there exist numerous interesting (left-linear) TRSs whose termination cannot be proven by simplification

[1] We point out that automatic techniques have been developed, for instance in [20, 1], which can handle self-embedded rewrite rules.

orderings. With the help of the freezing technique, we are able to show that many among these TRSs can be transformed into those whose termination can be shown by some methods based on simplification orderings.

We present some preliminaries in Section 2. Then in Section 3 we illustrate the basic idea behind the freezing technique by a simple example. We formally introduce the freezing technique in Section 4, and prove the termination of the freezing technique when it is applied to a left-linear TRS. We then present some examples in Section 5 to illustrate the effectiveness of the freezing technique, compare our work with some related work in Section 6, and conclude.

2 Preliminaries

In general, we shall stick close to the notations in [9] though some modifications may occur. We assume that the reader is familiar with term rewriting. The following is a brief summary of the notations we shall use later.

We fix a countably infinite set of variables: $x, y, z, \ldots$, which is denoted by $\mathcal{X}$ throughout the paper. We use $\mathcal{F}$ for a (finite) set of function symbols: $f, g, h, F, G \ldots$, where each function symbol has a fixed arity $\mathcal{A}r$; $\mathcal{T}(\mathcal{F})$ for the set of (first-order) terms: $l, r, s, t, \ldots$ over $\mathcal{F}$ and $\mathcal{X}$; $\mathcal{V}ar(t)$ for the set of variables in t; $\mathbf{t}_{m,n}$ for a sequence of terms: $t_m, t_{m+1}, \ldots, t_n (m > n$ means this is an empty sequence); ρ, σ for substitutions, and ϵ for the empty substitution; $\langle \mathcal{F}, \mathcal{R} \rangle$ for a TRS, where $\mathcal{R}$ consists of the rewriting rules of the form $l \to r$ such that $l, r \in \mathcal{T}(\mathcal{F})$ and $\mathcal{V}ar(r) \subseteq \mathcal{V}ar(l)$; $\langle \mathcal{F}, \mathcal{R} \rangle$ is a left-linear rewrite system if for every rewrite rule $l \to r$ in $\mathcal{R}$ there exists no variable which occurs more than once in l. Note that we may use f, g, h for some specific function symbols in certain parts of our presentation. As a consequence, we use F, G ranging over $\mathcal{F}$ when this happens, avoiding possible confusion.

We feel that it is convenient to use the notion of *contexts* in reasoning, and we present a definition. See [8] for some similar use of contexts.

Definition 1. Contexts C are defined as follows.

1. $[]$ is a context, and
2. $F(t_1, \ldots, t_{i-1}, C, t_{i+1}, \ldots, t_n)$ is a context if F is a function symbol with arity $\mathcal{A}r(F) = n \geq 1$ and $t_1, \ldots, t_{i-1}, t_{i+1}, \ldots, t_n$ are terms and C is a context.

$C[t]$ is the term obtained from replacing the "hole" $[]$ in C with term t, and $C[C']$ is the context obtained from replacing the "hole" $[]$ in C with context C'.

Given a TRS $\langle \mathcal{F}, \mathcal{R} \rangle$, we write $t \to_{\mathcal{R}} t'$ for $t, t' \in \mathcal{T}(\mathcal{F})$ if $t = C[l\sigma]$ and $t' = C[r\sigma]$ for some C and σ, where $l \to r \in \mathcal{R}$. If $t \to_{\mathcal{R}} t'$, we say t rewrites to t' (in one step). We use $\to^+_{\mathcal{R}}$ for the transitive closure of $\to_{\mathcal{R}}$ and $\to^*_{\mathcal{R}}$ for the reflexive and transitive closure of $\to_{\mathcal{R}}$. A notation of the following form stands for a finite or infinite $\to_{\mathcal{R}}$-rewriting sequence (from t_0):

$$t_0 \to_{\mathcal{R}} t_1 \to_{\mathcal{R}} t_2 \to_{\mathcal{R}} \cdots$$

We say $\mathcal{R}$ or $\to_{\mathcal{R}}$ is terminating if there exists no infinite $\to_{\mathcal{R}}$-rewriting sequence.

3 The Basic Idea

It is a well-known technique to prove the termination of a TRS via "simulation", as formally shown below.

Definition 2. Let $\langle \mathcal{F}, \mathcal{R} \rangle$ and $\langle \mathcal{F}_1, \mathcal{R}_1 \rangle$ be two TRSs such that $\mathcal{F} \subseteq \mathcal{F}_1$, and $\sim \subseteq \mathcal{T}(\mathcal{F}) \times \mathcal{T}(\mathcal{F}_1)$ be a relation satisfying $t \sim t$ for all $t \in \mathcal{T}(\mathcal{F})$. We write $\mathcal{R} \stackrel{\sim}{\Longrightarrow} \mathcal{R}_1$ if for any given $t \sim t_1$ and $t \to_{\mathcal{R}} t'$ there exists t'_1 such that $t_1 \to^+_{\mathcal{R}_1} t'_1$ and $t' \sim t'_1$.

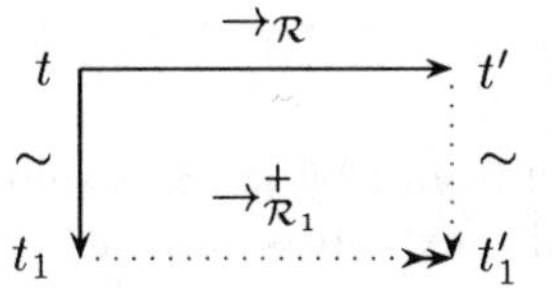

The need of $t \sim t$ for all $t \in \mathcal{T}(\mathcal{F})$ is to *start* the simulation as shown in the proof of next lemma.

Lemma 3. *If $\mathcal{R} \stackrel{\sim}{\Longrightarrow} \mathcal{R}_1$, then $\to_{\mathcal{R}}$ is terminating if $\to_{\mathcal{R}_1}$ is terminating.*

Proof. Assume that there exists an infinite $\to_{\mathcal{R}}$ rewriting sequence as follows.

$$t = t^{(0)} \to_{\mathcal{R}} t^{(1)} \to_{\mathcal{R}} t^{(2)} \to_{\mathcal{R}} t^{(3)} \to_{\mathcal{R}} \cdots$$

We can then construct an infinite $\to_{\mathcal{R}_1}$ rewriting sequence as shown in the diagram below.

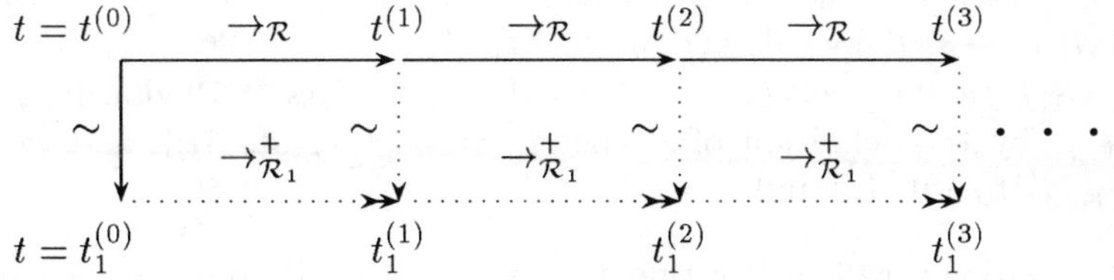

Therefore, there exists no infinite $\to_{\mathcal{R}}$ rewriting sequence since $\to_{\mathcal{R}_1}$ is terminating, i.e., $\to_{\mathcal{R}}$ is also terminating.

There exist many variations of the above approach. We use this formulation since it suffices for the development of our technique.

Note that Lemma 3 can yield an approach to termination proofs as follows. Suppose we want to prove that $\to_{\mathcal{R}}$ is terminating for some given TRS $\langle \mathcal{F}, \mathcal{R} \rangle$. We first construct another TRS $\langle \mathcal{F}_1, \mathcal{R}_1 \rangle$ with $\mathcal{F} \subseteq \mathcal{F}_1$ and a relation $\sim \subseteq \mathcal{T}(\mathcal{F}) \times \mathcal{T}(\mathcal{F}_1)$ satisfying $t \sim t$ for all $t \in \mathcal{T}(\mathcal{F})$, and prove $\mathcal{R} \stackrel{\sim}{\Longrightarrow} \mathcal{R}_1$. We then prove that $\to_{\mathcal{R}_1}$ is terminating. This yields that $\to_{\mathcal{R}}$ is terminating by Lemma 3. Clearly, we have to be able to construct $\mathcal{R}_1$ and $\sim$ in some way so that a termination proof for $\mathcal{R}_1$ can be given "more easily" than for $\mathcal{R}$, and this is the main subject of the paper. We now present a simple example before going into further details.

Example 1. Let $\mathcal{F} = \{f, g\}$ and $\mathcal{R}$ consist of the following rule[6].

$$f(f(x)) \to f(g(f(x)))$$

Clearly, it cannot be proven with simplification orderings that $\to_{\mathcal{R}}$ is terminating since the left-hand side of the rule is self-embedded in the right-hand side. Now we introduce the notion of *freezing*. Let $\underline{fg}$ be a new unary function symbol, and we define $\sim$ as follows.

$$\begin{array}{ll} x \sim x & \text{for all } x \in \mathcal{X}; \\ f(t) \sim f(t_1), g(t) \sim g(t_1), \text{ and } f(g(t)) \sim \underline{fg}(t_1) & \text{if } t \sim t_1; \end{array}$$

In other words, if t_1 is obtained from t by *freezing* some occurrences of $f(g(\cdots))$ into $\underline{fg}(\cdots)$, then $t \sim t_1$. We also extend $\sim$ to contexts as follows.

$$\begin{array}{ll} \Box \sim \Box & ; \\ f(C) \sim f(C_1), g(C) \sim g(C_1), \text{ and } f(g(C)) \sim \underline{fg}(C_1) & \text{if } C \sim C_1; \end{array}$$

Clearly, $C \sim C_1$ and $t \sim t_1$ implies $C[t] \sim C_1[t_1]$. Let $\mathcal{R}_1$ be the TRS consisting of the following rules, which can be generated automatically as shown in Section 4.

$$(1).\quad f(f(x)) \to \underline{fg}(f(x)) \qquad\qquad (2).\quad f(\underline{fg}(x)) \to \underline{fg}(f(g(x)))$$

Assume $t \sim t_1$ and $t \to_{\mathcal{R}} t'$. Then t is of the form $C[f(f(s))]$ and t' of the form $C[f(g(f(s)))]$. We do a case analysis on the form of t_1, showing $\mathcal{R} \overset{\sim}{\Longrightarrow} \mathcal{R}_1$.

- $t_1 = C_1[f(f(s_1))]$, where $C \sim C_1$ and $s \sim s_1$. Let $t_1' = C_1[\underline{fg}(f(s_1))]$, and we have $t_1 \to_{\mathcal{R}_1} t_1'$ by rule (1) and $t' \sim t_1'$.
- $t_1 = C_1[f(\underline{fg}(s_1))]$, where $C \sim C_1$ and $f(s) \sim \underline{fg}(s_1)$. Obviously, $f(s) \sim f(g(s_1))$ by the definition of $\sim$. Let $t_1' = C_1[\underline{fg}(f(g(s_1)))]$, and we have $t_1 \to_{\mathcal{R}_1} t_1'$ by rule (2) and $t' \sim t_1'$.

Later, Proposition 6 will justify that the case analysis is complete. Therefore, $\mathcal{R} \overset{\sim}{\Longrightarrow} \mathcal{R}_1$. Note that a termination proof for $\mathcal{R}_1$ can be given using rpo with the precedence: $f \succ \underline{fg} \succ g$. By Lemma 3, $\to_{\mathcal{R}}$ is terminating.

We now make an important observation. In the proof of $\mathcal{R} \overset{\sim}{\Longrightarrow} \mathcal{R}_1$, if we replace $\mathcal{R}_1$ with any $\mathcal{R}_*$ consisting of the following rules

$$f(f(x)) \to t_1 \qquad\qquad f(\underline{fg}(x)) \to t_2,$$

where t_1 and t_2 are any terms satisfying $f(g(f(x))) \sim t_1$ and $f(g(f(g(x)))) \sim t_2$, then $\mathcal{R} \overset{\sim}{\Longrightarrow} \mathcal{R}_*$ can be proven similarly. This means that we can construct a family of $\mathcal{R}_*$ such that $\mathcal{R} \overset{\sim}{\Longrightarrow} \mathcal{R}_*$ can be proven uniformly, and therefore the termination of any $\mathcal{R}_*$ in this family implies the termination of $\mathcal{R}$. Now our objective is to generate this family of TRSs automatically.

4 The Freezing Technique

In this section, we first give a formal presentation of the freezing technique for left-linear TRSs, and prove the termination of this technique. We also show with an example that the termination of the freezing technique can no longer be guaranteed if it is extended directly to a non-left linear rewriting system.

4.1 Left-linear TRSs

We fix a left-linear TRS $\langle \mathcal{F}, \mathcal{R} \rangle$ such that there exists a function symbol $f \in \mathcal{F}$ with arity $\mathcal{A}r(f) = n_f > 0$. Let g be some function symbol in $\mathcal{F}$ with arity $\mathcal{A}r(g) = n_g$. We may choose g to be the same as f. Also let $\mathcal{F}_* = \mathcal{F} \cup \{\underline{fg}\}$, where $\underline{fg}$ is a new function symbol with arity $\mathcal{A}r(\underline{fg}) = n_f + n_g - 1$.

Definition 4. Let m be a natural number between 1 and n_f. The $(f, g, m, \underline{fg})$-freezing relation $\sim \subseteq \mathcal{T}(\mathcal{F}) \times \mathcal{T}(\mathcal{F}_*)$ is defined by the following derivation rules: $t \sim t^*$ if and only if the judgement $t \sim t^*$ is derivable.

$$\frac{}{x \sim x} \qquad \frac{t_1 \sim t_1^* \quad \cdots \quad t_{\mathcal{A}r(F)} \sim t^*_{\mathcal{A}r(F)}}{F(t_1, \ldots, t_{\mathcal{A}r(F)}) \sim F(t_1^*, \ldots, t^*_{\mathcal{A}r(F)})}$$

$$\frac{t_1 \sim t_1^* \quad \cdots \quad t_{m-1} \sim t^*_{m-1} \quad t_m \sim g(\mathbf{s}^*_{1,n_g}) \quad t_{m+1} \sim t^*_{m+1} \quad \cdots \quad t_{n_f} \sim t^*_{n_f}}{f(\mathbf{t}_{1,m-1}, t_m, \mathbf{t}_{m+1,n_f}) \sim \underline{fg}(\mathbf{t}^*_{1,m-1}, \mathbf{s}^*_{1,n_g}, \mathbf{t}^*_{m+1,n_f})}$$

Similarly, $C \sim C_*$ can be defined by treating $[]$ as a variable. Note that F ranges over $\mathcal{F}$ in the definition.

Proposition 5. *We have the following.*

1. *If $C \sim C_*$ and $t \sim t^*$, then $C[t] \sim C_*[t^*]$.*
2. *If $t \sim t_1$ then $t\sigma \sim t_1\sigma$ for every substitution σ.*
3. *If $t \sim C[t_1]$ and $t_1 \sim t_2$, then $t \sim C[t_2]$.*

Proposition 6. *Suppose $t = C[s] \sim t^*$. Then one of the following two cases holds.*

- *$t^* = C_*[s^*]$ for some C_* and s^* such that $C \sim C^*$ and $s \sim s^*$.*
- *C is of the form $C'[f(\mathbf{t}_{1,m-1}, [], \mathbf{t}_{m+1,n_f})]$ and s is of the form $g(\mathbf{s}_{1,n_g})$. $t^* = C_*[s^*]$ for some C_* and s^* is of the form $\underline{fg}(\mathbf{t}^*_{1,m-1}, \mathbf{s}^*_{1,n_g}, \mathbf{t}^*_{m+1,n_f})$ such that $C' \sim C_*$ and $f(\mathbf{t}_{1,m-1}, s, \mathbf{t}_{m+1,n_f}) \sim s^*$.*

Proof. This follows from a structural induction on the derivation of $t \sim t^*$.

The use of Proposition 6 is to do case analysis in proofs.

The following definition of $\rhd$ is crucial for the development of the freezing technique. We will list some properties of $\rhd$ to ease the understanding as well as presenting an simple example. Also a comparison of the definition with *narrowing* may be helpful.

Definition 7. We define the relation $\rhd$ as follows, which is parameterized over $(f, g, m, \underline{fg})$.

$$\frac{}{x \rhd \langle x, \epsilon \rangle} \ (var)$$

$$\frac{t_1 \rhd \langle t_1^*, \sigma_1 \rangle \quad \cdots \quad t_{Ar(F)} \rhd \langle t_{Ar(F)}^*, \sigma_{Ar(F)} \rangle}{F(\mathbf{t}_{1,Ar(F)}) \rhd \langle F(\mathbf{t}_{1,Ar(F)}^*), \bigcup_{i=1}^{Ar(F)} \sigma_i \rangle} \ (fun)$$

$$\frac{t_i \rhd \langle t_i^*, \sigma_i \rangle \text{ for } 1 \le i \ne m \le n_f}{f(\mathbf{t}_{1,m-1}, x, \mathbf{t}_{m+1,n_f}) \rhd \langle \underline{fg}(\mathbf{t}_{1,m-1}^*, x_1, \ldots, x_{n_g}, \mathbf{t}_{m+1,n_f}^*), \sigma \rangle}, \ (inner)$$

where $x_1, \ldots, x_{n_g}$ are fresh variables, and

$$\sigma = \Big(\bigcup_{1 \le i \ne m \le n} \sigma_i \Big) \cup \{ x \mapsto g(x_1, \ldots, x_{n_g}) \}.$$

$$\frac{t_i \rhd \langle t_i^*, \sigma_i \rangle \text{ for } 1 \le i \ne m \le n_f \quad t_m \rhd \langle g(\mathbf{s}^*), \sigma_m \rangle}{f(\mathbf{t}_{1,m-1}, t_m, \mathbf{t}_{m+1,n_f}) \rhd \langle \underline{fg}(\mathbf{t}_{1,m-1}^*, \mathbf{s}_{1,n_g}^*, \mathbf{t}_{m+1,n_f}^*), \bigcup_{i=1}^{n_f} \sigma_i \rangle} \ (freeze)$$

$$\frac{l \rhd \langle l^*, \sigma \rangle \quad r^* = r\sigma}{l \to r \rhd \langle l^* \to r^*, \sigma \rangle}$$

$$\frac{l \rhd \langle g(\mathbf{l}_{1,n_g}^*), \sigma \rangle \quad r^* = f(x_1, \ldots, x_{m-1}, r\sigma, x_{m+1}, \ldots, x_{n_f})}{l \to r \rhd \langle \underline{fg}(x_1, \ldots, x_{m-1}, \mathbf{l}_{1,n_g}^*, x_{m+1}, \ldots, x_{n_f}) \to r^*, \sigma \rangle} \ (outer),$$

where $x_1, \ldots, x_{m-1}, x_{m+1}, \ldots, x_{n_f}$ are fresh variables.

The *master* $(f, g, m, \underline{fg})$-frozen version $\mathcal{R}_*$ of $\mathcal{R}$ is defined below.

$$\mathcal{R}_* = \{ l^* \to r^* \mid l \to r \rhd \langle l^* \to r^*, \sigma \rangle \text{ for some } l \to r \in \mathcal{R} \}$$

Some properties of $\rhd$ include (i) $t \to \langle t^*, \sigma \rangle$ implies $t\sigma = t^*$, (ii) $t \to \langle t, \epsilon \rangle$ for all every term t, and (iii) $l \to r \rhd \langle l \to r, \epsilon \rangle$. It can be readily verified that $\mathcal{R}_*$ is left-linear. If $\mathcal{R}$ is finite, then the finiteness of $\mathcal{R}_*$ clearly holds modulo renaming, i.e. we treat every pair of rules $l \to r$ and $l' \to r'$ in $\mathcal{R}_*$ as the same if the former can be obtained from renaming some variables in the latter and vice versa.

Example 2. This example is due to L. Bachmair. Let $\mathcal{R}$ consist of the following rules.

$$1.\ f(h(x)) \to f(i(x)) \quad 2.\ g(i(x)) \to g(h(x)) \quad 3.\ h(a) \to b \quad 4.\ i(a) \to b$$

Let us compute the master $(g, h, 1, \underline{gh})$-freezing version $\mathcal{R}_*$ of $\mathcal{R}$. It is straightforward to obtain the following,

$$f(h(x)) \rhd \langle f(h(x)), \epsilon \rangle \qquad g(i(x)) \rhd \langle g(i(x)), \epsilon \rangle$$
$$h(a) \rhd \langle h(a), \epsilon \rangle \qquad i(a) \rhd \langle i(a), \epsilon \rangle$$

which include all the possibilities since there is no application of the *(inner)* rule in this example. Hence $\mathcal{R}_*$ consists of rule 1,2,3,4 and rule 5: $\underline{gh}(a) \to g(b)$. Let $\mathcal{R}_1$ be a TRS consisting of rule $1, 3, 4, 5$ and rule $2'$: $g(i(x)) \to \underline{gh}(x)$, then $\mathcal{R}_1$ is a $(g, h, 1, \underline{gh})$-freezing version of $\mathcal{R}$. The termination of $\mathcal{R}_1$ can be shown by recursive path ordering with precedence: $h \succ i \succ \underline{gh} \succ g, b$. Notice that $\mathcal{R}_1$ is the generated TRS $\mathcal{S}$ in Example 14 in [20].

Lemma 8. *Let l be a linear term and ρ be a substitution. If $t = l\rho$ and $t \sim t^*$, then there exists a term l^* and a substitution ρ^* such that $t^* = l^*\rho^*$ and $l \rhd \langle l^*, \sigma\rangle$ and $\rho = \sigma\rho^*$.*

Proof. We proceed by a structural induction on t. If t is a variable then the case is trivial. Otherwise, we have the following cases.

- $t = F(\mathbf{t}_{1,n})$, where $F \neq f$. If l is a variable x, then $\rho = \{x \mapsto t\}$. Let $l^* = x$, $\sigma = \epsilon$ and $\rho^* = \{x \mapsto t^*\}$, and we are done. We now assume that l is not a variable. Since l is linear, l is of the form $F(\mathbf{l}_{1,n})$, and for $1 \leq i \leq n$, $t_i = l_i\rho_i$ for some ρ_i and $\rho = \bigcup_{i=1}^n \rho_i$. Since $F \neq f$, t^* is of the form $F(\mathbf{t}^*_{1,n})$ and $t_i \sim t_i^*$ for $1 \leq i \leq n$. By induction hypothesis, for $1 \leq i \leq n$, there exist l_i^* and ρ_i^* such that $t_i^* = l_i^*\rho_i^*$ and $l_i \rhd \langle l_i^*, \sigma_i\rangle$ and $\rho_i = \sigma_i\rho_i^*$. Let $l^* = F(\mathbf{l}^*_{1,n})$ and $\rho^* = \bigcup_{i=1}^n \rho_i^*$, then $l \rhd \langle l^*, \sigma\rangle$ for $\sigma = \bigcup_{i=1}^n \sigma_i$. It can be readily verified that $t^* = l^*\rho^*$ and $\rho = \sigma\rho^*$ since l is linear.
- $t = f(\mathbf{t}_{1,n_f})$. If l is a variable or t^* is of the form $f(\mathbf{t}^*_{1,n_f})$, where $t_i \sim t_i^*$ for $1 \leq i \leq n_f$, then this case can be proven as in the previous one. We now assume that l is of the form $f(\mathbf{l}_{1,n_f})$ and t^* is of form $\underline{fg}(\mathbf{t}^*_{1,m-1}, \mathbf{s}^*, \mathbf{t}^*_{m+1,n_f})$, where $t_i \sim t_i^*$ for $1 \leq i \neq m \leq n_f$ and $t_m \sim t_m^* = g(\mathbf{s}^*)$. If l_m is not a variable, then this case can also be proven as the previous one. We now assume $l = f(\mathbf{l}_{1,m-1}, x, \mathbf{l}_{m+1,n_f})$. Then for $1 \leq i \neq m \leq n$, $t_i = l_i\rho_i$ for some ρ_i and

$$\rho = (\bigcup_{1 \leq i \neq m \leq n_f} \rho_i) \cup \{x \mapsto t_m\}.$$

By induction hypothesis, for $1 \leq i \neq m \leq n_f$, there exist l_i^* and ρ_i^* such that $t_i^* = l_i^*\rho_i^*$ and $l_i \rhd \langle l_i^*, \sigma_i\rangle$ and $\rho_i = \sigma_i\rho_i^*$. Then

$$l \rhd \langle \underline{fg}(\mathbf{l}^*_{1,m-1}, x_1, \ldots, x_{n_g}, \mathbf{l}^*_{m+1,n_f}), \sigma\rangle,$$

where $\sigma = (\bigcup_{1 \leq i \neq m \leq n_f} \sigma_i) \cup \{x \mapsto g(x_1, \ldots, x_{n_g})\}$. Since $t_m \sim g(\mathbf{s}^*)$, $t_m = g(\mathbf{s})$ follows for some $\mathbf{s}$ and $\mathbf{s} \sim \mathbf{s}^*$. Let $l^* = \underline{fg}(\mathbf{l}^*_{1,m-1}, x_1, \ldots, x_{n_g}, \mathbf{l}^*_{m+1,n_f})$ and $\rho_m = \{x_1 \mapsto s_1^*, \ldots, x_{n_g} \mapsto s_{n_g}^*\}$ and $\rho = \cup_{i=1}^{n_f} \rho_i$, then it can be readily verified that $t^* = l^*\rho^*$ and $l \rhd \langle l^*, \sigma\rangle$ and $\rho = \sigma\rho^*$.

Lemma 9. *Let $l \rightarrow r$ be a left-linear rewrite rule and ρ be a substitution. If $t = C[l\rho] \sim t^*$, then $l \rightarrow r \rhd \langle l^* \rightarrow r^*, \sigma\rangle$ for some l^*, r^*, σ, and $t^* = C_*[l^*\rho^*]$ for some context C_* and $C[r\rho] \sim C_*[r^*\rho^*]$.*

Proof. Let $s = l\rho$. We do a case analysis on the form of t^* according to Proposition 6.

- $t^* = C_*[s^*]$, where $C \sim C_*$ and $s \sim s^*$. By Lemma 8, we have l^* and ρ^* such that $s^* = l^*\rho^*$ and $l \rhd \langle l^*, \sigma\rangle$ and $\rho = \sigma\rho^*$. Therefore, $l \rightarrow r \rhd \langle l^* \rightarrow r^*, \sigma\rangle$ for $r^* = r\sigma$, and this yields $r^*\rho^* = r\sigma\rho^* = r\rho$. Clearly, $C[r\rho] \sim C_*[r^*\rho^*]$.

– $t^* = C_*[s^*]$, where $C = C'[f(\mathbf{t}_{1,m-1}, [], \mathbf{t}_{m+1,n_f})]$ and $C' \sim C^*$ and

$$s = f(\mathbf{t}_{1,m-1}, \mathbf{s}_{1,n_g}, \mathbf{t}_{m+1,n_f}) \sim s^* = \underline{fg}(\mathbf{t}^*_{1,m-1}, \mathbf{s}^*, \mathbf{t}^*_{m+1,n_f}).$$

As the previous case, we have $l_0^* = g(\mathbf{l}^*)$, r_0^* and ρ^* such that $g(\mathbf{s}^*) = l_0^*\rho_0^*$ and $l \rhd \langle l_0^*, \sigma\rangle$ and $r\rho = r_0^*\rho_0^*$. Note $l \to r \rhd \langle l^* \to r^*, \sigma\rangle$, where

$$l^* = \underline{fg}(x_1, \ldots, x_{m-1}, \mathbf{l}^*_{1,n_g}, x_{m+1}, \ldots, x_{n_f})$$

and

$$r^* = f(x_1, \ldots, x_{m-1}, r_0^*, x_{m+1}, \ldots, x_{n_f}),$$

where $x_1, \ldots, x_{m-1}, x_{m+1}, \ldots, x_{n_f}$ are fresh.
Let $\rho^* = \rho_0^* \cup \{x_1 \mapsto t_1^*, \ldots, x_{m-1} \mapsto t^*_{m-1}, x_{m+1} \mapsto t^*_{m+1}, \ldots, x_{n_f} \mapsto t^*_{n_f}\}$, then

$$C[r\rho] = C'[f(\mathbf{t}_{1,m-1}, r\rho, \mathbf{t}_{m+1,n_f})] \sim C_*[r^*\rho^*].$$

Definition 10. Let $\mathcal{R}_1$ be a TRS such that $l_1 \to r_1 \in \mathcal{R}_1$ if and only if there exists $l^* \to r^* \in \mathcal{R}_*$ satisfying $l_1 = l^*$ and $r^* \sim r_1$. We call $\mathcal{R}_1$ a $(f, g, m, \underline{fg})$-frozen version of $\mathcal{R}$.

In general, there exist a (finite) family of $(f, g, m, \underline{fg})$-frozen versions of $\mathcal{R}$. For instance, the following is another $(f, g, 1, \underline{fg})$-frozen version of the $\mathcal{R}$ given in Example 1.

(1). $f(f(x)) \to \underline{fg}(f(x))$ (2). $f(\underline{fg}(x)) \to \underline{fg}(\underline{fg}(x))$

Theorem 11. *$\mathcal{R} \stackrel{\sim}{\Longrightarrow} \mathcal{R}_1$ holds for every $(f, g, m, \underline{fg})$-frozen version $\mathcal{R}_1$ of $\mathcal{R}$.*

Proof. Assume $t \sim t^*$ and $t \to_{\mathcal{R}} t'$. Then $t = C[l\rho]$ for some C and $l \to r \in \mathcal{R}$ and $t' = C[r\rho]$. By Lemma 9, there exists $l \to r \rhd \langle l^* \to r^*, \sigma\rangle$ such that $t^* = C_*[l^*\rho^*]$ for some C_* and $C[r\rho] \sim C_*[r^*\rho^*]$. By the definition of $\mathcal{R}_*$, $l^* \to r^* \in \mathcal{R}_*$. Since $l_1 \to r_1 \in \mathcal{R}_1$ for some l_1, r_1 such that $l^* = l_1$ and $r^* \sim r_1$, we have $t^* \to_{\mathcal{R}_1} C_*[r_1\rho^*]$, and therefore $C[r\rho] \sim C_*[r_1\rho^*]$ holds by (2) and (3) of Proposition 5. This yields $\mathcal{R} \stackrel{\sim}{\Longrightarrow} \mathcal{R}_1$.

Remark. We emphasize that the freezing technique as presented above can only be applied to left-linear TRSs. A straightforward extension to non-left linear TRSs may end with non-termination as shown by the following examples.

Example 3. Let $\mathcal{R}$ consist of the non-left linear rule $g(x, f(x)) \to c$, where c is a constant. Then the master $(f, f, 1, \underline{ff})$-frozen version $\mathcal{R}$, i.e., the TRS generated according to the above procedure for left-linear TRSs, consists of infinitely many rules (modulo renaming). For instance, the following rules are in $\mathcal{R}_*$:

$$g(x, f(x)) \to c, g(f(x), \underline{ff}(x)) \to c, g(\underline{ff}(x), \underline{ff}(f(x)) \to c, \ldots$$

The study on how to extend *freezing* to non-left linear TRSs is our immediate research topic.

4.2 Freezing more function symbols

Given a TRS $\langle \mathcal{F}, \mathcal{R} \rangle$ in which there exist f, g and h such that $\mathcal{A}r(f) \geq 1$ and $\mathcal{A}r(f) + \mathcal{A}r(g) - 1 \geq 1$. Note that f, g, h do not have to be distinct from each other. Let $\mathcal{R}_*$ be the master $(f, g, m_1, \underline{fg})$-frozen version of $\mathcal{R}$ and $\sim_1$ the freezing relation. Also let $\mathcal{R}_{**}$ be the master $(\underline{fg}, h, m_2, \underline{fgh})$-frozen version of $\mathcal{R}_*$ and $\sim_2$ the freezing relation. Then the master $(f, g, h, m_1, m_2, \underline{fgh})$-frozen version of $\mathcal{R}$ is defined as the TRS $\mathcal{R}_{**}^-$ which consists of all the rules $l \to r^-$ such that $l \to r \in \mathcal{R}_{**}$ for some $r^- \sim_1 r$ and $\underline{fg}$ has no occurrences in l. The corresponding freezing relation $\sim$ is defined as $t \sim t_2$ if $t \sim_1 t_1$ and $t_1 \sim_2 t_2$ for some t_1 and $\underline{fg}$ has no occurrence in t_2. Then $\mathcal{R}_1$ is a $(f, g, h, m_1, m_2, \underline{fgh})$-frozen version of $\mathcal{R}$ if $l_1 \to r_1 \in \mathcal{R}_1$ if and only if there exists $l \to r \in \mathcal{R}_{**}^-$ such that $l = l_1$ and $r \sim r_1$.

Theorem 12. *If $\mathcal{R}_1$ is a $(f, g, h, m_1, m_2, \underline{fgh})$-frozen version of a left-linear $\mathcal{R}$ and $\sim$ is the freezing relation, then $\mathcal{R} \overset{\sim}{\Longrightarrow} \mathcal{R}_1$.*

Proof. This simply follows from Theorem 11 with the explanation above.

This idea can certainly be generalized to freezing more function symbols, and we leave out the details.

We say that $\mathcal{R}$ is a 0-level frozen version of itself and $\mathcal{R}_{n+1}$ is an $n+1$-level frozen version of $\mathcal{R}$ if there exists a TRS $\mathcal{R}_n$ such that $\mathcal{R}_{n+1}$ is a $(f, g, m, \underline{fg})$-frozen version of $\mathcal{R}_n$ for some $f, g, m, \underline{fg}$ and $\mathcal{R}_n$ is an n-level frozen version of $\mathcal{R}$. We say to freeze

$$f(\bullet_1, \ldots, \bullet_{m-1}, g(\bullet_m, \ldots, \bullet_{m+\mathcal{A}r(g)-1}), \bullet_{m+\mathcal{A}r(g)}, \ldots, \bullet_{\mathcal{A}r(f)+\mathcal{A}r(g)-1})$$

into

$$\underline{fg}(\bullet_1, \ldots, \bullet_{m-1}, \bullet_m, \ldots, \bullet_{m+\mathcal{A}r(g)-1}, \bullet_{m+\mathcal{A}r(g)}, \ldots, \bullet_{\mathcal{A}r(f)+\mathcal{A}r(g)-1})$$

in $\mathcal{R}$ to mean constructing a $(f, g, m, \underline{fg})$-frozen version of $\mathcal{R}$.

4.3 Towards automated termination proofs

A straightforward combination of the freezing technique with others can be described as follows. Let $\mathcal{P}roc$ be a procedure which implements some approach to automated termination proofs. Given a TRS $\mathcal{R}$, we can enumerate all the n-level frozen versions of $\mathcal{R}$ for some n and use $\mathcal{P}roc$ to decide if one of them is terminating. This approach is impractical when the number of n-level frozen versions of $\mathcal{R}$ is too large. We use the following example to demonstrate a way to cope with this problem.

Example 4. This example is taken from [4]. Let $\mathcal{R}$ consist of the following rules.

1.	$Div2(\emptyset) \to \emptyset$
2.	$Div2(S(\emptyset)) \to \emptyset$
3.	$Div2(S(S(x))) \to S(Div2(x))$
4.	$LastBit(\emptyset) \to 0$
5.	$LastBit(S(\emptyset)) \to 1$
6.	$LastBit(S(S(x))) \to LastBit(x)$
7.	$Conv(\emptyset) \to \epsilon\&0$
8.	$Conv(S(x)) \to Conv(Div2(S(x)))\&LastBit(S(x))$

Suppose that we want to obtain some frozen version $\mathcal{R}_*$ of $\mathcal{R}$ such that $\mathcal{R}_*$ can be proven using some rpo. Then rule $1, 2, 3, 6$ are well-oriented under any (total) precedence relation. Since none of ϵ, 0, 1 and & occur in the left-hand side of any rule, we assign to them the lowest precedence, and the rule $4, 5, 7$ are then well-oriented. In order to orient rule 8, we freeze $Conv(Div2(\bullet_1))$ into $\underline{ConvDiv2}(\bullet_1)$, generating a TRS $\mathcal{R}_1$ consisting rule 1. – 7., and the following ones.

$$
\begin{array}{ll}
8'. & Conv(S(x)) \rightarrow \underline{ConvDiv2}(S(x))\&LastBit(S(x)) \\
9. & \underline{ConvDiv2}(S(S(x))) \rightarrow Conv(S(Div2(x))) \\
10. & \underline{ConvDiv2}(S(\emptyset)) \rightarrow Conv(\emptyset) \\
11. & \underline{ConvDiv2}(\emptyset) \rightarrow Conv(\emptyset)
\end{array}
$$

Now rule 8′, 9, and 10 are well-oriented under the following precedence:

$$S \succ Conv \succ \underline{ConvDiv2}, Div2, LastBit.$$

We then freeze $Conv(\emptyset)$ into $\underline{Conv\emptyset}$, generating a TRS $\mathcal{R}_2$ consisting of rule 1—7, 8′, 9, 10 and the following ones.

$$11'. \quad \underline{ConvDiv2}(\emptyset) \rightarrow \underline{Conv\emptyset} \qquad\qquad 12. \quad \underline{Conv\emptyset} \rightarrow \epsilon\&0$$

Then rule 11′ is well-oriented under $\underline{ConvDiv2} \succ \underline{Conv\emptyset}$, and rule 12 is well-oriented since ϵ, 0, and & have been given the lowest precedence. Hence $\mathcal{R}$ is terminating. Note that $\mathcal{R}_2$ is the $\mathcal{T} \cup \mathcal{S}$ in [4], which shows the termination of $\mathcal{R}$.

5 Examples

In this section we present some examples to show the effectiveness of the freezing technique. We say that $\mathcal{R}_1$ is a frozen version of $\mathcal{R}$ if $\mathcal{R}_1$ is a n-level frozen version of $\mathcal{R}$ for some n.

We will use the recursive path ordering (with status) approach (rpo(s)) *formulated in [19]* to prove the termination of TRSs. Given a function symbol F, the status $\tau(F)$ is either ⓜ for multiset status or $(i_1, \ldots, i_{\mathcal{A}r(F)})$, a permutation of $1, \ldots, \mathcal{A}r(F)$, for lexicographic status. If the status of F is not presented, then it is assumed to be $\tau(F) =$ ⓜ .

Example 5. This example is taken from [10], where it is proven terminating by dummy elimination. The TRS consists of the following rule

$$x * (y + z) \rightarrow (a(x, y) * y) + (x * a(z, x))$$

modulo associativity and commutativity of +.

We freeze $a(\bullet_1, \bullet_2) * \bullet_3$ into $\underline{a*}(\bullet_1, \bullet_2, \bullet_3)$ and $\bullet_1 * a(\bullet_2, \bullet_3)$ into $\underline{*a}(\bullet_1, \bullet_2, \bullet_3)$.

$$
\begin{aligned}
x * (y + z) &\rightarrow \underline{a*}(x, y, y) + \underline{*a}(x, z, x) \\
\underline{a*}(x_1, x_2, y + z) &\rightarrow \underline{a*}(a(x_1, x_2), y, y) + \underline{*a}(a(x_1, x_2), z, a(x_1, x_2))
\end{aligned}
$$

Let $\oplus$ and $\otimes$ be the usual addition and multiplication on positive integers. The following polynomial interpretation proves that the system is terminating modulo associativity and commutativity of $+$.

$$x + y = x \oplus y \oplus 2 \qquad a(x,y) = x \oplus y$$
$$\underline{a*}(x,y,z) = ((x \oplus y) \otimes z) \oplus (z \otimes z \otimes z)$$
$$\underline{*a}(x,y,z) = x \oplus y \oplus z \qquad x * y = x \otimes y \otimes y \otimes y$$

Example 6. This example is taken from [1], where it is claimed that the example eludes all the techniques in [21, 20, 16, 11], which aim for generating automatic termination proofs for TRSs.

$$minus(x,0) \to x \qquad minus(s(x),s(y)) \to minus(x,y)$$
$$quot(0,s(y)) \to 0 \qquad quot(s(x),s(y)) \to s(quot(minus(x,y),s(y)))$$

We freeze $quot(minus(\bullet_1,\bullet_2),\bullet_3)$ into $\underline{quotminus}(\bullet_1,\bullet_2,\bullet_3)$.

$$minus(x,0) \to x \qquad minus(s(x),s(y)) \to minus(x,y)$$
$$quot(0,s(y)) \to 0 \qquad quot(s(x),s(y)) \to s(\underline{quotminus}(x,y,s(y)))$$
$$\underline{quotminus}(x,0,y) \to quot(x,y)$$
$$\underline{quotminus}(s(x),s(z),y) \to \underline{quotminus}(x,z,y)$$

The termination of the above TRS can be proven using rpos with:

quasi precedence: $quot \approx \underline{quotminus} \succ s$
status: $\tau(quot) = (1,2), \tau(\underline{quotminus}) = (1,3,2)$

Example 7. (Greatest Common Divisor)

$$x - 0 \to x \qquad s(x) - s(y) \to x - y$$
$$0 < s(x) \to true \quad x < 0 \to false \quad s(x) < s(y) \to x < y$$
$$if(true,x,y) \to x \quad if(false,x,y) \to y \quad gcd(x,0) \to x \quad gcd(0,x) \to x$$
$$gcd(s(x),s(y)) \to if(x < y, gcd(s(x), y - x), gcd(x - y, s(y)))$$

Note that the left-hand side of the last rule becomes self-embedded in the right-hand side if y gets substituted with $s(x)$. We freeze $\mathrm{gcd}(\bullet_1 - \bullet_2, \bullet_3)$ into $\underline{gcd{-}L}(\bullet_1,\bullet_2,\bullet_3)$ and $\mathrm{gcd}(\bullet_1, \bullet_2 - \bullet_3)$ into $\underline{gcd{-}R}(\bullet_1,\bullet_2,\bullet_3)$, obtaining the following TRS $\mathcal{R}_1$.

$$x - 0 \to x \qquad s(x) - s(y) \to x - y$$
$$0 < s(x) \to true \quad x < 0 \to false \quad s(x) < s(y) \to x < y$$
$$if(true,x,y) \to x \quad if(false,x,y) \to y \quad gcd(x,0) \to x \quad gcd(0,x) \to x$$
$$gcd(s(x),s(y)) \to if(x < y, \underline{gcd{-}R}(s(x),y,x), \underline{gcd{-}L}(x,y,s(y)))$$
$$\underline{gcd{-}L}(x,y,0) \to x - y \qquad \underline{gcd{-}R}(0,x,y) \to x - y$$
$$\underline{gcd{-}L}(s(x),s(y),z) \to \underline{gcd{-}L}(x,y,z)$$
$$\underline{gcd{-}R}(x,s(y),s(z)) \to \underline{gcd{-}R}(x,y,z)$$
$$\underline{gcd{-}L}(x,0,y) \to gcd(x,y) \qquad \underline{gcd{-}R}(x,y,0) \to gcd(x,y)$$

The termination of $\mathcal{R}_1$ can be proven using the rpos with:

quasi precedence: $gcd \approx \underline{gcd-L} \approx \underline{gcd-R} \succ -, <, if \succ true, false$
status: $\tau(gcd) = (1,2), \tau(\underline{gcd-L}) = (1,3,2)$, and $\tau(\underline{gcd-R}) = (1,2,3)$

We point out that this example is much harder than Example 6 in [20], where if, $<$ and $-$ are simply treated as constructors.

Example 8. This example is taken from [2], which cannot be proven with a simplification ordering since the last rule is self-embedded:

$$app(nil,k) \to k \qquad app(l,nil) \to l \qquad app(x.l,k) \to x.app(l,k)$$
$$sum(x.nil) \to x.nil \qquad sum(x.(y.l)) \to sum((x+y).l)$$
$$sum(app(l,x.(y.k))) \to sum(app(l,sum(x.(y.k))))$$

We freeze $app(\bullet_1, sum(\bullet_2.\bullet_3))$ into $\underline{appsumdot}(\bullet_1,\bullet_2,\bullet_3)$ and $app(\bullet_1,\bullet_2.nil)$ into $\underline{appdotnil}(\bullet_1,\bullet_2)$:

$$app(nil,k) \to k \qquad app(l,nil) \to l \qquad app(x.l,k) \to x.app(l,k)$$
$$sum(x.nil) \to x.nil \qquad sum(x.(y.l)) \to sum((x+y).l)$$
$$sum(app(l,x.(y.k))) \to sum(\underline{appsumdot}(l,x,y.k))$$
$$\underline{appsumdot}(nil,x,y) \to sum(x.y)$$
$$\underline{appsumdot}(x.y,z,k) \to x.\underline{appsumdot}(y,z,k)$$
$$\underline{appsumdot}(l,x,y.k) \to \underline{appsumdot}(l,x+y,k)$$
$$\underline{appsumdot}(l,x,nil) \to \underline{appdotnil}(l,x)$$
$$\underline{appdotnil}(x.l,y) \to x.\underline{appdotnil}(l,y) \qquad \underline{appdotnil}(nil,x) \to x.nil$$

The termination of the above TRS can be proven by the rpos:

quasi precedence: $app \approx \underline{appsumdot} \succ \underline{appdotnil} \succ . \succ +$
status:
$\tau(app) = (1,2), \tau(\underline{appsumdot}) = (1,3,2), \tau(\underline{appdotnil}) = (1,2), \tau(.) = (2,1)$

The freezing technique *cannot* transform the following example of Kamin and Lévy into any TRS which can be proven terminating using rpos.

Example 9. Let $\mathcal{R}$ be the TRS consisting of the following rules.

$$p(s(s(x))) \to s(p(s(x))) \quad p(s(0)) \to 0 \quad fac(s(x)) \to fac(p(s(x))) * s(x)$$

Suppose that $\mathcal{R}_1$ is some frozen version of $\mathcal{R}$. If $\mathcal{R}_1$ can be proven terminating with some rpos, then $\mathcal{R}_2 = \mathcal{R}_1 \cup \{p(s(0)) \to s(0)\}$ can also be proven terminating using the same rpos. Clearly this is impossible since $\mathcal{R}_2$ is not terminating. Therefore, none of the frozen versions of $\mathcal{R}$ can be proven using rpos.

6 Related Work and Conclusion

Our work closely relates to *transformation orderings* [4], with which we give some comparison.

Given two relations R_1 and R_2, we write $R_1 \cdot R_2$ for the relation R such that xRz if and only if xR_1y and yR_2z for some y. Let $\mathcal{R}$ be a TRS for which we intend to find a termination proof. We then try to construct two TRSs $\mathcal{T}$ and $\mathcal{S}$ such that $\rightarrow_{\mathcal{R}} \subseteq \equiv_{\mathcal{T}} \cdot \rightarrow_{\mathcal{S}}^{+} \cdot \equiv_{\mathcal{T}}$,[2] where $\equiv_{\mathcal{T}} = (\rightarrow_{\mathcal{T}} \cup \leftarrow_{\mathcal{T}})^*$. In other words, the following diagram commutes.

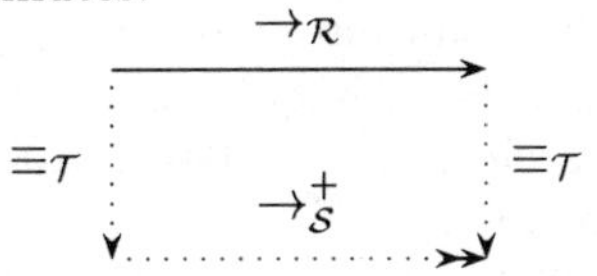

Notice the similarity and difference between this and the definition of $\overset{\sim}{\Longrightarrow}$. [3] Also we need to establish the following:

- $\mathcal{T}$ is confluent, and
- $\mathcal{S} \cup \mathcal{T}$ is terminating, and
- $\mathcal{S}$ locally cooperates with $\mathcal{T}$, namely, $\leftarrow_{\mathcal{T}} \cdot \rightarrow_{\mathcal{S}} \subseteq \rightarrow_{\mathcal{S}/\mathcal{T}}^{*} \cdot {}^{*}\!\!\leftarrow_{\mathcal{T}}$.

By Lemma 5 in [3], ${}^{*}\!\!\leftarrow_{\mathcal{T}} \cdot \rightarrow_{\mathcal{S}}^{*} \subseteq \rightarrow_{\mathcal{S}/\mathcal{T}}^{*} \cdot {}^{*}\!\!\leftarrow_{\mathcal{T}}$ holds since $\mathcal{S}$ locally cooperates with $\mathcal{T}$, where $\rightarrow_{\mathcal{S}/\mathcal{T}}$ stands for $\rightarrow_{\mathcal{T}}^{*} \cdot \rightarrow_{\mathcal{S}} \cdot \rightarrow_{\mathcal{T}}^{*}$. Now suppose that there exists an infinite $\rightarrow_{\mathcal{R}}$-rewriting sequence. We can then construct an infinite $\rightarrow_{\mathcal{S}/\mathcal{T}}$-rewriting sequence as shown below, yielding a contradiction since $\mathcal{S} \cup \mathcal{T}$ is terminating. Therefore, $\rightarrow_{\mathcal{R}}$ is terminating.

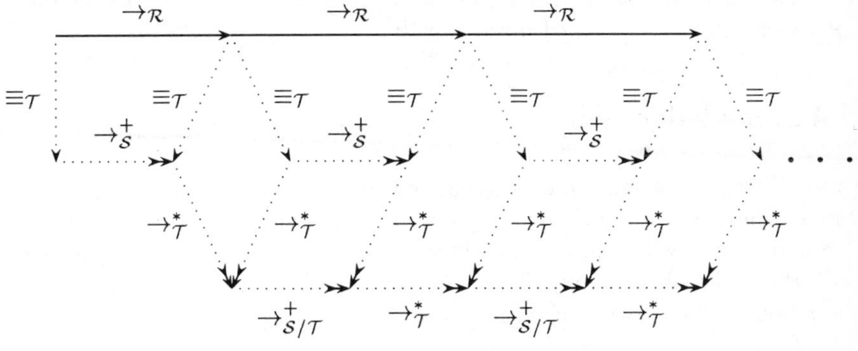

Note that this is very similar to Theorem 4 in [3]. The transformation ordering (TO) technique is the method which proves the termination of $\mathcal{R}$ by constructing of $\mathcal{T}$ and $\mathcal{S}$.

Although there exist many similarities between the freezing technique and the transformation ordering technique, we point out a significant difference as follows. After freezing $f(\ldots, g(\ldots), \ldots)$ into $\underline{fg}(\ldots)$, it attempting to relate freezing

[2] This condition can be weaken to $\rightarrow_{\mathcal{R}} \subseteq \rightarrow_{\mathcal{S}/\mathcal{T}}^{+} \cdot {}^{*}\!\!\leftarrow_{\mathcal{T}}$.

[3] This makes neither the transformation ordering technique nor the freezing technique encompass the other.

to transformation ordering by setting $\mathcal{T} = \{f(\mathbf{x}_{1,m-1}, g(\mathbf{y}_{1,\mathcal{A}r(g)}), \mathbf{x}_{m+1,\mathcal{A}r(f)}) \rightarrow \underline{fg}(\mathbf{x}_{1,m-1}, \mathbf{y}_{1,\mathcal{A}r(g)}, \mathbf{x}_{m+1,\mathcal{A}r(f)})\}$. However, this $\mathcal{T}$ may not be confluent(e.g., in the case where $f = g$). Moreover, if one applies freezing repeatedly, which is often the case in practice, the chance of obtaining a confluent $\mathcal{T}$ diminishes further.

In [20], Steinbach presented an algorithm to generate a transformation ordering for a given $\mathcal{R}$. The freezing technique closely resembles the projection technique in his algorithm. On the other hand, there also exist some significant differences.

1. When implementing the transformation ordering technique, one has to guarantee that the generated $\mathcal{T}$ is confluent, and this could be a rich source for the divergence of the algorithm. For a left-linear TRS, generating a frozen version of $\mathcal{R}$ always terminates.
2. Overall the algorithm in [20] seems to be quite involved and highly heuristic, while the freezing technique is conceptually simple and easily implementable.

In [3], a *quasi-commutation* method is introduced which does not require that the transformer system be confluent, either. However, there is no particular method given in [3] to facilitate the generation of such a transformer system. In contrast, the presentation of the freezing technique, an approach to transforming left-linear TRSs, is precisely the main contribution of this paper.

In summary, we have developed a technique called *freezing*, which can be applied to a left-linear TRS $\mathcal{R}$ to transform it into a family of left-linear TRSs such that the termination of any TRS in this family implies the termination of $\mathcal{R}$. We have shown that the transformation terminates for all (finite) left-linear TRSs. The effectiveness of this technique is demonstrated by many examples, all of which have been verified mechanically.

7 Acknowledgement

I gratefully acknowledge some electronic discussion with Joachim Steinbach regarding transformation orderings. I thank the anonymous referees for their detailed comments, which have undoubtedly enhanced the quality of the paper. Also I thank Frank Pfenning, Peter Andrews and Richard Statman for their support and for providing me with such a nice working environment.

References

1. T. Arts and J. Giesl. Termination of constructor systems. In Proceedings of the Seventh Conference on Rewriting Techniques and Applications, vol. 1103 of Lecture Notes in Computer Science, pp. 63-77, New Brunswick, USA, 1996.
2. T. Arts and J. Giesl. Automatically Proving Termination Where Simplification Orderings Fail. *In Proc. Colloquium on Trees in Algebra and Programming,* vol. 1214 of Lecture Notes in Computer Science, pp. 261-272, Lille, France, 1997.

3. L. Bachmair and N. Dershowitz. Communication, transformation and termination. *In Proceedings of the Eighth CADE*, vol. 230 of Lecture Notes in Computer Science, pp. 52-60, Oxford, July 1986.
4. F. Bellegarde and P. Lescanne. Termination by completion. *Applicable Algebra in Engineering, Communication and Computing*, vol. 1, pp. 79-96, 1990.
5. N. Dershowitz. Orderings for term rewriting systems. *TCS*, 17(3), pp. 279-301, 1982.
6. N. Dershowitz. Termination of rewriting. *Journal of Symbolic Computation*, vol. 3, pp.69-116, 1987.
7. N. Dershowitz and C. Hoot. Natural Termination, *TCS*, 142(2), pp. 179-207, 1995.
8. N. Dershowitz and Jean-Pierre Jouannaud. Rewrite Systems. In I. van Leeuwen, editor, *Handbook of Theoretical Computer Science*, vol. B, pp. 243-320.
9. N. Dershowitz and Jean-Pierre Jouannaud. Notations for rewriting. *EATCS*, vol. 43, pp. 162-172, 1991.
10. M. Ferreira. Dummy Elimination in Equational Rewriting. In Proceedings of the Seventh Conference on Rewriting Techniques and Applications, vol. 1103 of Lecture Notes in Computer Science, pp. 78-92, New Brunswick, USA, 1996.
11. M. Ferreira and H. Zantema. Dummy Elimination: making termination easier. In Proceedings of the 10th International Conference on Fundamentals of Computation Theory, LNCS 965, Dresden, 1995.
12. J. Giesl. Automated termination proofs with measure function. In *Proceedings of 19th Annual German Conference on AI*, LNAI 981, Bielefeld, 1995.
13. M. Hanus. The integration of functions into logic programming: From theory to practice. In *JLOGP*, vol. 19-20, pp. 583-628.
14. G. Huét and D. Lankford. On the uniform halting problem for term rewriting systems. Technical Report 283, INRIA, Le Chesnay, France, 1978.
15. D. Knuth and P. Bendix. Simple word problems in universal algebras. *Computational Problems in Abstract Algebra*, edited by J. Leech, Pergamon Press, pp. 263-297, 1970.
16. R. Kennaway. Complete term rewrite systems for decimal arithmetic and other total recursive functions. *Presented at the 2nd International Workshop on termination*, La Bresse, 1995
17. D. Lankford. On proving term rewriting systems are noetherian. Tech. Report Memo MTP-3, Louisiana Tech. University, 1979.
18. D. Plaisted. A recursively defined ordering for proving termination of term rewriting systems. Tech. report UIUC DCS-R-78-943, Univ. of Illinois, Urbana, 1978.
19. J. Steinbach. Simplification Orderings: History of Results, *Fundamenta Informaticae*, vol. 24, pp. 47-87, 1995.
20. J. Steinbach. Automatic termination proofs with transformation orderings. *In Proceedings of the Sixth RTA*, vol. 914 of Lecture Notes in Computer Science, pp. 11-25, 1995.
21. H. Zantema. Termination of term rewriting: interpretation and type elimination. *Journal of Symbolic Computation*, vol. 17, pp. 23-50, 1994.
22. H. Zantema. Termination of Term Rewriting by Semantic Labelling, *Fundamenta Informaticae*, vol. 24, pp 89-105, 1995.

Higher-Order Rewriting and Partial Evaluation

Olivier Danvy
BRICS,* Dept. of Computer Science, University of Aarhus†
Kristoffer Høgsbro Rose
LIP,‡ Ecole Normale Supérieure de Lyon§

Abstract

We demonstrate the usefulness of higher-order rewriting techniques for specializing programs, *i.e.*, for *partial evaluation*. More precisely, we demonstrate how casting *program specializers* as *combinatory reduction systems* (CRSs) makes it possible to formalize the corresponding program transformations as *meta-reductions*, *i.e.*, reductions in the internal "substitution calculus." For partial-evaluation problems, this means that instead of having to prove on a case-by-case basis that one's "two-level functions" operate properly, one can concisely formalize them as a combinatory reduction system and obtain as a corollary that static reduction does not go wrong and yields a well-formed residual program.

We have found that the CRS *substitution calculus* provides an adequate expressive power to formalize partial evaluation: it provides sufficient termination strength while avoiding the need for additional restrictions such as types that would complicate the description unnecessarily (for our purpose).

In addition, partial evaluation provides a number of *examples* of higher-order rewriting where being higher order is a central (rather than an occasional or merely exotic) property. We illustrate this by demonstrating how standard but non-trivial partial-evaluation examples are handled with higher-order rewriting.

1 Introduction

Most programs are overly general: they usually run with some invariants (*e.g.*, part of their input is constant). Partial evaluation aims at specializing programs with respect to these invariants [3, 11]. According to Kleene's S_n^m-theorem [13], specializing a program with respect to [an invariant] part of its input is computable, and running the corresponding specialized program

*Basic Research in Computer Science (Centre of the Danish Research Foundation).
†Building 540, Ny Munkegade, DK–8000 Aarhus C, Denmark; ⟨danvy@brics.dk⟩.
‡Laboratoire de l'Informatique du Parallélisme.
§46, Allée d'Italie, F–69364 Lyon 07, France; ⟨Kristoffer.Rose@ens-lyon.fr⟩.

on the remaining input should yield the same result as running the source program on the complete input, provided of course that the source program, the partial evaluator, and the residual program all terminate. The practical appeal of partial evaluation is that specialized programs are usually more efficient, so that running them amortizes the cost of partial evaluation.

What we find curious is that while in effect the renewal of partial evaluation originates in the area of rewriting techniques [12], there has been virtually no work to continue bridging the two areas. In this article we address this by formalizing the rewriting technique underlying standard partial-evaluation examples. Our prime target is the removal of interpretive overhead. In that view, any program traversing a data structure is an "interpreter" for that data structure (Abelson makes this point comprehensively in his foreword to "Essentials of Programming Languages" [9]). We thus focus on inductive data types and the associated functionals (or "derivors").

Our starting-point is the following analogy:

- In *partial evaluation* (PE) we specialize programs by performing part of them "in advance." This is achieved by using "two-level functions" that perform a mix of static evaluation and dynamic code generation.

- In rewriting we model (functional) programs by rewrite systems with a special "application" operator which is the root symbol of all (functional) rewrite rules. Since each application means we have to do work, we are interested in reducing the number of applications that are built, ideally to zero. We can do this using higher-order techniques in the framework of *combinatory reduction systems* (CRSs).

By recasting PE in the CRS formalism we should thus be able to exploit the meta-reductions in CRSs to perform the static reductions of our two-level functions.

Road map: The rest of the article is organized as follows: Section 2 reviews the two-level programming technique used in partial evaluation. Section 3 provides the necessary background in combinatory reduction systems, showing how simple program transformations can be cast as higher-order rewrite systems. Section 4 presents our synthesis between rewriting and partial evaluation, culminating with the detailed description of how to derive a specializer, and exemplified by a formal treatement of a classical example of specialization, a continuation-passing style (CPS) transformation for the λ-calculus. Section 5 concludes and briefly mentions some related work and future directions.

2 Partial Evaluation

In this section we illustrate partial evaluation through two examples: a first-order one and the higher-order one of traversing binary trees. The latter one hints at the style of transformation techniques used in the following sections.

2.1 A first-order example

So what is specialization? Specializing a program amounts to parameterizing it with code-generating (two-level) functions and running it. Let us demonstrate this with a first-order example: natural numbers and the exponentiation program, which is standard in partial evaluation. It is expressed as a simple conditional recursive equation as follows:

$$x^n = \begin{cases} x & \text{if } n = 1 \\ x^{n/2} \times x^{n/2} & \text{if } n \text{ even} \\ x \times x^{n-1} & \text{otherwise} \end{cases}$$

Seen as a function definition, *i.e.*, reading the equation from left to right, the exponentiation program is a *derivor*: it decomposes – or interprets – the exponent, in a trail of multiplications. Our aim is to specialize this derivor with respect to a particular exponent, which we achieve by interpreting the static exponent, in a trail of residual multiplications.

Here is the annotated derivor, where we have overlined the static parts that we can compute immediately when n becomes available:

$$x^{\overline{n}} = \begin{cases} x & \overline{\text{if } n = 1} \\ x^{\overline{n/2}} \times x^{\overline{n/2}} & \overline{\text{if } n \text{ even}} \\ x \times x^{\overline{n-1}} & \overline{\text{otherwise}} \end{cases}$$

Such a program is variously known as a *generating extension* [11] or a *back-quote interpreter* [9] in the literature.

In any case, repeatedly evaluating the overlined subexpressions of $x^{\overline{5}}$ gives the *residual* expression $x \times \big((x \times x) \times (x \times x)\big)$, which is in normal form since there are no further overlined expressions to evaluate. The trail of residual multiplications is all that remains of the static reductions.

2.2 A higher-order example

Consider the data type αBT of binary trees over some (unspecified) type α: αBT ::= $\mathit{Leaf}(\alpha) \mid \mathit{Node}(\alpha BT, \alpha BT)$ and its associated fold functional *Fold* typed as follows: $\mathit{Fold} : (\alpha \rightarrow \beta) \times (\beta \times \beta \rightarrow \beta) \rightarrow \alpha BT \rightarrow \beta$ As is customary in functional programming, we instantiate this fold functional with two functions: one for processing the leaves, and one for processing the nodes. For example, the application $\mathit{Fold}\,(\,\lambda x.1\,,\,\lambda\langle l, r\rangle.(l + r)\,)$ yields a function computing the number of leaves in a binary tree.

As is customary in partial evaluation, we instantiate this fold functional with three two-level functions: one for processing the leaves and one for processing the nodes, plus one to initialize the static computation. As before we overline static parts, here λs and applications.[1] Supplying a given binary tree to this fold function yields a residual program where the interpretive overhead of the fold function has been eliminated. For example, given unspecified function names L and N, the expression $\mathit{Fold}^{-}\,(\overline{\lambda}x.\,L\,x\,,\,\overline{\lambda}\langle l,r\rangle.\,N\,l\,r\,)$ yields a residual program combining L and N in a way that is isomorphic to the structure of the given binary tree. Applying the above to a binary tree such as $\mathit{Node}(\mathit{Node}(\mathit{Leaf}\,(1),\mathit{Leaf}\,(2)),\mathit{Leaf}\,(3))$ yields the residual program $N\,(\,N\,(L\,1)\,(L\,2))\,(L\,3)$ which is well-formed since neither overlines nor $\langle\cdot,\cdot\rangle$-pairs remain.

3 Combinatory Reduction Systems and Functional Programming

In this article, we use Klop's *Combinatory Reduction System* (CRS) formalism. In this section we first summarize the definition of CRSs [14, 16] before relating them to functional programming with a simple example demonstrating the use of higher-order rewriting to express improvements to functional programs.

3.1 Combinatory Reduction Systems

The following is a brief summary of the definition of CRSs. To avoid notational overloading of ordinary parentheses, we slightly modify the standard presentation of CRSs [16, §11–12]. We write $\cdot.\cdot$ and $\cdot[\cdot]$ instead of $[\cdot]\cdot$ and $\cdot(\cdot)$ for abstraction and meta-application, respectively.

3.1.1. Definition (many-sorted CRS). Assume a signature Σ of ranked symbols F^n, variables x, and ranked *meta-variables* Z^n (in both cases the superscript n is the rank).

1. CRS *terms* have the form $t ::= x \mid x.t \mid F^n(t_1,\ldots,t_n)$ and must be *closed* (that is, $\mathrm{fv}(t) = \{\}$ where $\mathrm{fv}(x) = \{x\}$, $\mathrm{fv}(x.t) = \mathrm{fv}(t) \setminus \{x\}$, and $\mathrm{fv}(F^n) = \bigcup_{i=1}^{n} \mathrm{fv}(t_i)$). The three forms are respectively called *variable*, *abstraction*, and *construction*.

2. CRS *meta-terms* extend CRS terms to $t ::= \ldots \mid \mathrm{Z}^n[t_1,\ldots,t_n]$. The new form is called a *meta-application*.

---

[1] An application is denoted by the space between two subexpressions, so $x\,\overline{\ }\,y$ is a static application whereas $x\,y$ is not.

3. An *assignment* σ specifies how to eliminate meta-applications that use specific meta-variables. It is a collection of pairs $(\mathrm{Z}^n[x_1, \ldots, x_n], t')$ with distinct x_i; $\sigma(t)$ is the resulting term where everywhere in t, $\mathrm{Z}^n[t_1, \ldots, t_n]$ is replaced by $t'\{x_1 := t_1, \ldots, x_n := t_n\}$ (which denotes an ordinary simultaneous substitution). All assignment must happen without name capture and clashes (this requires judicious renaming of bound variables following rather complicated but still intuitive "safeness" rules [16, Definition 12.3]).
4. CRS *rules*, written $\ell \to r$, are constructed from two meta-terms ℓ and r with the following additional restrictions:
   (a) ℓ (the left-hand side) must be a "pattern:" a construction where all meta-applications have the form $\mathrm{Z}^n[x_1, \ldots, x_n]$ with distinct x_i, and
   (b) r (the right-hand side) can only contain meta-applications with meta-variables occurring in the left-hand side.
5. We say that a term t *matches* a pattern ℓ if an assignment σ exists such that $t = \sigma(\ell)$ (the intuition being that each pattern meta-application $\mathrm{Z}^n[x_1, \ldots, x_n]$ becomes part of the assignment of σ).
6. The rules define the CRS *rewrite relation*: $s \to t$ iff s and t are identical except for one subterm: In s, it must be $\sigma(\ell)$ for some assignment σ, the "redex." In t, it must be $\sigma(r)$, the "contractum."
7. A *sorted CRS* is the subsystem obtained by restricting terms to be "well-sorted" according to some syntax specification, and assigning to each meta-variable a sort that it must match.

We use the usual abbreviations for CRSs. In particular, we omit the rank superscript and abbreviate $F^1(x.F^1(y.t))$, $F^0()$, and $\mathrm{Z}^0[]$, as $Fxy.t$, F, and Z, respectively. We also exploit conventions introduced in syntax productions to bind meta-variables to sorts and introduce infix binary constructors.

We do not delve further into the exciting details of the properties of rewriting systems in general and CRS in particular but refer the reader to the comprehensive literature on the subject [15, 16]. Instead we go straight to our basic example.

3.1.2. Example (2-level λ-calculus). The *2-level λ-calculus*, denoted $\overline{\lambda}$, is the single-sorted CRS over the *$\overline{\lambda}$-terms*

$$\mathrm{E} ::= x \mid \lambda x.\mathrm{E} \mid \mathrm{E}_0\ \mathrm{E}_1 \mid \overline{\lambda} x.\mathrm{E} \mid \mathrm{E}_0 \mathbin{\overline{\ }} \mathrm{E}_1$$

where concatenation denotes "application" (the invisible infix application function symbol sometimes written as @), $\mathrm{E}_0 \,\overline{\ }\, \mathrm{E}_1$ is "overlined application" (also $\overline{@}$), and both associate to the left as usual. Its rewrite rules read

$$(\lambda x.\mathrm{E}[x])\ \mathrm{E}' \to \mathrm{E}[\mathrm{E}'] \qquad (\beta)$$

$$(\overline{\lambda} x.\mathrm{E}[x])\,\overline{\ }\,\mathrm{E}' \to \mathrm{E}[\mathrm{E}']. \qquad (\overline{\beta})$$

(The subset with no overlines and with just β as reduction is the usual $\lambda\beta$-calculus denoted $\boldsymbol{\lambda}$ [1].)

3.1.3. Definition (abstract rewriting). Binary relations are denoted by arrows. Relational *composition* is written $\underset{1}{\to} \cdot \underset{2}{\to}$; the *inverse* of $\to$ is $\leftarrow$, its *transitive reflexive closure* is $\twoheadrightarrow$, and its *normalisation function* is $\twoheadrightarrow\!\!\!|$ (the restriction of $\twoheadrightarrow$ to just the reductions ending in a normal form). Two relations *commute* if $(\underset{1}{\leftarrow} \cdot \underset{2}{\to}) \subseteq (\underset{2}{\to} \cdot \underset{1}{\leftarrow})$; a relation $\to$ is *confluent* if $\twoheadrightarrow$ self-commutes. Finally a relation is *convergent* if it is confluent and *terminating*, *i.e.*, has no infinite reduction sequences.

3.1.4. Definition (CRS restrictions). A CRS is *left-linear* if all meta-variables occurring in each left-hand side are distinct. A CRS is *non-overlapping* if it is impossible for a symbol in a term to be part of two redexes in the term. A CRS is *orthogonal* if it is left-linear and non-overlapping. A *constructor* CRS's symbols are in two disjoint sets: *functions* that occur at the root of left-hand sides and *constructors* that do not. Finally, a CRS is a *term-rewriting system* (TRS) if all meta-variables used in rules are nullary.

3.1.5. Theorem. *Orthogonal CRSs are confluent [14, 16].*

3.2 Comparing to functional programs

First-order functional programs are usually said to correspond to left-linear constructor TRSs. We observe that untyped higher-order functional programming corresponds to adding β to the underlying formalism, thus interpreting the special relationship between the application function symbol and the λ constructor.

3.2.1. Example (binary tree folding, functional style). A *binary tree* of integers has two sorts: trees, $\mathrm{T} ::= \mathit{Leaf}(\mathrm{I}) \mid \mathit{Node}(\mathrm{T}_1, \mathrm{T}_2)$ and integers, I. "Folding" over the tree means replacing each $\mathit{Node}(\mathrm{T}_1, \mathrm{T}_2)$ with an application $N\ \mathrm{T}_1\ \mathrm{T}_2$ and each leaf $\mathit{Leaf}(\mathrm{I})$ with $L\ \mathrm{I}$, as discussed in section 2.2. A typical "functional program" rewrite system to do this is the following (left-linear constructor) TRS over trees extended with the symbol *Fold*:

$$Fold(\text{L}, \text{N}, Leaf(\text{I})) \to \text{L I} \tag{1}$$

$$Fold(\text{L}, \text{N}, Node(\text{T}_1, \text{T}_2)) \to \text{N}\,\big(Fold(\text{L}, \text{N}, \text{T}_1)\big)\,\big(Fold(\text{L}, \text{N}, \text{T}_2)\big) \tag{2}$$

For any tree T this system rewrites $Fold(L, N, \text{T})$ to the folding of T with L and N. Running it[2] on an example term gives

$$\begin{aligned} &Fold(L, N, Node(Node(Leaf(1), Node(Leaf(2), Leaf(3))),\\ &\qquad\qquad Node(Node(Leaf(4), Leaf(5)), Leaf(6))))\\ &\qquad\qquad \twoheadrightarrow N\,(N\,(L\,1)(N\,(L\,2)(L\,3)))(N\,(N\,(L\,4)(L\,5))(L\,6)) \end{aligned}$$

as should be expected.

Assume now that we wish to *flatten* trees just as we programmed it in the previous section, converting the tree to a list of the leaf integers. The following makes $Flatten(\text{T})$ rewrite to this effect when combined with *Fold* and λ to achieve the function reduction (thus this program is higher-order):

$$Flatten(\text{T}) \to Fold(\ \lambda ia.Cons(i, a)\ ,\ \lambda c_1 c_2 a.c_1(c_2\ a)\ ,\ \text{T}\ )\ Nil \tag{3}$$

Running the system on an example term gives

$$\begin{aligned} &Flatten(Node(Node(Leaf(1), Node(Leaf(2), Leaf(3))),\\ &\qquad\qquad Node(Node(Leaf(4), Leaf(5)), Leaf(6))))\\ &\qquad \twoheadrightarrow Cons(1, Cons(2, Cons(3, Cons(4, Cons(5, Cons(6, Nil)))))). \end{aligned}$$

The inconvenience of the above system (and of functional programming in general) is that the only rule doing actual rearrangement is β: the folding itself does nothing but build applications. This can be fixed by exploiting the possibilities of the CRS formalism for matching functions in any rule rather than just in β. In the next example, we thus reconsider flattening.

3.2.2. Example (binary tree folding, CRS). Consider binary trees as before with the additional auxiliary symbols L^1 and N^2. We then define folding by pattern matching on the functions to let the CRS formalism do the work of unfolding the leaf- and node-functions; furthermore we add a root function applied by the wrapper *Fold'*:

$$Fold'(\lambda t.\text{R}[t], \text{L}, \text{N}, \text{T}) \to \text{R}[Fold(\text{L}, \text{N}, \text{T})] \tag{4}$$

$$Fold(\lambda i.\text{L}[i], \text{N}, Leaf(\text{T})) \to \text{L}[\text{T}] \tag{5}$$

$$\begin{aligned} &Fold(\text{L}, \lambda st.\text{N}[s,t], Node(\text{T}_1, \text{T}_2)) \to\\ &\qquad \text{N}\big[Fold(\text{L}, \lambda st.\text{N}[s,t], \text{T}_1), Fold(\text{L}, \lambda st.\text{N}[s,t], \text{T}_2)\big] \end{aligned} \tag{6}$$

---

[2]All examples were run with the CRS implementation of the second author's PhD thesis [21, chapter6], adapted to the present syntax.

Then flattening is merely dispersing yet another "continuation" wrapped in special $\langle\cdot,\cdot\rangle$ brackets which are pattern-matched by the fold functions:

$$\mathit{Flatten}(\mathrm{T}) \to \mathit{Fold}'(\ \lambda x.x(\lambda z.\langle z, \mathit{Nil}\rangle),\ \lambda i.L\ i,\ \lambda lr.N\ l\ r,\ \mathrm{T}\ ) \quad (7)$$

$$L\ \mathrm{A}\ (\lambda z.\langle z, \mathrm{B}\rangle) \to \mathit{Cons}(\mathrm{A}, \mathrm{B}) \quad (8)$$

$$N\ \mathrm{A}\ \mathrm{B}\ \mathrm{C} \to \mathrm{A}\ (\lambda v.\langle v, \mathrm{B}\ \mathrm{C}\rangle) \quad (9)$$

Notice that there are two levels of functions involved: the function constructed with an explicit λ in the arguments to *Fold'*, and the function encoded by the L and N rules. The solution is elegant and very efficient since all the constructed abstractions are known to have the form $\lambda x.\langle x, \mathrm{A}\rangle$ where x is not free in A, so substitution is not costly – in fact we exploit this in the pattern of (8) where B is not written as $\mathrm{B}[z]$ because we know it does not contain a free occurrence of z. Folding the sample tree gives the same result as in the previous example, of course. The only remaining inconvenience is the fact that we still have to prove that all $\langle\cdot\rangle$ brackets are eliminated.

4 Synthesis

In this section we apply the higher-order rewriting technology discussed in the last section to partial evaluation, and we formalise the notion of a program specializer accordingly. We use this to prove a general theorem of well-annotatedness of specializers when given in the form of two-level derivors.

4.1 Derivors

We only consider syntax-directed program transformations. They are usually specified compositionally in the following sense.

4.1.1. Definition. A constructor CRS is *compositional* if each function symbol is compositional in one of its arguments, *i.e.*, if it has a distinguished argument such that the distinguished argument of all function constructions in the right-hand side of rules is always a strict subterm of the distinguished argument of the function construction on the left-hand side. Only one exception is permitted (to facilitate "root" rules): if a function symbol occurs only on left-hand sides, then the rules where it occurs are exempted from the constraint.

4.1.2. Lemma. *A compositional constructor CRS is terminating.*

Proof. Interpret a term into a multiset of the size of subterms in compositional positions (with left-hand-side symbols adding all subterms and multiplying by the largest number of copies): rewriting always decreases this multiset in the well-founded *multiset ordering* [8] (details in the full version of this article [6]). □

Now we can characterize the program transformers under consideration.

4.1.3. Definition. A *derivor* is an orthogonal and compositional constructor CRS where the normal forms contain only constructors.

4.1.4. Theorem. *Derivors are convergent.*

Proof. Use Theorem 3.1.5 and Lemma 4.1.2. □

With this we can express an interesting class of systems, which is directly relevant to partial evaluation.

4.1.5. Definition. A *two-level derivor* is a derivor producing $\overline{\boldsymbol{\lambda}}$-terms (of Definition 3.1.2) with the restrictions that $\overline{@}$ does not occur on any left-hand side and that $\overline{\lambda}$ is a constructor. A *well-annotated* two-level derivor is one producing $\overline{\boldsymbol{\lambda}}$-terms for which the $\overline{\beta}$-normal form is a $\boldsymbol{\lambda}$-term, *i.e.*, all overlines are eliminated.

Two-level derivors have interesting properties: first of all it is easy to see that "static reduction does not go wrong," which is mandatory in partial evaluation [11].

4.1.6. Proposition. *For any two-level derivor, $\mathcal{D}$, $\mathcal{D} \cup \boldsymbol{\lambda}$ is confluent.*

Proof. The restrictions on the occurrences of $\overline{@}$ and $\overline{\lambda}$ ensure that the combined system remains orthogonal. □

However, it remains difficult to prove well-annotatedness and termination of a two-level derivor because $\overline{\beta}$ has the full Turing-complete power in it. Both can be proven if one restricts the permitted $\overline{\beta}$ to a subset known to terminate, such as the simply typed λ-calculus. Then the entire construction of $\overline{\boldsymbol{\lambda}}$-terms has to be shown well-typed, a property that is easy to lose by even minute changes to the system. (This is related to why we have chosen CRSs as our basis formalism; we comment on this in the conclusion.)

Both properties can be shown for the first-order flatten in Example 3.2.1 but are trivial for the higher-order flatten of Example 3.2.2 since it produces no overlines at all. We exploit this property in the following example.

4.2 The call-by-value CPS transform

Continuation-passing style (CPS) is a sublanguage of the λ-calculus that is *insensitive to evaluation order* [20]. It is interesting to compiler writers because transforming a program into CPS makes it "sequential" in a way that facilitates code generation. Here we merely formalize one published CPS transformation [4] (details in the full version of this article [6]).

4.2.1. Definition (Call-by-Value CPS transformation). The eager, or *Call-by-Value*, CPS transformation can be expressed as a derivor over the two-sorted syntax

$$\mathrm{V} ::= x \tag{10}$$

$$\mathrm{E} ::= \mathrm{V} \mid \lambda x.\mathrm{E} \mid \mathrm{E}_0\ \mathrm{E}_1 \mid \overline{\lambda} x.\mathrm{E} \mid \mathrm{E}_0 \,\overline{\ }\, \mathrm{E}_1 \mid \mathit{CPS1}(\mathrm{E}) \mid \langle \mathrm{E} \rangle \tag{11}$$

(the first sort just contains variables) with rules

$$\mathit{CPS1}(\mathrm{E}) \to \lambda k.\langle \mathrm{E} \rangle \,\overline{\ }\, (\overline{\lambda} m.km) \tag{12}$$

$$\langle \mathrm{V} \rangle \to \overline{\lambda} k.k \,\overline{\ }\, \mathrm{V} \tag{13}$$

$$\langle \lambda x.\mathrm{E}[x] \rangle \to \overline{\lambda} k.k \,\overline{\ }\, (\lambda x.\lambda k.\langle \mathrm{E}[x] \rangle \,\overline{\ }\, (\overline{\lambda} m.km)) \tag{14}$$

$$\langle \mathrm{E}_0 \mathrm{E}_1 \rangle \to \overline{\lambda} k.\langle \mathrm{E}_0 \rangle \,\overline{\ }\, (\overline{\lambda} m.\langle \mathrm{E}_1 \rangle \,\overline{\ }\, (\overline{\lambda} n.mn(\lambda a.k \,\overline{\ }\, a))) \tag{15}$$

(where we exploit the sorting to ensure that (13) is only applied to variables).

The *CPS1* system is obviously a two-level derivor. It is possible to prove its well-annotatedness and termination directly using a typing argument [4, 19]. Instead, let us integrate the "administrative" $\overline{\beta}$-contractions in the transformation, making it truly one-pass in a rewriting sense; this will mechanically lead us to Sabry and Felleisen's "compacting" CPS transformation [22]. That this integration is well-defined is clear from Proposition 4.1.6. What remains is to express the transformation as a derivor that does not require "post-processing" in the form of static reductions or erasure to make it obvious that it cannot generate static applications or static abstractions.

4.3 Deriving specializers

4.3.1. Definition. A two-level derivor is a *specializer* if its normal forms are λ-terms.

A specializer thus encodes static reductions into the derivor (so specializers are trivially well-annotated). The "Holy Grail of Partial Evaluation" follows:

4.3.2. Corollary. *Specializers are convergent (since they are derivors).*

So merely *expressing* a program transformation as a two-level derivor whose normal forms are $\boldsymbol{\lambda}$-terms ensures both that "static reduction does not go wrong" and that static normal forms (*i.e.*, specialized programs) exist and are unique.

The remainder of this section is devoted to show how one can mechanically obtain a specializer from a well-annotated two-level derivor and vice versa.

4.3.3. Theorem. *Let $\mathcal{D}$ be a two-level derivor. Then there is a specializer realizing $\underset{\mathcal{D}}{\rightarrow} \cdot \underset{\beta}{\twoheadrightarrow}$ if and only if $\mathcal{D}$ is well-annotated.*

Before we prove this by actually constructing the specializer, let us illustrate the method for our example system. First we observe that the problematic function symbol is $\langle\cdot\rangle$ because it has the (normalization function) type $\boldsymbol{\lambda} \to \overline{\boldsymbol{\lambda}} \to \overline{\boldsymbol{\lambda}}$ which is of order 2, and we want to change it to make the result belong to $\boldsymbol{\lambda}$. The technique we use is to *uncurry* the uses of $\langle\cdot\rangle$ to obtain something which has the type $\boldsymbol{\lambda} \times (\overline{\boldsymbol{\lambda}} \to \overline{\boldsymbol{\lambda}}) \to \boldsymbol{\lambda}$ and thus creates no overlines anywhere. This is expressed by the *representation shift* from the curried $\langle \mathrm{E}_1 \rangle \overline{\ } (\overline{\lambda} m.\mathrm{E}_2[m])$ to the uncurried $\langle \mathrm{E}_1, \overline{\lambda} m.\mathrm{E}_2[m] \rangle$ which gives the following set of rules (the first one again for initialisation):

$$CPS2(\mathrm{E}) \to \lambda k.\langle \mathrm{E}, \overline{\lambda} m.km \rangle \tag{16}$$

$$\langle \mathrm{V}, \overline{\lambda} k.\mathrm{F}[k] \rangle \to \mathrm{F}[\mathrm{V}] \tag{17}$$

$$\langle \lambda x.\mathrm{E}[x], \overline{\lambda} k.\mathrm{F}[k] \rangle \to \mathrm{F}[\lambda x.\lambda k.\langle \mathrm{E}[x], \overline{\lambda} m.km \rangle] \tag{18}$$

$$\langle \mathrm{E}_0\mathrm{E}_1, \overline{\lambda} k.\mathrm{F}[k] \rangle \to \langle \mathrm{E}_0, \overline{\lambda} m.\langle \mathrm{E}_1, \overline{\lambda} n.mn(\lambda a.\mathrm{F}[a]) \rangle \rangle \tag{19}$$

This is sufficient since there are now no $\overline{@}$s on any right-hand side and all $\overline{\lambda}$s are eliminated by $\langle\cdot,\cdot\rangle$.

4.3.4. Proposition. *CPS2 is a specializer.*

Proof. *CPS2* is clearly a compositional derivor, hence $\langle\cdot,\cdot\rangle$ is well-defined as a normalisation function with with type $\boldsymbol{\lambda} \times (\overline{\boldsymbol{\lambda}} \to \overline{\boldsymbol{\lambda}}) \to \boldsymbol{\lambda}$; from this the proposition follows, which is easy since in the constructor CRS with $\langle\cdot,\cdot\rangle$, the only function symbol is closed with respect to the sub-$\overline{\boldsymbol{\lambda}}$ system with terms

$$\mathrm{A} ::= x \mid \lambda x.\mathrm{A} \mid \mathrm{A}_0\mathrm{A}_1 \mid \langle \mathrm{A}, \overline{\lambda} x.\mathrm{A} \rangle \tag{20}$$

which degenerates to $\boldsymbol{\lambda}$ for normal forms because all possible variations of $\langle \mathrm{A}, \overline{\lambda} x.\mathrm{A} \rangle$ match one of (17–19). □

This integration of administrative reductions into the CPS transformation is known for several years now [4, 22, 23]. What we have done here is to derive it using our rewriting account of partial evaluation. In particular, the resulting term need no post-processing (such as erasing remaining annotations).

Even systems with higher-order types can be handled by first reducing the type order with supercombinator extraction [10], which one could call "meta-λ-lifting" since it is targeted at lifting out all higher-order applications of $\overline{@}$. This is, in fact, what we did with the tree flattening Example 3.2.2, and what we use in the following general construction.

Proof of Theorem 4.3.3.

Case $\Rightarrow$. Given the two-level derivor, $\mathcal{D}$, and a specializer, $\mathcal{S}$, such that $\underset{\mathcal{S}}{\rightarrow} = \underset{\mathcal{D}}{\rightarrow} \cdot \underset{\overline{\beta}}{\twoheadrightarrow\!\!\shortmid}$. Then $\mathcal{D}$ is well-annotated because we know static reduction will finish with a term containing no $\overline{@}$s.

Case $\Leftarrow$. Given $\mathcal{D}$ a well-annotated two-level derivor. Then $\underset{\mathcal{D}}{\rightarrow} \cdot \underset{\overline{\beta}}{\twoheadrightarrow\!\!\shortmid}$ is a function into λ. Let us specify how to transform the rules $\mathcal{D}$ into a specializer. Clearly, the problem is to get rid of $\overline{@}$s on the right-hand sides. Thus we have three subcases:

Base subcase: If there are no $\overline{@}$s at all then the system is already a specializer because all $\overline{\lambda}$s must be eliminated in some way by the system even without $(\overline{\beta})$ because no $\overline{\beta}$-redexes are created.

Uncurry: If the system has a rule of the form $F^n(\vec{t}) \rightarrow \overline{\lambda} k.s$ then add the new rule $F^n(\vec{\mathrm{T}}) \rightarrow F_1^{n+1}(\vec{\mathrm{T}}, \overline{\lambda} k.k)$ with F_1 a fresh function symbol, and replace the rule with $F_1^{n+1}(\vec{t}, \overline{\lambda} k.\mathrm{E}[k]) \rightarrow s'$ where s' is obtained from s by replacing

- all occurrences of $k \,\overline{\ }\, t'$, for some t', by $\mathrm{E}[t']$, and
- all occurrences of $(F(\vec{t'}))\,\overline{\ }\, t''$, for some $\vec{t'}, t''$, by $F_1(\vec{t'}, t'')$.

with E a new meta-variable (of the appropriate sort).

$\overline{\lambda}$-lift step: If the system has a rule of the form $F^n(\vec{t}) \rightarrow C\{\overline{\lambda} k.s\}$ which was not generated by uncurrying and where $C\{\cdot\}$ is a non-empty context, then add the (generic) rule $A^2(\overline{\lambda} x.\mathrm{Z}[x], \mathrm{T}) \rightarrow \mathrm{Z}[x]$ and replace the rule with $D^n(\vec{t}) \rightarrow C\{\overline{\lambda} k.s'\}$ where s' is obtained by replacing in s all occurrences of $k \,\overline{\ }\, t$ by $A^2(k, t)$.

The iteration terminates (with a number of iterations corresponding to the order of the involved two-level types). The resulting system has no $\overline{@}$s left because the well-annotated $\mathcal{D}$ cannot have other instances of $\overline{@}$ which would not be $\overline{\beta}$-reducible.

From the two cases we conclude that $\underset{\mathcal{D}}{\rightarrow} \cdot \underset{\beta}{\twoheadrightarrow}$ is a specializer if and only if the two-level derivor $\mathcal{D}$ is well-annotated. □

5 Conclusion

Foremost we report a success: using higher-order rewriting, we have been able to formalize the partial-evaluation technique of two-level programming, and we have illustrated it with two non-trivial examples: flattening a binary tree in Section 3 and the so-called "one-pass" CPS transformation in Section 4. The immediate benefit, from a partial-evaluation point of view, is obvious: the formalization comes with a generic proof technique to establish the correctness of program specialization. A dual benefit also holds, from the rewriting point of view: the idea of tapping into a source of examples where being higher-order is what makes the examples work.

Why CRSs? One question immediately arises: Why have we used CRSs rather than any of the other formalisms for higher-order rewriting? In particular the complexity of Definition 3.1.1, due to the fact that it is "standalone," seems excessive. The major reason was that we have found CRSs were very easy to understand in an informal and intuitive way, first of all due to Klop, van Oostrom, and van Raamsdonk's survey [16]. Once we had worked with a few examples, CRSs have posed few problems. It is perhaps an significant factor here that program transformation is a very "syntactic" activity and the purpose of CRSs was to provide a syntactic theory of systems with binding [14].

One could instead use a formalism founded on known systems: such are usually much more concisely defined (for better and worse). A good candidate for this is HRS [18] where the "substitution calculus" used to describe the mechanics of rewrite steps is Church's λ^{τ} (simply typed λ-terms with β). Two difficulties need to be overcome: (1) The notion of "binding" in HRSs is more semantic, which makes it nonobvious to work with free variables as we did in (10), something most program transformers do. (2) The syntactic constructors of the source language need to be typed in HRSs to ensure that the notion of substitution is well-defined. The (weaker) "calculus of developments" used by CRSs guarantees that substitution terminates no matter which constructions are used so no special considerations are needed [25]. One could see the demand for typing as an advantage, in particular in our last proof where the "uncurrying" is type directed: it would be nice if the underlying formalism provided support for system transformations involving type changes.

An even more drastic approach would be to use a formalism where the substitution calculus is a "plug-in" such as HORS [26]: this could provide for more advanced notions of "static"reduction, for example including arithmetic as needed by the first-order "power" example. One worry remains, however: the typed systems (including HRSs) work on $\beta\eta$-long normal forms. It is not clear to which extent this interferes with the transformations and syntactic constraints we have discussed.

Related work: We only know of three lines of work relating rewriting and partial evaluation, and none that establish a common ground between them. (They focus more on highlighting the fact that TRSs can be seen as a fully functional programming language but did not exploit rewriting technology for the formalization of partial evaluation.) In his M.Sc. thesis [2], Bondorf investigated the (self-applicable) partial evaluation of TRSs. He thus wrote a partial evaluator for TRSs, using a TRS. Sherman and Strandh [24] use partial evaluation to optimize the implementation of term-rewriting systems. Dershowitz [7] uses rewriting as the basic mechanism for abstracting and instantiating program schemas.

Higher-order systems such as λ-prolog or Elf can also be used for program transformation. For example, Danvy and Pfenning have formalized the CPS transformation in Elf [5].

Future work: In addition to the understanding better the rôle of types in higher-order rewriting, we plan to investigate the relation to specific published notions of reduction and λ-lifting, specifically 2-level λ-lifting [17].

Acknowledgements: To the anonymous referees for perceptive comments, and to Tobias Nipkow for encouraging us not to stop at CRSs to formalize partial evaluation.

References

[1] Henk Barendregt. *The Lambda Calculus — Its Syntax and Semantics.* North-Holland, 1984.

[2] Anders Bondorf. Towards a self-applicable partial evaluator for term rewriting systems. In Dines Bjørner, Andrei P. Ershov, and Neil D. Jones, editors, *Partial Evaluation and Mixed Computation*, pages 27–50. North-Holland, 1988.

[3] Charles Consel and Olivier Danvy. Tutorial notes on partial evaluation. In Susan L. Graham, editor, *Proceedings of the Twentieth Annual ACM Symposium on Principles of Programming Languages*, pages 493–501, Charleston, South Carolina, January 1993. ACM Press.

[4] Olivier Danvy and Andrzej Filinski. Representing control, a study of the CPS transformation. *Mathematical Structures in Computer Science*, 2(4):361–391, December 1992.

[5] Olivier Danvy and Frank Pfenning. The occurrence of continuation parameters in CPS terms. Technical report CMU-CS-95-121, School of Computer Science, Carnegie Mellon University, Pittsburgh, Pennsylvania, February 1995.

[6] Olivier Danvy and Kristoffer Høgsbro Rose. Higher-order rewriting and partial evaluation. Technical Report BRICS RS-97-46, Department of Computer Science, University of Aarhus, Aarhus, Denmark, December 1997.

[7] Nachum Dershowitz. Program abstraction and instantiation. *ACM Transactions on Programming Languages and Systems*, 7(3):446–477, 1985.

[8] Nachum Dershowitz and Zohar Manna. Proving termination with multiset orderings. *Communications of the ACM*, 22(8):465–476, 1979.

[9] Daniel P. Friedman, Mitchell Wand, and Christopher T. Haynes. *Essentials of Programming Languages.* The MIT Press and McGraw-Hill, 1991.

[10] John Hughes. Super combinators: A new implementation method for applicative languages. In Daniel P. Friedman and David S. Wise, editors, *Conference Record of the 1982 ACM Symposium on Lisp and Functional Programming*, pages 1–10, Pittsburgh, Pennsylvania, August 1982.

[11] Neil D. Jones, Carsten K. Gomard, and Peter Sestoft. *Partial Evaluation and Automatic Program Generation.* Prentice Hall International Series in Computer Science. Prentice-Hall, 1993.

[12] Neil D. Jones, Peter Sestoft, and Harald Søndergaard. An experiment in partial evaluation: The generation of a compiler generator. In Jean-Pierre Jouannaud, editor, *Rewriting Techniques and Applications*, number 202 in Lecture Notes in Computer Science, pages 124–140, Dijon, France, May 1985.

[13] Stephen C. Kleene. *Introduction to Metamathematics.* D. van Nostrand, Princeton, New Jersey, 1952.

[14] Jan Willem Klop. *Combinatory Reduction Systems.* Mathematical Centre Tracts 127. Mathematisch Centrum, Amsterdam, 1980.

[15] Jan Willem Klop. Term rewriting systems. In Samson Abramsky, Dov M. Gabby, and T. S. E. Maibaum, editors, *Handbook of Logic in Computer Science, Vol. 2*, chapter 1, pages 2–116. Oxford University Press, Oxford, 1992.

[16] Jan Willem Klop, Vincent van Oostrom, and Femke van Raamsdonk. Combinatory reduction systems: Introduction and survey. *Theoretical Computer Science*, 121:279–308, 1993.

[17] Flemming Nielson and Hanne Riis Nielson. 2-level λ-lifting. In Harald Ganzinger, editor, *Proceedings of the Second European Symposium on Programming*, number 300 in Lecture Notes in Computer Science, pages 328–343, Nancy, France, March 1988.

[18] Tobias Nipkow. Orthogonal higher-order rewrite systems are confluent. In M. Bezem and J. F. Groote, editors, *Typed Lambda Calculi and Applications*, number 664 in Lecture Notes in Computer Science, pages 306–317, Utrecht, The Netherlands, March 1993.

[19] Jens Palsberg. Correctness of binding-time analysis. *Journal of Functional Programming*, 3(3):347–363, July 1993.

[20] Gordon D. Plotkin. Call-by-name, call-by-value and the λ-calculus. *Theoretical Computer Science*, 1:125–159, 1975.

[21] Kristoffer Høgsbro Rose. *Operational Reduction Models for Functional Programming Languages.* PhD thesis, DIKU, Computer Science Department, University of Copenhagen, Universitetsparken 1, DK-2100 København Ø, February 1996. DIKU report 96/1, available from ⟨URL: http://www.diku.dk/research/published/96-1.ps.gz⟩.

[22] Amr Sabry and Matthias Felleisen. Reasoning about programs in continuation-passing style. *LISP and Symbolic Computation*, 6(3/4):289–360, December 1993.

[23] Amr Sabry and Philip Wadler. Compiling with reflections. In R. Kent Dybvig, editor, *Proceedings of the 1996 ACM SIGPLAN International Conference on Functional Programming*, pages 13–24, Philadelphia, Pennsylvania, May 1996. ACM Press.

[24] David Sherman and Robert Strandh. Optimization of equational programs using partial evaluation. In Paul Hudak and Neil D. Jones, editors, *Proceedings of the ACM SIGPLAN Symposium on Partial Evaluation and Semantics-Based Program Manipulation*, SIGPLAN Notices, Vol. 26, No 9, pages 72–82, New Haven, Connecticut, June 1991. ACM Press.

[25] Vincent van Oostrom and Femke van Raamsdonk. Comparing combinatory reduction systems and higher-order rewrite systems. In *HOA-93*, volume 816 of *LNCS*, pages 276–304. Springer-Verlag, 1993.

[26] Vincent van Oostrom and Femke van Raamsdonk. Weak orthogonality implies confluence: the higher-order case. Technical Report CS-R9501, CWI, 1995.

SN Combinators and Partial Combinatory Algebras

Yohji AKAMA

Department of Information Science, Tokyo University, Bunkyou-ku, Tokyo, 113, (e-mail)akama@is.s.u-tokyo.ac.jp

Abstract. We introduce an intersection typing system for combinatory logic. We prove the soundness and completeness for the class of partial combinatory algebras. We derive that a term of combinatory logic is typeable iff it is SN. Let $\mathcal{F}$ be the class of non-empty filters which consist of types. Then $\mathcal{F}$ is an extensional non-total partial combinatory algebra. Furthermore, it is a fully abstract model with respect to the set of SN terms of combinatory logic. By $\mathcal{F}$, we can solve Bethke-Klop's question; "find a suitable representation of the finally collapsed partial combinatory algebra of $\mathcal{P}$". Here, $\mathcal{P}$ is a partial combinatory algebra, and is the set of closed SN terms of combinatory logic modulo the inherent equality. Our solution is the following: the finally collapsed partial combinatory algebra of $\mathcal{P}$ is representable in $\mathcal{F}$. To be more precise, it is isomorphically embeddable into $\mathcal{F}$.

1 Introduction

Combinatory logic (CL, for short) is a simple rewriting system where the terms (CL-terms, for short) are built up from variables and the two basic combinators $\mathbf{S}, \mathbf{K}$ by means of term application (MN). The rewriting rules are

$$\mathbf{S}LMN \to_{cl} LN(MN). \quad \mathbf{K}MN \to_{cl} M.$$

The set of closed SN CL-terms and the set of CL-terms will be denoted by $\mathcal{SN}^0$ and $\mathcal{T}erm$, respectively.

Once the set of SN CL-terms is characterized, then so is the set of SN terms of various λ-calculi $\lambda\#$, through a feasible translation $(-)^\#$ such that A is SN in the λ-calculus $\lambda\#$ iff $A^\#$ is SN in CL. By "feasible", we mean that the translation simultaneously replaces each $\lambda x, \lambda y, \ldots$ with an abstraction algorithm $\lambda^\# x, \lambda^\# y, \ldots$ such that $(\lambda^\# x. M)N \to^+_{cl} M[x := N]$.

Example 1. – The weak λ-calculus λW. The redexes are exactly outside abstractions (Most functional programming languages are subsystems of this calculus). Here $\lambda^W x. MN \equiv \mathbf{S}(\lambda^W x. M)(\lambda^W x. N)$; $\lambda^W x. x \equiv \mathbf{SKK}$; $\lambda^W x. a \equiv \mathbf{K}a$, otherwise.

– Howard-Hindley-Cağman's weak λ-calculus [7] λH. The redexes are exactly the *free* β-redexes (A functional programming language which is a subsystem of this calculus is implemented by *super combinators* due to Turner).

Here $\lambda^H x.\, x \equiv \mathbf{SKK}$; $\lambda^H x.\, M \equiv \mathbf{K}M$, if x is not (free) in M; $\lambda^H x.\, M'x \equiv M'$, if x is not (free) in M'; $\lambda^H x.\, MN \equiv \mathbf{S}(\lambda^H x.\, M)(\lambda^H x.\, N)$, otherwise.
- $\lambda\beta(\eta)$-calculus. Here $\lambda^\beta x.\, M \equiv \mathbf{K}(\lambda^T x.\, M)(M[x := \circledast])$. See Akama [1].

So, we characterize the set of SN CL-terms, by introducing an intersection type assignment system for CL, such that for every CL-term A, A is SN iff A has a type. The system enjoys soundness and completeness for the class of PCA.

According to Bethke-Klop [6],

- a quotient structure $\mathcal{SN}^0/ =_{cl}$, (abbreviated to $\mathcal{P}$) is a *partial combinatory algebra* (PCA, for short).
- The finally collapsed PCA, which is obtained from a PCA $\mathcal{A}$ by equating elements as much as possible with keeping the structure of a PCA, is a quotient structure $\mathcal{A}/ \approx_{\mathcal{A}}$. Here, $\approx_{\mathcal{A}}$ is an equivalence relation over $\mathcal{A}$. The definition of $\approx_{\mathcal{A}}$ resembles that of the strongest consistent equality $=_{\mathcal{K}^*}$ between λ-terms (See Barendregt [2]).

We represent Bethke-Klop's final collapse $\mathcal{P} \;/ \approx_{\mathcal{P}}$ of a PCA $\mathcal{P}$ [6], in the so-called *filter domain* of our intersection type assignment system. The representation result is a characterization of $\approx_{\mathcal{P}}$ in the filter domain.

Intersection type discipline was introduced by Coppo, Dezani-Ciancaglini and Salle in 1970s originally for the λ-calculus. The reason why we introduce an intersection type assignment system for CL is the following features of the intersection type assignment system for the λ-calculus:

1. It characterizes strong/weak normalization properties of λ-terms (See van Bakel [20], for example).
2. It characterizes the strongest consistent equality $=_{\mathcal{K}^*}$ between λ-terms. The characterization of $=_{\mathcal{K}^*}$ involves Böhm out technique (Barendregt [2]); given any "non-equivalent" λ-terms M, N, we should construct a context $C[\,]$ such that $C[M]$ has a head normal form but $C[N]$ not.

The technical originality of our paper will be following:

1. A λ-term does not have the arity, and is eager (e.g. $(\lambda xyz.\, xz(yz))1 \to_\beta \lambda yz.\, 1z(yz)$). On the other hand a CL-term is lazy (e.g. $\mathbf{S}1 \not\to_{cl}$). The arity of $\mathbf{S}$ is three. To cope with the laziness of CL, we used the maximum type ω ($A \le \omega$ for all type A) and equipped our type assignment system with an axiom $\mathbf{S} : \omega \to \omega \to \omega$. The axiom reflects the partiality of the application operator of the PCA's.
2. When we represent Bethke-Klop's final collapse of a PCA $\mathcal{P}$, we will introduce another Böhm out technique $\mathcal{P} \;/ \approx_{\mathcal{P}}$ for CL: Given any "non-equivalent" CL-terms M, N, we should construct a context $C[\,]$ such that $C[M]$ is SN but $C[N]$ not. The $C[\,]$ is much more simpler than the $C[\,]$ of original Böhm out technique for the λ-calculus.

Variants of our type assignment system for CL are also sound and complete for more general, recently studied partial combinatory algebras, such as *conditional partial combinatory algebras* introduced by Hyland-Ong [13].

The organization of the rest of this paper is the following: In Section 2, we will introduce an intersection type assignment system which characterizes SN. In Section 3, we will prove the soundness and the completeness for the class of PCA's. The completeness will be proved by verifying that the system's filter domain $\mathcal{F}$ is indeed a PCA. In Section 4, we will embed $\mathcal{P} \mathbin{/} \approx_{\mathcal{P}}$ into $\mathcal{F}$. In Section 5, we will conclude this paper with some remark.

2 Characterization of SN

Before we introduce an intersection type assignment system such that a closed CL-term M is SN iff $\vdash M : \omega$, let's observe some properties of SN CL-terms. Let $\Delta \equiv \lambda^W x.\, xx \equiv \mathbf{S}(\mathbf{SKK})(\mathbf{SKK})$. Then, $\mathbf{S}(\mathbf{K}\Delta)(\mathbf{K}\Delta)$ is SN, but not SN if applied to any CL-term M; $\mathbf{S}(\mathbf{K}\Delta)(\mathbf{K}\Delta)M \to_{cl} \mathbf{K}\Delta M(\mathbf{K}\Delta M) \to_{cl} \Delta\Delta \to_{cl} \Delta\Delta$. This forces $\vdash \mathbf{S}(\mathbf{K}\Delta)(\mathbf{K}\Delta) : \omega$ and $\not\vdash \mathbf{S}(\mathbf{K}\Delta)(\mathbf{K}\Delta) : \omega \to \omega$; hence in our typing system, ω is not subsumed by $\omega \to \omega$. To cope with above property of $\mathbf{S}$, we include a typing axiom

$$\frac{}{\mathbf{S} : \omega \to \omega \to \omega}\ (\mathbf{S_2}).$$

Our system is similar to Dezani-Hindley's [8] intersection type assignment system for CL, except that ours lacks a type-subsumption axiom $\omega \leq \omega \to \omega$ but has $(\mathbf{S_2})$. Our type structure is exactly the same as Egidi-Honsell-Ronchi's intersection type assignment system for Plotkin lazy call-by-value λ-calculus [19].

Definition 1. *1. The* types *are generated by $T_\wedge ::= \omega \mid T_\wedge \wedge T_\wedge \mid T_\wedge \to T_\wedge$. Here ω is a unique type constant.*
The type-subsumption $\leq \subseteq T_\wedge \times T_\wedge$ *is the* preorder *such that*

$$A \leq \omega. \qquad A \leq A \wedge A. \qquad A \wedge B \leq A. \qquad A \wedge B \leq B.$$
$$A \leq A', B \leq B' \Rightarrow A \wedge B \leq A' \wedge B'. \quad A \leq A', B \leq B' \Rightarrow A' \to B \leq A \to B'.$$
$$(A \to B) \wedge (A \to B') \leq A \to (B \wedge B').$$

2. The typing rules *are* $(\mathbf{S_2})$ *and the following:*

$$\frac{M : A}{M : B}\ (\leq)\ \text{if } A \leq B. \qquad \frac{M : A \quad M : B}{M : A \wedge B}\ (\wedge i). \qquad \frac{M : A \to B \quad N : A}{MN : B}\ (\to e).$$

$$\frac{}{\mathbf{S} : (A \to B \to C) \to (A \to B) \to A \to C}\ (\mathbf{S}). \qquad \frac{}{\mathbf{K} : A \to B \to A}\ (\mathbf{K}).$$

Let Γ be a possibly infinite list of the form $x : A$, where x is any variable and $A \in T_\wedge$. When we can derive $M : A$ from Γ, we write $\Gamma \vdash M : A$.

We note that if $\Gamma \vdash M : A$ then $\{x : B \mid x \text{ occurs in } M\} \vdash M : A$.

We will define a computability predicate for the typing.

Definition 2. *We let*

$$\begin{aligned}
Comp(\Gamma, M, \omega) &\iff \Gamma \vdash M : \omega \text{ and } M \text{ is SN} \\
Comp(\Gamma, M, A \to B) &\iff \forall \Gamma', N.\ (Comp(\Gamma', N, A) \Rightarrow Comp(\Gamma \cup \Gamma', MN, B)) \\
Comp(\Gamma, M, A_1 \wedge A_2) &\iff Comp(\Gamma, M, A_1)\ \&\ Comp(\Gamma, M, A_2).
\end{aligned}$$

Lemma 1. *1.* $\Gamma \vdash x\vec{M}: A$ *&* $x\vec{M}$ *is* SN $\Rightarrow$ *Comp*$(\Gamma, x\vec{M}, A)$.
2. *Comp*$(\Gamma, M, A) \Rightarrow \Gamma \vdash M : A$ *and* M *is* SN.

Proof. By induction on A.

Lemma 2. *1. If* *Comp*(Γ', N, B) *and* *Comp*$(\Gamma, C[M], A)$, *then* *Comp*$(\Gamma \cup \Gamma', C[\mathbf{K}MN], A)$.
2. *Comp*$(\Gamma, C[LN(MN)], A) \Rightarrow$ *Comp*$(\Gamma, C[\mathbf{S}LMN], A)$.

Proof. By induction on A. **Case** $A \equiv \omega$**:** As for the claim (1): Lemma 1(2) implies that $\Gamma \vdash C[M] : A$, $C[M]$ is SN, $\Gamma' \vdash N : B$, and N is SN. So, by the Subject Expansion Lemma, $\Gamma \cup \Gamma' \vdash C[\mathbf{K}MN] : A$. By Lemma 7, $C[\mathbf{K}MN]$ is SN. Therefore, OK. As for the claim (2): By Lemma 1(2), $\Gamma \vdash C[LN(MN)] : A$, and $C[LN(MN)]$ is SN. By the Subject Expansion Lemma, $\Gamma \vdash C[\mathbf{S}LMN] : A$. By Lemma 7, $C[\mathbf{S}LMN]$ is SN. Therefore, we have the desired conclusion. **The other case:** Calculate according to the definition of *Comp* for higher types.

Theorem 1. *The computability predicate is sound and complete for the typing. To be precise, the following are equivalent*

1. $x_1 : B_1, \ldots, x_n : B_n \vdash M : A$.
2. *For each* N_i *such that* *Comp*(Γ_i, N_i, B) *with* $1 \le i \le n$,

$$Comp(\cup_i \Gamma_i, M[x_1 := N_1, \ldots, x_n := N_n], A).$$

Proof. The implication from (1) to (2) is already proved by Lemma 1(1). The converse will be proved by induction on $\Gamma \vdash M : A$.

Case $\dfrac{}{S : \omega \to \omega \to \omega}(\mathbf{S_2})$: Assume *Comp*$(\Gamma_i, M_i, \omega)$ $(i = 1, 2)$. By Lemma 1(2), M_i is SN and $\Gamma_i \vdash M_i : \omega$. So, $\mathbf{S}M_1M_2$ is SN, and $\Gamma_1 \cup \Gamma_2 \vdash \mathbf{S}M_1M_2 : \omega$. By the definition, *Comp*$(\Gamma_1 \cup \Gamma_2, \mathbf{S}M_1M_2, \omega)$, and thus *Comp*$(\emptyset, \mathbf{S}, \omega \to \omega \to \omega)$. From it, we can prove the desired conclusion.

Case $\dfrac{}{\mathbf{S} : (D \to B \to E) \to (D \to B) \to D \to E}(\mathbf{S})$: Suppose *Comp*$(\Gamma_1, M_1, D \to B \to E)$, *Comp*$(\Gamma_2, M_2, D \to B)$, and *Comp*$(\Gamma_3, M_3, D)$. By the definition, *Comp*$(\cup_i \Gamma_i, M_1M_3(M_2M_3), E)$. By Lemma 2(2), *Comp*$(\cup_i \Gamma_i, \mathbf{S}M_1M_2M_3, E)$. Therefore *Comp*$\big(\emptyset, \mathbf{S}, (D \to B \to E) \to (D \to B) \to D \to E\big)$. From it, we can prove *Comp*$(\Gamma, \mathbf{S}, A)$.

Case $\dfrac{L : B \to A \quad N : B}{LN : A}$: By I.H.'s, *Comp*$(\cup_i \Gamma_i, L[\ldots], B \to A)$ and *Comp*$(\cup_i \Gamma_i, N[\ldots], B)$. By the definition of *Comp*, *Comp*$\big(\cup_i \Gamma_i, (LN)[\ldots], A\big)$.

Case $\dfrac{M : B}{M : A}$ $(\le)$ if $B \le A$: By I.H., *Comp*$(\cup_i \Gamma_i, M[...], B)$. By induction on $B \le A$, we have the desired consequence. **The other cases** are easy.

From Lemma 1(1) and the previous Theorem, we derive

Corollary 1. *If* $\Gamma \vdash M : A$, *then* M *is* SN.

Below, we will prove the converse of the Corollary.

Lemma 3. *For each M in normal form, there exist Γ such that $\Gamma \vdash M : \omega$.*

Proof. M is either $x\,\vec{N}$, $\mathbf{S}, \mathbf{S}M'$, $\mathbf{S}M'M''$, $\mathbf{K}$, or $\mathbf{K}M'$ with $\vec{N}, M'$ and M'' being normal and x being a variable. By induction on the M, we can prove the statement. Use $\vdash \mathbf{S}, \mathbf{K} : \omega \to \omega \to \omega$.

Lemma 4 (Generation). $\Gamma \vdash MN : A \Rightarrow \exists B. \Gamma \vdash M : B \to A \;\&\; \Gamma \vdash N : B$.

The proof is by induction on the premise.

Lemma 5 (Subject Expansion). *Let $C \in \mathrm{Contexts}_{Term}$.*

1. *If $\Gamma \vdash C[ML(NL)] : D$, then $\Gamma \vdash C[\mathbf{S}MNL] : D$.*
2. *If $\Gamma \vdash C[M] : D$ and $\Gamma' \vdash N : E$ for some E, then $\Gamma \cup \Gamma' \vdash C[\mathbf{K}MN] : D$.*

Proof. The proof is by induction on the length of C. **Case $C \equiv []$:** For claim (1), we first repeatedly apply the Generation Lemma. Some A_1, A_2, B satisfy $\Gamma \vdash M : A_1 \to B \to D$, $\Gamma \vdash L : A_1$, $\Gamma \vdash N : A_2 \to B$, and $\Gamma \vdash L : A_2$. By $(\leq)$, we have the above four with each A_i replaced with $A_1 \wedge A_2$. By repeatedly applying $(\to e)$, we have $\Gamma \vdash \mathbf{S}MNL : D$. For claim (2), by $(\to e)$, $\Gamma \cup \Gamma' \vdash \mathbf{K}MN : D$. **Case $C \equiv C'M'$:** For claim (1), by the Generation Lemma, for some D' $\Gamma \vdash C'[ML(NL)] : D' \to D \;\&\; \Gamma \vdash M' : D'$. By I.H., $\Gamma \vdash C'[\mathbf{S}MNL] : D' \to D$. So, $\Gamma \vdash C[\mathbf{S}MNL] : D$. For claim (2), the proof is similar. **Case $C \equiv M'C'$:** As above.

Lemma 6. *If M is* SN, *then $\Gamma \vdash M : \omega$ for some Γ.*

Proof. Because M is SN, there is a reduction sequence $M \equiv M_0 \to_{cl} M_1 \to_{cl} \cdots \to_{cl} M_n$ such that the redex of each step does not contain properly a redex, and M_n is normal. By Lemma 3, $\Gamma \vdash M_n : \omega$. When the redex of $M_{n-1} \to_{cl} M_n$ is $\mathbf{K}N_1N_2$, then N_2 is normal by the premise. By Lemma 3, $\Gamma' \vdash N_2 : \omega$ for some Γ'. By the previous Lemma, we have $\Gamma \cup \Gamma' \vdash M_{n-1} : \omega$. When the redex of $M_{n-1} \to_{cl} M_n$ is $\mathbf{S}N_1N_2N_3$, the previous Lemma implies $\Gamma \vdash M_{n-1} : \omega$. By iterating this argument along the reduction sequence, we will establish the conclusion.

To sum up, we have the following:

Theorem 2 (Main). *M is* SN, *iff there is Γ such that $\Gamma \vdash M : \omega$.*

Let Σ be $\{x : A \mid x$ is a variable and $A \in T_\wedge\}$. We will have the following:

Corollary 2. *M is* SN, *iff $\Sigma \vdash M : \omega$.*

3 Soundness and Completeness for the PCA's

In Section 3.1, we review partial combinatory algebras from term rewriting point of view. In Subsection 3.2, we prove the *soundness* of our intersection type assignment system for the class of PCA's. The completeness is proved by constructing a PCA $\mathcal{F}$ canonical to our typing system. The construction is rudimentary; $\mathcal{F}$ is the class of filters consisting of types. Actually $\mathcal{F}$ is extensional (Subsection 3.3). From this completeness, the *subject reduction property* of our typing system follows.

3.1 Partial Combinatory Algebra

A *partial combinatory algebra* (PCA) is a structure $\langle A, \cdot, \mathbf{s}, \mathbf{k}\rangle$ such that $\cdot$ is a partial binary operator on A, and $\mathbf{s}, \mathbf{k}$ are distinct elements of A that satisfies the conditions

$$(\mathbf{S}_2): (\mathbf{s}\cdot f)\cdot g\downarrow, \quad ((\mathbf{s}\cdot f)\cdot g)\cdot a \simeq (f\cdot a)\cdot(g\cdot a), \quad (\mathbf{k}\cdot f)\cdot a = f.$$

The symbol $\simeq$ means that if one side is defined, so is the other side, and that both are equal, while the symbol $=$ means that both sides are defined with equal values.

It is investigated by Beeson [3]. The motivating example of a PCA is the set of natural numbers with the application operator $\{n\}(m)$ ("the value of the n-th unary partial recursive function applied to m"). However, from the viewpoint of TRS, a more relevant example will be the PCA constructed from the set $\mathcal{SN}^0$ of closed SN CL-terms and the intrinsic equality $=_{cl}$ of CL.

Theorem 3 (Bethke-Klop[6]). *Let $\mathcal{P}$ be $\langle \mathcal{SN}^0/=_{cl}, \cdot/=_{cl}, [\mathbf{S}]_{=_{cl}}, [\mathbf{K}]_{=_{cl}}\rangle$ with $\cdot$ being term application. Then, $\mathcal{P}$ is a* PCA.

The well-definedness as a quotient structure is assured by the following:

Lemma 7. *Suppose (a) N is obtained from M by contracting a redex R and (b) if R is $\mathbf{K}M_1M_2$ then M_2 is* SN. *Then the* SN *of N implies the* SN *of M.*

The previous lemma is due to Klop [16], Akama [1, Lemma 2.10], or Gramlich [10, Lemma 3.4.30]:

3.2 Soundness for PCA's

Definition 3 (Models). *Given a* PCA *$\mathcal{D} := \langle D, \cdot, \mathbf{s}, \mathbf{k}\rangle$. Let ξ be a valuation of term variables in $\mathcal{D}$.*

- *Let $[\![-]\!]^{\mathcal{D}} : T_\wedge \to 2^D$ take ω and $\wedge$ respectively to D and the intersection operator. And $[\![A \to B]\!]^{\mathcal{D}} = \{\mathbf{a} \in D \mid \forall \mathbf{b} \in [\![A]\!]^{\mathcal{D}}(\mathbf{a}\cdot\mathbf{b} \in [\![B]\!]^{\mathcal{D}})\}$.*
- *$\mathcal{D}, \xi \models M : A$, iff $[\![M]\!]^{\mathcal{D}}_\xi \in [\![A]\!]^{\mathcal{D}}$.*
  Here $[\![x]\!]^{\mathcal{D}}_\xi = \xi(x)$, $[\![\mathbf{S}]\!]^{\mathcal{D}}_\xi = \mathbf{s}$, $[\![\mathbf{K}]\!]^{\mathcal{D}}_\xi = \mathbf{k}$, $[\![MN]\!]^{\mathcal{D}}_\xi = [\![M]\!]^{\mathcal{D}}_\xi \cdot [\![N]\!]^{\mathcal{D}}_\xi$.
- *$\mathcal{D}, \xi \models \Gamma$, iff $\mathcal{D}, \xi \models x : A$ for every $(x : A) \in \Gamma$.*
- *$\Gamma \models M : A$ iff for all $\mathcal{D}, \xi$ such that $\mathcal{D}, \xi \models \Gamma$, we have $\mathcal{D}, \xi \models M : A$.*

By induction on the derivation of $A \leq B$, we can show that $A \leq B$ implies $[\![A]\!]^{\mathcal{D}} \subseteq [\![B]\!]^{\mathcal{D}}$.

Theorem 4 (Soundness). *If $\Gamma \vdash M : A$, then $\Gamma \models M : A$.*

Proof. By induction on the derivation $\Gamma \vdash M : A$. Let $\mathcal{D} = \langle D, \mathbf{s}, \mathbf{k}, \cdot\rangle$ be an arbitrary PCA. **Case It is inferred by ($\leq$):** Then for some B, $\Gamma \vdash M : B$ with $B \leq A$. By I.H., $\Gamma \models M : B$. By the observation above, we also have $\Gamma \models M : A$. **Case It is an instance of (S):** Let $\mathbf{f} \in [\![A \to B \to C]\!]^{\mathcal{D}}$, $\mathbf{g} \in [\![A \to B]\!]^{\mathcal{D}}$, and $\mathbf{a} \in [\![A]\!]^{\mathcal{D}}$. By the interpretation for arrow type, $\mathbf{fa(ga)} \in [\![C]\!]^{\mathcal{D}}$. By an axiom $\mathbf{sabc} \simeq \mathbf{ac(bc)}$, $\mathbf{sabc} \in [\![C]\!]^{\mathcal{D}}$. Therefore, $[\![\mathbf{S}]\!]^{\mathcal{D}}_\xi \in [\![(A \to B \to C) \to (A \to B) \to A \to C]\!]^{\mathcal{D}}$. **Case It is ($\mathbf{S}_2$):** Let $\mathbf{a}, \mathbf{b} \in [\![\omega]\!]^{\mathcal{D}}$. Because of the definition of PCA, $\mathbf{sab}\downarrow$, i.e., $\mathbf{sab} \in [\![\omega]\!]^{\mathcal{D}}$. Hence, OK. **The other cases:** Clear.

3.3 The Filters of Types as a PCA

Definition 4 (Filter Domain).

1. $\mathbf{a} \in \mathcal{F}$ *if* $\mathbf{a}$ *is a non-empty filter consisting of types. That is,* $\mathbf{a}$ *contains* ω, *is closed w.r.t* $\wedge$, *and is upper-closed w.r.t.* $\leq$. *The filter generated from types* $X, Y, \ldots$ *is denoted by* $\{X, Y, \ldots\}^{\uparrow}$.
2. $\mathbf{a} \cdot \mathbf{b} \ni A :\Longleftrightarrow \exists B.(\mathbf{a} \ni B \rightarrow A$ *and* $\mathbf{b} \ni B)$
3. $\mathbf{s} \ni A :\Longleftrightarrow \vdash \mathbf{S} : A, \qquad \mathbf{k} \ni A :\Longleftrightarrow \vdash \mathbf{K} : A$

Here, we say that a PCA is *extensional*, if $(\forall \mathbf{x}.\ \mathbf{a}_0\mathbf{x} \simeq \mathbf{a}_1\mathbf{x})$ implies $\mathbf{a}_0 = \mathbf{a}_1$. Non-total PCA will be abbreviated to NCA.

Theorem 5. $\langle \mathcal{F}, \cdot, \mathbf{s}, \mathbf{k} \rangle$ *is an extensional* NCA.

The rest of this subsection devotes to the proof of the above theorem.

Proof. We observe that if $\mathbf{a} \cdot \mathbf{b}$ is not empty then it is in $\mathcal{F}$. Because a filter $\{\omega\}^{\uparrow}$ contains no arrow type, it applied to itself is undefined. Therefore, $\mathcal{F}$ is a *non-total* applicative structure.

Furthermore, $\mathbf{s}\mathbf{a}_1\mathbf{a}_2$ is defined for $\mathbf{a}_1, \mathbf{a}_2 \in \mathcal{F}$, because $\omega \rightarrow \omega \rightarrow \omega \in \mathbf{s}$ by definition and the filters $\mathbf{a}_i$ necessarily contain ω.

We observe that $\mathbf{a} \simeq \mathbf{b}$ iff $\mathbf{a}$ and $\mathbf{b}$ are the same as sets. The extensionality of $\mathcal{F}$ can be proved as follows: Suppose $(\forall \mathbf{x} \in \mathcal{F}.\ \mathbf{a}_0\mathbf{x} \subseteq \mathbf{a}_1\mathbf{x})$. If $\omega \in \mathbf{a}_0$, then $\omega \in \mathbf{a}_1$, since each $\mathbf{a}_i$ is a non-empty filter. If $X \rightarrow A \in \mathbf{a}_0$, then $A \in \mathbf{a}_0 \cdot \{X\}^{\uparrow}$. Because of the premise, we have $A \in \mathbf{a}_1 \cdot \{X\}^{\uparrow}$. This means that for some $B \geq X$, $B \rightarrow A \in \mathbf{a}_1$. Since $B \rightarrow A \leq X \rightarrow A$ and $\mathbf{a}_1$ is a filter, we have $X \rightarrow A \in \mathbf{a}_1$. If $A_1 \wedge A_2 \in \mathbf{a}_0$, then $A_i \in \mathbf{a}_0$ for each i. By induction hypotheses, $A_i \in \mathbf{a}_1$, which implies $A_1 \wedge A_2 \in \mathbf{a}_1$. Therefore $\mathbf{a}_0 \subseteq \mathbf{a}_1$.

We have only to prove $\mathbf{sabc} \simeq \mathbf{ac}(\mathbf{bc})$ and $\mathbf{kab} = \mathbf{a}$. For the verification of the fist claim, let $D \in \mathbf{sabc}$. Then $\exists A \in \mathbf{a}, \exists B \in \mathbf{b}, \exists C \in \mathbf{c}\,(A \rightarrow B \rightarrow C \rightarrow D \in \mathbf{s})$. We observe that

$$A \rightarrow B \rightarrow C \rightarrow D \geq \bigwedge_{i \in I} (P_i \rightarrow Q_i \rightarrow R_i) \rightarrow (P_i \rightarrow Q_i) \rightarrow P_i \rightarrow R_i \wedge \omega \rightarrow \omega \rightarrow \omega.$$

for some I. By Lemma 8(1) presented subsequently to this proof, we have for some E, $A \leq C \rightarrow E \rightarrow D$ and $B \leq C \rightarrow E$. Since $\mathbf{a}, \mathbf{b}, \mathbf{c}$ are upper-closed, we have $\mathbf{a} \ni C \rightarrow E \rightarrow D$ and $\mathbf{b} \ni C \rightarrow E$. Therefore $\mathbf{ac}(\mathbf{bc}) \ni D$.

The converse, that is $D \in \mathbf{ac}(\mathbf{bc}) \Rightarrow \mathbf{sabc} \ni D$, is easy.

The second claim $\mathbf{kab} = \mathbf{a}$ is proved similarly to the first claim.

To complete the proof, we prove the following lemma, which is close to Kurata's equivalence condition [17, Theorem 3.5] for *a* filter domain to be a combinatory algebra (CA, for short).

Lemma 8. *1. Suppose*

$$A \rightarrow B \rightarrow C \rightarrow D \geq \bigwedge_{i \in I} (P_i \rightarrow Q_i \rightarrow R_i) \rightarrow (P_i \rightarrow Q_i) \rightarrow P_i \rightarrow R_i \wedge \omega \rightarrow \omega \rightarrow \omega.$$

Then, for some E, $A \leq C \rightarrow E \rightarrow D$ *and* $B \leq C \rightarrow E$.

2. If $A \to B \to C \geq \bigwedge_{i \in I}(P_i \to Q_i \to P_i)$, then $C \geq A$.

To prove the lemma, we first define the *arity* of a type.

Definition 5. *Given $A \in T_\wedge$, define* $\mathrm{arity}(A)$ *as follows:* $\mathrm{arity}(\omega) = 0$, $\mathrm{arity}(A' \to A'') = \mathrm{arity}(A'') + 1$, *and* $\mathrm{arity}(A_1 \wedge A_2) = \max_i(\mathrm{arity}(A_i))$.

The induction on the derivation of $A \leq B$ establishes the following.

Lemma 9. $A \leq B$ *implies* $\mathrm{arity}(A) \geq \mathrm{arity}(B)$.

Therefore, in Lemma 8, the arity of the lhs $A \to B \to C \to D$ of the premise is not greater than that of $\omega \to \omega \to \omega$ in the rhs. We first verify that the lhs is greater than $\bigwedge_{i \in I}(P_i \to Q_i \to R_i) \to (P_i \to Q_i) \to P_i \to R_i$. We then try to draw the conclusion of Lemma 8.

The main tool here to handle the relation $\leq$ is a translation $(-)^*$ from $T_\wedge$ to the set NT of *normal types*. It is defined as follows.

1. If $A_1, \ldots A_n \in \mathsf{NT}$ with $n \geq 0$, then $A_1 \to A_2 \to \cdots \to A_n \to \omega \in \mathsf{NT}$.
2. If $A, B \in \mathsf{NT} - \{\omega\}$, then $A \wedge B \in \mathsf{NT}$.

When a normal type is of the first form, we say that it is an $\to$-*normal type* and write $\mathsf{NT}^{\to}$ for the set of all $\to$-normal types.

$$\omega^* \equiv \omega. \quad (E \to F)^* \equiv \bigwedge_{1 \leq i \leq m} (E^* \to F_i). \quad (E_0 \wedge E_1)^* \equiv \begin{cases} E_i^* & \text{if } E_{1-i}^* \equiv \omega; \\ E_1^* \wedge E_2^* & \text{otherwise.} \end{cases}$$

Here, if $F^* \equiv \omega$, then $m = 1$ and $F_i \equiv \omega$. Otherwise, $F^* \equiv \bigwedge_{1 \leq i \leq m} F_i$ with $F_i \in \mathsf{NT}^{\to} - \{\omega\}$.

The properties of $(-)^*$ useful in verifying Lemma 8 are the following

Proposition 1. $A \leq A^* \leq A$.

Proposition 2. *Suppose $A_1, \ldots, A_m, B_1, \ldots, B_n \in NT^{\to} - \{\omega\}$. Then $\bigwedge_{i=1}^m A_i \leq \bigwedge_{j=1}^n B_j$, iff $\forall j \in \{1, \ldots, n\}, \exists i \in \{1, \ldots, m\}(A_i \leq B_j)$.*

Proposition 3. *For all $A \to B, C \to D \in NT^{\to}$, $A \to B \leq C \to D$ implies $A \geq C$ and $B \leq D$.*

The proofs are omitted. Now we are ready to prove Lemma 8.

As for Claim (1). By Proposition 1, the premise implies

$$\bigwedge_{k \in K}(A^* \to B^* \to C^* \to D_k) \geq \bigwedge_{i \in I,\ j \in J_i} (P_i \to Q_i \to R_i)^* \to (P_i \to Q_i)^* \to P_i^* \to R_i^{(j)} \wedge \omega \to \omega \to \omega, \tag{1}$$

where

$$D^* \equiv \bigwedge_{k \in K} D_k \text{ with } D_k \in \mathsf{NT}^{\to} - \{\omega\}. \qquad R_i^* \equiv \bigwedge_{j \in J_i} R_i^{(j)} \text{ with } R_i^{(j)} \in \mathsf{NT}^{\to} - \{\omega\}.$$

By Proposition 2, for each k, $A^* \to B^* \to C^* \to D_k$ is a supertype of some conjunct of the rhs of Ineq.(1). If $A^* \to B^* \to C^* \to D_k \geq \omega \to \omega \to \omega$, then by comparing the arities of the both side, we derive a contradiction with Lemma 9. Therefore,

$$\forall k \in K\, \exists i_k \in I\, \exists j_k \in J_{i_k}$$
$$A^* \to B^* \to C^* \to D_k \;\geq\; (P_{i_k} \to Q_{i_k} \to R_{i_k})^* \to (P_{i_k} \to Q_{i_k})^* \to P_{i_k}^* \to R_{i_k}^{(j_k)}.$$

By Proposition 3, for each k, we have $A^* \leq (P_{i_k} \to Q_{i_k} \to R_{i_k})^*$, $B^* \leq (P_{i_k} \to Q_{i_k})^*$, $C^* \leq P_{i_k}^*$, and (¶) : $D_k \geq R_{i_k}^{(j_k)}$. By using Proposition 1,

$$A \leq P_{i_k} \to Q_{i_k} \to R_{i_k} \tag{2}$$
$$B \leq P_{i_k} \to Q_{i_k} \tag{3}$$
$$C \leq P_{i_k}\;. \tag{4}$$

We first note that $R_{i_k} \overset{\textit{by Proposition 1}}{\leq} (R_{i_k})^* \equiv \bigwedge_{j \in J_{i_k}} R_{i_k}^{(j)} \leq R_{i_k}^{(j_k)} \overset{\textit{by (¶)}}{\leq} D_k$.
Then,

$$\bigwedge_{k \in K} R_{i_k} \leq \bigwedge_{k \in K} D_k \equiv D^* \overset{\textit{by Proposition 1}}{\leq} D. \tag{5}$$

Let $E \equiv \bigwedge_k Q_{i_k}$. Then $A \overset{\textit{by Ineq(2)}}{\leq} P_{i_k} \to E \to R_{i_k} \overset{\textit{by Ineq.(4)}}{\leq} C \to E \to R_{i_k}$.

So, $A \leq C \to E \to \bigwedge_k R_{i_k} \overset{\textit{by Ineq.(5)}}{\leq} C \to E \to D$.

On the other hand, $B \leq C \to Q_{i_k}$ by Ineq.(4). So, $B \leq C \to E$. This completes the proof of the claim (1) of Lemma 8.

The proof of the claim (2) is similar. This completes Lemma 8. Hence the proof of Theorem 5 is finished.

3.4 Completeness for the PCA's

Lemma 10. $[\![A]\!]^{\mathcal{F}} = \{\mathbf{a} \in \mathcal{F} \mid \mathbf{a} \ni A\}$.

Proof. By induction on A. **Case** $A = \omega$**:** Since every $\mathbf{a} \in \mathcal{F}$ contains ω, rhs is $[\![A]\!] = \mathcal{F}$. **Case** $A = A_1 \to A_2$**:** Then $\mathbf{a} \in [\![A]\!]$ iff $\forall \mathbf{b} \in [\![A_1]\!].\,(\mathbf{ab} \in [\![A_2]\!])$, iff $\forall \mathbf{b}.\,(A_1 \in \mathbf{b} \Rightarrow A_2 \in \mathbf{ab})$ by I.H.'s. Let's denote the last formula by (¶). Since $A_1 \in \{A_1\}^{\uparrow}$, (¶) implies $A_2 \in \mathbf{a} \cdot \{A_1\}^{\uparrow}$. This means that for some $A_1' \geq A_1$, we have $A_1' \to A_2 \in \mathbf{a}$. Therefore $A_1 \to A_2 \in \mathbf{a} \in \mathcal{F}$. Conversely, let $A = A_1 \to A_2 \in \mathbf{a} \in \mathcal{F}$. In view of the application operator of the filters, it is easy to see that this implies (¶). **Case** $A = A_1 \wedge A_2$**:** Then, $\mathbf{a} \in [\![A]\!]$ iff $\mathbf{a} \in [\![A_1]\!]$ and $\mathbf{a} \in [\![A_2]\!]$, iff $A_1 \in \mathbf{a}$ and $A_2 \in \mathbf{a}$ by I.H.'s. Therefore, $A_1 \wedge A_2 \in \mathbf{a}$.

Definition 6. – *Given a finite Γ, define a valuation ξ_Γ in $\mathcal{F}$ by $\xi_\Gamma(x) = \{A \mid \Gamma \vdash x : A\}$. It is easy to see that ξ is indeed a valuation in $\mathcal{F}$.*
– *Given a valuation ξ in $\mathcal{F}$, Let Γ_ξ be $\{x : A \mid A \in \xi(x)\}$.*

Lemma 11. *1. $\mathcal{F}, \xi_\Gamma \models \Gamma$.*

2. $\Gamma \vdash M : A$ *iff* $\Gamma_{\xi_\Gamma} \vdash M : A$.
3. Given a CL*-term* M *and a valuation* ξ *in* $\mathcal{F}$, $[\![M]\!]^{\mathcal{F}}_{\xi} = \{A \mid \Gamma_\xi \vdash M : A\}$.

Proof. The proof is fairly standard. See van Bakel [20].

Theorem 6 (Completeness). *If* $\Gamma \models M : A$*, then* $\Gamma \vdash M : A$.

Proof. The premise implies $\mathcal{F}, \xi_\Gamma \models M : A$ by Lemma 11(2), which is $[\![M]\!]^{\mathcal{F}}_{\xi_\Gamma} \in [\![A]\!]^{\mathcal{F}}$. By Lemma 10, $A \in [\![M]\!]^{\mathcal{F}}_{\xi_\Gamma}$, which implies $\Gamma_{\xi_\Gamma} \vdash M : A$ by Lemma 11(3). By Lemma 11(1), $\Gamma \vdash M : A$.

Theorem 7 (Subject Reduction). $\Gamma \vdash M : A \;\&\; M \to_{cl} M' \Rightarrow \Gamma \vdash M' : A$.

Proof. By Soundness, for all $\mathcal{D}, \xi$ such that $\mathcal{D}, \xi \models \Gamma$, we have $[\![M]\!]^{\mathcal{D}}_{\xi} \in [\![A]\!]^{\mathcal{D}}$. Since $[\![M]\!]^{\mathcal{D}}_{\xi}\downarrow$, we can prove $[\![M]\!]^{\mathcal{D}}_{\xi} = [\![M']\!]^{\mathcal{D}}_{\xi}$, by simple induction on M. Therefore, $\Gamma \models M' : A$. By Completeness, $\Gamma \vdash M' : A$.

4 Bethke-Klop's Problem as Full-Abstraction Problem

4.1 The Final Collapse of PCA

The *finally collapsed* PCA of a PCA $\mathcal{A}$ is obtained from $\mathcal{A}$ by equating elements as much as possible with keeping the structure of PCA. To be precise, the finally collapsed PCA is stated in terms of a *homomorphism* from $\mathcal{A}$ to a PCA.

Definition 7 (Final Collapse (Bethke-Klop[6])). *Let* $\mathcal{A} = \langle A, \mathbf{s}, \mathbf{k}, \cdot\rangle$ *and* $\mathcal{B} = \langle B, \mathbf{s}', \mathbf{k}', \cdot'\rangle$ *be* PCA*'s.*

1. A homomorphism *from* $\mathcal{A}$ *to* $\mathcal{B}$ *is a mapping* $\varphi : A \to B$ *such that* $\varphi(\mathbf{s}) = \mathbf{s}'$, $\varphi(\mathbf{k}) = \mathbf{k}'$, *and* $\varphi(\mathbf{a} \cdot \mathbf{b}) \simeq \varphi(\mathbf{a}) \cdot' \varphi(\mathbf{b})$ *for all* $\mathbf{a}, \mathbf{b} \in A$.
If φ *is bijective, then* φ *is called an* isomorphism.
2. φ *is called a collapse of* $\mathcal{A}$ *if* φ *is a surjective homomorphism from* $\mathcal{A}$ *to a* PCA.
3. A collapse $\varphi : \mathcal{A} \to \mathcal{B}$ *of* $\mathcal{A}$ *is called* final, *if for all collapses* φ' *of* $\mathcal{A}$ *there exists a unique homomorphism* ψ *with* $\psi \circ \varphi' = \varphi$. *We call* $\mathcal{B}$ *the* finally collapsed PCA *of* $\mathcal{A}$.

The final collapse of CA's is investigated by Jacopini[14], Jacopini-Zilli[15], and Berarducci-Intrigila[4]. Here the leading question is whether given λ-terms M and N, the equation $M = N$ can be added consistently to the λ-calculus. Not every extensional CA has a final collapse (Jacopini[14]). However, every non-total PCA has a final collapse.

Definition 8 (Observational Equivalence on $\mathcal{A}$). *Let* $\mathcal{A}$ *be* $\mathcal{SN}^0$, *or a* PCA.

$$\mathbf{a} \approx_{\mathcal{A}} \mathbf{b} :\iff \forall C \in \mathrm{Contexts}_{\mathcal{A}},\; C[\mathbf{a}] \in \mathcal{A} \iff C[\mathbf{b}] \in \mathcal{A}.$$

Here, $\mathrm{Contexts}_{\mathcal{A}} ::= \square \mid \mathbf{a}\mathrm{Contexts}_{\mathcal{A}} \mid \mathrm{Contexts}_{\mathcal{A}}\mathbf{a}$, *with* $\square$ *begin a distinct formal constant and* $\mathbf{a} \in \mathcal{A}$. *For* $C \in \mathrm{Contexts}_{\mathcal{A}}$, $C[\mathbf{b}]$ *denotes the expression obtained from* C *by replacing* $\square$ *by* $\mathbf{b} \in \mathcal{A}$.

Theorem 8 (Bethke-Klop[6]). *Given an* NCA $\mathcal{A} = \langle A, \cdot, \mathbf{s}, \mathbf{k}\rangle$, *the quotient structure* $\mathcal{A}/\approx_{\mathcal{A}}$, *which is defined by* $\langle A/\approx, \cdot/\approx_{\mathcal{A}}, [\mathbf{s}]_{\approx_{\mathcal{A}}}, [\mathbf{k}]_{\approx_{\mathcal{A}}}\rangle$, *is indeed well-defined. Moreover,* $\theta : \mathcal{A} \to \mathcal{A}/\approx_{\mathcal{A}}; \mathbf{a} \mapsto [\mathbf{a}]_{\approx_{\mathcal{A}}}$ *is the final collapse of* $\mathcal{A}$.

In [6], Bethke-Klop showed the following, by an ingenious diagram chase of the descendants of redexes in combinatory logic (recall that $\mathcal{P}$ is an abbreviation of $\mathcal{SN}^0/=_{cl}$).

Definition 9. *Let* ker($\mathcal{C}$) *be the* PCA *which is generated from* **s** *and* **k** *by means of the application operator. We say that* $\mathcal{C}$ *is* minimal, *if* $\mathcal{C}$ = ker($\mathcal{C}$).

Theorem 9. *The finally collapsed* PCA $\mathcal{P}/\approx_{\mathcal{P}}$ *of* $\mathcal{P}$ *is extensional and minimal.*

The extensionality of $\mathcal{P}/\approx_{\mathcal{P}}$ implies $\mathbf{sk} = \mathbf{k}(\mathbf{skk})$ in $\mathcal{P}/\approx_{\mathcal{P}}$, although $\mathbf{SK} \neq_{cl} \mathbf{K}(\mathbf{SKK})$ in CL. Other unusual equations [1] may hold in $\mathcal{P}/\approx_{\mathcal{P}}$. The situation about the structure of $\mathcal{P}/\approx_{\mathcal{P}}$ is delicate, since the extensionality of NCA is difficult to achieve.

Theorem 10 (Bethke [5]).

1. *No extensional* NCA *is embedded in a* CA.
2. *No extensional* NCA *has an extensional* CA *as a substructure.*

The known construction of extensional NCA is ingenious, as listed below:

Example 2 (of extensional NCA's).

1. A CPO model $\mathcal{R}$ (Bethke [5]).
   A modification of Plotkin-Scott-Engeler's graph model. $\mathcal{R}$ has $[\mathcal{R} \to_\perp \mathcal{R}]$ as a *retract*. $\mathcal{Q}$ of a similar domain equation $\mathcal{Q} = [\mathcal{Q} \to_\perp \mathcal{Q}]_\perp$ is not extensional [9, Lemma20].
2. PCA's modulo an applicative bisimulation (Egidi-Honsell-Ronchi [9], Pino Perez [18]).

Because of the complication of $\approx_{\mathcal{P}}$, Bethke-Klop's extensional NCA $\mathcal{P}/\approx_{\mathcal{P}}$ is difficult; for example, it is difficult to choose the representatives from the quotient structure $\mathcal{P}/\approx_{\mathcal{P}}$. So, it has a good reason to tackle Bethke-Klop's problem [6]:

Question 1. determine the structure of the finally collapsed PCA $\mathcal{P}/\approx_{\mathcal{P}}$ *of* $\mathcal{P}$ *is, or how to find a suitable representation of its elements.*

---

[1] The extensionality of $\mathcal{P}/\approx_{\mathcal{P}}$ has a flavor of ω-rule, because the extensionality of $\mathcal{P}/\approx_{\mathcal{P}}$ is: given closed SN CL-terms M, N, if ML *is* NL for every *closed* SN CL-term L, then M *is* N.

4.2 Representation

We will represent $\mathcal{P} / \approx_{\mathcal{P}}$ in $\mathcal{F}$. Recall that $\mathcal{F}$ is an extensional NCA of filters consisting of types (see Section 3.3).

Definition 10 (Representing function). *Each element of $\mathcal{P}/\approx_{\mathcal{P}}$ is of the form $[[M]_{=_{cl}}]_{\approx_{\mathcal{P}}}$ for some $M \in \mathcal{SN}^0$. Let a representing function φ take it to $[\![M]\!]^{\mathcal{F}}$.*

Then we will prove that φ is indeed a well-defined, injective homomorphism from $\mathcal{P} / \approx_{\mathcal{P}}$ to $\mathcal{F}$. Let's rephrase the well-definedness and the injectivity of φ.

Lemma 12. *For $M_1, M_2 \in \mathcal{SN}^0$, $[[M_1]_{=_{cl}}]_{\approx_{\mathcal{P}}} = [[M_2]_{=_{cl}}]_{\approx_{\mathcal{P}}}$ iff $M_1 \approx_{\mathcal{SN}^0} M_2$*

Proof. We have only to verify $M_1 =_{cl} M_2 \Rightarrow M_1 \approx_{\mathcal{SN}^0} M_2$. Let $C \in \text{Contexts}_{\mathcal{SN}^0}$. By the confluence of $\rightarrow_{cl}$, for some P, $M_i \twoheadrightarrow_{cl} N$. Since the reduction sequence erases only SN terms, so does $C[M_i] \twoheadrightarrow_{cl} C[N]$. By Lemma 7, $C[M_i]$ is SN iff $C[N]$ is SN. Therefore $C[M_1]$ is SN iff $C[M_2]$ is SN. Hence, $M_1 \approx_{\mathcal{SN}^0} M_2$.

Corollary 3. *1. These are equivalent:*
- *(Full abstractness) $M \approx_{\mathcal{SN}^0} N \Rightarrow [\![M]\!]^{\mathcal{F}} = [\![N]\!]^{\mathcal{F}}$.*
- *φ is well-defined.*

2. These are equivalent:
- *(Adequacy) $M \approx_{\mathcal{SN}^0} N \Leftarrow [\![M]\!]^{\mathcal{F}} = [\![N]\!]^{\mathcal{F}}$.*
- *φ is injective.*

The proof of the well-definedness(or full abstractness) consists of the proof of the contraposition; for any M, N such that $[\![M]\!]^{\mathcal{F}} \neq [\![N]\!]^{\mathcal{F}}$, we will construct a context that discriminate M from N in a sense that $C[M] \in \mathcal{SN}^0$ but $C[N] \notin \mathcal{SN}^0$.

This construction is a kind of Böhm out technique which is used for the characterization of $=_{\kappa^*}$ by Böhm Trees and D_∞ (Hyland [12], Wadsworth [21]) and the filter domain of an intersection type assignment system for the λ-calculus.

The context we construct is much more simpler than that in the original Böhm out technique. Actually, our construct is an *applicative context*, i.e, a context of the form $[\ ]N'_1 \cdots N'_n$. It is because our construction uses the computability predicate of Definition 2.

Lemma 13. *For every $A \in T_\wedge$, there is a term c_A such that $\vdash c_A : A$.*

Proof. Since the only atomic type ω is the greatest among types, the induction on a type A proves $\omega^{\text{arity}(A)} \rightarrow \omega \leq A$. By the definition of *Comp*, $Comp(\emptyset, \lambda^H x_1 \ldots x_{\text{arity}(A)}.\ x_1,\ \omega^{\text{arity}(A)} \rightarrow \omega)$. By Lemma 1(2), we have $\vdash \lambda^H x_1 \ldots x_{\text{arity}(A)}.\ x_1 : \omega^{\text{arity}(A)} \rightarrow \omega$.

Theorem 11. *$M \approx_{\mathcal{SN}^0} N$ implies $[\![M]\!]^{\mathcal{F}} = [\![N]\!]^{\mathcal{F}}$; hence φ is well-defined.*

Proof. We have only to derive from ($\vdash M : A$ & $\not\vdash N : A$) that $M \not\approx_{\mathcal{SN}^0} N$, by induction on A. First we will consider the case where $A \equiv A_1 \to \cdots \to A_n \to \omega$ ($n \geq 0$). Then $\vdash M : A$ implies $Comp(\emptyset, M, A)$ by Lemma 1. By the definition of $Comp$, $\&_{i=1}^{n} Comp(\Gamma_i, N_i, A_i)$ implies $Comp(\cup_i \Gamma_i, MN_1 \cdots N_n, \omega)$.

On the other hand, we can derive $\Sigma \not\vdash N : A$ from $\not\vdash N : A$. So, $Comp(\Sigma, N, A)$ cannot be true. Because $\Sigma \supseteq \cup_i \Gamma_i$, there exist $N_1, \cdots, N_n, \Gamma_1, \cdots, \Gamma_n$ such that $\&_{i=1}^{n} Comp(\Gamma_i, N_i, A_i)$ but not $Comp(\Sigma, NN_1 \cdots N_n, \omega)$. We claim that

> (‡) for some $N'_1, \cdots N'_n \in \mathcal{SN}^0$, $\&_{i=1}^{n} Comp(\emptyset, N'_i, A_i)$ but $NN'_1 \cdots N'_n$ is not SN.

(∵) Let N'_i be $N_i[x := c_B, \ldots]$, where $x : B, \ldots \vdash N_i : A_i$ and c_B is constructed in Lemma 13. Then N'_i is closed. So, $Comp(\emptyset, c_B, B)$ by Lemma 1, and $Comp(\emptyset, N_i[x := c_B, \ldots], A_i)$ again by the same lemma. I.e., $Comp(\emptyset, N'_i, A_i)$. By Lemma 1(2), $N'_i \in \mathcal{SN}^0$. We have only to prove $NN'_1 \cdots N'_n$ is not SN. Since $Comp(\Sigma, NN_1 \cdots N_n, \omega)$ does not hold, by the definition of $Comp$, we have either $\Sigma \not\vdash NN_1 \cdots N_n : \omega$ or N is not SN. The first alternative implies the second, by Corollary 2. So, $NN_1 \cdots N_n$ is not SN. Neither is $NN'_1 \cdots N'_n$, since it is a substitution instance of $NN_1 \cdots N_n$. This ends the proof of the Claim.

Let C be the context $[\,]N'_1 \cdots N'_n$. Then, it is in $\text{Contexts}_{\mathcal{SN}^0}$, because $N'_i \in \mathcal{SN}^0$. Then, the claim (‡) implies $Comp(\emptyset, C[M], \omega)$. Therefore $C[M]$ is SN, by the Main Theorem. On the other hand, $C[N]$ is not SN. Thus the context C separates M from N, i.e., $M \not\approx N$.

In the other case, A is of the form $A_1 \wedge A_2$. Then, for some i, $\vdash M : A_i$ but $\not\vdash N : A_i$. By I.H., $M \not\approx N$.

Theorem 12. *$[\![M_1]\!]^{\mathcal{F}} = [\![M_2]\!]^{\mathcal{F}}$ implies $M_1 \approx_{\mathcal{SN}^0} M_2$; hence φ is injective.*

Proof. $[M]_{=_{cl}} \mapsto [\![M]\!]^{\mathcal{F}}$ is a collapse of $\mathcal{P}$. From the finality of a collapse $[M]_{=_{cl}} \mapsto [M]_{\approx_{\mathcal{SN}^0}}$, the statement follows. A type theoretic proof is possible.

Theorem 13. *φ is a homomorphism from $\mathcal{P} \,/ \approx_{\mathcal{P}}$ to $\mathcal{F}$.*

Proof. It is easy to show that the $\mathbf{s}, \mathbf{k}$ of $\mathcal{P} \,/ \approx_{\mathcal{P}}$ is mapped to those of $\mathcal{F}$. Let's verify that $\varphi(\mathbf{a} \cdot \mathbf{b}) \simeq \varphi(\mathbf{a}) \cdot \varphi(\mathbf{b})$.

Let $\mathbf{a} = \left[[M]_{=_{cl}}\right]_{\approx}$ and $\mathbf{b} = \left[[N]_{=_{cl}}\right]_{\approx}$. Then $\mathbf{ab} \simeq \left[[M]_{=_{cl}}[N]_{=_{cl}}\right]_{\approx} \simeq \left[[MN]_{=_{cl}}\right]_{\approx}$. On the other hand, $\varphi(\mathbf{a})\varphi(\mathbf{b}) \simeq [\![M]\!]^{\mathcal{F}}[\![N]\!]^{\mathcal{F}} = \{B \mid \vdash M : A \to B \;\&\; \vdash N : A\}$. Here $\{B \mid \vdash M : A \to B \;\&\; \vdash N : A\} \subseteq \{B \mid \vdash MN : B\}$, and vise versa by the Generation Lemma. Therefore, $\varphi(\mathbf{a})\varphi(\mathbf{b}) \simeq [\![MN]\!]^{\mathcal{F}}$.

If $\mathbf{ab}$ is defined, then $\left[[MN]_{=_{cl}}\right]_{\approx}$ is defined. So, MN is an SN CL-term. By the Main Theorem, $[\![MN]\!]^{\mathcal{F}}$ is not empty. Then $\varphi(\mathbf{a})\varphi(\mathbf{b})$ is defined and is equal to $\varphi(\mathbf{ab})$. If $\varphi(\mathbf{a})\varphi(\mathbf{b})$ is defined, then MN has a type. By the Main Theorem, MN is an SN CL-term. Therefore $\mathbf{ab}$ is defined. Thus $\varphi(\mathbf{ab})$ is defined and is equal to $\varphi(\mathbf{a})\varphi(\mathbf{b})$.

In view of Theorem 9, this subsection's theorems amount to the following:

Theorem 14. *$\mathcal{P} \,/ \approx_{\mathcal{P}}$ is isomorphic to* $\ker(\mathcal{F})$.

5 Conclusion

Our type assignment system characterizes the SN CL-terms, and it enjoys semantical properties, unlike the intersection type assignment system which characterizes the set of SN λ-terms. In the latter system, a modest β-expansion can lose the types, although it is sound and complete for the λI-models [11],

Let's recall that Hyland-Ong [13] defined a *conditional partial combinatory algebra* (CPCA) as a PCA minus $(\mathbf{S}_2)$. Let's define a *1-partial combinatory algebra* (1-PCA) as a CPCA such that $(\mathbf{S}_1)$: $\mathbf{sa}\downarrow$. So, we have CPCA, 1-PCA, PCA, and CA, to list them from the most general to the least.

Theorem 15. *1. Let* $\mathbf{s}_0$ *be* $\{(A \to B \to C) \to (A \to B) \to A \to C \mid A, B, C \in T_\wedge\}^\uparrow$. *Then,* $\langle \mathcal{F}, \cdot, \mathbf{s}_0, \mathbf{k}\rangle$ *is an extensional* CPCA.
2. Let $\mathbf{s}_1$ *be* $\{\omega \to \omega,\ (A \to B \to C) \to (A \to B) \to A \to C \mid A, B, C \in T_\wedge\}^\uparrow$. *Then,* $\langle \mathcal{F}, \cdot, \mathbf{s}_1, \mathbf{k}\rangle$ *is an extensional* 1-PCA.

We have benefited from the existence of the type-subsumption axiom $A \leq \omega$. We can describe as a typing axiom the partiality of the constant $\mathbf{s}$ of $\mathcal{F}$. Because ω is also the unique atomic type, we can prove that each type is inhabited (Lemma 13) and that the homomorphism φ is well-defined (Theorem 11, a full abstraction result).

We hope our treatment can be applied to describe with filter domains various partial combinatory algebras. For example,

- The CPCA in Hyland-Ong [13]. That is the set of closed SN λ-terms modulo a weak β-equality [7]. Our conjectures are
  - it has the finally collapsed CPCA, and
  - the finally collapsed CPCA can be embedded into $\langle \mathcal{F}, \cdot, \mathbf{s}_0, \mathbf{k}\rangle$.
- The other extensional NCA's in Example 2.

Acknowledgement. The author thanks Masami Hagiya, Toshihiko Kurata, Mariangiola Dezani-Ciancaglini, J.R. Hindley, and anonymous referees. The idea of the system is brewed through discussion with Toshihiko Kurata.

References

1. Y. Akama. A λ-to-CL translation for strong normalization. In Ph. de Groote and J.R. Hindley, eds., *TLCA, Proceedings*, vol. 1210 of *LNCS*, pp. 1–10. 1997.
2. H.P. Barendregt. *The Lambda Calculus, Its Syntax and Semantics.* North-Holland, second edition, 1984.
3. M. Beeson. *Foundations of Constructive Mathematics.* 1984.
4. A. Berarducci and B. Intrigila. Some new results on easy lambda-terms. *TCS*, 121:71–88, 1993.
5. I. Bethke. On the existence of extensional partial combinatory algebras. *JSL*, 52(3):819–833, 1987.
6. I. Bethke and J. W. Klop. Collapsing partial combinatory algebras. In G. Dowek et al., eds., *HOA, Proceedings*, vol. 1074 of *LNCS*, pp. 57 – 73. 1996.

7. N. Cağman and J.R. Hindley. Combinatory weak reduction in lambda calculus. Technical report, University of Wales Swansea, Swansea, U.K., 1997.
8. M. Dezani-Ciancaglini and J.R. Hindley. Intersection types for combinatory logic. *TCS*, 100(2):303–324, 1992.
9. L. Egidi, F. Honsell, and S. Ronchi Della Rocca. Operational, denotational and logical descriptions: A case study. *Fundamenta Informaticae*, 16:149–169, 1992.
10. B. Gramlich. *Termination and Confluence Properties of Structured Rewrite Systems.* PhD thesis, Universität Kaiserslautern, 1996.
11. F.Honsell and S.Ronchi della Rocca. Models for theories of functions strictly depending on all their arguments. Internal report, Department of Computer Science, Turin, Italy, 1984.
12. J. Martin E. Hyland. A syntactic characterization of the equality in some models of the lambda calculus. *J. London Math. Soc. (2)*, 12:361–370, 1976.
13. J.M.E. Hyland and C.-H. L. Ong. Modified realizability toposes and strong normalization proofs (extended abstract). In M. Bezem and J.F. Groote, eds., *TLCA, Proceedings*, vol. 664 of *LNCS*, pp. 179–194. 1993.
14. G.Jacopini. A condition for identifying two elements of whatever model of combinatory logic. In C.Böhm, editor, *λ-calculus and computer science theory*, vol. 37 of *LNCS*, pp. 213–219. 1975.
15. G.Jacopini and M. Venturini Zilli. Easy terms in the lambda-calculus. *Fundamenta Informaticae*, 8(2):225–233, 1985.
16. J.W. Klop. Term rewriting systems. In S. Abramsky et al., editors, *Handbook of Logic in Computer Science*, volume 2, pages 2–117. Oxford UP, 1992.
17. T. Kurata. Subtype relations yielding filter lambda-models. In T. Ida, A. Ohori, and M. Takeichi, eds., *Second Fuji International Workshop on Functional and Logic Programming, Proceedings*, pp. 210–229. World Scientific, 1997.
18. R. Pino Perez. An extensional partial combinatory algebra based on λ-terms. In A. Tarlecki, editor, *MFCS, Proceedings*, vol. 520 of *LNCS*, pp. 387–396. 1991.
19. G. Plotkin. Call-by-name, call-by-value, and the λ-calculus. *TCS*, 1:125–159, 1975.
20. S. van Bakel. Complete restrictions of the intersection type discipline. *TCS*, 102:135–163, 1992.
21. C. P. Wadsworth. The relation between computational and denotational properties for scott's D_∞-models of the lambda calaulus. *SIAM J. Computing*, 5:488–521, 1976.

Coupling Saturation-Based Provers by Exchanging Positive/Negative Information

Dirk Fuchs

Fachbereich Informatik, Universität Kaiserslautern
Postfach 3049, 67653 Kaiserslautern
Germany
E-mail: dfuchs@informatik.uni-kl.de

Abstract. We examine different possibilities of coupling saturation-based theorem provers by exchanging positive/negative information. Positive information is given by facts that should be employed for proving a proof goal, negative information is represented by facts that do not appear to be useful. We introduce a basic model for cooperative theorem proving employing both kinds of information. We present theoretical results regarding the exchange of positive/negative information as well as practical methods that allow for a gain of efficiency in comparison with sequential provers. Finally, we report on experimental studies conducted in the areas unfailing completion and superposition.

1 Introduction

In general, automated theorem proving is based on the solution of search problems which usually comprise huge search spaces. Thus, a skillful control of the search conducted by the prover is especially important to deal with hard search problems. Therefore, for first-order theorem proving many different calculi, each of them controllable by various heuristics, have been developed. However, it is problematic in this context to decide which calculus and which heuristic should be employed when tackling a given problem.

Another main problem is that usually many specialized theorem provers for sub-logics of first-order logic exist that are able to solve problems stemming from their specialized logic very efficiently, but cannot deal with problems specified in full first-order logic. E.g., the availability of high-performance equational provers like DISCOUNT ([ADF95]) is somewhat limited although many problems contain a large number of formulas where equality is involved in.

The first problem, the lack of knowledge about the quality of certain provers for certain problems, could be solved by using competitive versions of different provers. But such a system can at most be as good as the best of its provers. Therefore, our approach is the development of *cooperative theorem provers*. We want to achieve that on the one hand some provers work in parallel and that on the other hand a further gain of efficiency—caused by synergetic effects—is obtained. Cooperation may also be the right way to deal with our second problem because by cooperation specialized provers can support universal provers.

In this paper we want to examine how cooperation between several *saturation-based theorem provers* could look like. Note that most rewrite-based provers in first-order logic with equality belong to this class, e.g. all superposition-based provers. We deal with the cooperation of both homogeneous and heterogeneous provers. In this context, we call a set of provers homogeneous if all provers employ the same calculus and differ from each other only in the heuristic they employ. Otherwise, we consider a set of provers to be heterogeneous.

There are some approaches that try to couple homogeneous (e.g., [AD93], [FD97]) and heterogeneous provers ([Sut92]). All approaches have in common that the provers only exchange *positive information* in form of deduced facts the receiving provers should use. We also exchange positive information in form of important facts (*lemmas*) that the cooperating provers have deduced. The main difference between our approach and the method presented in [Sut92] is that we do not exchange each deduced fact but only a few facts selected by *referees*. The main difference between our approach and the TEAMWORK method ([AD93]) is that we employ a different communication model (see section 3) that allows us also to deal with heterogeneous provers. Another difference is that we use a more complicated selection process of facts so as to integrate more knowledge into this process. Furthermore, our extension to all approaches that deal with cooperating provers is that we also use *negative information*, i.e. information on deduced facts that do not appear to be needed for a proof ("bad" facts) and that the receiving provers should hence not use.

Thus, our two main topics regarding both the exchange of good and bad facts are: On the one hand we design techniques for *determining good and bad facts*, on the other hand we develop methods for *processing the received information.*

The paper is organized in the following way: At first we introduce some basics of saturation-based theorem provers and introduce the application domains unfailing completion ([BDP89]) and superposition ([BG94]). After that, we discuss in section 3 how the exchange of facts can be organized. Section 4 deals in more detail with the exchange of positive information, section 5 deals with the exchange of negative information. Results of our experiments are presented in section 6. Finally, an outlook at possible future work concludes the paper.

2 Basics of Automated Deduction

The general problem in theorem proving is given as follows: Given a set of facts Ax (axioms), is a further fact λ_G (goal) a logical consequence of the axioms? Commonly, automated theorem provers utilize certain *calculi* for accomplishing the task mentioned above. *Saturation-based calculi* go the way to producing logic consequences from Ax until a fact covering the goal appears. Typically a saturation-based calculus contains inference rules of an inference system $\mathcal{I}$ which can be applied to a set of facts. Expansion inference rules are able to generate new facts from known ones, contraction inference rules allow for the deletion of facts or replacing facts by other ones. We write $\Lambda \vdash_{\mathcal{I}} \Lambda'$ if we can derive the fact set Λ' from Λ through the application of one inference rule. A sequence of sets $(\Lambda_i)_{i\geq 0}$ is called an $\mathcal{I}$-derivation if $\Lambda_i \vdash_{\mathcal{I}} \Lambda_{i+1}$ for all i. In order to solve proof problems by utilizing an inference system $\mathcal{I}$-derivations $Ax = \Lambda_0 \vdash_{\mathcal{I}} \Lambda_1 \vdash_{\mathcal{I}} \ldots$ have to be performed until a fact set Λ_n is derived which contains a fact covering the goal. For all calculi described in the following subsections *fairness* of derivations implies completeness, i.e. each valid proof goal can be proven by performing fair derivations (see, e.g., [BDP89] for unfailing completion, and [BG94] for superposition).

Definition 1 Persistent Facts, Fairness of an $\mathcal{I}$-derivation.
Let $\mathcal{I}$ be an inference system, let I_e be the set of expanding inference rules. $I_e(M)$ denotes the set of all facts derivable from M by applying an inference from I_e to some facts from M. Furthermore, let $\Lambda_0 \vdash_{\mathcal{I}} \Lambda_1 \vdash_{\mathcal{I}} \ldots$ be an $\mathcal{I}$-derivation.

1. We define the *persistent facts* of this $\mathcal{I}$-derivation by $\Lambda^\infty = \bigcup_{i \geq 0} \bigcap_{j \geq i} \Lambda_j$.
2. The $\mathcal{I}$-derivation is called *fair* iff $I_e(\Lambda^\infty) \subseteq \bigcup_{i \geq 0} \Lambda_i$.

Our framework is as follows: A sequential theorem prover (SP) maintains a set $\mathcal{F}^P$ of *potential* or *passive facts* from which it selects and removes one fact λ at a time. After the application of some contraction inference rules on λ, it is put into the set $\mathcal{F}^A$ of *activated facts* (i.e. it is *activated*) or discarded if it was deleted by a contraction rule. Activated facts are, unlike passive facts, allowed to produce new passive facts via exhaustively applying expanding inference rules. Initially, $\mathcal{F}^A = \emptyset$ and $\mathcal{F}^P = Ax$. Note that we call in the following the pair $(\mathcal{F}^A, \mathcal{F}^P)$ the *search state* of the prover. The indeterministic selection or *activation step* is realized by heuristic means: After the generation of a new fact and before its insertion into the set of passive facts, a heuristic weight $\omega_\lambda \in \mathbb{N}$ is associated with the fact. This heuristic weight remains unchanged during the proof process. It is needed for the activation step since the fact with the smallest heuristic weight is selected. Conflicts are handled using the FIFO strategy.

The heuristic weight of a fact is computed by employing a heuristic $\mathcal{H}$. A heuristic can utilize following information as input: Firstly, a heuristic can use the syntactic structure of a fact, e.g. its number of symbols. Secondly, it can depend on the systems of active and passive facts $\mathcal{F}^A$ and $\mathcal{F}^P$ of the prover at the moment of the generation of λ. E.g., the goal-oriented heuristics defined in [DF94] fall back on this information so as to compute structural differences between a fact and current goals. Finally, a heuristic can also use the *derivation tree* (see [Fuc97a]) T_λ of a fact λ. Such derivation trees can be used in order to prefer facts that are descendants of certain facts, e.g., the original proof goal. All in all, for a fact λ we obtain $\omega_\lambda = \mathcal{H}(\lambda, \mathcal{F}^A, \mathcal{F}^P, T_\lambda)$.

We call a heuristic *fair* if it guarantees for each start set Λ that each fact being derived from Λ by employing algorithm SP and hence being passive at a certain moment is either activated or discarded after a finite period of time. It is easy to recognize that the use of fair heuristics implies that the algorithm conducts fair derivations.

For our experiments we restricted ourselves to the area of first-order theorem proving, i.e., in the following facts correspond to clauses. We experimented with the provers SPASS and DISCOUNT. SPASS ([WGR96]) is an automatic prover for first-order logic with equality. It is based on the superposition calculus (see [BG94]). The inference rules of SPASS can be divided into expansion (e.g. superposition) and contraction rules (e.g. subsumption). DISCOUNT ([ADF95]) is based on the *unfailing completion* procedure (see [BDP89]). This procedure offers possibilities to develop high-performance provers in pure equational logic. The inference system underlying unfailing completion is in main parts a restricted version of the superposition calculus and contains additional contraction rules.

3 A Basic Model for Cooperation

Architecture and Behavior: On each computer in a computer network a prover conducts a search for the common proof goal. The network consists of either homogeneous or heterogeneous provers. Since possibly not all provers can deal with problems specified in the same logic, we must sometimes transform the axiomatization of different provers. We assume that the network is fully intermeshed, i.e. each prover can communicate with all other provers.

It is important that each prover has enough time to tackle the problem without being permanently interrupted by others. Thus, we decided to let the provers work independently for a while and only cooperate periodically. The provers try to solve the same problem independently during so-called *working phases* P_w^i, cooperation takes place during *cooperation phases* P_c^i. The phases are alternating, i.e. the sequence of phases is $P_w^0, P_c^0, P_w^1, P_c^1, \ldots$ In a working phase we require at least one real activation, i.e. the insertion of a new fact into $\mathcal{F}^A$. During a cooperation phase P_c^i $(i \geq 0)$ each prover performs the following activities: Firstly, a set of positive lemmas $\mathcal{P}_j^i$ and a set of bad facts $\mathcal{N}_j^i$ are determined and transmitted to each receiving prover j. Secondly, the facts $\lambda \in \mathcal{P}^i \cup \mathcal{N}^i$ which the prover has received from other provers are processed.

Classification of Facts: In order to determine lemmas and bad facts the concepts "good fact" and "bad fact" have to be defined. This, however, is a serious problem when dealing with saturation-based provers. Due to the general undecidability, it is impossible to predict whether or not a fact is part of a proof. Thus, a semantic classification of facts into good or bad regarding their ability to contribute to solutions of the proof problem is not sensible. Hence, we choose a more pragmatic approach: Instead of calling facts good or bad for the *proof* of the goal we classify facts as good or bad w.r.t. their usefulness for the *search* for a proof. Thus, we try to estimate whether facts may be more or less useful for the *process of finding a proof* instead of whether they are *part of a proof*.

We make these ideas more precise: Often there is not only one proof of a goal but many different proofs exist. A proof can be considered to be better than another proof w.r.t. the process of finding it (proof search) if less unnecessary inferences are performed and less unnecessary facts are generated during the search. In this context inferences or facts are unnecessary if they do not contribute to the final proof of the goal which is found by the prover. Our aim is to find "good" proofs in this sense, i.e. proofs that can be found in a short proof run. From this point of view facts are bad, i.e. not so useful for proving the goal, if they entail a large amount of unnecessary inferences and hence possibly a long proof run. Otherwise, we consider facts to be good. As we will see in the following, it is possible to design criteria that are well-suited for estimating whether facts are good or bad regarding this concept. These criteria are mainly based on the fact that we consider expanding inferences to be negative because the generation of too many facts entails—since usually a main part is not needed in a proof—a high demand for memory and unnecessary computation. In contrast, contracting inferences are considered to be positive. This is surely justified if we recall that mainly simplification, a contracting inference, is necessary in order to develop efficient rewrite-based provers.

In the following, we firstly have to answer the question of how to recognize that facts belong to the sets $\mathcal{P}_j^i$ or $\mathcal{N}_j^i$. As discussed later in more detail, we employ for each receiver j selection functions $\varphi_{\mathcal{P},j}$ and $\varphi_{\mathcal{N},j}$ that determine in each cooperation phase P_c^i sets $\mathcal{P}_j^i$ and $\mathcal{N}_j^i$, respectively. Secondly, we have to determine how to process the received information.

4 Cooperation by Exchanging Positive Facts

In order to exchange important lemmas it is necessary to identify them, to transmit them from one prover to all other provers of the network, and to process the information on good facts received from other provers. We start with the description of the selection and transmission techniques. After that we describe the processing of the received lemmas.

4.1 Selection and Transmission of Lemmas

The most important problem of each cooperative system is to provide for each prover enough lemmas that allow for a reduction of the search effort but not to exchange too much unnecessary information. The latter does not only produce unnecessary communication (remember that typically communication is much more expensive than computation and should therefore be restricted) but also additional computation effort. This is due to the fact that unnecessary facts have to be included into the search state thus entailing more and more unnecessary computation in further steps of the search. Therefore, possible facts to be exchanged should be judged by its sender to reduce the amount of communication and computation. It may also be useful to judge the remaining facts on the side of the receiver if information available only on this side can support the selection of facts. Hence, we employ referees on both sides of a communication channel, namely *send-referees* at the sender site and *receive-referees* at the receiver site.

Send-Referees A send-referee for a receiving prover is a pair $(P_{\mathcal{P}}, \varphi_{\mathcal{P}})$ of a *filter predicate* $P_{\mathcal{P}}$ and a *selection function* $\varphi_{\mathcal{P}}$. When employing a referee the selection of facts during a cooperation phase takes place in the following manner: At first the active facts of the sender are filtered by the filter predicate. Only facts λ for that $P_{\mathcal{P}}(\lambda)$ holds are allowed to pass through the filter. After that, important lemmas $\mathcal{P}$ are selected from the remaining facts with the help of $\varphi_{\mathcal{P}}$.

Since we possibly couple heterogeneous provers only such facts should pass through the filter $P_{\mathcal{P}}$ which the receiving prover can integrate into its set of facts. Hence, the task of the filter predicate is to prevent facts of a certain logic from being transmitted to other provers which are only able to deduce facts from another logic. Moreover, facts are filtered out that are redundant for the receiving prover. Redundant are all axioms and facts selected in an earlier phase.

The selection function $\varphi_{\mathcal{P}}$ is used to choose among the facts that passed $P_{\mathcal{P}}$. $\varphi_{\mathcal{P}}$ employs several judgment functions $\psi_1, \ldots, \psi_n$ in order to measure different aspects of the quality of a fact, mainly regarding our abstract concept from section 3. These functions can employ different kinds of knowledge for judging facts. The easiest way is to integrate only local knowledge into the judgment functions, i.e. knowledge about the sender. Moreover, also knowledge about the receiver can be employed: Firstly, one can have knowledge about the identity of the receiver, i.e. about its calculus and underlying logic. Secondly, one can have further knowledge about the concrete status of the receiver at a certain point in time: This kind of knowledge can be divided into knowledge about the current search state of the prover (i.e. its active and passive facts) and knowledge about its method to change the search state in future (i.e. its heuristic).

Judgment functions employing local knowledge define the quality of a fact according to the *success* of this fact recorded by the prover that generated it. So, the quality of a fact is determined by its history. The judgment function ψ_S ("statistical judgment") employs for each fact λ and for each inference type, the inferences λ was involved in so far. ψ_S rates "good" inferences (i.e., contracting inferences) positive, "bad" inferences (i.e., expansion inferences) negative (see [FD97]).

If we fall back on knowledge about the heuristic of the receiver the quality of a fact should be determined by the effects this fact will have on the future search process of the receiver. Here, facts are of interest that with a high probability have not already been found by the receiver but are nevertheless rated as very good (by the receiver). Judgment function ψ_H ("heuristical judgment"), described in [FD97], favors facts that have a small heuristic weight regarding the part of the heuristic of the receiver that is only based on syntactic properties of a fact, but have at least one ancestor with a high heuristic weight. Thus, it is probable that it is difficult for the receiver to deduce these facts and we hence avoid the transmission of redundant information.

Knowledge about the current search state of the receiver cannot be employed by judgment functions of a send-referee because the needed information is not available at the sender site.

Receive-Referees A receive-referee consists of a selection function $\varphi_{\mathcal{P}}$ which employs again different judgment functions. These functions employ knowledge about the search state of the receiver. In general, there are two main principles how receive-referees can be designed, a contraction-based and an expansion-based principle (see [FD97] for details).

Function ψ_U follows the first principle. Basically, it favors facts that might be often involved in contracting inferences in future. In order to estimate this, $\psi_U(\lambda)$ counts how often λ could be used in order to contract active facts of the receiver if it was integrated into its search state. Note that $\psi_U(\lambda)$ can efficiently be computed because usually the set of active facts is rather small.

Since we consider expansion inferences to be negative we require that expansion inferences involving λ should at least with a high probability contribute to a proof. Therefore $\psi_{SG}(\lambda)$ sums the value $\gamma(\bar{\lambda})$ for each active fact $\bar{\lambda}$ of the receiver that would be involved in an expanding inference with λ if λ was integrated into the receiver's search state. $\gamma(\bar{\lambda})$ is the higher the more likely $\bar{\lambda}$ contributes to a proof and hence possibly also a descendant of λ and $\bar{\lambda}$. E.g., the similarity of $\bar{\lambda}$ to the goal regarding the definitions from [DF94] could be a hint that $\bar{\lambda}$ contributes to a proof. For exact definitions see [FD97].

4.2 Processing received Lemmas

Now, we will examine aspects of the processing of received lemmas. We consider a prover whose underlying calculus is complete in a certain logic. As we have mentioned in section 2 each valid proof goal can be proven by performing a fair derivation (using a fair heuristic) until a fact covering the goal appears.

If we couple our prover with others by exchanging lemmas its internal algorithm SP must be modified to algorithm CP (cooperative prover): It is necessary to activate facts not only from the set of potential facts during the working phases but also periodically lemmas received from other provers must be activated during the cooperation phases. We assume that all received lemmas are activated during a cooperation phase in an arbitrary order. Now, the question arises whether the prover remains complete, i.e., we are interested in the question whether it is possible that completeness is lost because of the periodical "disturbance" through others.

First, we must extend our notion of $\mathcal{I}$-derivations to $\mathcal{I}_{Dist}$-derivations.

Definition 2 Inference System with Disturbance $\mathcal{I}_{Dist}$.
Let $\mathcal{I}$ be an inference system, let I be the set of inference rules of $\mathcal{I}$. Let Λ_0 be an initial set of facts. If M is a set of facts and A a fact, $M \models A$ denotes that A is a logic consequence of M. Then the *inference system with disturbance* $\mathcal{I}_{Dist}$ contains the inference rules $I_{Dist} = I \cup \{\mathtt{Dist}\}$, with

$$(\mathtt{Dist}) \qquad \Lambda \vdash \Lambda \cup \{\lambda\}; \text{if } \Lambda_0 \models \lambda$$

This describes exactly what happens during a cooperation phase: Facts derived from other provers, i.e. logic consequences from the initial system of facts, can be integrated into the system of facts without an additional proof. A (infinite) sequence of fact sets $(\Lambda_i)_{i\geq 0}$ is an $\mathcal{I}_{Dist}$*-derivation* iff $\Lambda_i \vdash_{\mathcal{I}_{Dist}} \Lambda_{i+1}$, $\forall i \geq 0$. We call an $\mathcal{I}_{Dist}$-derivation $(\Lambda_i)_{i\geq 0}$ fair iff $I_{Dist,e}(\Lambda^\infty) \subseteq \cup_{i\geq 0}\Lambda_i$. (We do not consider Dist to be an expanding inference.) Again, regarding our calculi each valid goal can be proven by performing a fair $\mathcal{I}_{Dist}$-derivation until the goal appears. Note that our algorithm CP produces $\mathcal{I}_{Dist}$-derivations. We call the heuristic that CP employs *fair despite disturbance* if it guarantees for each start set Λ that each fact being derived from Λ by employing algorithm CP and hence being passive at a certain moment is either activated or discarded after a finite period of time. If CP employs a heuristic which is fair despite disturbance it produces only fair $\mathcal{I}_{Dist}$-derivations.

Now, we want to examine whether the notions of fairness and fairness despite disturbance are the same. If this is true then each prover which performs fair $\mathcal{I}$-derivations with a fair heuristic would also remain complete if it is coupled with others via the exchange of lemmas. Unfortunately, this is not the case in general. But at least if we employ a *strong fair* heuristic (see [AD93]), i.e. a heuristic where the set $M_z = \{\lambda : \exists$ fact sets $\mathcal{C}$ and $\mathcal{D}$, and a derivation tree $T_\lambda : \mathcal{H}(\lambda, \mathcal{C}, \mathcal{D}, T_\lambda) = z\}$ is finite for all numbers z, we can preserve completeness.

Theorem 3. *([Fuc97a]) Let $\mathcal{I}$ be an inference system, $\mathcal{I}_{Dist}$ be the respective inference system with disturbance. Then it holds:*

1. *Each heuristic which is fair despite disturbance is fair.*
2. *There are heuristics which are fair but not fair despite disturbance.*
3. *Each strong fair heuristic is fair despite disturbance.*

As we can see it might be the case that a fair heuristic for SP is not fair for CP, i.e. coupling a prover with others can cause incompleteness. Fortunately, most heuristics are strong fair and hence also fair despite disturbance. Nearly all heuristics we used in SPASS and DISCOUNT have this very property.

5 Cooperation by Exchanging Negative Facts

In the following, we describe a method for detecting facts that behave badly in a certain way. Moreover, we present in which way a prover can employ received bad facts so as to improve its search.

5.1 Determination of Bad Facts

In order to determine a set $\mathcal{N}$ of bad facts the active facts of a prover are filtered by a filter predicate $P_\mathcal{N}$ and then some facts are selected from the remaining ones by a selection function $\varphi_\mathcal{N}$.

Filtering facts is again necessary due to the (possible) heterogeneity of the provers. It is wise to let only such facts pass through the filter which are syntactically correct regarding the logic of the receiving prover. Moreover, the filter predicate is used to avoid that a certain fact is sent twice to receiving provers.

In general, we can select as before bad facts at the sender site as well as at the receiver site. Due to efficiency reasons, however, we did not select facts at the receiver site and employed only local knowledge, i.e. knowledge about the sender, and knowledge about the heuristic of the receiver.

In our concrete realization we employed only one kind of judgment functions ψ_B for judging facts and integrated local knowledge and knowledge about the heuristic of the receiver in it. We split ψ_B in a part $\psi_\mathcal{L}$ based on local knowledge and a part $\psi_\mathcal{H}$ based on knowledge about the heuristic of the receiver. Thus, we obtain $\psi_B(\lambda) = \psi_\mathcal{L}(\lambda) + \psi_\mathcal{H}(\lambda)$. Facts λ are the worse the higher $\psi_B(\lambda)$ is.

At first we consider local knowledge in order to estimate if a fact contributes with a high probability only to long proof runs. We employ $\psi_L = -\psi_S$, i.e. facts are the worse the higher their number of expansion and the lower their number of contraction inferences is. The second part $\psi_\mathcal{H}$ is simply given as $\psi_\mathcal{H} = -\mathcal{H}_R$, $\mathcal{H}_R$ being the part of the heuristic of the receiving prover which is based on syntactic properties of a fact. Hence, facts that have a high weight regarding the heuristic of a receiver are considered to be less negative because it is quite improbable that the receiver has activated them or will activate them in future. Thus, function $\psi_\mathcal{H}$ aims at minimizing the transmission of information the receiver cannot use.

5.2 Processing Bad Facts

We process the set of bad facts received from other theorem provers in two different manners: On the one hand, if received bad facts are already in the system of active facts of the prover, a *restructuring of the search state* that prevents the prover to work with bad facts in future might be the right way. Hence our aim is—similarly to the idea of dynamic programming—that paths in the search space whose exploration would lead to high costs are postponed or even neglected. Thus, possibly short proof runs occur if the bad facts are really not needed and a proof can be found quickly when exploring a proof path not employing the bad facts. Nevertheless, it must be guaranteed that eventually all paths are explored so as to preserve completeness if a bad fact is really necessary. Even in such a case a restructuring might be sensible: Parts of a proof that are not based on this fact can possibly be exploited faster without using the fact and then remaining parts of the proof can be searched for. Thus, we can gain efficiency, too. On the other hand, bad facts of other provers can even have a control aspect: By using this information one can *modify the heuristic weights* of facts so as to postpone the activation of bad facts. Thus, a posteriori knowledge of other provers (bad facts) can be transformed into a priori knowledge of the receiver (heuristic weights). Now, we discuss both approaches in more detail:

Modification of Heuristic Weights A modification of the heuristic weights of a prover regarding a set $\mathcal{N}^j$ of bad facts received in cooperation phase P_c^j is only sensible if some facts $\mathcal{N}_P^j \subseteq \mathcal{N}^j$ are not already activated. The aim of such a modification is then to postpone or avoid the activation of facts $\lambda \in \mathcal{N}_P^j$.

Definition 4 Modified Heuristic Weights $\widehat{\omega_\lambda}$.
Let γ be a real-valued parameter, $\mathcal{N}^j$ be the set of all bad facts the prover has obtained in cooperation phase P_c^j $(j \geq 0)$. Then the modified heuristic weights $\widehat{\omega_\lambda}$ of passive facts λ (with original heuristic weights ω_λ) equal in each working phase P_w^i the following weights ω_λ^i: If $i = 0$ then $\omega_\lambda^0 = \omega_\lambda$. If $i > 0$ then

$$\omega_\lambda^i = \begin{cases} \gamma \cdot \omega_\lambda & , \lambda \in \mathcal{N}^{i-1} \\ \omega_\lambda & , \text{otherwise} \end{cases}$$

If γ is high then the activation of bad facts may be completely impossible during the next working phase, otherwise it may only be delayed.

Finally, we want to examine if a prover that originally uses a fair heuristic or a heuristic which is fair despite disturbance might become incomplete when modifying the heuristic weights as described. Under certain restrictions on the original heuristic of a prover this is impossible:

Theorem 5. *([Fuc97a]) Let $\mathcal{I}_{Dist}$ be an inference system with disturbance. Let $\mathcal{H}$ be a strong fair heuristic. Furthermore, let CP be realized in such a way that it uses $\mathcal{H}$ and modifies heuristic weights $\widehat{\omega_\lambda}$ of passive facts λ as described in the preceding definition 4. Then CP produces only fair $\mathcal{I}_{Dist}$-derivations.*

Thus, a prover cannot become incomplete if it starts with a strong fair heuristic.

Restructuring the Search State By restructuring the search state we want to achieve that a prover does not work with already activated bad facts or at least postpone inferences where bad facts take part in. Hence we try to correct a posteriori a wrong decision we made due to our low a priori knowledge.

As already mentioned, such a restructuring is only reasonable if bad facts are already in the system of active facts of the receiver. However, such a restricted application of the restructuring technique entails some problems: Remember that we want to exchange positive and negative information among homogeneous and heterogeneous provers. Now, it is well-known (see, e.g., [DF96]) that methods based on the exchange of important lemmas between homogeneous provers require the heuristics of the cooperating provers to be quite different. Therefore, in our context we have to employ different heuristics, too. But this entails that different provers do not have so many active facts in common. Moreover, this is also true when coupling heterogeneous provers. Thus, in the most cases bad facts as suggested by the sender are not in the systems of active facts of the receivers. Nevertheless, surely a lot of other unnecessary facts are activated by the receivers. Thus, it would be desirable for a prover to derive from the badness of received facts that also some of its own facts are possibly unnecessary.

The solution of this problem is to consider a kind of "similarity": According to our definition of badness a fact is bad if it slows down the search for the proof goal. Since this often depends on certain syntactical properties of a fact we could infer from the *syntactical similarity* between a fact $\bar{\lambda}$ and a bad fact λ that also $\bar{\lambda}$ is a bad fact. Regarding our definition of the judgment function ψ_B one can see that facts λ and $\bar{\lambda}$ that are syntactically very similar would have with a high probability similar judgments $\psi_B(\lambda)$ and $\psi_B(\bar{\lambda})$ (if the facts were activated at the same time). Thus, it is surely justified to generalize the information on bad facts to syntactically similar facts $\mathcal{S}$. So, λ represents a *scheme* $\mathcal{S}_\lambda$ of bad facts. In our realization such a scheme is a subset of the active facts of the receiver. For details concerning this generalization see [Fuc97a].

Now, if we are able to generalize information on bad facts we do not try to restructure our search state w.r.t. the bad facts $\mathcal{N}^i$ received in cooperation phase P_c^i but we use the set $\mathcal{N}_S^i = \cup_{\lambda \in \mathcal{N}^i} \mathcal{S}_\lambda$. A restructuring of the search state employing the set $\mathcal{N}_S^i$ should now guarantee that the examination of search paths is postponed where facts $\lambda \in \mathcal{N}_S^i$ are involved in. In order to do this one can use the so-called *inference rights* as described in [Fuc97b]. Inference rights are essentially annotations to facts that describe rights to take part in certain inferences. The main idea of these rights is to forbid during some time intervals certain facts that tend to produce much offspring to take part in expanding inferences by detracting the respective inference rights. Thus, we gain a flexible control over the expansion of the fact set without losing simplification power.

When using inference rights we do not employ any longer algorithm SP (if we do not exchange lemmas) or algorithm CP (if we exchange lemmas) using the conventional inference systems $\mathcal{I}$ or $\mathcal{I}_{Dist}$, respectively, but we employ algorithms $SP^{\mathcal{R}}$ or $CP^{\mathcal{R}}$ employing inference systems $\mathcal{I}^{\mathcal{R}}$ or $\mathcal{I}_{Dist}^{\mathcal{R}}$ working on facts with rights (λ, R). Such a right R is a set of inference types λ can be involved in. The inference system $\mathcal{I}^{\mathcal{R}}$ ($\mathcal{I}^{\mathcal{R}}_{Dist}$) contains the rules of $\mathcal{I}$ ($\mathcal{I}_{Dist}$) but considers the rights in such a way that an inference can only be performed if all facts involved in it have the respective inference right. Additionally, $\mathcal{I}^{\mathcal{R}}$ ($\mathcal{I}^{\mathcal{R}}_{Dist}$) contains a rule for detracting rights. Then, by utilizing inference rights such a postponing as suggested is realized in $SP^{\mathcal{R}}$ ($CP^{\mathcal{R}}$): Expanding inferences with facts $\lambda \in \mathcal{N}^i_S$ are forbidden for a while by detracting the inference rights for all expanding inferences ("deactivation") in the cooperation phase. Thus, the examination of proof paths where such facts are involved in is postponed. Moreover, we do not need to perform so many time-consuming expanding inferences and are able to examine other proof paths with less computation effort. Note that by utilizing such a fine-grained control via inference rights we do not loose simplification power because contraction inferences are still allowed. Thus, we can indeed find proofs very quickly if the "deactivated" facts are really not needed for any proof. Nevertheless, it might be the case that such a detected bad fact is necessary for the proof. Thus, we employ a *recover mechanism* for inference rights (by using an additional recover rule in $\mathcal{I}^{\mathcal{R}}$ ($\mathcal{I}^{\mathcal{R}}_{Dist}$)) that can give inference rights back to a fact and performs all inferences that were postponed by deactivation. Also in this case it is possible that we gain efficiency: Parts of the proof that are "parallel" to the bad fact, i.e. the facts in this part do not have the bad fact as an ancestor, can possibly be found faster if the bad fact is deactivated. The remaining parts of the proof can be found after the recovery of rights. Thus, the restructuring of the search state can entail a rearrangement of the proof parts found by the prover and thus a gain of efficiency. Such a recovery of rights is performed by $SP^{\mathcal{R}}$ ($CP^{\mathcal{R}}$) periodically in the cooperation phases.

Finally, we want to examine if a prover that performs fair $\mathcal{I}$-derivations or fair $\mathcal{I}_{Dist}$-derivations with a strong fair heuristic might become incomplete when considering the received information for restructuring the search state. First, we define fairness of $\mathcal{I}^{\mathcal{R}}$- ($\mathcal{I}^{\mathcal{R}}_{Dist}$-)derivations analogously to section 2, i.e. all expansion inferences must be performed to persistent facts without considering the inference right. Again, fairness implies in our context the completeness of the provers. Then it holds:

Theorem 6. *([Fuc97a]) Let $\mathcal{I}$ ($\mathcal{I}_{Dist}$) be an inference system (with disturbance). Let $\mathcal{H}$ be a strong fair heuristic. Furthermore, let $SP^{\mathcal{R}}$ ($CP^{\mathcal{R}}$) be realized in such a way that it uses $\mathcal{H}$, that each deactivated fact can recover all inference rights after a finite period of time, and that no facts are deactivated infinitely often. Then $SP^{\mathcal{R}}$ ($CP^{\mathcal{R}}$) produces only fair $\mathcal{I}^{\mathcal{R}}$- ($\mathcal{I}^{\mathcal{R}}_{Dist}$-)derivations.*

Hence, easily realizable conditions on the heuristic and the algorithm can guarantee completeness. In our experiments we limited the number of deactivations to the value 10. Moreover, we recovered inference rights after 5 cooperation phases.

6 Experimental Results

In order to examine the potential of our cooperation concepts we conducted our experimental studies in different domains of the problem library TPTP v.1.2.1

(see [SSY94]). We experimented on the one hand with cooperating homogeneous provers, i.e. provers that are based on the same calculus and differ from each other only in the heuristic they use. On the other hand, we let also heterogeneous provers cooperate. Firstly, we evaluate our two cooperation concepts separately and use either only positive information or negative information. Secondly, we let the provers cooperate by the exchange of important lemmas and bad facts. All experiments that are described in the following were performed on SPARCstations-20/712.

6.1 Cooperation of Homogeneous Provers

We examined cooperation of homogeneous provers in the area of superposition based theorem proving employing the prover SPASS. As our test set we chose problems stemming from the CADE-13 ATP system competition. A minimal requirement in order to couple different incarnations of a prover is that it has different heuristics at its disposal. SPASS, however, uses only one fixed strategy that counts a weighted sum of the number of variables and two times the number of function symbols of a fact. In order to get different heuristics we allowed different weightings of certain function symbols.

In order to let the provers exchange lemmas we filtered facts with $P_{\mathcal{P}}$, selected 25 facts with the functions ψ_S and ψ_H at the sender site, and selected again from these facts 10 facts at the receiver site with ψ_U and ψ_{SG}. In order to detect bad facts we used the predicate $P_{\mathcal{N}}$ and selection function $\varphi_{\mathcal{N}}$ as described in section 5. We selected in each cooperation phase 5 bad facts. We employed for all experiments fixed parameters of the selection functions.

In order to investigate the potential of our cooperation concepts we compare them with the standard version of SPASS. We let two heuristics cooperate: The SPASS standard heuristic and an automatically generated heuristic.

Table 1 displays an excerpt of our results. We list here only the results of such problems the standard heuristic of SPASS was able to solve, running at least 5 seconds. Columns 2 and 3 display the run times needed by the two coupled heuristics, columns 4–6 the runtimes of the cooperating system when employing our cooperation concepts. Column 7 presents the results when using OTTER ([McC94]).

The results reveal that cooperation by exchanging important lemmas allows for an obvious gain of efficiency in comparison with the standard version of the prover ($\mathcal{H}_1$). Moreover, our cooperative provers solve the most problems about two times faster compared to the best of the coupled heuristics. It is to be emphasized that due to our inefficient implementation (exchange of facts via files) the results can further be improved. Coupling of superposition provers by using negative information does not entail results as good as before: we achieve on the average a speed-up of 1.6 compared to the best of the coupled heuristics. But there are difficult examples (e.g., `RNG018-6`, `GRP148-1`) where the exchange of negative information leads to much better results than the exchange of positive information. Employing both kinds of information entails in the prevailing number of cases a further gain of efficiency. Particularly when tackling difficult problems the results are very encouraging.

problem	$\mathcal{H}_1$	$\mathcal{H}_2$	pos	neg	pos/neg	OTTER
LCL196-1	292.4s	311.7s	83.8s	272.1s	80.2s	34.9s
LCL163-1	10.0s	11.9s	7.5s	9.8s	7.0s	–
GRP048-2	23.3s	17.4s	8.7s	12.9s	8.3s	101.9s
GRP148-1	951.2s	253.1s	184.7s	97.4s	91.1s	70.3s
GRP169-1	80.1s	15.7s	13.7s	9.9s	10.8s	9.9s
GRP169-2	56.1s	26.2s	9.7s	14.7s	12.8s	9.5s
GRP174-1	8.0s	8.3s	5.8s	3.9s	4.0s	9.6s
RNG018-6	639.9s	199.7s	152.3s	87.1s	79.1s	0.5s
NUM009-1	8.1s	6.3s	2.6s	8.7s	4.2s	21.5s

Table 1. Coupling incarnations of SPASS by exchanging positive/negative information

6.2 Cooperation of Heterogeneous Provers

We examine cooperation among heterogeneous provers by coupling SPASS and DISCOUNT. If we are interested in solving problems in full first-order logic with equality we have to take care of the problem that DISCOUNT can only cope with unit clauses.[1] This is not a problem during the proof run because facts DISCOUNT cannot work with are not allowed to pass through the filter of SPASS. But, we must at least transform the initial axiomatization. Our solution of this problem is following decomposition of a proof problem, represented as a set of clauses $\mathcal{C}$ whose inconsistency should be shown: Each positive equation $P(t_1, \ldots, t_n) = true$ or $s = t$ of $\mathcal{C}$ is chosen as an axiom for DISCOUNT, each negative equation acts as a proof goal. We only considered examples where we could isolate enough positive equations such that DISCOUNT could not find a complete system of facts. If no negative equation was an element of $\mathcal{C}$ DISCOUNT worked in a completion mode. Obviously, by this kind of problem decomposition DISCOUNT is often not able to solve its proof problem. However, it can produce many lemmas or bad facts from the part of the search space it can traverse.

In order to exchange important lemmas we selected 10 facts via ψ_S and ψ_H at the sender site, then an additional selection of 5 facts using ψ_U and ψ_{SG} took place at the receiver site. The same number of bad facts was exchanged.

Generally, we let SPASS work in its standard mode. DISCOUNT activated facts with a goal-oriented heuristic as described in [DF94]. In order to measure the strength of our cooperation concepts we experimented in the light of various problems taken from TPTP. We dealt particularly with the domains ROB and HEN. In these domains our cooperative system was either better than each of the coupled provers or at least as good as the best. We present in table 2 a small excerpt from these experiments, enriched with some highlights taken from other domains. More problems can be found in [Fuc97a].

The best results of SPASS when working alone can be found in column 2. Column 3 displays the run times when using DISCOUNT. The entry "–" denotes

[1] Note that unit literals $P(t_1, \ldots, t_n)$ can easily be transformed into equations $P(t_1, \ldots, t_n) = true$.

problem	SPASS	DISCOUNT	pos	neg	pos/neg	OTTER
BOO007-4	403.4s	–	19.3s	373.4s	19.3s	10.9s
GRP169-1	80.1s	–	41.9s	50.5s	40.3s	9.9s
GRP177-2	–	–	58.3s	–	58.3s	–
GRP179-1	–	–	80.7s	–	80.7s	–
LCL163-1	10.0s	12.0s	7.7s	7.4s	6.3s	–
ROB005-1	–	109.6s	39.9s	74.2s	35.2s	44.9s
ROB008-1	–	98.8s	33.3s	58.4s	17.0s	0.4s
ROB011-1	105.3s	–	21.9s	82.4s	20.4s	0.6s
ROB016-1	9.8s	–	4.6s	3.8s	3.4s	0.7s
ROB022-1	15.1s	–	6.9s	12.2s	4.7s	5.0s
ROB023-1	204.6s	–	4.4s	204.8s	4.1s	2.2s
LDA010-2	–	–	682.7s	–	682.7s	–
HEN009-5	309.9s	–	105.9s	99.4s	109.3s	119.7s
HEN011-5	41.2s	–	26.2s	41.5s	23.3s	–

Table 2. Coupling SPASS and DISCOUNT by exchanging positive/negative information

that the problem could not be solved within 1000 seconds. Note that DISCOUNT is often not able to solve the problem because it has only parts of the proof problem as input. Columns 4–6 show the run times when employing one or both of our cooperation concepts, column 7 the run time when using OTTER.

If we take a closer look at the results we can observe that they are much better than in the homogeneous case. By exchanging important lemmas we are able to solve all of the listed problems, whereas SPASS is only able to solve 64%, DISCOUNT 21%, and OTTER 64%. Even when we take into account the fact that SPASS *and* DISCOUNT work as cooperative provers there are still problems neither SPASS nor DISCOUNT can cope with, but that can be solved if the provers cooperate (GRP177-2, GRP179-1, LDA010-2). When coupling provers only by exchanging negative information the speed-ups are lower but in almost all cases we could gain efficiency. Moreover, there are also some examples where the speed-ups are rather high. As we have already emphasized in the preceding experiments we can find problems where the use of negative information outperforms the use of positive information. Very hard problems, however, seem to be out of reach. Combining both exchange of important lemmas and bad facts, is again the best way for coupling theorem provers.

7 Discussion

We have presented two different cooperation techniques well-suited for coupling homogeneous and heterogeneous provers: On the one hand the exchange of important lemmas. On the other hand the exchange of certain "bad facts".

We examined these two cooperation concepts theoretically and discovered weak conditions on the heuristic of a prover and the processing of the received information that are sufficient for completeness. Thus, in most cases our cooperation concepts do not destroy completeness. Besides theoretical aspects, we pointed out practical ways to realize cooperative provers and introduced some concrete techniques and heuristics. The experiments have shown that our methods indeed enabled the cooperative provers to outperform sequential provers.

It would be interesting to examine whether other positive or negative information than good or bad facts might be useful for coupling different provers. E.g., the exchange of control information in order to couple different provers would be an interesting topic for further research.

Acknowledgments. I would like to thank Jürgen Avenhaus, Marc Fuchs, and the referees for constructive criticism and many helpful comments.

References

[AD93] J. Avenhaus and J. Denzinger. Distributing equational theorem proving. In *Proc. 5th RTA*, pages 62–76, Montreal, 1993. LNCS 690.

[ADF95] J. Avenhaus, J. Denzinger, and M. Fuchs. DISCOUNT: A System For Distributed Equational Deduction. In *Proc. 6th RTA*, pages 397–402, Kaiserslautern, 1995. LNCS 914.

[BDP89] L. Bachmair, N. Dershowitz, and D.A. Plaisted. Completion without Failure. In *Coll. on the Resolution of Equations in Algebraic Structures*. Academic Press, Austin, 1989.

[BG94] L. Bachmair and H. Ganzinger. Rewrite-based equational theorem proving with selection and simplification. *Journal of Logic and Computation*, 4(3):217–247, 1994.

[DF94] J. Denzinger and M. Fuchs. Goal oriented equational theorem proving. In *Proc. 18th KI-94*, pages 343–354, Saarbrücken, 1994. LNAI 861.

[DF96] J. Denzinger and D. Fuchs. Referees for teamwork. In *Proc. FLAIRS '96*, pages 454–458, Key West, 1996.

[FD97] D. Fuchs and J. Denzinger. Cooperation in theorem proving by loosely coupled heuristics. Technical Report SR-97-03 (ftp://ftp.uni-kl.de/reports_uni-kl/computer_science/SEKI/1997/Fuchs.SR-97-03.ps.gz), University of Kaiserslautern, Kaiserslautern, 1997.

[Fuc97a] D. Fuchs. Coupling saturation-based provers by exchanging positive/negative information. Technical Report SR-97-07 (ftp://ftp.uni-kl.de/reports_uni-kl/computer_science/SEKI/1997/Fuchs.SR-97-07.ps.gz), University of Kaiserslautern, Kaiserslautern, 1997.

[Fuc97b] D. Fuchs. Inference Rights for Controlling Search in Generating Theorem Provers. In *Proc. EPIA '97*, pages 25–36, Coimbra, 1997. LNAI 1323.

[McC94] W. McCune. Otter 3.0 reference manual and guide. Technical Report ANL-94/6, Argonne National Laboratory, Argonne, 1994.

[SSY94] G. Sutcliffe, C.B. Suttner, and T. Yemenis. The TPTP Problem Library. In *CADE-12*, pages 252–266, Nancy, 1994. LNAI 814.

[Sut92] G. Sutcliffe. A heterogeneous parallel deduction system. In *Proc. FGCS'92 Workshop W3*, 1992.

[WGR96] C. Weidenbach, B. Gaede, and G. Rock. Spass & Flotter Version 0.42. In *Proc. CADE-13*, pages 141–145, New Brunswick, 1996. LNAI 1104.

An On-line Problem Database

Nachum Dershowitz[1]* and Ralf Treinen[2]**

[1] Department of Computer Science, University of Illinois, Urbana, IL 61801, U.S.A.
[2] L.R.I., Université de Paris-Sud, F91405 Orsay cedex, France

Preface

The RTA list of open problems was created in 1991 by Jean-Pierre Jouannaud, Jan Willem Klop, and the first author [21] on the occasion of the *Fourth International Conference on Rewriting Techniques and Applications* (RTA). Updated lists have since been published at RTA '93 [22] and RTA '95 [23]. Bending to these electronic times, we have recently placed a combined list on the world-wide web at

`http://www.lri.fr/~rtaloop`

This list can also be accessed from the "Rewriting Home Page", currently at

`http://www.loria.fr/~vigneron/RewritingHP`

The RTA list seeks to summarize open problems and subsequent solutions in fields of interest to this conference. For the current proceedings, the main subjects were

Term rewriting systems	Symbolic and algebraic computation
Unification and matching	Completion techniques
String and graph rewriting	Conditional and typed rewriting
Rewriting-based theorem proving	Parallel rewriting and deduction
Constrained rewriting and deduction	Constraint solving
Higher-order rewriting	Lambda calculi
Functional and logic programming languages	

We continue to solicit electronic contributions of new problems, progress reports, solutions, and comments. These should be mailed electronically to

`rtaloop@lri.fr`

Of the 87 problems included in previous lists (problems 1–44 in [21]; 45–77 in [22]; and 78–87 in [23]), seventeen have been solved (specifically 7, 20, 33, 39, 42, 76, and 77 in [22]; 4, 44, and 68 in [23]; and 3, 41, 51, 52, 78, 81, and 87 here), and many others have seen progress. Thus, the lists have indeed helped promote and focus research in rewriting. In this report, we provide a brief update on problems about which we know of significant progress since the appearance of [23].

* Supported in part by the National Science Foundation under grant CCR-97-00070.
** Supported in part by the Esprit working group CCL II (22457).

Solved Problems

Problem #3 [Deepak Kapur]

A term t is *ground reducible* with respect to a rewrite system R if all its ground (variable-free) instances contain a redex. Ground reducibility is decidable for ordinary rewriting (and finite R) [11,34,61], but n^{n^n} is the best known upper bound in general, $2^{dn \log n}$ and $2^{cn/\log n}$ are the best upper and lower bounds, respectively, for left-linear systems, where n is the size of the system R and c, d are constants [34]. Can these bounds be improved?

Ground-reducibility is EXPTIME-complete [13].

Problem #41 [Participants at Unif Val d'Ajol]

The complexity of the theory of finite trees when there are finitely many symbols is known to be PSPACE-hard [44]. Is it in PSPACE? The same question applies to infinite trees.

The problem is non-elementary [83].

Problem #51 [Hubert Comon, Max Dauchet]

For an arbitrary finite term rewriting system R, is the first-order theory of one-step rewriting ($\rightarrow_R$) decidable? Decidability would imply the decidability of the first-order theory of encompassment (that is, being an instance of a subterm) [9], as well as several known decidability results in rewriting. (It is well known that the theory of $\rightarrow_R^*$ is in general undecidable.)

This has been answered negatively in [78]. Sharper undecidability results have been obtained for the following subclasses of rewrite systems: linear, shallow, $\exists^*\forall^*$-fragment [69]; linear, terminating, $\exists^*\forall^*\exists^*$-fragment [84]; right-ground, terminating, $\exists^*\forall^*$-fragment [46]. Decidability results have been obtained for the positive existential theory [52], the case of unary signatures [32], and for left-linear right-ground systems [73].

Problem #52 [Richard Statman]

It has been remarked by C. Böhm [5] that Y is a fixed point combinator if and only if $Y \leftrightarrow^* (SI)Y$ (Y and SIY are convertible). Also, if Y is a fixed point combinator, then so is $Y(SI)$. Is there a fixed point combinator Y for which $Y \leftrightarrow^* Y(SI)$?

This was solved by B. Intrigila [31] who showed that there is no such fixed point combinator.

Problem #78 [Pierre Lescanne]

There are confluent calculi of explicit substitutions, but these do not preserve termination (strong normalization) [16,48], and there are calculi that are not confluent on open terms, but which do preserve termination [42]. Is there a calculus of explicit substitution that is both confluent and preserves termination?

The calculus presented in [51] enjoys both properties.

Problem #81 [Andreas Weiermann]

If the termination of a finite rewrite system over a finite signature can be proved using a simplification ordering, then the derivation lengths are bounded by a Hardy function of ordinal level less than the small Veblen number $\phi_{\Omega^\omega}0$. (See [85].) Is it possible to lower this bound by replacing the Hardy function by a slow growing function? That is, is it possible to bound the derivation lengths by a multiply recursive function?

H. Touzet [74] has shown in her thesis that the answer is negative, exhibiting a simplifying rewrite system that has derivation bounds "longer" than multiply recursive. What now remains open is what complexity can be achieved using simplifying rewrite systems.

Problem #87 [Hans Zantema]

Termination of string-rewriting systems is known to be undecidable [30]. Termination of a single term-rewriting rule was proved undecidable in [18,41]. It is also undecidable whether there exists a simplification ordering that proves termination of a single term rewriting rule [49] (cf. [33]). Is it decidable whether a single term rewrite rule can be proved terminating by a monotonic ordering that is total on ground terms? (With more rules it is not [87].)

A negative solution has been provided in [28].

Significant Progress

Problem #13 [Jean-Jacques Lévy]

By a lemma of G. Huet [29], left-linear term-rewriting systems are confluent if, for every critical pair $t \approx s$ (where $t = u[r\sigma] \leftarrow u[l\sigma] = g\tau \rightarrow d\tau = s$, for some rules $l \rightarrow r$ and $g \rightarrow d$), we have $t \rightarrow^{\|} s$ (t reduces in one parallel step to s). (The condition $t \rightarrow^{\|} s$ can be relaxed to $t \rightarrow^{\|} r \leftarrow^{\|} s$ for some r when the critical pair is generated from two rules overlapping at the roots; see [76].) What if $s \rightarrow^{\|} t$ for every critical pair $t \approx s$? What if for every $t \approx s$ we have $s \rightarrow^{=} t$? (Here $\rightarrow^{=}$ is the reflexive closure of $\rightarrow$.) What if for every critical pair $t \approx s$, either $s \rightarrow^{=} t$ or $t \rightarrow^{=} s$? In the last case, especially, a confluence proof would be interesting; one would then have confluence after critical-pair completion without regard for termination. If these conditions are insufficient, the counterexamples will have to be (besides left-linear) non-right-linear, nonterminating, and nonorthogonal (have critical pairs). See [36].

Significant progress is reported in [60].

Problem #21 [Max Dauchet] Is termination of one linear (left and right) rule decidable? Left linearity alone is not enough for decidability [17].

In [22], the following remark was added:

A less ambitious, long-standing open problem (mentioned in [20]) is decidability for *one* (length-increasing) monadic (string, semi-Thue) rule. Termination is undecidable for nonlength-increasing monadic systems of rules [8]. For one monadic rule, confluence is decidable [40,86]. What about confluence of one nonmonadic rule?

Termination and uniform termination of one string rule of the form $0^p1^q \rightarrow v$, *where* $p, q > 0, v \in \{0,1\}^*$, *has been shown decidable [70]. For a fixed system of this form the termination problem is of linear complexity. A simple characterisation of the systems of the above form which are not uniformly terminating has been givem in [37]. It would be nice to extend these results to more general non-overlapping left sides. See also Problem #87.*

Problem #32 [John Pedersen]

Is there a finite term-rewriting system of some kind for free lattices?

As mentioned in Problem #77 [23], it has been shown in [25] that there is no finite, normal form, associative-commutative term-rewriting system for lattices.

Problem #43 [Jean-Pierre Jouannaud]

Design a framework for combining constraint solving algorithms.

The combination approach of [1] has been extended in [2] to constraints involving predicate symbols other that equality, and [3] in turn extends this approach to constraint-solving over solution domains that are not free structures. These results are presented in a uniform framework by [4].

The work of [66] has been extended to the case of "shared constructors" by [24].

Problem #59 [M. Kurihara, M. Krishna Rao]

One of the earliest results established on modularity of combinations of term-rewriting systems is the confluence of the union of two confluent systems which share no symbols [75]; if symbols are shared modularity is not preserved by union [39]. Some sufficient conditions for modularity of confluence of constructor-sharing systems that are terminating have been found [39,50]. Are there interesting sufficient conditions that are independent of termination?

Left-linearity is a sufficient condition, as shown long ago in [65]. In [55], it is established that confluence is modular in the presence of the weak normalization property. (This result has been extended in [64,63] for hierarchical combinations.) In [19], some results are given when only one of the systems is terminating.

There are other sufficient conditions for modularity of confluence that do not require termination of the combined system even when function symbols are shared. One set of conditions, viz., "persistence", "relative termination", and lr-disjointness, is given in [80,81]. An abstract confluence theorem without termination is given in [27].

Problem #61 [Tobias Nipkow, M. Takahashi] For higher-order rewrite formats as given by combinatory reduction systems [35] and higher-order rewrite systems [53,71], confluence has been proved in the restricted case of orthogonal systems. Can confluence be extended to such systems when they are weakly orthogonal (all critical pairs are trivial)? When critical pairs arise only at the root, confluence is known to hold.

Weakly orthogonal higher-order rewriting systems are confluent. This has been shown both via the Tait-Martin-Löf method and via finite developments [79].

Details and extensions similar to Huet's parallel closure condition can be found in [57,58,62].

Problem #70 [Jean-Claude Raoult]

There exist finite automata for words, trees, and dags. No really good comparable notion is available for graphs. (Perhaps there is one akin to the ideas in [43] on label rewriting.)

A well motivated notion of "graph acceptor" has been presented in [72].

Problem #71 [Jean-Claude Raoult]

There are good algorithms for pattern-matching for words and trees, but not yet for graphs.

An algorithm for finding the rules of a graph grammar that are applicable to a graph has been given in [7].

Problem #72 [Jean-Claude Raoult]

Graph rewritings, like term or word rewritings, are usually finitely branching. There are relations that are not finitely branching, yet satisfy good properties: rational transductions of words, tree-transductions. A good definition of graph transduction, that extends rational word transductions is still lacking.

See [14,15].

Problem #73 [Jean-Claude Raoult]

Termination is, as we know, undecidable. Yet, there are several sufficient conditions ensuring termination for word and term rewritings. Most are suitable extensions of Higman's or Kruskal's embeddings [38]. Robertson and Seymour [68] have achieved a similar theorem for undirected graphs. However, no embedding theorem has yet been proved for directed graphs, and (consequently?) powerful termination orderings remain to be designed.

In [67], embedding theorems are proved for directed wqo-labelled graphs and hypergraphs.

Problem #79 [Mizuhito Ogawa]

Does a system that is non-overlapping under unification with infinite terms (unification without "occur-check" [47]) have unique normal forms? This conjecture was originally proposed in [54] with an incomplete proof, as an extension of the result on strongly non-overlapping systems [35,10]. Related results appear in [59,77,45], but the original conjecture is still open. This is related to Problem #58. This problem is also related with modularity of confluence of systems sharing constructors, see [56].

The answer is yes if the system is also nonduplicating [81]. A direct technique is given in [81]. The nonduplicating condition can be relaxed under a certain technical condition [81]. Some extensions to handle root overlaps are given in

[82] and a restricted version of the result in [10] is also proved using the direct technique in [82].

Acknowledgements

We couldn't have produced this report without the advice we received from many people: Franz Baader, Witold Charatonik, Adam Cichon, Bruno Courcelle, Miki Herman, Benedetto Intrigila, Jean-Pierre Jouannaud, Pierre Lescanne, Krishna Rao Madala, Albert Meyer, Aart Middeldorp, Cesar Munoz, Tobias Nipkow, Enno Ohlebusch, Vincent van Oostrom, Michio Oyamaguchi, Detlef Plump, Michael Rusinowitich, Géraud Senizergues, Wayne Snyder, Richard Statman, Hélène Touzet, Rakesh Verma, and Hans Zantema.

References

1. F. Baader and K. Schulz. Unification in the union of disjoint equational theories: Combining decision procedures. In D. Kapur, editor, *Proceedings of the Eleventh International Conference on Automated Deduction (Saratoga Springs, NY)*, volume 607 of *Lecture Notes in Artificial Intelligence*, Berlin, June 1992. Springer-Verlag.
2. F. Baader and K. Schulz. Combination of constraint solving techniques: An algebraic point of view. In *Proceedings of the 6th International Conference on Rewriting Techniques and Applications*, volume 914 of *Lecture Notes in Artificial Intelligence*, pages 352–366, Kaiserslautern, Germany, 1995. Springer Verlag.
3. F. Baader and K. Schulz. On the combination of symbolic constraints, solution domains, and constraint solvers. In *Proceedings of the International Conference on Principles and Practice of Constraint Programming, CP95*, volume 976 of *Lecture Notes in Artificial Intelligence*, Cassis, France, 1995. Springer Verlag.
4. F. Baader and K. Schulz. Combination of constraint solvers for free and quasi-free structures. *Theoretical Computer Science*, 1998. To appear.
5. H. P. Barendregt. *The Lambda Calculus, its Syntax and Semantics.* North-Holland, Amsterdam, second edition, 1984.
6. M. Bidoit and M. Dauchet, editors. *Theory and Practice of Software Development*, Lecture Notes in Computer Science, vol. 1214, Lille, France, April 1997. Springer-Verlag.
7. H. Bunke, T. Glauser, and T.-H. Tran. An efficient implementation of graph grammars based on the RETE-matching algorithm. In H. Ehrig, H.-J. Kreowski, and G. Rozenberg, editors, *Graph Grammars and Their Application to Computer Science*, volume 532 of *Lecture Notes in Computer Science*, pages 174–189, 1991.
8. A.-C. Caron. Linear bounded automata and rewrite systems: Influence of initial configurations on decision properties. In *Proceedings of the International Joint Conference on Theory and Practice of Software Development, volume 1: Colloquium on Trees in Algebra and Programming (Brighton, U.K.)*, volume 493 of *Lecture Notes in Computer Science*, pages 74–89, Berlin, April 1991. Springer-Verlag.
9. A.-C. Caron, J.-L. Coquidé, and M. Dauchet. Encompassment properties and automata with constraints. In C. Kirchner, editor, *Proceedings of the Fifth International Conference on Rewriting Techniques and Applications*, volume 690 of *Lecture Notes in Computer Science*, Montreal, Canada, 1993. Springer-Verlag.

10. P. Chew. Unique normal forms in term rewriting systems with repeated variables. In *Proceedings of the Thirteenth Annual Symposium on Theory of Computing*, pages 7–18. ACM, 1981.
11. H. Comon. *Unification et Disunification: Théorie et Applications.* PhD thesis, l'Institut National Polytechnique de Grenoble, 1988.
12. H. Comon, editor. *8th International Conference on Rewriting Techniques and Applications*, Lecture Notes in Computer Science, vol. 1232, Sitges, Spain, June 1997. Springer-Verlag.
13. H. Comon and F. Jacquemard. Ground reducibility is EXPTIME-complete. In *Twelfth Annual IEEE Symposium on Logic in Computer Science*, pages 26–34, Warsaw,Poland, June 1997. IEEE.
14. B. Courcelle. Monadic-second order graph transductions: A survey. *Theoretical Computer Science*, 126:53–75, 1994.
15. B. Courcelle. The expression of graph properties and graph transformations in monadic second-order logic. In G. Rozenberg, editor, *Handbook of graph grammars and computing by graph transformations, vol. 1: Foundations*, chapter 5, pages 313–400. World Scientific, New-Jersey, London, 1997.
16. P.-L. Curien, T. Hardin, and J.-J. Lévy. Confluence properties of weak and strong calculi of explicit substitutions. RR 1617, Institut National de Rechereche en Informatique et en Automatique, Rocquencourt, February 1992.
17. M. Dauchet. Simulation of Turing machines by a left-linear rewrite rule. In N. Dershowitz, editor, *Proceedings of the Third International Conference on Rewriting Techniques and Applications (Chapel Hill, NC)*, volume 355 of *Lecture Notes in Computer Science*, pages 109–120, Berlin, April 1989. Springer-Verlag.
18. M. Dauchet. Simulation of Turing machines by a regular rewrite rule. *Theoretical Computer Science*, 103(2):409–420, 1992.
19. N. Dershowitz. Innocuous constructor-sharing combinations. In Comon [12], pages 202–216.
20. N. Dershowitz and J.-P. Jouannaud. Rewrite systems. In J. van Leeuwen, editor, *Handbook of Theoretical Computer Science*, volume B: Formal Methods and Semantics, chapter 6, pages 243–320. North-Holland, Amsterdam, 1990.
21. N. Dershowitz, J.-P. Jouannaud, and J. W. Klop. Open problems in rewriting. In R. Book, editor, *Proceedings of the Fourth International Conference on Rewriting Techniques and Applications (Como, Italy)*, volume 488 of *Lecture Notes in Computer Science*, pages 445–456, Berlin, April 1991. Springer-Verlag.
22. N. Dershowitz, J.-P. Jouannaud, and J. W. Klop. More problems in rewriting. In C. Kirchner, editor, *Proceedings of the Fifth International Conference on Rewriting Techniques and Applications (Montreal, Canada)*, volume 690 of *Lecture Notes in Computer Science*, pages 468–487, Berlin, June 1993. Springer-Verlag.
23. N. Dershowitz, J.-P. Jouannaud, and J. W. Klop. Problems in rewriting III. In *Proceedings of the Sixth International Conference on Rewriting Techniques and Applications (Kaiserslautern, Germany)*, Lecture Notes in Computer Science, Berlin, April 1995. Springer-Verlag.
24. E. Domenjoud, F. Klay, and C. Ringeissen. Combination techniques for non-disjoint equational theories. In A. Bundy, editor, *CADE 94*, volume 814 of *Lecture Notes in Artificial Intelligence*, pages 267–281, Nancy, France, 1994. Springer-Verlag.
25. R. Freese. personal communication, 1993.
26. H. Ganzinger, editor. *7th International Conference on Rewriting Techniques and Applications*, Lecture Notes in Computer Science, vol. 1103, New Brunswick, NJ, USA, July 1996. Springer-Verlag.

27. A. Geser. *Relative Termination.* PhD thesis, Universität Passau, Passau, Germany, 1990.
28. A. Geser, A. Middeldorp, E. Ohlebusch, and H. Zantema. Relative undecidability in the termination hierarchy of single rewrite rules. In Bidoit and Dauchet [6], pages 237–248.
29. G. Huet. Confluent reductions: Abstract properties and applications to term rewriting systems. *J. of the Association for Computing Machinery*, 27(4):797–821, October 1980.
30. G. Huet and D. S. Lankford. On the uniform halting problem for term rewriting systems. Rapport laboria 283, Institut de Recherche en Informatique et en Automatique, Le Chesnay, France, March 1978.
31. B. Intrigila. Non-existent statman's double fixed point combinator does not exist, indeed. *Information and Computation*, 137(1):35–40, 1997.
32. F. Jacquemard. *Automates d'arbres et Réécriture de termes.* PhD thesis, Université de Paris-Sud, 1996. In French.
33. J.-P. Jouannaud and H. Kirchner. Construction d'un plus petit ordre de simplification. *RAIRO Theoretical Informatics*, 18(3):191–207, 1984.
34. D. Kapur, P. Narendran, and H. Zhang. On sufficient completeness and related properties of term rewriting systems. *Acta Informatica*, 24(4):395–415, August 1987.
35. J. W. Klop. *Combinatory Reduction Systems*, volume 127 of *Mathematical Centre Tracts.* Mathematisch Centrum, Amsterdam, 1980.
36. J. W. Klop. Term rewriting systems. In S. Abramsky, D. M. Gabbay, and T. S. E. Maibaum, editors, *Handbook of Logic in Computer Science*, volume 2, chapter 1, pages 1–117. Oxford University Press, Oxford, 1992.
37. Y. Kobayashi, M. Katsura, and K. Shikishima-Tsuji. Termination and derivational complexity of confluent one-rule string-rewriting systems. Technical report, Toho university, Japan, 1997.
38. J. B. Kruskal. Well-quasi-ordering, the Tree Theorem, and Vazsonyi's conjecture. *Transactions of the American Mathematical Society*, 95:210–225, May 1960.
39. M. Kurihara and A. Ohuchi. Modularity of simple termination of term rewriting systems with shared constructors. *Theoretical Computer Science*, 103:273–282, 1992.
40. W. Kurth. *Termination und Konfluenz von Semi-Thue-Systems mit nur einer Regel.* PhD thesis, Technische Universitat Clausthal, Clausthal, Germany, 1990. In German.
41. P. Lescanne. On termination of one rule rewrite systems. *Theoretical Computer Science*, 132:409–420, 1992.
42. P. Lescanne and J. Rouyer-Degli. The calculus of explicit substitutions $\lambda\upsilon$. Technical Report RR-2222, INRIA-Lorraine, January 1994.
43. I. Litovski, Y. Métivier, and E. Sopena. Definitions and comparisons of local computations on graphs. *Mathematical Systems Theory*, to appear. Available as internal report 91-43 of LaBRI, University of Bordeaux 1.
44. M. J. Maher. Complete axiomatizations of the algebras of the finite, rational and infinite trees. In *Proceedings of the Third IEEE Symposium on Logic in Computer Science*, pages 348–357, Edinburgh, UK, July 1988. IEEE, Computer Society Press.
45. K. Mano and M. Ogawa. A new proof of Chew's theorem. Technical report, IPSJ PRG94-19-7, 1994.
46. J. Marcinkowski. Undecidability of the first order theory of one-step right ground rewriting. In Comon [12], pages 241–253.

47. A. Martelli and G. Rossi. Efficient unification with infinite terms in logic programming. In *International conference on fifth generation computer systems*, pages 202–209, 1984.
48. P.-A. Melliès. Typed λ-calculi with explicit substitutions may not terminate, 1995. To appear.
49. A. Middeldorp and B. Gramlich. Simple termination is difficult. *Applicable Algebra in Engineering, Communication and Computing*, 6(2):115–128, 1995.
50. A. Middeldorp and Y. Toyama. Completeness of combinations of constructor systems. In R. Book, editor, *Proceedings of the Fourth International Conference on Rewriting Techniques and Applications (Como, Italy)*, volume 488 of *Lecture Notes in Computer Science*, pages 174–187, Berlin, April 1991. Springer-Verlag.
51. C. Muñoz. Confluence and preservation of strong normalisation in an explicit substitutions calculus (extended abstract). In *Eleventh Annual IEEE Symposium on Logic in Computer Science*, pages 440–447, New Brunswick, New Jersey, July 1996. IEEE Computer Society Press.
52. J. Niehren, M. Pinkal, and P. Ruhrberg. On equality up-to constraints over finite trees, context unification and one-step rewriting. In *14th International Conference on Automated Deduction*, Townsville, Australia, July 1997. To appear.
53. T. Nipkow. Higher-order critical pairs. In *Proceedings of the Sixth Symposium on Logic in Computer Science*, pages 342–349, Amsterdam, The Netherlands, 1991. IEEE.
54. M. Ogawa and S. Ono. On the uniquely converging property of nonlinear term rewriting systems. Technical report, IEICE COMP89-7, 1989.
55. E. Ohlebusch. On the modularity of confluence of constructor-sharing term rewriting systems. In *19th Colloquium on Trees in Algebra and Programming*, volume 787 of *LNCS*, pages 261–275. Springer-Verlag, 1994.
56. E. Ohlebusch. On the modularity of confluence of constructor-sharing term rewriting systems. In *Proceedings of the Colloquium on Trees in Algebra and Programming*, 1994.
57. V. v. Oostrom. *Confluence for Abstract and Higher-Order Rewriting.* PhD thesis, Vrije Universiteit, Amsterdam, 1994.
58. V. v. Oostrom. Developing developments. *Theoretical Computer Science*, 175:159–181, 1997.
59. M. Oyamaguchi and Y. Ohta. On the confluent property of right-ground term rewriting systems. *Trans. IEICE*, J76-D-I:39–45, 1993.
60. M. Oyamaguchi and Y. Ohta. A new parallel closed condition for Church-Rosser of left-linear term rewriting systems. In Comon [12], pages 187–201.
61. D. A. Plaisted. Semantic confluence tests and completion methods. *Information and Control*, 65(2/3):182–215, May/June 1985.
62. F. v. Raamsdonk. *Confluence and Normalization for Higher-Order Rewriting.* PhD thesis, Vrije Universiteit, Amsterdam, 1996.
63. M. K. Rao. Modular aspects of term graph rewriting. *Theoretical Computer Science.* To appear. Special issue on Rewriting Techniques and Applications conference RTA'96.
64. M. K. Rao. Semi-completeness of hierarchical and super-hierarchical combinations of term rewriting systems. In *Theory and Practice of Software Development, TAPSOFT'95*, volume 915 of *Lecture Notes in Computer Science*, pages 379–393. Springer-Verlag, 1995.
65. J.-C. Raoult and J. Vuillemin. Operational and semantic equivalence between recursive programs. *J. of the Association for Computing Machinery*, 27(4):772–796, October 1980.

66. C. Ringeissen. Unification in a combination of equational theories with shared constants and its application to primal algebras. In A. Voronkov, editor, *Proceedings of the Conference on Logic Programming and Automated Reasoning (St. Petersburg, Russia)*, volume 624 of *Lecture Notes in Artificial Intelligence*, Berlin, July 1992. Springer-Verlag.
67. N. Robertson and P. Seymour. Graph minors 23: Nash-williams' immersion conjecture. Preprint, March 1996.
68. N. Robertson and P. D. Seymour. Graph minors IV. Tree-width and well-quasi-ordering. Submitted 1982; revised January 1986.
69. F. Seynhaeve, M. Tommasi, and R. Treinen. Grid structures and undecidable constraint theories. In Bidoit and Dauchet [6], pages 357–368.
70. G. Sénizergues. On the termination problem for one-rule semi-Thue system. In Ganzinger [26], pages 302–316.
71. M. Takahashi. λ-calculi with conditional rules. In M. Bezem and J. F. Groote, editors, *Proceedings of the International Conference on Typed Lambda Calculi and Applications (Utrecht, The Netherlands)*, volume 664 of *Lecture Notes in Computer Science*, pages 406–417, Berlin, 1993. Springer-Verlag.
72. W. Thomas. Automata theory on trres and partial orders. In Bidoit and Dauchet [6], pages 20–38.
73. S. Tison. Automates comme outil de décision dans les arbres. Dossier d'habilitation à diriger des recherches, December 1990. In French.
74. H. Touzet. *Propriétés combinatoires pour la terminaison de systèmes de réécriture.* PhD thesis, Université Henri Poincaré – Nancy 1, Nancy, France, September 1991. In French.
75. Y. Toyama. On the Church-Rosser property for the direct sum of term rewriting systems. *J. of the Association for Computing Machinery*, 34(1):128–143, January 1987.
76. Y. Toyama. Commutativity of term rewriting systems. In K. Fuchi and L. Kott, editors, *Programming of Future Generation Computers II*, pages 393–407. North-Holland, 1988.
77. Y. Toyama and M. Oyamaguchi. Church-Rosser property and unique normal form property of non-duplicating term rewriting systems. In N. Dershowitz and N. Lindenstrauss, editors, *Workshop on Conditional Term Rewriting Systems (Jerusalem, July 1994)*, Lecture Notes in Computer Science. Springer-Verlag, to appear.
78. R. Treinen. The first-order theory of one-step rewriting is undecidable. In Ganzinger [26], pages 276–286.
79. V. v. Oostrom and F. v. Raamsdonk. Weak orthogonality implies confluence: the higher-order case. In A. Nerode and Y. V. Matiyasevich, editors, *Third International Symposium on the Logical Foundations of Computer Science*, volume 813 of *Lecture Notes in Computer Science*, pages 379–392, St. Petersburg, Russia, July 1994. Springer-Verlag.
80. R. M. Verma. Unique normal forms and confluence for rewrite systems. In *Int'l Joint Conf. on Artificial Intelligence*, pages 362–368, 1995.
81. R. M. Verma. Unicity and modularity of confluence for term rewriting systems. Technical report, University of Houston, 1996.
82. R. M. Verma. Unique normal forms for nonlinear term rewriting systems: Root overlaps. In *Symp. on Fundamentals of Computation Theory*, volume 1279 of *Lecture Notes in Computer Science*, pages 452–462. Springer-Verlag, September 1997.

83. S. Vorobyov. An improved lower bound for the elementary theories of trees. In M. A. McRobbie and J. K. Slaney, editors, *13th International Conference on Automated Deduction*, Lecture Notes in Computer Science, pages 275–287, New Brunswick, NJ, july/august 1996. Springer-Verlag.
84. S. Vorobyov. The first-order theory of one step rewriting in linear noetheran systems is undecidable. In Comon [12], pages 254–268.
85. A. Weiermann. Bounding derivation lengths with functions from the slow growing hierarchy. Preprint Münster, 1993.
86. C. Wrathall. Confluence of one-rule Thue systems. In *Proceedings of the First International Workshop on Word Equations and Related Topics (Tübingen)*, volume 572 of *Lecture Notes in Computer Science*, pages 237–246, Berlin, 1990. Springer-Verlag.
87. H. Zantema. Total termination of term rewriting is undecidable. Technical Report UU-CS-1994-55, Utrecht University, December 1994.

Author Index

Lecture Notes in Computer Science

For information about Vols. 1–1300

please contact your bookseller or Springer-Verlag

Vol. 1301: M. Jazayeri, H. Schauer (Eds.), Software Engineering - ESEC/FSE'97. Proceedings, 1997. XIII, 532 pages. 1997.

Vol. 1302: P. Van Hentenryck (Ed.), Static Analysis. Proceedings, 1997. X, 413 pages. 1997.

Vol. 1303: G. Brewka, C. Habel, B. Nebel (Eds.), KI-97: Advances in Artificial Intelligence. Proceedings, 1997. XI, 413 pages. 1997. (Subseries LNAI).

Vol. 1304: W. Luk, P.Y.K. Cheung, M. Glesner (Eds.), Field-Programmable Logic and Applications. Proceedings, 1997. XI, 503 pages. 1997.

Vol. 1305: D. Corne, J.L. Shapiro (Eds.), Evolutionary Computing. Proceedings, 1997. X, 307 pages. 1997.

Vol. 1306: C. Leung (Ed.), Visual Information Systems. X, 274 pages. 1997.

Vol. 1307: R. Kompe, Prosody in Speech Understanding Systems. XIX, 357 pages. 1997. (Subseries LNAI).

(Eds.), Database
eedings, 1997.

ls.), Interactive
communication
ages. 1997.

e Analysis and
e I. XXII, 722

ge Analysis and
e II. XXII, 794

(Eds.), Rules in
VII, 214 pages.

ucas (Eds.), FME
ened Foundations
. XIII, 685 pages.

ve Logic Program-
97 pages. 1997.

k (Eds.), Algebraic
ycle. Proceedings,

Vol. 1316: M. Li, A. Maruoka (Eds.), Algorithmic Learning Theory. Proceedings, 1997. XI, 461 pages. 1997. (Subseries LNAI).

Vol. 1317: M. Leman (Ed.), Music, Gestalt, and Computing. IX, 524 pages. 1997. (Subseries LNAI).

Vol. 1318: R. Hirschfeld (Ed.), Financial Cryptography. Proceedings, 1997. XI, 409 pages. 1997.

Vol. 1319: E. Plaza, R. Benjamins (Eds.), Knowledge Acquisition, Modeling and Management. Proceedings, 1997. XI, 389 pages. 1997. (Subseries LNAI).

Vol. 1320: M. Mavronicolas, P. Tsigas (Eds.), Distributed Algorithms. Proceedings, 1997. X, 333 pages. 1997.

Vol. 1321: M. Lenzerini (Ed.), AI*IA 97: Advances in Artificial Intelligence. Proceedings, 1997. XII, 459 pages. 1997. (Subseries LNAI).

Vol. 1322: H. Hußmann, Formal Foundations for Software Engineering Methods. X, 286 pages. 1997.

Vol. 1323: E. Costa, A. Cardoso (Eds.), Progress in Artificial Intelligence. Proceedings, 1997. XIV, 393 pages. 1997. (Subseries LNAI).

Vol. 1324: C. Peters, C. Thanos (Eds.), Research and Advanced Technology for Digital Libraries. Proceedings, 1997. X, 423 pages. 1997.

Vol. 1325: Z.W. Raś, A. Skowron (Eds.), Foundations of Intelligent Systems. Proceedings, 1997. XI, 630 pages. 1997. (Subseries LNAI).

Vol. 1326: C. Nicholas, J. Mayfield (Eds.), Intelligent Hypertext. XIV, 182 pages. 1997.

Vol. 1327: W. Gerstner, A. Germond, M. Hasler, J.-D. Nicoud (Eds.), Artificial Neural Networks – ICANN '97. Proceedings, 1997. XIX, 1274 pages. 1997.

Vol. 1328: C. Retoré (Ed.), Logical Aspects of Computational Linguistics. Proceedings, 1996. VIII, 435 pages. 1997. (Subseries LNAI).

Vol. 1329: S.C. Hirtle, A.U. Frank (Eds.), Spatial Information Theory. Proceedings, 1997. XIV, 511 pages. 1997.

Vol. 1330: G. Smolka (Ed.), Principles and Practice of Constraint Programming – CP 97. Proceedings, 1997. XII, 563 pages. 1997.

Vol. 1331: D. W. Embley, R. C. Goldstein (Eds.), Conceptual Modeling – ER '97. Proceedings, 1997. XV, 479 pages. 1997.

Vol. 1332: M. Bubak, J. Dongarra, J. Waśniewski (Eds.), Recent Advances in Parallel Virtual Machine and Message Passing Interface. Proceedings, 1997. XV, 518 pages. 1997.

Vol. 1333: F. Pichler. R.Moreno-Díaz (Eds.), Computer Aided Systems Theory – EUROCAST'97. Proceedings, 1997. XII, 626 pages. 1997.

Vol. 1334: Y. Han, T. Okamoto, S. Qing (Eds.), Information and Communications Security. Proceedings, 1997. X, 484 pages. 1997.

Vol. 1335: R.H. Möhring (Ed.), Graph-Theoretic Concepts in Computer Science. Proceedings, 1997. X, 376 pages. 1997.

Vol. 1336: C. Polychronopoulos, K. Joe, K. Araki, M. Amamiya (Eds.), High Performance Computing. Proceedings, 1997. XII, 416 pages. 1997.

Vol. 1337: C. Freksa, M. Jantzen, R. Valk (Eds.), Foundations of Computer Science. XII, 515 pages. 1997.

Vol. 1338: F. Plášil, K.G. Jeffery (Eds.), SOFSEM'97: Theory and Practice of Informatics. Proceedings, 1997. XIV, 571 pages. 1997.

Vol. 1339: N.A. Murshed, F. Bortolozzi (Eds.), Advances in Document Image Analysis. Proceedings, 1997. IX, 345 pages. 1997.

Vol. 1340: M. van Kreveld, J. Nievergelt, T. Roos, P. Widmayer (Eds.), Algorithmic Foundations of Geographic Information Systems. XIV, 287 pages. 1997.

Vol. 1341: F. Bry, R. Ramakrishnan, K. Ramamohanarao (Eds.), Deductive and Object-Oriented Databases. Proceedings, 1997. XIV, 430 pages. 1997.

Vol. 1342: A. Sattar (Ed.), Advanced Topics in Artificial Intelligence. Proceedings, 1997. XVII, 516 pages. 1997. (Subseries LNAI).

Vol. 1343: Y. Ishikawa, R.R. Oldehoeft, J.V.W. Reynders, M. Tholburn (Eds.), Scientific Computing in Object-Oriented Parallel Environments. Proceedings, 1997. XI, 295 pages. 1997.

Vol. 1344: C. Ausnit-Hood, K.A. Johnson, R.G. Pettit, IV, S.B. Opdahl (Eds.), Ada 95 – Quality and Style. XV, 292 pages. 1997.

Vol. 1345: R.K. Shyamasundar, K. Ueda (Eds.), Advances in Computing Science - ASIAN'97. Proceedings, 1997. XIII, 387 pages. 1997.

Vol. 1346: S. Ramesh, G. Sivakumar (Eds.), Foundations of Software Technology and Theoretical Computer Science. Proceedings, 1997. XI, 343 pages. 1997.

Vol. 1347: E. Ahronovitz, C. Fiorio (Eds.), Discrete Geometry for Computer Imagery. Proceedings, 1997. X, 255 pages. 1997.

Vol. 1348: S. Steel, R. Alami (Eds.), Recent Advances in AI Planning. Proceedings, 1997. IX, 454 pages. 1997. (Subseries LNAI).

Vol. 1349: M. Johnson (Ed.), Algebraic Methodology and Software Technology. Proceedings, 1997. X, 594 pages. 1997.

Vol. 1350: H.W. Leong, H. Imai, S. Jain (Eds.), Algorithms and Computation. Proceedings, 1997. XV, 426 pages. 1997.

Vol. 1351: R. Chin, T.-C. Pong (Eds.), Computer Vision – ACCV'98. Proceedings Vol. I, 1998. XXIV, 761 pages. 1997.

Vol. 1352: R. Chin, T.-C. Pong (Eds.), Computer Vision – ACCV'98. Proceedings Vol. II, 1998. XXIV, 757 pages. 1997.

Vol. 1353: G. BiBattista (Ed.), Graph Drawing. Proceedings, 1997. XII, 448 pages. 1997.

Vol. 1354: O. Burkart, Automatic Verification of Sequential Infinite-State Processes. X, 163 pages. 1997.

Vol. 1355: M. Darnell (Ed.), Cryptography and Coding. Proceedings, 1997. IX, 335 pages. 1997.

Vol. 1356: A. Danthine, Ch. Diot (Eds.), From Multimedia Services to Network Services. Proceedings, 1997. XII, 180 pages. 1997.

Vol. 1357: J. Bosch, S. Mitchell (Eds.), Object-Oriented Technology. Proceedings, 1997. XIV, 555 pages. 1998.

Vol. 1358: B. Thalheim, L. Libkin (Eds.), Semantics in Databases. XI, 265 pages. 1998.

Vol. 1360: D. Wang (Ed.), Automated Deduction in Geometry. Proceedings, 1996. VII, 235 pages. 1998. (Subseries LNAI).

Vol. 1361: B. Christianson, B. Crispo, M. Lomas, M. Roe (Eds.), Security Protocols. Proceedings, 1997. VIII, 217 pages. 1998.

Vol. 1362: D.K. Panda, C.B. Stunkel (Eds.), Network-Based Parallel Computing. Proceedings, 1998. X, 247 pages. 1998.

Vol. 1363: J.-K. Hao, E. Lutton, E. Ronald, M. Schoenauer, D. Snyers (Eds.), Artificial Evolution. XI, 349 pages. 1998.

Vol. 1364: W. Conen, G. Neumann (Eds.), Coordination Technology for Collaborative Applications. VIII, 282 pages. 1998.

Vol. 1365: M.P. Singh, A. Rao, M.J. Wooldridge (Eds.), Intelligent Agents IV. Proceedings, 1997. XII, 351 pages. 1998. (Subseries LNAI).

Vol. 1367: E.W. Mayr, H.J. Prömel, A. Steger (Eds.), Lectures on Proof Verification and Approximation Algorithms. XII, 344 pages. 1998.

Vol. 1368: Y. Masunaga, T. Katayama, M. Tsukamoto (Eds.), Worldwide Computing and Its Applications — WWCA'98. Proceedings, 1998. XIV, 473 pages. 1998.

Vol. 1370: N.A. Streitz, S. Konomi, H.-J. Burkhardt (Eds.), Cooperative Buildings. Proceedings, 1998. XI, 267 pages. 1998.

Vol. 1372: S. Vaudenay (Ed.), Fast Software Encryption. Proceedings, 1998. VIII, 297 pages. 1998.

Vol. 1373: M. Morvan, C. Meinel, D. Krob (Eds.), STACS 98. Proceedings, 1998. XV, 630 pages. 1998.

Vol. 1375: R. D. Hersch, J. André, H. Brown (Eds.), Electronic Publishing, Artistic Imaging, and Digital Typography. Proceedings, 1998. XIII, 575 pages. 1998.

Vol. 1376: F. Parisi Presicce (Ed.), Recent Trends in Algebraic Development Techniques. Proceedings, 1997. VIII, 435 pages. 1998.

Vol. 1377: H.-J. Schek, F. Saltor, I. Ramos, G. Alonso (Eds.), Advances in Database Technology – EDBT'98. Proceedings, 1998. XII, 515 pages. 1998.

Vol. 1378: M. Nivat (Ed.), Foundations of Software Science and Computation Structures. Proceedings, 1998. X, 289 pages. 1998.

Vol. 1379: T. Nipkow (Ed.), Rewriting Techniques and Applications. Proceedings, 1998. X, 343 pages. 1998.

Vol. 1380: C.L. Lucchesi, A.V. Moura (Eds.), LATIN'98: Theoretical Informatics. Proceedings, 1998. XI, 391 pages. 1998.

Vol. 1381: C. Hankin (Ed.), Programming Languages and Systems. Proceedings, 1998. X, 283 pages. 1998.

Vol. 1382: E. Astesiano (Ed.), Fundamental Approaches to Software Engineering. Proceedings, 1998. XII, 331 pages. 1998.

Vol. 1383: K. Koskimies (Ed.), Compiler Construction. Proceedings, 1998. X, 309 pages. 1998.